AA004479

eTELEMED 2016

The Eighth International Conference on eHealth, Telemedicine, and Social Medicine

DIGITAL HEALTHY LIVING 2016

A Multidisciplinary View on Digital Support for Healthy Living and Self-management for Health

MATH 2016

The International Symposium on Mobile and Assistive Technology for Healthcare

April 24 - 28, 2016

Venice, Italy

eTELEMED 2016 Editors

Marike Hettinga, Windesheim University of Applied Sciences, the Netherlands

Alma Leora Culén, University of Oslo, Norway

Lisette Van Gemert-Pijnen, University of Twente - Enschede, the Netherlands

Anne G. Ekeland, Norwegian Centre for Integrated Care and Telemedicine / University Hospital of North Norway | University of Tromsø, Norway

Trine Bergmo, University Hospital of North Norway, Norway

Paolo Perego, Politecnico di Milano, Italy

Maurizio Caon, University of Applied Sciences and Arts Western Switzerland, Switzerland

Cristiana Degano, Gruppo SIGLA S.r.l., Italia

Piero Giacomelli, Spac SPA, Italy

Kari Dyb, Norwegian Centre for Integrated Care and Telemedicine / University Hospital of North Norway | University of Tromsø, Norway

Printed from e-media with permission by:

Curran Associates, Inc.
57 Morehouse Lane
Red Hook, NY 12571

Some format issues inherent in the e-media version may also appear in this print version.

Printed by Curran Associates, Inc. (2016)

International Academy, Research, and Industry Association (IARIA)
412 Derby Way
Wilmington, DE 19810

Phone: (408) 893-6407
Fax: (408) 527-6351

petre@iaria.org

Additional copies of this publication are available from:

Curran Associates, Inc.
57 Morehouse Lane
Red Hook, NY 12571 USA
Phone: 845-758-0400
Fax: 845-758-2633
Email: curran@proceedings.com
Web: www.proceedings.com

TABLE OF CONTENTS

ETELEMED 6: TELEMEDICINE/EHEALTH APPLICATIONS

ETELEMED 7: PERSONALIZED EHEALTH

ETELEMED 8: EHEALTH DATA RECORDS

ETELEMED 9: EHEALTH SYSTEMS AND COMMUNICATIONS

ETELEMED 10: EHEALTH INFORMATION PROCESSING

ETELEMED 11: LARGE-SCALE COST-EFFECTIVE EHEALTH SYSTEMS

DIGITAL HEALTHY LIVING 1: ASSISTIVE TECHNOLOGIES

DIGITAL HEALTHY LIVING 2: HEALTH-ORIENTED INTERACTION DESIGN

MATH 1: MOBILE AND ASSISTIVE TECHNOLOGY FOR HEALTHCARE I

MATH 2: MOBILE AND ASSISTIVE TECHNOLOGY FOR HEALTHCARE II

eTELEMED 2016

Forward

The Eighth International Conference on eHealth, Telemedicine, and Social Medicine (eTELEMED 2016), held between April 24 and April 28, 2016 in Venice, Italy, continued a series of events considering advances in techniques, services, and applications dedicated to a global approach to eHealth, including a regard on federated aspects considering the mobility of the population, the cross-nations agreements, and the new information technology tools.

We are facing the generalization of digital society across multiple social areas. The globalization imposes the revision of the health costs a society can support. The progress in different domains, such as image processing, wireless communications, computer vision, cardiology, and information storage and management assure a virtual team to access online to the latest achievements.

The processing of medical data benefits now from advanced techniques for color imaging, visualization of multi-dimensional projections, Internet imaging localization archiving as well as from a higher resolution of medical devices.

Collecting, storing, and handling patient data requires robust processing systems, safe communications and storage, and easy and authenticated online access.

We assist at an unprecedented and rapid deployment of the use of electronic imagery, navigation portals, positive attitude on telemedicine, distributed surgery teams, tele-cardiology, and remote medicine. Development of wireless homecare, of special types of communications with patient data, of videoconferencing and tele-presence, and the progress in image processing and date protection increased the eHealth applications and services, and extended Internet-based patient coverage areas. Social and economic aspects as well as the integration of classical systems with the telemedicine systems are still challenging issues.

The conference had the following tracks:

- Sociological, sociotechnical and multi-disciplinary perspectives
- Preventive and emergency health systems
- Telemedicine/eHealth services
- eHealth technology and devices
- Telemedicine/eHealth applications
- Personalized eHealth
- eHealth data records
- eHealth systems and communications
- eHealth information processing
- Challenges of large-scale, cost-effective eHealth systems

The conference also featured the following symposia:

- ***DIGITAL HEALTHY LIVING 2016, A Multidisciplinary View on Digital Support for Healthy Living and self-management of Health***
- ***MATH 2016, The International Symposium on Mobile and Assistive Technology for Healthcare***

We take here the opportunity to warmly thank all the members of the eTELEMED 2016 technical program committee, as well as the numerous reviewers. The creation of such a high quality conference program would not have been possible without their involvement. We also kindly thank all the authors that dedicated much of their time and effort to contribute to eTELEMED 2016. We truly believe that, thanks to all these efforts, the final conference program consisted of top quality contributions.

Also, this event could not have been a reality without the support of many individuals, organizations and sponsors. We also gratefully thank the members of the eTELEMED 2016 organizing committee for their help in handling the logistics and for their work that made this professional meeting a success.

We hope eTELEMED 2016 was a successful international forum for the exchange of ideas and results between academia and industry and to promote further progress in the fields of eHealth, telemedicine and social medicine. We also hope that Venice, Italy, provided a pleasant environment during the conference and everyone saved some time to enjoy the unique charm of the city.

eTELEMED Advisory Committee

Lisette Van Gemert-Pijnen, University of Twente - Enschede, The Netherlands
Amir Hajjam-El-Hassani, University of Technology of Belfort-Montbéliard, France
Päivi Hämäläinen, The National Institute for Health and Welfare - Helsinki, Finland
Frank Lievens, Med-e-Tel, Belgium
Åsa Smedberg, Stockholm University, Sweden
Piero Giacomelli, SPAC SPA -Arzignano (Vicenza), Italia
Marike Hettinga, Windesheim University of Applied Sciences, Netherlands
Ramesh Krishnamurthy, Health Systems and Innovation Cluster, World Health Organization - Geneva, Switzerland

eTELEMED 2016 Special Area Chairs

Sociological and Socio-Technical Approaches

Anne G. Ekeland, Norwegian Centre for Integrated Care and Telemedicine / University Hospital of North Norway | University of Tromsø, Norway
Kari Dyb, Norwegian Centre for Integrated Care and Telemedicine / University Hospital of North Norway | University of Tromsø, Norway

eTELEMED 2016

Committee

eTELEMED Advisory Committee

Lisette Van Gemert-Pijnen, University of Twente - Enschede, The Netherlands
Amir Hajjam-El-Hassani, University of Technology of Belfort-Montbéliard, France
Päivi Hämäläinen, The National Institute for Health and Welfare - Helsinki, Finland
Frank Lievens, Med-e-Tel, Belgium
Åsa Smedberg, Stockholm University, Sweden
Piero Giacomelli, SPAC SPA -Arzignano (Vicenza), Italia
Marike Hettinga, Windesheim University of Applied Sciences, Netherlands
Ramesh Krishnamurthy, Health Systems and Innovation Cluster, World Health Organization - Geneva, Switzerland

eTELEMED 2016 Special Area Chairs

Sociological and Socio-Technical Approaches

Anne G. Ekeland, Norwegian Centre for Integrated Care and Telemedicine / University Hospital of North Norway | University of Tromsø, Norway
Kari Dyb, Norwegian Centre for Integrated Care and Telemedicine / University Hospital of North Norway | University of Tromsø, Norway

eTELEMED 2016 Technical Program Committee

Sasan Adibi, Research In Motion (RIM), Canada
Hussain Al-Rizzo, University of Arkansas at Little Rock, USA
Giner Alor Hernández, Instituto Tecnológico de Orizaba - Veracruz, México
Flora Amato, University of Naples "Federico II", Italy
Tomaz Amon, Center for Scientific Visualization - Ljubljana, Slovenia
Ennio Amori, Parma University Hospital, Italy
Adriano O. Andrade, Federal University of Uberlandia, Brazil
Hege Andreassen, Norwegian Center for Integrated Care and Telemedicine, Norway
Adina Astilean, Technical University of Cluj-Napoca, Romania
Bruno Azevedo, Pontifícia Universidade Católica do Rio de Janeiro (PUC-Rio), Brazil
Omar Badreddin, Northern Arizona University, USA
Piotr Bała, ICM University of Warsaw, Poland
Luigi Benedicenti, University of Regina, Canada
Kazi S. Bennoor , National Institute of Diseases of Chest & Hospital - Mohakhali, Bangladesh
Emmanuel Benoist, University of Applied Sciences Bern (BFH-TI), Switzerland

Ilias Maglogiannis, University of Piraeus, Greece
Asta Maknickaite, Technical University of Denmark, Denmark
Flora Malamateniou, University of Piraeus, Greece, Digital Health Services Laboratory (DHSL), Greece
Salah H. Mandil, eStrategies & eHealth for WHO and ITU - Geneva, Switzerland
Cherisse Mark, MED-EL, Austria
Roque Marin Morales, Universidad de Murcia, Spain
Rodrigo Mariño, University of Melbourne, Australia
Begoña Martinez-Salvador, Universitat Jaume I, Spain
Elio Masciari, ICAR-CNR, Italy
Cezary Mazurek, Poznan Supercomputing and Networking Center, Poland
Teresa Meneu, Universidad Politécnica de Valencia, Spain
Yoshitoshi Murata, Iwate Prefectural University, Japan
Radha Nagarajan, University of Kentucky, USA
Tadashi Nakano, Osaka University, Japan
Stuart J. Nelson, US National Library of Medicine - Bethesda, USA
Goran Nenadic, University of Manchester, UK
Lambert Nieuwenhuis, University of Twente, The Netherlands
Harm op den Akker, Roessingh Research and Development - Telemedicine group Enschede, Netherlands
Eiring Øystein, Norwegian Knowledge Center for the Health Services, Norway
Guy Pare, HEC Montreal, Canada
Constantinos Pattichis, University of Cyprus, Cyprus
Oscar Peters, University of Twente, The Netherlands
Barbara Pierscionek, Kingston University London, UK
Vitor Pinheiro, IBM Research, Brazil
Isabel Pirillo, Università di Camerino, Italy
Rainer Planinc, Vienna University of Technology, Austria
Andriana Prentza, University of Piraeus, Greece
Markela Psymarnou, VIDAVO SA - Thessaloniki, Greece
Juha Puustjärvi, Univerity of Helsinki, Finland
João Manuel R. S. Tavares, Universidade do Porto, Portugal
Elisabeth Rakus-Andersson, Blekinge Institute of Technology, Sweden
David Riaño, Universitat Rovira i Virgili, Spain
Joel Rodrigues, University of Beira Interior, Portugal
Valter Roesler, Universidade Federal do Rio Grande do Sul (UFRGS), Brazil
Robert J. Rosati, Center for Home Care Policy and Research - New York, USA
Uwe Roth, Luxembourg Institute of Science and Technology, Luxembourg
David Russell, Center for Home Care Policy & Research - New York, USA
Sanna Salanterä, University of Turku, Finland
Abdel-Badeeh M. Salem, Ain Shams University - Cairo, Egypt
Majid Sarrafzadeh, University of California - Los Angeles, USA
Matthieu-Patrick Schapranow, Hasso Plattner Institute, Germany
Jacob Scharcanski, UFRGS - Universidade Federal do Rio Grande do Sul, Brazil

Christos N. Schizas, University of Cyprus, Cyprus
Michael Schumacher-Debril, University of Applied Sciences Western Switzerland (HES-SO), Switzerland
Sandra Sendra Compte, Polytechnic University of Valencia, Spain
Clauirton Siebra, Federal University of Paraiba, Brazil
Åsa Smedberg, Stockholm University, Sweden
Terje Solvoll, University hospital of North Norway - Tromsø, Norway
Sanjay Sood, Centre for Development of Advanced Computing (C-DAC), Mohali, India
Marijke Span, Windesheim University of Applied Sciences, Netherlands
Hrvoje Stancic, University of Zagreb, Croatia
Frederick Steinke, Humboldt Universität zu Berlin, Germany
Maciej Stroinski, Poznan Supercomputing and Networking Center, Poland
Irene Tagliente, Pediatric Hospital Bambino Gesù, Rome / University of Rome 'La Sapienza' / University of Cassino and the Southern Lazio, Cassino, Italy
Toshiyo Tamura, Osaka Electro-communication University, Japan
Svetlana Taneva, Healthcare Human Factors, Canada
Víctor Torres Padrosa, Polytechnic School / Faculty of Medicine | Universitat de Girona, Spain
Vicente Traver Salcedo, ITACA - Universitat Politècnica de València, Spain
Ceara Treacy, Regulated Software Research Centre & Lero - Dundalk Institute of Technology, Ireland
Masatsugu Tsuji, University of Hyogo, Japan
Gary Ushaw, Newcastle University, UK
Aristides Vagelatos, CTI - Athens, Greece
Bert-Jan van Beijnum, University of Twente Enschede, Netherlands
Lisette Van Gemert-Pijnen, University of Twente - Enschede, The Netherlands
Dimitrios D. Vergados, University of Piraeus, Greece
Miriam Vollenbroek Hutten, University of Twente – Enschede, The Netherlands
Xiaohong Wang Gao, Middlesex University - London, UK
Jobke Wentzel, University of Twente - Enschede, The Netherlands
Lixia Yao, UNC Charlotte, USA
Huiyu Zhou, Queen's University Belfast, UK
André Zúquete, Instituto de Engenharia Electrónica e Telemática de Aveiro (IEETA), Portugal

DIGITAL HEALTHY LIVING 2016

Advisory Committee

Marike Hettinga, Windesheim University of Applied Sciences, Netherlands
Alma Leora Culén , University of Oslo, Norway
Ramesh Krishnamurthy, Health Systems and Innovation Cluster, World Health Organization - Geneva, Switzerland

Program Committee Members

Eric Brangier, Université de Lorraine, France
Maurizio Caon, HumanTech Institute HES-SO | Haute école d'ingénierie et d'architecture | University of Applied Sciences and Arts of Western Switzerland, Switzerland
Cristiana Degano, Gruppo SIGLA, Italy
Sisse Finken, Linnaeus University, Sweden
Philippe Gorce, HandiBio, Toulon University, France
Yiannis Koumpouros, Technological Educational Institute of Athens, Greece
Ramesh Krishnamurthy, Health Systems and Innovation Cluster, World Health Organization - Geneva, Switzerland
Margaret Machniak, University of Oslo, Norway
Santiago Gil Martinez, University of Agder, Norway
Mart Min, Tallinn University of Technology, Estonia
Britt Östlund, Royal Institute of Technology, Sweden
Paolo Perego, Politecnico di Milano, Italy
Caterina Rizzi, Università degli Studi di Bergamo, Italy
Pietro Siciliano, Institute for Microelectronics and Microsystems IMM-CNR, Italy
Neeraj Kumar Singh, University of Toulouse - INP/ IRIT, France
Carolien Smits, Windesheim University of Applied Sciences, The Netherlands
Leonor Teixeira, Institute of Electronics and Telematics Engineering of Aveiro (IEETA) / Department of Economics, Management and Industrial Engineering (DEGEI) - University of Aveiro, Aveiro, Portugal
Nadine Vigouroux, IRIT - UMR CNRS 5505 - Toulouse | Université Fédérale de Toulouse, France

MATH 2016

Program Committee Members

Maurizio Caon, HES-SO Haute Ècole Spécialisée de Suisse Occidentale, Switzerland
Cristiana Degano, Gruppo SIGLA s.r.l., Italy
Paolo Perego, Politecnico di Milano, Italy
Luca Bianconi, Gruppo SIGLA s.r.l., Italy
Renata Guarnieri, Politecnico di Milano, Italy
Mauro Migliardi, Università di Padova, Italy
Luca Manfredini, Centro Regionale di Riferimento di Cure Palliative e Terapia del Dolore Pediatrico - Istituto G. Gaslini, Italy
Marco Tettamanti, Ospedale S. Raffaele Milano, Italy
Riccardo Berta, Elios Lab, Dipartimento DITEN, Università di Genova, Italy
Giuseppe Andreoni, Politecnico di Milano, Italy
Alessandro De Gloria, Elios Lab, Dipartimento DITEN, Università di Genova, Italy

Supporting eHealth Innovations: an Insurer's Perspective

Ruud Janssen, Sikke Visser, Elles Gyaltsen-Lohuis, Marike Hettinga
Research group IT-innovation in Health Care
Windesheim University of Applied Sciences
Zwolle, The Netherlands
Email: {TJWM.Janssen}{S.Visser}{EGE.Lohuis}{M.Hettinga}@windesheim.nl

Abstract—**Health care insurers play a vital role in the implementation of eHealth and other care innovations within the health care sector. Yet, very little is said in the literature about how innovations are being evaluated by insurers, and what criteria are being used to do so. This paper describes the results of a case study into the evaluation process and criteria of a large health care insurer in The Netherlands. The results show that experts from several departments within the insurer are involved, that these experts each bring their own set of criteria to the table, and that the decision to provide support for an innovation (e.g., by funding its wider implementation or by reimbursing its use) is based on consensus between these experts. Based on these results, an interactive website was developed to better inform entrepreneurs, project managers, researchers and other people active in the field of eHealth innovation about the insurer's role and criteria for supporting eHealth innovations.**

Keywords-eHealth; innovation; health care; insurer; support; evaluation; criteria.

I. INTRODUCTION

This paper describes a follow-up study of the research project *Successful Entrepreneurship in eHealth* [1]. During this project an *eHealth innovation map* for the Dutch health care system was developed: a diagram showing entrepreneurs in eHealth (a) which health care parties to involve during the implementation process, (b) their roles and their mutual relations, (c) their interests in eHealth innovation, and (d) the kinds of evidence that can be used to convince these parties of the added value of an eHealth innovation [2]. Furthermore, a set of corresponding fact sheets and an interactive website were developed to provide eHealth entrepreneurs with concise information for choosing an appropriate *innovation route* within the health care system [3].

At the outset of the project, but also afterwards while disseminating the results to the wider audience of eHealth entrepreneurs, it was found that the role and interests of the health care insurer within the Dutch system were largely unclear to the entrepreneur. Based on this finding and facilitated by an additional research grant, the decision was made to conduct a follow-up study with the aim to shed more light on the insurer's process and criteria for evaluating eHealth innovations. Such evaluation takes place, for instance, when an entrepreneur requests the support of an insurer by funding the implementation of an eHealth innovation by health care providers, or by reimbursing its use by health care providers or patients. It was expected that the current confusion among entrepreneurs about the insurer's role (and the resulting mismatch of expectations when requesting support from an insurer) could thus be reduced.

To the best of the authors' knowledge, this is the first study into the process and criteria used by a health care insurer to evaluate health care innovations such as eHealth. For instance, the authors have searched extensively in the MEDLINE database [4] and found no results mentioning specific processes or criteria. The study was carried out in close cooperation with two innovation experts from a large Dutch health care insurer, who in this way were hoping to make the insurer's role and interests more transparent to entrepreneurs.

The remainder of this paper is structured as follows: Section II describes the methods used for data collection and analysis, Section III reports on the main findings with respect to the insurer's evaluation process and evaluation criteria, and Section IV summarizes the conclusions and next steps.

II. METHOD

A. *Data collection*

During the study, data were collected in six ways:

1) Interviews with two innovation experts

A semi structured interview was held with two innovation experts from a large Dutch health insurer. The interview focused on the evaluation process and the evaluation criteria of this insurer.

2) Screening of the insurer's innovation portal

The insurer's innovation portal (a website and accompanying web form where health care innovations can be submitted by entrepreneurs and care providers to request the support from the insurer) was screened for any information pertaining to the process and criteria.

3) Documentation of submissions and assessments

The research team was provided with detailed documentation of all submitted innovations over the last quarter (17 in total), including written evaluations by all experts involved in the evaluation process.

4) Observation of one expert meeting

One researcher was present during the meeting where these submissions and their evaluations were discussed among all involved experts, and where decisions were made whether or not to provide support. Due to the confidential nature of the discussions only field notes were taken; no audio recordings were made and no pictures were taken.

5) Documentation of decisions and considerations

Two days after the meeting, the research team was provided with the written decisions and accompanying considerations that had been sent to the applicants.

6) Consultation of two other experts

Two other experts were frequently consulted during the study to check that the results were sufficiently representative of the processes and criteria of other insurers in The Netherlands: the first an innovation expert working at another insurer, and the second an eHealth program officer at a large national research funding agency.

B. Data analysis

Analysis of the collected data was performed in five stages:

1) Extracting and assigning remarks, questions etc.

All remarks, questions, suggestions etc. that were made as part of the evaluation of the 17 submitted innovations were extracted from the field notes, written assessments and decisions, and then assigned to the individual expert who made them.

2) Grouping experts into expert roles

Based on their job title and the department they worked for within the insurer, the individual experts were grouped into expert roles.

3) Identifying the criteria per expert role

All remarks and questions assigned to the experts within a single expert role were clustered (using the affinity diagramming technique [5]) to identify the criteria per role.

4) Prioritization of the criteria

The resulting criteria were then prioritized per role by all involved experts to arrive at a set of three main criteria per role.

5) Fine-tuning the roles and criteria

Throughout steps 2-4 the intermediate results were discussed with the two innovation experts that had been interviewed as part of the data collection process. Together with the research team they fine-tuned the roles and criteria based on feedback they obtained from their colleagues.

C. Website development

Based on the resulting expert roles and corresponding sets of main criteria, an interactive website was developed. The website had to document the insurer's role in the health care system and its interests in eHealth innovation, the insurer's evaluation process for eHealth innovations, and the criteria used within this process. As part of the development process, an evaluation of the website's content and usability was carried out with three entrepreneurs. This was done by means of a structured user interface walkthrough [5].

III. FINDINGS AND RESULTS

A. Evaluation process

The insurer participating in this research has developed an innovation portal which is part of its website and where, year round, entrepreneurs and other health care innovators are invited to submit their ideas by means of a form. Among the questions asked in this form are: *What patient needs are addressed by your idea? How does your idea improve the quality of care? How does your idea lower the cost of care?* and *What kind of support do you request from the insurer*?

All ideas submitted in this way are first assessed in writing, typically by three to five experts working within the insurer. Once every three to four months, the submitted ideas and the written assessments are discussed in a meeting between all involved experts. During the meeting that was observed, 17 ideas were discussed. Once consensus about an idea has been reached between the experts, the decision whether or not to support the idea is made and the applicant is notified. Applicants whose ideas have been accepted are invited for a follow-up meeting at the insurer. It may also happen that an applicant is requested to come up with additional information.

B. Experts and expert roles

Overall, 28 experts were involved in the evaluation of the 17 submitted innovations. Among them were care and cure purchasers (8), innovation consultants (5), policy coordinators (5), pharmaceutical and medical advisors (4), commercial consultants (4), IT experts (1), and department managers (1). The experts worked for the departments purchasing (10), innovation (7), policy (4), commerce (4), and medical advice (3). Of these experts, 21 were involved in the written evaluation of the submitted innovations whereas 11 experts were present during the meeting where the submitted innovations were discussed.

Based on job title and department and on feedback provided by the experts themselves, the experts were grouped into seven expert roles: the *medical advisor*, the *innovation consultant*, the *policy coordinator*, the *proposition manager*, the *purchaser*, the *market consultant*, and the *technology consultant*. The first and second columns of Table I list the expert roles and the respective scopes when evaluating innovations. For instance, the medical advisor role focuses on the medical quality of an innovation: does it conform to the state-of-the art in medical evidence and professional standards?

C. Evaluation criteria

Using the affinity diagramming technique [5] all questions corresponding to a single expert role were clustered to identify criteria. For instance, all questions and remarks made by the four medical and pharmaceutical experts in the written assessments and/or during the expert meeting, were all assigned to the *medical expert* role and then clustered. Affinity diagramming is particularly suited for this task since it allows topics to emerge from the data rather than from predefined categories. The topics that emerged in this way were *financing*, *target group*, *patient problem*, *uniqueness*, *added value*, and *implementation*, with each role contributing questions and remarks to one or more topics.

Next, criteria were identified per topic. This was done in several iterations and in close co-operation with the experts themselves. As a final step, the generated criteria were prioritized to arrive at a set of three main criteria per role. See the third column of Table I. For instance, the medical advisor role uses cost effectiveness, medical evidence and substitution of existing care as the three main criteria when evaluating an innovation.

D. *Interactive website*

Based on the results of the study, an interactive website [6] was developed for entrepreneurs, project managers, researchers and other people active in the field of eHealth innovation. The website consists of five pages: (1) a general introduction about the aim of the website, (2) a brief description of how to submit an innovation for evaluation by the insurer, (3) the main page where entrepreneurs can read about the insurer's criteria and complete an assessment to learn how their own innovation might score when evaluated by the insurer, (4) background information about the insurer and its role within the health care system, and (5) links to relevant websites and organizations.

The result of the assessment (see Figures 1 and 2) is explicitly not shown in terms of a 'fail' or 'pass' as this would be an unrealistic simplification. Rather, the result is indicated by seven color-coded avatars that each stand for one of the expert roles. An avatar's color represents whether the respective expert role is likely to be convinced (green), undecided (orange) or unconvinced (red), and is determined by the answers given to three questions (either 'yes', 'no' or 'unsure'). For instance, for a green avatar the answers should include at least 2 yes's and 0 no's. Overall, this reflects the actual evaluation process, where the individual experts might be convinced, undecided or unconvinced and where the final decision is made based on a discussion where consensus between the experts is reached.

IV. CONCLUDING REMARKS

Health care insurers play a crucial role in the implementation of innovations within the health care sector. Yet, very little (if anything) is said in the literature about how innovations are being evaluated by insurers, and what criteria are being used to do so. Although the results presented here are based on the process and criteria of a single insurer, the study has delivered useful results and clearly deserves a follow-up.

Such a follow-up study should, preferably, also be done in other countries with different health care systems. The process and criteria we found, must be determined by the characteristics of the health care system and the role of insurers within the system. For instance, the Dutch system can be characterized as a universal health care system with compulsory insurance carried out by competing private insurers [7]. This may explain why some criteria address the insurer's image or customer retention. We expect that the criteria of insurers acting within other types of systems will to some extent be different.

Besides contributing to the transparency about the insurer's evaluation process and criteria, the study has also proven beneficial for the insurer itself. The co-operating experts were surprised to learn, after the first stages of the analysis, how much overlap existed between the remarks made and questions asked by every one of them. The ensuing iterations to fine-tune the expert roles and evaluation criteria has led to a greater awareness and a more streamlined evaluation process within the insurer. It also led to a strong involvement among the experts, who clearly saw the benefits of a greater transparency about the insurer's evaluation process, such as fewer misunderstandings among entrepreneurs and a higher quality of future submissions.

During the last stages of the study, three representatives of the target group (all eHealth entrepreneurs) were invited to evaluate the website by means of a structured user interface walkthrough [5]. This has led to several improvements of the website. More importantly, it showed that the information presented on the website and the assessment are indeed useful.

Figure 1. Part of the main page of the developed website. The avatars of four out of seven expert roles are visible.

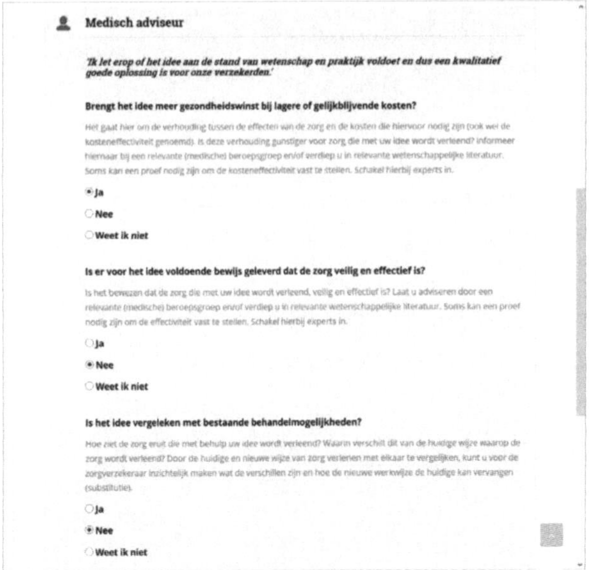

Figure 2. Part of the assessment. The three questions for the expert role *medical advisor* are visible.

The participating entrepreneurs stressed the value of adding descriptions to the website of successful cases where insurer and entrepreneur have indeed co-operated in the wider implementation of an innovation.

Overall, the study and website have received a warm welcome among entrepreneurs, insurers and government officials. Removing obstacles for health innovation is an important issue on the agenda of the Ministry of Public Health. Nevertheless, the actual added value of the website remains to be seen (for instance by studying the question: is there any difference in success rates among entrepreneurs who did and who didn't consult the website before submitting their ideas to the insurer?). The Ministry of Public Health has, in the meantime, provided an additional grant to extend the study to more insurers, and to incorporate the description of several successful cases as requested by the entrepreneurs. The results will be reported at this conference in due time.

ACKNOWLEDGMENT

The authors wish to thank the experts who contributed to this study: Matthijs van der Bijl and Tanya Levi (innovation consultants at health care insurer VGZ), Mirande Groener (eHealth program manager at health care insurer Achmea) and Dr. Inge Valstar (program officer at research funding agency ZonMw). The research described in this paper was made possible by a research grant of Stichting Innovatie Alliantie (NVG-3-2011-19-14M).

REFERENCES

[1] R. Janssen et al., "Developing evidence guidelines for eHealth small and medium-sized enterprises," Proc. eTELEMED 2013, Feb. 2013, pp. 92-95.

[2] R. Janssen et al., "Innovation routes and evidence guidelines for eHealth Small and Medium-sized Enterprises," Int. J. Adv. Life Sc., 2013, 5(3&4), pp. 188-203.

[3] R. Janssen et al., "An eHealth innovation map for small and medium-sized enterprises," Proc. eTELEMED 2014, Mar. 2014, pp. 292-301.

[4] See http://www.pubmed.gov. [Retrieved March, 2016]

[5] J. Hackos and J. Redish, "User and task analysis for interface design," New York: Wiley, 1998.

[6] See http://www.watwildezorgverzekeraar.nl (in Dutch). [Retrieved March, 2016]

[7] See http://en.wikipedia.org/wiki/Universal_health_coverage_by_country#Netherlands. [Retrieved March, 2016]

TABLE I. EXPERT ROLES, EVALUATION SCOPES AND MAIN EVALUATION CRITERIA

Expert role	Evaluates whether the innovation...	Main evaluation criteria
Medical advisor	satisfies the state-of-the art in medical evidence and professional standards	1. cost effectiveness (does it contribute to better health at equal or lower cost) 2. medical evidence (is there any proof that it is safe and effective) 3. substitution (can it replace a current treatment)
Innovation consultant	contributes to cost reduction while maintaining or improving quality of care	1. business case (are costs balanced by benefits) 2. stakeholder support (has it been developed in co-creation with stakeholders) 3. pilot readiness (is it ready for a trial to evaluate its added value)
Policy coordinator	fits the insurer's responsibilities and strategy	1. patient problem (does it address an existing patient problem) 2. insurer role (does it fit with the insurer's role within the health care system) 3. insurer strategy (does it fit with the insurer's policy and strategy)
Proposition manager	has added value for the insurer and its subscribers	1. insurer image (will it aid the insurer to distinguish itself from its competitors) 2. customer retention (will it help to retain or attract customers) 3. customer value (will it improve the insurer's customer offerings)
Purchaser	can be reimbursed within the public health insurance regulations	1. substitution (will it replace, not supplement, existing care) 2. care regulations (can it be made to fit within current regulations) 3. reimbursement regulations (is it reimbursable under current regulations)
Market consultant	addresses an existing patient need	1. patient problem (does it address an existing patient problem) 2. patient involvement (have patients been involved during its development) 3. patient need (does it fit with patient needs and desires)
Technology consultant	is truly new	1. scalability (can it be scaled to more care providers) 2. uniqueness (is it unique compared to its alternatives) 3. privacy & compliancy (is information handled according to relevant standards)

Promoting Patient Voices on the Internet

Ethical considerations about web-based dissemination of research on patient narratives

Marianne V. Trondsen[1], Line Lundvoll Warth[2], Kari Dyb[3], Torsten Risør[4], Hege K. Andreassen[5]

[1, 2, 3, 5]Norwegian Center for E-health Research
Tromsø, Norway

[4]Faculty of Health Sciences
UiT - The Arctic University of Norway
Tromsø, Norway

E-mail: marianne.trondsen@telemed.no; line.lundvoll.warth@telemed.no; kari.dyb@telemed.no; torsten.risor@uit.no; hege.andreassen@telemed.no

Abstract—**Web-based promotion of qualitative research has a potential to reform traditional scientific dissemination, as well as it challenges ethical norms of research participation and anonymity. In this poster, we discuss ethical considerations of web-based dissemination of qualitative research on patient narratives through video recordings. Including subjective un-anonymized patient stories in research dissemination to inform and support, might have benefits for the field of science, patients, health professionals and the general population. However, ethical norms and guidelines for anonymity needs to be considered with care when promoting patient voices on the Internet.**

Keywords-ethics; patient narratives; qualitative methods; video recordings; Internet.

I. Introduction

There is a growing body of qualitative research aimed to explore in depth how people experience and manage their illness or health challenges [1][2]. The Internet provides new opportunities for dissemination of such research, within not only the established field of science, but also to lay people, health professionals and the general population. However, ethical norms and guidelines for anonymity needs to be considered with care when promoting patient voices on the Internet.

The traditional way of disseminating research has been through academic channels, such as conferences and scientific journals. In line with ethical norms and guidelines for scientific practice, the voice and the story of the informant usually is anonymized in written publications, striving for anonymity and the protection of privacy. Internet opens a new avenue of public dissemination for qualitative researchers, challenging the established ethical norms and guidelines.

In Norway, all health research projects involving patient and health information need to be evaluated by the regional committees for medical and health research ethics, who assess whether the project plan include a detailed strategy for handling data and ethical reflections with regard to patient involvement.

In this poster, we will exemplify some of the ethical considerations with regard to web dissemination of patient narratives using our project "Helsesnakk.no" ("Health talk Norway") as case. In this project to begin in 2016, we will develop a Norwegian website, to disseminate qualitative research on peoples' experiences of health and illness, through use of video- or audio-recorded in-depth interviews with patients and family carers [3]. The purpose is to share information based on qualitative research to inform and support people who suffer from similar illnesses or health challenges, as well for the benefit for caregivers, health professionals, students and the public. In "Health talk Norway", researchers interviewing people about their health experiences, analyze the interviews using thorough scientific methodology, and select and distribute short extracts from the results on the website. The extracts may appear on the website as videos with texts (transcripts), as audio with text, or as anonymous text. In the following, we concentrate on the video clips since they are most sensitive regarding patients anonymity in research. The aim of this poster is to discuss ethical consideration of web-based dissemination of qualitative research on patient narratives through such un-anonymized video clips.

The poster is organized as follows; Section II describes the various methods of web-based dissemination of the research. Section III discusses ethical considerations of dissemination of patients' narratives using video recordings. The conclusions are presented in Section IV, and the acknowledgement closes the poster.

II. Video-Recorded Narratives on the Website

The informants are recruited based on their diagnosis or medical condition. Before the interview, they receive written information about the study. Moreover, they must sign an agreement selecting which of the following forms they agree to be interviewed for dissemination on the website: video-recorded with picture, sound and written text; audio-recorded with sound and texts; or audio-recorded for dissemination only in anonymously text.

The interviews are transcribed, and the informants have the opportunity to read the transcripts and remove parts or sections. A process of analyzing the data material follows, and the researchers carefully select illustrative video clips for the website according to the informants' consent. Video clips are uploaded using the Vimeo service, with settings blocking all access to the videos except when played through the domain "Helsesnakk.no." The process follows standard research norms. The informant may withdraw his or her consent and have the extracts published on the website removed at any time. However, it may not be possible to remove all existing copies from circulation.

III. Ethical Considerations of Patient Voices

Traditionally, in qualitative research researchers are striving for anonymity. Dissemination of video clips from the interviews at the website implies that the informants' names will not appear, but because we can see their faces, these stories cannot be anonymized. Such a break with the traditional way of disseminating research prompts the need to pay extra attention to the ethical considerations.

Thorough information about the study and the web dissemination is important. Hence, we must be sure that the informant gives written consent and is clearly capable of giving consent. As researchers, we need to be "objective" with respect to patient involvement, introducing the opportunity to participate in the research without pushing. Our experience is that recruitment for video-recorded interviews is more difficult than for audio-recorded interviews. Informants who sign up for the website might take on a whole range of attitudes from skepticism to total openheartedness as they tell their stories.

Published materials on websites will be visible to the public, family and friends for years. Young people grow up, and their attitude toward being on the website might change. Their stories might affect how people evaluate them, i.e. some are at the beginning of their work careers. The researcher must pay attention to factors such as age, how people talk about others (relatives, health professionals), and whether they are too openhearted in telling their stories. Should we limit these stories or should we let them speak? What is the most "correct" consideration to ensure ethics?

The website aims to communicate a maximum variation of health experiences. This means that we might receive stories from informants who want to be a public voice for tabooed, shamed, or sensitive themes. They might be eager to use their stories to inform or help others, to break taboos in society, or to promote voices we are seldom able to hear. In the methodological literature addressing ethical considerations, it is an ongoing discussion of the dilemma between the benefit and the risk associated with research participation about sensitive issues or for vulnerable populations [4][5]. Instead of causing distress for the informants, a growing body of research have point out that research participation might be educational, enriching, therapeutic or empowering for vulnerable populations [4]. Nevertheless, research addressing health and illness experiences, potential sensitive topics, and to select clips to be published on a website requires particular awareness. This will imply a researchers' good judgement about people's capabilities, their basis for consent, and how thorough the information given about the study. Does the researcher need to protect some voices? Will it be ethically correct to publish, or unethical not to publish the stories, since the informants have shared their stories for the purposed of being published on the website? Not being identified as the data source or given opportunity to face a distinctive health condition can also cause disadvantages for a person or a group for example by maintaining of taboos or "invisibility" [6]. It is argued that in some cases the default of anonymization might be replaced by a careful liberation, together with the informants, of how to handle the issues of identification and confidentiality [6].

IV. Conclusion

The Internet holds a potential to reform traditional research dissemination. The expansion of the Internet allows us to reach out broader and faster, not only to the scientific community, but also to the general population. For qualitative researchers, interested in peoples' health experiences, the use of the Internet as dissemination channel impose some new ethical considerations. Whilst the researchers' obligation to ensure informant anonymity has previously been the accepted norm, new technology is now challenging this ideal. When establishing a web site for disseminating insights based in research on patient narratives, anonymizing the individual stories will not always be the ethical thing to do. Rather one can claim that anonymizing the informant in this kind of research dissemination will contribute to silence and hide the patient voice from scientific, as well as public discourse, and thus be un-ethical. Including subjective un-anonymized stories in research dissemination must however be handled with care. Hence, informed consent and the right to withdraw at any time are fundamental.

Acknowledgment

The study is supported by grants from Northern Norway Regional Health Authority; project number HST 1270-15.

References

[1] D. Ridge and S. Ziebland, "Understanding depression through a 'coming out' framework," Sociology of Health & Illness, 34(5), 2012, pp. 730-745.

[2] I. Grønning, G. Scambler, and A. Tjora, "From fatness to badness: The modern morality of obesity", Health, 17(3), 2013, pp. 266-283.

[3] H. K. Andreassen, K. Dyb, M. V. Trondsen, and L. Lundvoll Warth, "Study protocol: Health talk Norway", Proceedings of the 13th Scandinavian Conference on Health Informatics, 2015, pp. 63-65.

[4] K. Dyregrov, "Bereaved parents' experience of research participation", Social Science & Medicine, 58(2), 2004, pp. 391-400.

[5] M. M. Stevens, B. A. Lord, M.-T. Proctor, S. Nagy, and E. O'Riordan, "Research With Vulnerable Families Caring for Children With Life-Limiting Conditions", Qualitative Health Research, 20(4), 2010, pp. 496-505.

[6] A.-L. Svalastog and S. Eriksson, "You can use my name; you don't have to steal my story – a critique of anonymity in indigenous studies", Developing World Bioethics, 10(2), 2010, pp. 104-110.

Online Information about Cancer Patient Pathways (CPP) in Norway

Heidi Gilstad
Health Informatics Research Group
Department of Neuroscience, Faculty of Medicine
The Norwegian University of Science and Technology (NTNU)
Trondheim, Norway
e-mai: heidi.gilstad@ntnu.no

Abstract—**A Cancer Patient Pathways (CPP)- program was implemented in Norway in January 2015. An important emphasis of CPP is that the patient should be ensured adequate information and involvement in relation to her/his cancer diagnosis and treatment. Updated information about CPPs in Norway is mainly available through two Web portals: helsedirektoratet.no and helsenorge.no. With reference to Critical Discourse Analysis and Fairclough's notions *knowledge exchange* and *activity exchange*, the online information about CPP is examined and discussed with an emphasis on what characterizes the communication exchange. The results show that the CPP online information has an unclear addressee, and that there is more activity exchange with directives to health personnel than knowledge exchange with health facts to patients on both health portals. The conclusion is that in order to develop targeted and adequate online information, health authorities need to be aware of citizens' eHealth literacy.**

Keywords- cancer patient pathways; CPP, critical discourse analysis; CDA; communication exchange; knowledge exchange; activity exchange; the ADIUVAT-model.

I. INTRODUCTION

The research-based knowledge about of how eHealth will influence healthcare delivery is limited. The potential of using eHealth services for health information exchange seems promising, since a large group of people can be reached efficiently. However, there are important issues to take into consideration when eHealth is adopted in healthcare, one of them being the digital divide [1]. Although eHealth has become widely available, accessible and affordable, there is a cultural and social gap between those who use the eHealth services and those who do not. This divide, caused, for example, by a lack of infrastructure, computer equipment, motivations or skills, affects the society on an overall level. Even more importantly, the eHealth literacy level of the citizens varies [2] [3] [4] [5] [6]. While some are perfectly able to understand and apply the information given, others are to be considered illiterate. We should acknowledge that these mechanisms also might have unintentional consequences within health. The problems related to the eHealth divide should therefore be taken into account in the communication and information distributed to patients of all diagnostic categories.

Healthcare authorities and healthcare institutions are responsible for providing health information relevant to the healthcare services they offer to the population and the patients. But what characterizes health information? How can we distinguish good quality health information from bad quality health information? The objective of this paper is to examine certain aspects of health information provided from health authorities and health institutions concerning cancer patient pathways (CPP). It is especially interesting to analyze how existing information is presented and to whom. The following research question guides the paper: *What characterizes the online information from the authorities concerning the cancer patient pathways in Norway?*

The expected outcome of this paper is threefold: 1) knowledge about the discursive characteristics of the online information about CPP from the Norwegian authorities available for health personnel and for patients and their caregivers, 2) a discussion of notions from Critical Discourse Analysis as tools for analysing online health information in general, 3) a model of the patients health information pathway. The subsequent part of Section I offers an introduction to the CPP program in Norway. Section II presents the theoretical and analytical framework of the paper, while Section III is the methodology section. Section IV offers the analysis and results, and these are discussed in Section V, where the ADIUVAT-model is also suggested. Section VI contains concluding remarks.

A. The introduction of Cancer Patient Pathways in Norway.

Inspired by Denmark, the Norwegian Government introduced Cancer Patient Pathways (CPP) (Norwegian: "pakkeforløp for kreft") in January 2015. The politically decided introduction of 28 CPPs is expected to have an impact on logistics and on information flow in patient care. The purpose of the CPPs is that "cancer patients will experience a well-organized, coherent and predictable pathway without unnecessary, non-medically justified delays in assessment, diagnosis, treatment and rehabilitation. An important aspect is that patients must be ensured adequate information and involvement." [7] [8] The Directorate for Health emphasizes that it is a logistics reform, and not new guidelines for diagnosis [9].

After 12 months of iterative implementation of 28 CPP and 31 diagnostic guidelines in primary and specialist healthcare, the overall experience is positive; the pathway is more predicable for the patients and their next- of kins [9]. However, some issues are raised that indicate that there is still room for improvement. One issue is that patients are more focused on time, and tend to have time as a driving force in their communication with healthcare. However, the CPP does not reduce medically justified delays. Moreover, experiences show that the bottlenecks in the CPP are imaging, endoscopy, and pathology. This is in coherence with the findings in Denmark [10]. This challenge calls for a structural change in the respective departments as well as a focus on communication across departments and towards patients. Moreover, the transition between primary and specialist healthcare is challenging, for example with regard to the quality of referrals from the general practitioners. Previously the GPs wrote a thorough description, whereas now many write only "pakke" (package). Also, the lack of coordinated data systems poses challenges. The use of contract specialists and private actors makes routines and information flow challenging due to different systems. The procedures of coding are challenging for both doctors and health secretaries due to different patient administration systems. Moreover, the preparation of reports to aid in improvement work is not yet initiated. Yet another issue is the coordination of the pathway for the patient. The role of the pathway coordinators is regarded as important. However, experiences show that the teaching and training of pathway coordinators and the definitions of their tasks are not adequately established. Finally, there is a challenge with regard to management, economy and redisposition of resources since the government has not allocated more financial support for the implementation of the reform.

The glue in the logistical chain of the CPP is communication and information flow. The Directorate of Health [7] emphasizes the importance of involving the patient and next-of-kin in decision-making throughout the pathway. The patient is given the *promise*, and the *right*, to be involved actively in the decision-making about own health. Moreover, ethics in the communication i.e. "respect and empathy, shall be maintained". In addition, the Directorate of Health emphasizes that social aspects, expectations and the "individual abilities" of the patient shall be taken into consideration, and the pathway shall be predictable at all stages. The promise, and right, to be involved in decision-making, to be treated ethically, to be addressed according to the individual health- and eHealth literacy level and to experience a predictable pathway demonstrates a good intention from the authorities about patient-centeredness in healthcare. But what are the implications of including patients in decision-making about their own health? Decision- making is the situation where a choice is made among many possible choice alternatives. The task of making decisions about his/her own health based on assessments of different alternatives requires an understanding of what we can expect from the healthcare system, and what kind of information is available at what time during the patient pathway.

Inclusion of patients can be done through mainly two ways of communication and information exchange. Firstly, through face-to-face communication where the specific condition of the patient is discussed and followed in iterative consultations, and secondly, through making information available in a written form for patients to consult to obtain knowledge about the condition in general as well as with regard to the condition of the specific patient. Subsequently, the online information available about CPP in Norway is discussed in relation to analytic notions from Critical Discourse Analysis.

II. THE THEORETICAL AND ANALYTIC FRAMEWORK

A. Discourse Analysis

The theoretical and analytic framework for this project is Discourse analysis. Discourse analysis is a perspective on social life that contains both methodological and conceptual elements [11]. There are various approaches to discourse analysis depending on the object of study. This paper draws on notions from Critical Discourse Analysis (CDA) [12].

According to Fairclough [12], discourse analysis is based on the assumption that language is an irreducible part of social life, dialectically interconnected with other elements of social life, so that social analysis and research always has to take account of language. Social life can, and should, be studied through a focus on language. CDA is concerned with continuity and change at a structural level in society as well as at a more narrow textual level. In this paper the focus is on a textual level, with the point of departure that texts are produced to bring changes in society, for example in healthcare, education, politics and trade.

In the analysis of the online information about CPP we make a distinction between two types of exchange in communication: *Knowledge exchange,* that has a focus on exchange of information, making claims, stating facts, and *activity exchange,* that has a focus on activity, on people doing things or getting people to do things (12). Based on this distinction, we analyse online information about CPP with the help of *speech functions* (for example instructions, solicitations, statements, questions, demands and offers), and look at what kind on sentence types are appearing (for example interrogative, and imperative sentences). The generalized speech functions can again be distinguished into speech acts. The speech function "Offer" could for example include the speech acts promising, threatening and thanking, while the speech function "Demand" would include for example the speech acts ordering and requesting.

III. Methodology

A keyword search on "pakkeforløp" and "pakkeforløp for kreft" was conducted. A list of hospitals, regional health authorities, interest organisations and "other" linked to the keyword appeared. The links were systematically examined, and the content described. Also external links were mapped. The majority of the institutions mentioned had brief information about CPP, but linked to the Webpages provided by the Directorate of Health for more information. Some hospitals had developed their own information on CPP, with overviews of CPP coordinators, leaders and timeframes for each cancer type. Since the Web portals from the Directorate of Health were linked to from most institutions, there are expectations at hospitals all over the country that the Directorate of Health should provide adequate information about CPP. This is the reason why we decided to study the content on CPP available at helsedirektoratet.no and helsenorge.no more closely. In this paper, the following data is studied: CPP online information provided by the Directorate of Health including: 1) Information at the health portal helsedirektoratet.no [13] and 2) Information at the health portal helsenorge.no [7]. Selected sections are studied in depth, and analyzed in a CDA perspective.

IV. Analysis

A. Information about CPP in Norway available online

The online information is examined theoretically and analytically through drawing on the following notions from CDA; 1) addresser and addressee, and 2) knowledge exchange versus activity exchange.

B.Addresser and addressee

The primary addresser of both helsedirektoratet.no and helsenorge.no is the Directorate of Health.

Helsedirektoratet.no information concerning CPP has a double addressee: first, GPs, healthcare professionals in primary- and specialist healthcare, and secondly, patients. The information for healthcare professionals as primary addressee is for example the contact information emphasizing that the GP and other healthcare professionals can contact the Directorate via email address if they have questions. In addition, topics of the Website are diagnostic guidelines and the role of the CCP coordinator. On the same page, there is a link called "CPP and patient information". The primary addressee is the healthcare professional who can subsequently advice the patient to access it herself/himself.

The primary addressee at *helsenorge.no* is the patient. Helsenorge.no is a health portal, and the idea is to gather information relevant for the patients (for example general health information, booking appointments, electronic health record at hospitals (pilot in two regions), ePrescriptions, referrals, deductibles, summary care record/ kjernejournal). The information about CPP is found under the categories diagnosis/cancer/CPP (updated 30.08.2015).

In the subsequent section we analyze the online health information on CPP in two cases, respectively: Case 1: assessment of selected CPP information at helsedirektoratet.no and Case 2 assessment of selected CPP information at helsenorge.no.

C. Case 1: helsedirektoratet.no

The information on the Web portal helsedirektoratet.no concerning CPP is organized in the following main categories: 1) Aim of CPP, 2) CPP and patient information, 3) Diagnostic guidelines- the role of the general practitioner (GP), 4) Implementation of CPP, 5) Experiences from Denmark and 6) Code Guidelines.

Category 2), "CPP and patient information", links to "general information for all CPPs", which is the same information as presented in the Action Plan and "general patient information about examination for suspected cancer". In addition, "CPP and patient information", links to information about all 28 types of cancer included in the CPP program. The information on each cancer type is organized with a) an introduction to the specific type of cancer, b) the entrance to the CPP (risk groups, symptoms, referral to CPP, decision about CPP, information and dialogue with the patient, responsibility for the referral, registrations, pathway times, c) examination about the cancer type, d) initial treatment, e) follow-up and control, pathway times, e) registrations of codes in the CPP (codes to be used at examination, at biopsy, at transferal to another hospital, at decision about initial treatment, at treatment, at the end of the CPP). With reference to the analytic categories of Fairclough [12], in the subsequent section we do an analysis of the communication exchange in category 2) "CPP and patient information".

a) Lung cancer as an illustratve example

All information about the 28 cancer types is structured similarly, so the information about lung cancer is used as an illustrative example [14]. The link: "The introduction to CPP for lung cancer" leads to the section on "General information about lung cancer" which contains a brief *knowledge exchange*, as manifest in for example the following two sentences with *statement of facts*: "Annually around 2,900 are diagnosed with lung cancer. In 2012 respectively 1,600 men and 1,300 women were diagnosed." The factual statements are followed by an *evaluation* "The prognosis is not good", with a support of a new *factual statement*: "Five years of relative survival is 19 % for women and 13 % for men". This worrying reading is tentatively *modified by a new statement of fact*: "Modern treatment resulted in somewhat better prognosis than previously". The next section "National action program" is brief, and links to an action plan (broken link).

The subsequent section, "Pathway coordination" refers to the coordination of the pathway. The main kind of communication exchange here is *activity exchange*, as exemplified in: "Coordination of the patient pathway shall ensure effective pathways from the time the referral is received by specialist health service until treatment is started or CPP is concluded, without undue delay and with close cooperation between all involved departments and specialists." This *instruction*, from the Directorate of Health to all departments and specialists, gives directions about how to proceed with CPP effectively and within an expected timeframe. Subsequently, a new *instruction* is given, this time specifically about how hospitals are expected to coordinate the pathway: "All hospitals that investigate and treat cancer should have course coordinators who have close contact with the patient and involved agencies." This is followed by an *instruction* on who shall be in charge of the particular CPP: "Pulmonary Physician will be the one that formally starts CPP for lung cancer, and has a central role in coordination with other specialists." The two latter *instructions* concern roles and responsibilities professionally, and within and across institutions.

Also in the section "Multidisciplinary teams", the *activity exchange* is the primary communication exchange mode, "All hospitals that treat lung cancer shall have regular multidisciplinary decision-making meetings. At meetings where lung surgery is considered, the lung specialist, the thorax surgeon, the oncologist, the radiologist, the pathologist, the nuclear medicine specialist and the pathway coordinator should participate". This demonstrates an *expectation* from the Directorate of Health about how the respective hospitals should organize decision-making and who are expected to participate. This requires thorough coordination of clinical personnel, and puts demands on their priorities. Interestingly, the patient is not mentioned. This is followed by a *knowledge exchange* formulated as an *evaluation*: "These meetings ensure quality control of diagnosis and surgical treatment as well as planning of further treatment."

The section "Information and dialogue with the patient" is also characterized by *activity exchange,* but is to a lesser extent dominated by instructions and more characterized by underlying *values* from the addresser as to how information *should* be conveyed. Fairclough [12] claims that there is a value-content in factual statements that links knowledge exchange to activity exchange. When promoting information and dialogue with the patient as an important tool in the CPP, the Directorate of Health expresses the objective that patients and next-of kins should experience good information, involvement, influence and dialogue throughout the pathway. In an *activity exchange,* the inclusion of the patient is not formulated as an instruction, but a *solicitation*, expressed in present tense: "The patient and the responsible doctor jointly decide the further pathway". However, the expectation about the patients being treated with respect and empathy is an *instruction*: "The communication with the patients and their families *shall* in all circumstances be based on respect and empathy. Information and dialogue should take place in a considerate way and be adapted to the recipient's individual abilities such as age, social situation, language, expressed wishes and needs." This instruction contains value-content about how patient communication is best conducted, but does not really specify what respect and empathy actually means in the context. However, subsequently we read a *value-content solicitation* about an aspect that the healthcare professional should emphasize in the communication, namely the patients expectations throughout the pathway: "Further, the communication with the patient *should* include a clarification concerning expectations about the pathway, including inclusion of patient and next-of kin". Note that the difference between the tense of the use of the verb *shall* and *should* also indicates the difference between instruction and solicitation here.

Finally, there are *instructions* about how communication and information should be: «Communication and information shall be consistent and coordinated. As part of the communication, the patient and next-of-kin shall iteratively be involved and informed about examination results and next step in the CPP." The addressee of this instruction is seemingly the coordinator since she is the one that is appointed to have all coordination and overall contact with the patient. The subsequent *instruction* is giving directions to the hospital managements about how they should organize for good communication: "The hospitals shall, in cooperation with relevant patient organizations, prepare for talks with authorized peers if the cancer patients and/or the next-of-kins wish this." The Directorate has *expectations* about information and dialogue that accompanies the implementation of CPP, but with unspecified criteria.

D. Case 2: helsenorge.no

The information online about CPP offered on the patient health portal helsenorge.no is introduced with a section concerning the goals of the CPP, i.e. that patients should experience a complete and predictable pathway without unnecessary delays caused by non-medical factors, and that patients shall receive information and be involved.

The pathway coordinator is mentioned briefly as contact point for the patients. The patients will be informed about a telephone number to call for information. In addition, on the right side of the Website, there is a telephone number, with opening hours during daytime. The information emphasizes that the results are being good so far, for example in keeping the predefined time schedules.

Helsenorge.no information concerning CPP links to information booklets for the 28 distinct types of cancer included in the CPP-program. The booklets introduce the respective CPP for the specific type of cancer with two pages concerning the examination about cancer and the phases for examination and treatment. Exanimation and treatment is illustrated in a table as illustrated in Table 1, where the expected time is particularly emphasized.

TABLE 1: ESTIMATED TIME IN EACH PHASE OF THE CPP

De forskjellige fasene i utredningen	Kommentarer	Anbefalt innen:	
Fase 1: **Dette er tiden fra sykehuset mottar henvisning til du møter i sykehus første gang**	Forløpskoordinatoren sørger for å sette opp timene du skal ha i utredningen.	**7 kalenderdager**	
Fase 2: **Dette er tiden fra du møter i sykehuset første gang til utredningen din er ferdig**	I løpet av dette tidsrommet blir det gjort undersøkelser av deg for å avklare om du har kreft eller ikke. Ved mistanke om kreft vil du undersøkes av lege. Det tas blodprøver, beinmargsprøve og ofte ulike røntgenundersøkelser. Når resultatene fra undersøkelsene og prøvene er klare, vil det som oftest kunne avklares om du har kreft eller ikke. Beslutning om diagnose tas. Har du ikke kreft, avsluttes pakkeforløpet.	**20 kalenderdager**	
Fase 3: **Dette er tiden fra det er fastslått at du har kreft og til behandlingen starter**	Har du kreft, vurderes nå om du skal behandles og eventuelt hvilken behandling som er best for deg. Beslutning om dette tas i samråd med deg. Myelomatose trenger noen ganger ikke behandling, men når behandling er aktuelt vil dette oftest være kjemoterapi, av og til med strålebehandling.	**Medikamentell behandling**	**3 kalenderdager**

All information about the CPP diagnostic categories is organized in the same way. Table 1 is an illustrative example borrowed from the booklet on myelomatosis. The table emphasizes *knowledge exchange* in *progress description* of the 3 different phases in the examination and the *estimated* time frame (in days) for each phase. The middle column contains comments. The first comment, "The pathway coordinator makes sure that you will undergo examinations in order to clarify if you have cancer or not", is an example *activity exchange* with a third party actor, the coordinator. The primary addressee is the patient. However, the indirect addressee here is the coordinator and the healthcare institution that has the responsibility to provide for this service.

The comments concerning phase 2, can be interpreted as a mix between knowledge exchange and activity exchange: depending on who the addressee is: "During this time frame examinations will be done to decide whether you have cancer or not. On suspicion of cancer you are examined by a doctor. Blood tests, bone marrow tests, and different x-rays are conducted. When the results are ready, usually it is possible to decide whether you have cancer or not. If you haven't got cancer, the CPP is terminated". Here we can identify a primary addressee; the patient, and an indirect addressee; the institutions in charge of providing care. The progress of the process is described. This is an example that shows that the information can be regarded as *knowledge exchange* to the primary addressee, the patient, as it explains the rationale behind what is going to happen throughout the pathway. However, to the indirect addressee, the information is rather *activity exchange*, and gives instructions to the healthcare professional about the procedures throughout the CPP.

V. DISCUSSION

We have seen that the information at both helsedirektoratet.no and helsenorge.no concerning CPP has multiple addressees; the healthcare institutions concerned with CPP, the healthcare professionals, including the management and the CPP coordinators, and the patients. At *helsedirektoratet.no*, the communication exchange about coordination, roles and responsibilities is characterized by activity exchange formulated as instructions concerning roles and responsibilities professionally and within and across institutions and expectations about multidisciplinary competencies required for professional decision-making. The communication exchange about the information with the patients is characterized by values. There are examples of instructions and solicitations containing value-content. Moreover, there are instructions and expectations about communicative responsibilities, but with unspecified criteria. At *helsenorge.no*, the knowledge exchange is characterized by the progress description and the estimation of time as factual information about the procedures in relation to the CPP. Whether the communication can be interpreted as knowledge exchange or activity exchange depends on the addressee: what can be interpreted as factual knowledge about procedures for the patient, can be interpreted as activity exchange for the healthcare professionals.

The theoretical terms from Critical Discourse Analysis *knowledge exchange* and *activity exchange* are valuable for analytically identifying what characterizes the communication exchange of online health information. The *knowledge exchange* is mainly relevant for expressing facts concerning cancer and CPP to the patients. However, from an analytic point of view, the term knowledge exchange is context-dependent, since the content interpretation may depend on who the main addressee is. The *activity exchange* is commonly used in the CPP information online, both as instructions and solicitations to the healthcare professionals and the institutions.

Online health information for the patients should be presented according to the needs of the patients at the different points in the pathway. Online information that pays attention to the heterogeneity of the patient group may be one of several measures to diminish the digital divide. Factual and targeted health information at each step of the pathway enables the patient to take part in decision-making, and make informed choices.

Health authorities that are responsible for ensuring adequate health information and communication, and consequently patients, would gain by taking the patient's eHealth literacies, i.e. prerequisites for and capabilities to access, understand and apply online information [2][3][4][5][6], into consideration when developing health information. Especially important is the consciousness about what information and communication exchange is relevant at what point during the pathway. Based on insights from the analysis of the CPP communication exchange, and informed by discussions on eHealth literacy (ibid), below is suggestion of a model of the patients' health information pathway. The model is called ADIUVAT (lat: helps), an acronym based on the patient's steps to finding, understanding and taking action upon the health information.

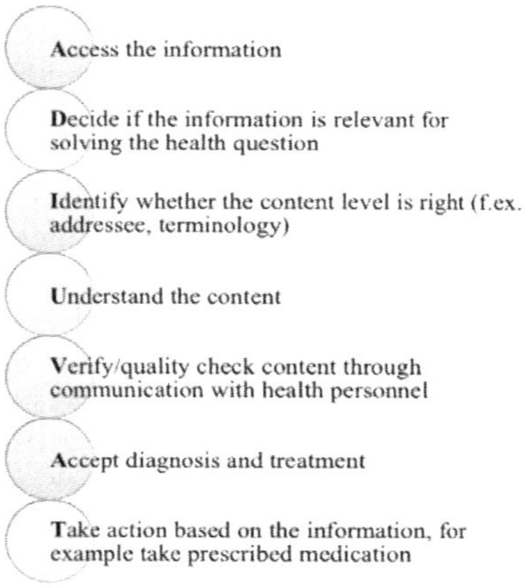

Figure 1: The ADIUVAT model

First, the citizen needs to know how to **A**ccess the information. After having overcome that first obstacle, the second step is to **D**ecide whether the information is relevant for solving the health question or not. This requires some basic knowledge about disease and medical notions. Thirdly, the citizen must **I**dentify whether the content level is right, if she/he is in the target group for the information and if the terminology is known. The fourth, and maybe the mort important step, is to actually **U**nderstand the content. Although the citizen can understand the information on a general level, she/he will probably not fully understand the implications for her/his personal condition. Thus, as the fifth step, she may need to **V**erify and qualitycheck the content through communication with health personnel. Having understood the content in relation to diagnostic results of the personal condition, the sixth step would be to **A**ccept the diagnosis and the treatment, and finally to **T**ake action based on the information, for example by taking the prescribed medication. Although being under development, the ADIUVAT model may be used as a starting point for discussions about what kind of information the patients need at what point in the pathway.

VI. Conclusion

The implementation of cancer patient pathways in the hospitals in Norway was a politically decided top-down process from the Norwegian authorities. The online information, which mainly consists of instructions, is perhaps a natural consequence of the top-down strategy, as it mirrors the instructions from the political leadership. Although the intentions of the Norwegian health authorities are changes in the patient-provider relationship, moving from a paternalistic towards a more participant model, this brief analysis demonstrates that the linguistic style of the online information is rather authoritarian. The online information about CPP would benefit by having more clearly defined addressees, more knowledge exchange of facts and more clearly defined visions and criteria for what constitutes high quality health communication and information exchange. Patients, healthcare professionals and healthcare institutions would profit of an overall and explicit elaboration about the defined criteria of high quality health communication and information exchange concerning cancer patient pathways, not only with regard to online information but also in verbal communication between patients and professionals in face-to face consultations. Knowledge-based and targeted information is an important measure to reduce the digital divide.

Acknowledgments

My gratitude goes to the Department of Neuroscience at NTNU and to the Operation Room of the Future at St.Olavs Hospital for providing resources to the postdoc project, and to Mikael Lyngstad and David Bowie for their inspiration.

References

[1] Wangberg, Andreassen, Kummervold, Wynn, & Sørensen (2009). Use of the internet for health purposes: trends in Norway 2000–2010. Scandinavian journal of caring sciences.

[2] Andreassen, Bujnowska-Fedak, Chronaki, Dumitru, Pudule, Santana, Wynn. (2007). European citizens' use of E-health services: A study of seven countries. BioMed Central Public health.

[3] Gilstad. (2014) Toward a comprehensive model of eHealth literacy. Toward a comprehensive model of eHealth literacy. CEUR Workshop Proceedings.

[4] Monkman and Kushniruk *(ACMI Fellow; CAHS Fellow)* eHealth literacy issues, constructs, models, and methods for health information technology design and evaluation, http://www.kmel-journal.org/ojs/index.php/online-publication/article/view/514/281

[5] Bautista(2015).From Solving a Health Problem to Achieving Qua lity of Life: Redefining eHealth Journal of Literacy and Technology, Volume 16, Number 2.

[6] Norgaard, Furstrand, Klokker, Karnoe, Batterham, Kayser, Osborne (2015): The e-health literacy framework: A conceptual framework for characterizing e-health users and their interaction with e-health.

[7] Helsenorge.no/sykdom/kreft/pakkeforlop-for-kreft. (2015)

[8] Cancer Package Pathway (CPP). A general introduction for all CPPs, Hdir 11/2014, https://helsedirektoratet.no/retningslinjer/pakkeforlop

[9] Tveit, Hdir, Experience Conference CPP, 2015, Stjørdal.

[10] Dyrop, Safwat, Vedsted, Maretty-Nielsen, Hansen, Jørgensen, et al. Cancer Patient Pathways shortens waiting times and accelerates the diagnostic process of suspected sarcoma patients in Denmark. Health Policy. 2013;113(1):110-7.

[11] Wood, Kroger.(2000): Doing discourse analysis: Methods for studying action in talk and text. Thousand Oaks, CA, US: Sage Publications, Inc

[12] Fairclough (2003): Analysing Discourse: Textual Analysis for Social Research

[13] Directorate of Health (2015): CPP https://helsedirektoratet.no/kreft/pakkeforlop-for-kreft and

[14] Directorate of Health (2015): CPP Lung cancer, https://helsedirektoratet.no/retningslinjer/pakkeforlop-for-lungekreft.

Designing in Rural Highland Contexts

Exploring the role of technology in facilitating human connections

Tara French, Jeroen Blom, Sneha Raman
Institute of Design Innovation, The Glasgow School of Art
Forres, UK
Email t.french@gsa.ac.uk, j.blom@gsa.ac.uk, s.raman@gsa.ac.uk

Abstract— **There is a need when introducing new technology in health and social care settings to involve those who will use the technology, or be affected by it, at an early stage of the design process. Experience Labs employ a participatory approach for different stakeholders to share lived experiences, and collaboratively create ideas and solutions for health and social care challenges. The Labs serve the whole of Scotland, including the remote Highland areas, and are designed bespoke to the project goal and context. In this paper, we propose that the participatory approach of the Experience Labs provides a valuable space for collaboration to explore the potential and impact of supportive technology in rural communities. Initial findings suggest that participating in Experience Labs leads to a better understanding of communities of care as established by individuals and support structures in place, and how these can be mediated by technology. We conclude that the introduction of technology in these rural Highland communities of care needs to support existing networks, unique to each community, and consider the impact on changing existing structures.**

Keywords- rural health; participatory design; community; technology; experience labs; care structures; collaboration space; experiential learning.

I. INTRODUCTION

A. *Health care in rural areas of Scotland*

The Highlands and Islands of Scotland has a population of approximately 466,00 [1] and is 'one of the most sparsely populated parts of the European Union' [2]. The existing care infrastructure is facilitated by NHS Highlands and the area is served by one general hospital in the Highland capital of Inverness. The National Health Service (NHS) in Scotland encompasses 14 regional health boards. Each board is responsible for providing services and resources to their local region. Health services in remote and rural contexts are served by GPs, community health care staff, Scottish Ambulance Service, and local council [3].

B. *Challenges for rural communities*

Several challenges exist for rural communities in relation to health and social care including access to services [4], geographic challenges [4], recruitment and retainment of professionals [3] [5] and lack of integrated care infrastructure [3]. Although the benefits of telehealth and digital innovations are recognised in terms of supporting communities and reducing hospital admissions [6], there is a scepticism surrounding the usefulness, organisational fit, and quality of communication of telemedicine [7]. To date telehealth impact has been small [3] and therefore a new approach to designing is required for rural communities.

The role of design in shaping future health and social care services is now increasingly recognised across the health service. Research in the field of rural primary healthcare is also beginning to embrace community participatory approaches and outcomes suggest that involving communities adds value to the process [3]. Within the NHS, design thinking and approaches such as user-centred health design and evidence based co-design have been employed within the context of health improvement [8] [9]. Online design tools and toolkits have also opened up design to those who are not formally trained designers [10]. The toolkits are useful for healthcare staff to involve patients in ongoing service improvement, however, their capacity to create transformative and sustainable innovations to public health and care challenges is not clear.

The scope of wellbeing and quality of life extends beyond enabling patients to experience a good state of health, to ensuring that they have dignity, positive relationships and feel included in the process. This cannot be created and maintained using digital means alone. There is a need to ensure that technology is designed to support both professionals and patients rather than replace existing services and resources, as this can contribute further to experiencing feelings of loneliness and isolation. This highlights the importance of 'community' when considering a rural context such as the Highlands and Islands of Scotland.

Design can play a key role in crafting technology to facilitate connections between individuals and communities by enabling communication, and overcoming some of the existing barriers to participation and access. We term this process creating 'communities of care' which aims to enable individuals to become active agents of their own health, and support each other in their community.

This paper seeks to contribute to the understanding of the role of technology in the rural health and care context, and the role of design in co-creating sustainable futures with health professionals and patients. We begin by introducing our design-led Experience Lab approach in section 2, highlighting the collaborative and temporal nature of the research approach to explore the potential and impact of technology in the delivery of care in rural communities. In section 3 we present three case studies of Experience Labs in the rural context to illustrate the approach and way in which the community of care is established for each. Finally, we discuss the knowledge we have gained when designing in the rural context and the resulting implications for technology innovations.

II. EXPERIENCE LABS: MOBILE TEMPORAL SPACES FOR HEALTH AND SOCIAL CARE INNOVATION

When considering the role of technology in health and social care innovation, there is a need to involve those who will use or benefit from the innovation at a much earlier stage of the process [11] to design and develop technology that is appropriate within the context.

Experience Labs are led and developed by the Institute of Design Innovation at The Glasgow School of Art. The Labs are currently applied within the context of health and social care, forming a core part of the Digital Health and Care Institute (DHI) [12]. The DHI is an innovation centre funded by the Scottish Funding council and is tasked to deliver value for the Scottish people and economy by innovating in the health and social care sector. The Experience Labs employ a participatory design approach to provide a space for collaboration and enable co-creation with different stakeholders across a range of projects. We are exploring the potential of our approach across the health and social care context in order to share our knowledge and learning with the wider design community and researchers in this context. We hypothesise that our participatory approach can respond to the challenges faced in this context through the space we create for collaboration and creativity, where we support participants towards creating "preferable futures" [13]; through designing bespoke tools and artefacts to support design activities to make ideas tangible; through the interdisciplinary skills of the Lab team; and an evaluation approach to communicate the value and impact, and share the knowledge and learning.

It is recognised that Experience Labs are not a new concept, however, there are elements which make the Experience Lab approach unique in their context. The location of the Lab team is distributed: based in Forres on the edge of the Highlands and Islands, and Glasgow in the central belt of Scotland. Experience Labs are mobile and serve the whole of Scotland with projects taking place in several locations, for example Wick and Thurso on the North coast, the Isle of Skye in the west, Fort William in the South and Inverness, the capital of the Highlands in the East (see Figure 1). As described previously, the Highlands present unique challenges for health and care services and rural communities. The Experience Lab team needs to have a presence in the area and engage with the local communities and organizations in order to gain an understanding of the context and landscape, and address the local care challenges appropriately. The mobility of the Labs provides the opportunity to create temporary spaces, bespoke to the project and context, in which concepts can be explored quickly.

The participatory nature of the Labs aims to support the development of collaborative relationships through creating new communicative spaces and experiential learning [14]. Participants are invited to design and evaluate solutions, together with the Lab team and with relevant stakeholders, with a focus on how proposed solutions would have an impact on their individual lives and context. The Lab team design the tools and artefacts to support participants to make their ideas tangible and use methods such as prototyping and role-play to test and iterate ideas during the Lab [15]. The collective skill of the team allows for a non-linear process that is flexible and can respond to the needs of participants. The participants' lived experience is at the core of the insights and concepts emerging from the Labs. The asset-based approach [16] [17] of the Labs focus on empowering participants to be creative, and to share their skills and experience towards identifying opportunities for healthcare innovation. This allows participants to tackle challenges and identify opportunities for change or improvement rather than focussing on problems and deficits [16] [17].

III. EXPERIENCE LABS IN THE RURAL CONTEXT: CASE STUDIES

The following section describes three case studies of projects carried out in the rural context with communities in the Highlands. The projects focus on the needs and expectations of people to ensure that health and social care services are accessible to all in ways that are relevant and desirable. In each case study, we describe the context of the project, the design of the Experience Labs and the community of care established in the context. Figure 1 provides a map of Scotland, including population density. The map illustrates areas in the Highlands where Experience Labs discussed in the case studies have taken place using red markers.

A. *Case study 1: Digital brokering*

1) Context: The Digital Brokering project explored the potential of a digital platform to connect those who need small services to those who can provide them, with a particular focus on rural communities. The need for a platform was identified by Albyn Housing Association and the University of the Highlands and Islands [18]. It was found that accessing and exchanging small services between residents would have a positive effect on rural communities.

2) Design: The Experience Lab provided the opportunity for multiple stakeholders, including Highland rural residents, public sector and third sector representatives to come together and map their experiences of accessing services in rural communities. The mapping method employed in the Lab [19] provided participants with the tools to identify people, places and things that were important in terms of accessing services. A second session then focussed on the potential impact of a digital platform in accessing these services in terms of the technological requirements and the effect on existing relationships and structures.

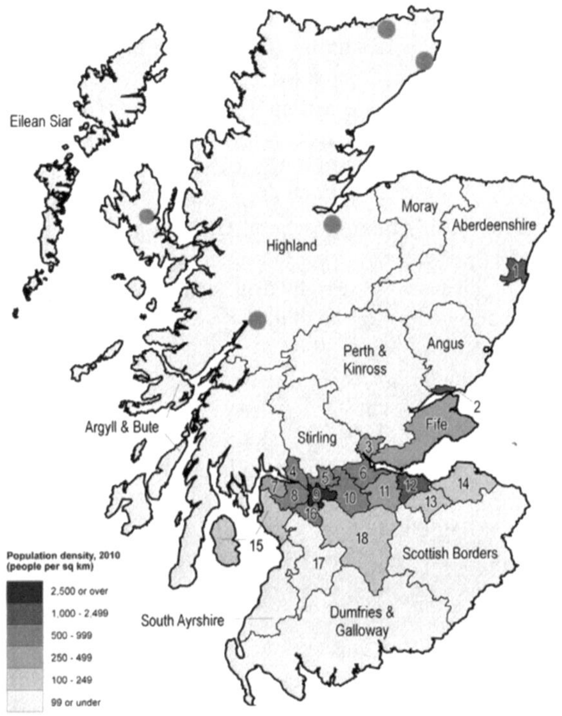

Figure 1. Areas of the Highlands where Experience Labs took place (red markers) and population density in Scotland [20]

3) Community of care: In order to explore the potential of a digital platform, the Experience Lab made the current experience of accessing services and the potential impact of technology support tangible. The community of care in this context described the relationships and structures within rural areas. The project specifically suggested how an envisioned technology would strengthen rather than replace structures through offline accessibility, accreditation of services and moderation of feedback. Furthermore, it was discussed that the platform could indicate opportunities for service providers to start/expand a business to benefit the community in supporting unmet needs. However, this would require careful consideration of an organization with local knowledge to prevent bypassing of local businesses and potentially damaging existing structures.

B. *Case study 2: Virtual hospice*

1) Context: A key objective of Hospice care is to empower individuals and carers to live with life-shortening illness in their own communities, by offering support and increasing their confidence. This project involved a collaboration with Highland Hospice with the aim of widening access for patients and professionals and enabling higher quality palliative care to be delivered through a 'Virtual Hospice.' The project involved designing a sequential series of Labs in order to define the various elements of the Virtual Hospice.

2) Design: The project involved a series of Experience Labs with Highland Hospice staff, healthcare professionals, patients and a carer. The aim of the Labs was to explore awareness, use and barriers to use of Highland Hospice services and identify opportunities for Hospice services that could be delivered at a distance, as well as ideas for potential future services. One of the Labs was replicated across four different locations in the Highlands (Inverness, Fort William, Isle of Skye, and Wick) using an iterative approach and focussed on identifying opportunities for delivery of hospice care tailored to the local context.

3) Community of care: The Virtual Hospice was initially envisioned by Highland Hospice as a room within the main Hospice building where digital services could be delivered. Over the course of the Labs, this definition expanded from a purely digital platform to involve a multi-layered and interconnected network of people and places supported by technology. Across locations a number of key barriers to accessing Highland Hospice services were identified. Some participants proposed that one of the ways to overcome these barriers would be for Highland Hospice to build on existing networks of people in the community, and for this to be mediated by technology; not for the solution to be digital. This led to ideas specific to the local context – e.g. collaborating with Men's Shed in Skye and a Care Home in Fort William. Both ideas focussed primarily on leveraging on existing community or care groups as points of delivery for Hospice services.

The community of care therefore involves the linking of networks involving people, e.g., local healthcare professionals, social and care groups, patients and families; community spaces, e.g., the Men's Shed and Care Home; and digital platforms supporting delivery of improved palliative care.

C. *Case study 3: Directory app*

1) Context: A third case study provides an example of a different type of community of care but highlights challenges in a rural context compared to more urban areas, and how these could be addressed. The Directory App project explored ways of developing a directory of service application providing information about alternative services and points of care for the Scottish Ambulance Service [21].

2) Design: The Experience Lab began by exploring the current services available to ambulance clinicians across a number of health boards through a mapping session. The mapping session involved marking health and care locations accessible to each clinician in their respective NHS regional boundary on a physical map. The Lab also involved a role-play session of an ambulance call out scenario where the ambulance clinician required access to information about alternative services in the area (see Figure 2). The role-play session provided insights, which led to the co-design of a paper directory of services. The first prototype was tested by role-play using a mock tablet with the paper screens inserted. In the final design session the prototype was further iterated to develop the 'ideal' directory of services (see Figure 2).

3) Community of care: In this case study, the directory of services was initially envisioned to serve as a digital directory of contact information for local services available to ambulance clinicians in their current healthcare boundary.

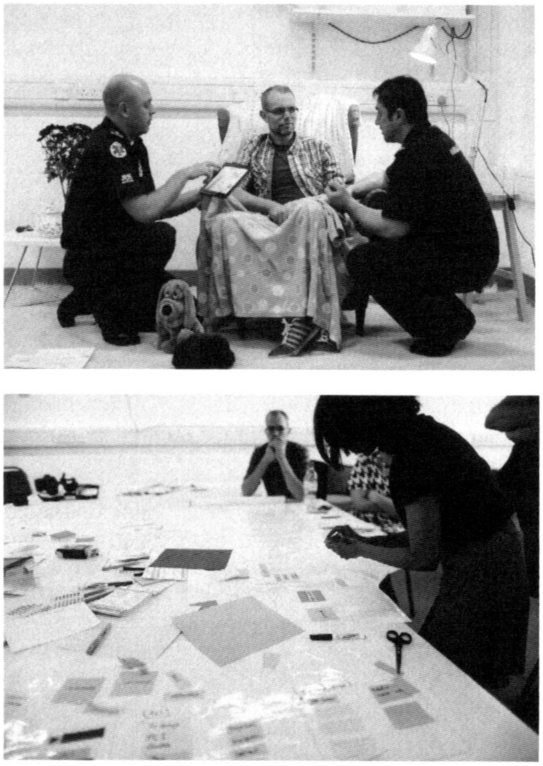

Figure 2. Images from the Directory App Experience Lab including role play and design session of the 'ideal' directory of services.

It became evident during the Lab that participants wished to have the ability to make contact with and refer patients to local services in the community, rather than simply have a directory of contact information. Therefore the community of care involves connecting ambulance clinicians with appropriate local services in their current healthcare boundary which best serves the needs of the patient, avoiding inappropriate transfer to A&E. This would provide ambulance clinicians with an awareness of points of care within their own regional boundary as well as other health care boundaries in Scotland and enable the patient to receive the most appropriate care.

IV. DISCUSSION

The findings and knowledge gained from the projects presented within the case studies, provide important considerations when designing in the rural care context, and highlight a number of implications for rural technology innovations.

A. Designing in the rural care context

1) Identifying needs through a participatory approach: The Experience Lab offers a participatory approach to designing for rural care contexts. The Lab provides the opportunity to ascertain the needs of individuals and communities living in rural contexts in order to ensure that the types of solutions proposed during the Lab meet the needs of the people involved. Early stages of the Experience Lab approach enable a deep understanding of participants' lived experience. It is important when designing in the rural care context to understand the experience of the participants, their communities and the wider rural landscape to have an awareness of existing support and infrastructure. The case studies presented support the participatory approach of the Lab in giving voice to individuals and communities living in rural areas, allowing them to feel a level of ownership over the outcomes.

2) Imagining future possibilities through tools and artefacts: The Labs allow participants to imagine and express their thoughts on technology through the use of artefacts and tools. These make ideas and solutions tangible and also help participants to see them in the context of use by weaving on their daily lives and relationships as part of the narrative. This helps to prioritise technology or solutions that are not only 'possible' or 'feasible,' but also preferable [13]. This is important for change, especially in health and social care where success of new systems rely on positive attitudes and behaviours. Participants in the Virtual Hospice case study could envisage how both the Men's Shed and Care Home hubs could work through a visual tool and props, which helped to develop the idea and create a narrative. In addition, the case study of the directory of service application for ambulance clinicians took this a step further by enabling participants to experience the co-designed application through role-play.

3) Creating value and enabling new collaborations: The collaborative nature of the sessions and the use of bespoke design tools bring issues and challenges to life between different stakeholders. In the case studies, rural participants expressed their appreciation of being heard and having the opportunity to share their experiences. Institutional stakeholders involved in the collaboration valued the depth in the scenarios being created through the tools and artefacts used in the Lab, and the level of detail expressed in relation to the needs and desires of people who may use the proposed solution. By focussing on communities and context of use, the relationships and exchanges are enhanced through the new ideas that emerge. This will potentially support the adoption of a new technology because it will be used by the same people in the same environment as those involved in the Lab.

The communicative spaces facilitated by the Experience Lab, enables collaborations and relationships to be established for new ways of creating communities of care. For example in the Virtual Hospice case study, a potential collaboration which emerged involved the Hospice engaging with a local care home in one Highland region. By being involved in designing and trialling this model in the Lab, a senior care home nurse could imagine part of being a system that does not yet exist and feel more inspired and open to try a new way of working.

B. Implications for technology innovations in the rural care context

1) Technology as an enabler: The case studies presented in this paper highlight the emerging theme across the Experience Labs conducted in rural contexts to date; the role of technology in enabling and mediating connections with others i.e. enabling a 'community of care.' In this case, technology is not the single solution to the challenges which are experienced in rural contexts, but is a way of facilitating support through existing networks and connections. We therefore suggest that participants in Highland communities have deeper incentives to rely on human contact potentially due to the rural context within which they live. In the case studies, participants talked about the human support, which in rural areas automatically means connecting people locally.

2) Low risk, iterative model for technology innovation: In all the three case studies, the technology proposed by participants were simple existing technology that people use in their everyday lives, or adaptations of the same. This provides an opportunity for trialling new health and social care models quickly and without very high set-up costs. At the same time it can significantly cut down the development time for completely new technology. In order to inspire individuals and clinicians to adopt new ways of living and working it is important that they use these everyday technologies they feel confident with and fit with their everyday life. As there are no high costs or lengthy development time associated with the iterative approach, this instils more confidence in rural health boards to invest and trial new models of health and social care delivery and enables them to gather evidence quickly, which can then inform further iterations and development of the technology-supported solutions.

3) Practical issues of technology in the rural context: A practicality of using technology which is evident in rural areas is internet connectivity. A service which relies on an online platform would quickly be rendered useless if people cannot access it when required. In the Digital Brokering case study, this issue was resolved by adding an offline directory to a platform aimed at connecting people to exchange services. In this situation the technology was primarily being viewed as a mediator with the reliability on other people being the main goal.

4) Long term impacts of technology on communities:

Participants are aware of the wider impacts of technology on the community they live in beyond its functional use. In the Digital Brokering case study, it was suggested that a platform for brokering of small services between residents would need careful moderation by an organization familiar with the community. Participants described the potential impact of negative feedback on the dynamics of a community, and how a platform might lead to bypassing existing local small businesses. When considering the introduction of a technological solution to a rural context it is important for participants from different backgrounds to collectively explore the impact of proposed technologies on their specific community and existing structures. The fragile nature of small Highland communities requires particular attention to the impact over time, not necessarily for the functioning of the technology, but for its position within a community.

There are also important implications in terms of ensuring people do not become isolated or lonely through the introduction of technology-focussed solutions. The case studies presented proposed solutions aiming to connect individuals. We anticipate that the resulting innovations may also provide ways to address social isolation and loneliness through the support provided by the community of care, enabled by technology.

V. CONCLUSION

Communities of care in the rural Highland context depend on providing support through existing networks of people and care infrastructure within that community. Technology is the enabler to ensure that people have access to care and services that meet their needs. The Experience Lab approach highlights the importance of the impact of new technologies and services on the people involved and in the context of use, rather than focussing on the functionality of technology or service alone. In this way, the Experience Lab approach allows people to imagine a wide variety of possibilities and identify those that are preferable. The collaborative nature of the sessions can assist in overcoming some of the scepticisms surrounding the introduction of new technologies.

The participatory research approach of the Experience Labs offers an insight into the potential for technology in the delivery of care from a community perspective. New collaborations are enabled through the participatory approach. The use of design tools and artefacts help to explore future possibilities and empower participants to create value within the community. The Labs help to better understand what a local community of care entails in a rural context, and how it can be supported by technology.

Based on the case studies presented, it is suggested that in order for technology to effectively enable delivery of care in rural contexts there is a need to consider existing networks of support within the community and adapt and tailor the technology based on the local context. The case studies suggest that adopting a low risk, fast and iterative approach ensures appropriateness and acceptance when designing and implementing new technologies in the rural context. However, there is also a need to consider practical issues as well as long-term impacts of technology on communities.

Traditionally when designing new technologies and services, there is a focus on designing to address the needs of the general population, which results in gaps when applied to rural contexts with their unique challenges [22]. The focus on the rural context and the mobile nature of Experience Labs ensures that the innovations are more applicable to the local needs.

Future work will explore whether the solutions, which address the unique challenges in rural Highlands are applicable and relevant to the development of similar solutions for the urban context.

ACKNOWLEDGMENT

We would like to thank all our participants who took part in the Experience Labs described in the case studies. We would also like to thank our project partners: Albyn Housing Association, The Glasgow School of Art, Highland Hospice, Scottish Ambulance Service and the University of the Highlands and Islands. We would like to acknowledge the support from our Experience Lab colleagues throughout the projects.

REFERENCES

[1] H. Smith, N. Sobey and C. Ross, Highlands and Islands Area Profile, Highlands and Islands Enterprise, May, 2014.

[2] The Scottish Government, Highlands and Islands Scotland: European Regional Development Fund 2007-2013: Structural Funds Operational Programme, July, 2008.

[3] J. Farmer and A. Nimegeer, "Community participation to design rural primary healthcare services," BMC Health Services Research, vol. 14, March, 2014, doi: 10.1186/1472-6963-14-130.

[4] L. Philip, A. Gilbert, N. Mauthner, and E. Phimister, "Scoping study of older people in rural Scotland," Research Findings, vol. 3, 2003, pp. 1-4,.

[5] J. Farmer, W. Lauder, H. Richards, and S. Sharkey, "Dr John has gone: assessing health professionals' contribution to remote rural community sustainability in the UK," Social Science Medicine, vol. 57, August, 2003, pp 673-686.

[6] A. Roberts, L. Garrett and D. J. Godden, "Can telehealth deliver for rural Scotland? Lessons from the Argyll and Bute Telehealth Programme," Scottish Medical Journal, vol. 57, February, 2012, pp. 33-37.

[7] G. King, H. Richards, and D. Godden, "Adoption of telemedicine in Scottish remote and rural general practices: a qualitative study," Journal of Telemedicine and Telecare, vol. 13, 2007, pp. 382-386.

[8] S. Bowen, K. McSeveny, E. Lockley, D. Wolstenholme, M. Cobb, and A. Dearden, "How was it for you? Experiences of participatory design in the UK health service", CoDesign, vol. 9, 2013, pp. 230-246.

[9] G. Robert, J. Cornwell, L. Locock, A. Purushotham, G. Sturmey, and M. Gager, "Patients and staff as codesigners of healthcare services", BMJ, 350, 2015, pp. 7714.

[10] H. Bevan, G. Robert, P. Bate, L. Maher, and J. Wells, "Using a Design Approach to Assist Large-Scale Organizational Change "10 High Impact Changes" to Improve the National Health Service in England", The Journal of Applied Behavioral Science, vol. 43, 2007, pp. 135-152.

[11] E. B. N. Sanders and P. J. Stappers, "Co-creation and the new landscapes of design," Co-Design: International Journal of CoCreation in Design and the Arts, vol. 4, 2008, pp. 5-18.

[12] Digital Health and Care Institute, Available from www.dhi-scotland.com. [retrieved: March, 2016]

[13] I. McAra-McWilliam, Available from: http://tinyurl.com /odual6h 2014. [retrieved: March, 2016]

[14] P. Reason and H. Bradbury, The Sage Handbook of Action Research. London: Sage Publications Inc., 2008.

[15] T. French, G. Teal and S. Raman, "Experience Labs: co-creating health and care innovations using design tools and artefacts", Design + Research + Society DRS 2016, In press.

[16] J. Foot and T. Hopkins, "A glass half full: how an asset approach can improve community health and wellbeing," London: Improvement and Development Agency, 2010.

[17] A. Morgan and E. Ziglio, "Revitalising the evidence base for public health: an assets model," Promotion and Education, vol. 14, 2007, pp. 17-22.

[18] G. Roger and S. Bradley, "Feasibility of delivering assisted living services to vulnerable Albyn Housing Society Customers," Centre for Rural Health, April 2012.

[19] M. P. Johnson, "Mapping design things: making design explicit in the discourse of change, "unpublished Ph.D. Thesis, 2015.

[20] Office for National Statistics, Scotland population density: by council area, 2010, August, 2012.

[21] T. French and G. Teal, "Co-designing a digital directory of services," Procedia Computer Science, vol. 63, September, 2015, pp. 445-450, doi:10.1016/j.procs.2015.08.366.

[22] R. Chapman and T. Slaymaker, "ICTs and Rural Development: Review of the Literature, Current Interventions and Opportunities for Action," Overseas Development Institute, Working Paper 192, 2002, pp. 5.

Patient-Centred Healthcare Team

Work Practice, Experiences, and Estimated Benefits

Trine S Bergmo
Norwegian Centre for e-Health Research
University Hospital of North Norway
Tromsø, Norway
Email: Trine.Bergmo@telemed.no

Anne Grethe Olsen, Johanne Aasen, Monika Dalbakk
Grete Jakobsen, Torunn I Johansen, Stian Kristoffersen, Veronica Sørensen
Division of Internal Medicine
University Hospital of North Norway
Tromsø, Norway
Email: {Anne.Grethe.Olsen, Monika.Dalbakk, Johanne.Aasen, Grete.Jakobsen, Torunn.Irene.Johansen, Stian.Kristoffersen, Veronica.Sørensen}@unn.no

Birgitte Aabotsvik, Monika Eriksen, Elisabeth Kollnes
Marte F Larsen, Hege Mathisen
Community Nursing Service
Tromsø Municipality
Tromsø, Norway
Email: {Birgitte.Aabotsvik, Monika.Eriksen, Elisabeth.Kollnes, Marte.Finanger.Larsen, Hege.Mathisen}@tromso.kommune.no

Stine Haustreis
Hospital Pharmacy North Norway
Tromsø, Norway
Email: Stine.Haustreis@sykehusapotek-nord.no

***Abstract*—The University Hospital of North Norway and the Community Nursing Service in Tromsø Municipality established an integrated patient-centred care model to improve the continuity and quality of care for multimorbid older patients living in the region. The model includes a proactive interdisciplinary team providing outreach services supported by m-health technologies. The team consists of healthcare personnel from the Community Nursing Service and the hospital. The main aim is to help the patients get the needed hospital care, ensure safe discharge, provide support in the home environment, and prevent unnecessary hospital admissions. This paper describes the mobile team's work practices and the information exchange process across two different levels of care. Furthermore, we analysed changes in treatment plans and estimated avoided hospitalisations. This is a prospective registration study supplemented with focus group discussions. This paper reports preliminary results from the first 98 patients the team handled.**

***Keywords*—Geriatrics, integrated care, patient centeredness, interdisciplinary teamwork, e-health, mobile health, safe discharge, avoided hospitalisation.**

I. INTRODUCTION

The Division of Internal Medicine at the University Hospital of North Norway (UNN) and the community service providers at the administration in the municipality of Tromsø initiated an integrated patient-centred outreach service to improve the continuity and quality of care for multimorbid older adults in the region. Mobile, proactive, and patient-centred interdisciplinary teams were established, which include personnel from the Community Nursing Service and the hospitals in Tromsø and Harstad. Their main task is to follow patients through the system, helping them receive appropriate, patient-centred, and timely healthcare services when needed at the most appropriate location. This also includes providing support at home to minimise hospital admissions. The patient-centred healthcare team (hereafter "the team") is funded as a cost-sharing collaboration between the municipalities and hospitals. As part of this large-scale intervention, a project evaluating effectiveness and cost effectiveness in a prospective matched control before-and-after study is underway. This study will start recruiting patients in June 2016. The protocol describing this research study is published elsewhere [1].

In this paper, we focus on the team established in Tromsø. The first test patients were referred to the team in October 2014. Approximately 300 patients were referred to the team in 2015. The team is now part of the everyday clinical practice. The team's work represents a new service model that is organised and works between primary and secondary care, producing new relations and work practices. The service model is not built around and driven by e-health technologies; rather, it uses ICT and mobile devices extensively to support team work by improving communication and facilitating information exchange between health providers and between patients and providers. We will describe the patient-centred outreach model, including the team's work practice, experiences, and challenges during its first year. Furthermore, we will report some preliminary results for the first 98 patients.

This paper is structured as follows: Section II provides the background and includes an overview of the local context and use of an interdisciplinary team in clinical work. Section III outlines the material and research setting. Section IV describes the method used in this study. Section V reports some preliminary results. Section VI discusses implications and limitations. Finally, conclusions and future work are discussed in Section VII.

II. BACKGROUND

The proportion of older people is rapidly increasing. It is estimated that the number of people over 67 years old in Norway will continue to increase by 13 000 annually. This demographic will nearly double from 624 000 in 2010 to 1.5 million in 2060, accounting for 22% of the total population [2]. The older population tends to have more health problems. More than 80% of individuals over 65 years old have one or more chronic conditions [3]. Individuals with multiple chronic conditions are also more likely to be hospitalised [4].

The effective management of acutely ill and complex patients with multiple conditions poses one of the greatest challenges in current hospital care [5]. Furthermore, healthcare utilization and costs increase significantly with an increasing number of chronic conditions [6]. One way to improve the management of multimorbid older adults is to implement integrated care models. Integrated care models were developed in response to fragmented and reactive care systems and a lack of patient involvement. One way to improve patient involvement and achieve more patient-centred care is to establish interdisciplinary teams [7].

Interdisciplinary teamwork offers an integrated approach to providing coordinated healthcare patients with complex long-term needs. This has been considered a good practice for more than two decades [8]. Evidence also suggests that interdisciplinary teamwork can improve the quality of care [9-13]. Well integrated models have been found to improve the process of care and could reduce hospitalisation and community service use [9]. Teamwork may also improve prescription and medication adherence [10].

The UNN and the community service providers at the municipality have jointly established an interdisciplinary patient-centred team to improve the quality of care for older multimorbid patients in the Tromsø area. This was mainly to improve the quality of care, reduce hospitalisations and cut costs. The team includes personnel from the Community Nursing Service and the hospital.

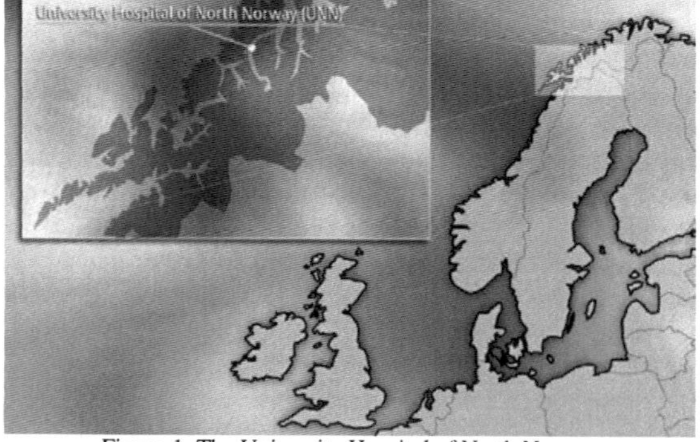

Figure 1. The University Hospital of North Norway

The UNN is the leading healthcare provider and health trust in North Norway. It serves as the local hospital for Troms County residents and other parts of Nordland, providing a full range of hospital functions (see Figure 1). The Municipality of Tromsø, which is the host for the hospital, is the largest in the area. It has a population of 72 000. The Municipality is responsible for ensuring good and proper health and care services at the primary care level to all its residents. These health and care services include childcare, preventive healthcare, nursing care, medical services, rehabilitation, and social services.

III. MATERIALS AND RESEARCH SETTING

In this section, we present the integrated patient-centred care model, its structure, and the team's work practice.

A. *The patient-centred healthcare team model*

The patient-centred healthcare team represents a mobile, seamless, and proactive model that aims to ensure safe discharge and prevent hospital admissions for older patients with multiple conditions. The team's main task is to follow patients through the system and help them receive appropriate and timely healthcare services when needed at the most appropriate location. The team identifies and assesses care needs early, provides support during discharge and follow-up, facilitates coordination and integrated services, and provides home-based services by actively monitoring, supporting, and caring for patients outside the hospital until adequate follow-up services are in place. This could reduce unplanned hospital admissions, reduce the need for community services, and improve or prevent deterioration in health and functional outcomes.

The patients will actively be involved in the care and self-management process. The team will identify patient goals, assess home situations, and facilitate individual tailored care plans and follow-up protocols. Such patient involvement and engagement in care has been shown to improve health and functional outcomes [14].

B. *The team structure*

The interdisciplinary patient-centred care model is a complex intervention that delivers collaborative care focused on care coordination. The team has two separate focus areas: a pre-hospital focus and a safe discharge focus. The former targets patients living at home. The latter addresses hospitalised patients. The core team consists of the team leader and two full-time coordinators (both geriatric nurses). One coordinator comes from the hospital, and the other is selected from the Community Nursing Service. The core team is responsible for daily teamwork management and ensuring patients receive appropriate and timely support and services. The core team seeks to:

- Understand which professionals providing care and services are currently available,
- Develop joint work to enable services integration,
- Encourage a proactive and structured assessment of patients goals and needs,
- Facilitate understanding and ensure the use of personal and advance care plans,

- Act as a central resource for the health and social care professionals, both within the team and across the municipality and hospital sectors.

The wider interdisciplinary team includes hospital and community personnel and consists of ten positions. These include one geriatric specialist (senior consultant), two nurses, two district nurses, two physiotherapists, two occupational therapists, and one pharmacist. Four of these are part-time positions.

C. Team work practice

The structure and organisation of the service model cover complex areas, such as patient and information logistics. Patient logistics include regulating the patient flow through the system, knowing when and how blockages occur and resolving them, using various instruments (such as case management, discharge protocols, capacity constraints, and home monitoring). Information logistics includes team communication, information exchange, and patient communication.

The patient-centred model includes four main tasks: to identify the patients with special needs for coordinated care; to conduct patient-centred need assessments; to facilitate individualised care plans and follow-up protocols; and to initiate meetings for the coordination and integration of patient care across service providers.

In the first phase after the start-up, the team actively promoted its existence and services by visiting GPs, district nurses, other community service providers, and hospital wards and clinics. The team has been assigned office space at the hospital and scheduled two daily meetings to plan activities and discuss patients. Between team meetings, the team conducts home visits, visits hospitalised patients, attends coordination meetings, assesses patients, and plans follow-up activities, writes case reports, and updates patient records. The patients are referred to the team by phone, electronic referrals, and ordinary mail. Most are inpatients at the hospital when they are referred to the team (see Table I).

The team assesses each patient's needs based on what the patients view as important. The team also assesses the fall risk and the need for special aid, reviews medication lists, and ensures that the patients' GPs are informed. Furthermore, the team ensures continuity of care in the hospital setting and during the transition phase from hospital to home. The team is responsible for the patients until adequate follow-up services are in place. Figure 2 illustrates the main steps in the patient-centred model.

D. Communication and information exchange

The team works with hospital inpatients, patients in nursing homes, and individuals living at home supported by home care. In Norway, the community services and the hospital sector have two distinct and separate patient information systems. Therefore, the patients referred to the team have two different patient records: the hospital's Distributed Information and Patient System for Hospitals (DIPS in Norwegian) and Profil, the community services' patient records.

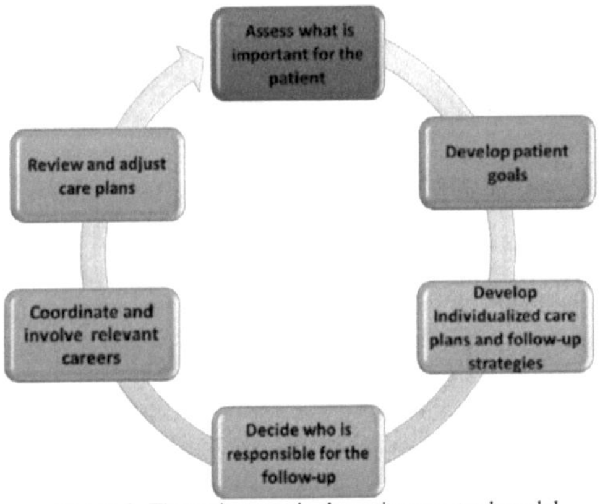

Figure 2. The main steps in the patient-centred model.

Ensuring effective day-to-day information exchange between team members across these sectors is one of the challenges in the project thus far. The team uses two different systems to access patient records. Two desktop computers are connected to the community service network and are used to access patient information in Profil. Three computers are connected to the hospital network and are used to access the hospital records. All team members have login credentials for both systems. To get all the needed patient information on one specific patient the team must log on to both systems. The team also employs laptops.

Furthermore, the team uses information and communication technologies to support patient treatment and follow-up outside of the hospital. The team members use mobile devices (phones and tablets) to communicate outside the office. The data security system does not allow them to include patient records on the tablets so they must be in regular contact with the office to receive updated patient information. The tablets are also used to film physiotherapy sessions. This facilitates patient exercises at home without in-person supervision. The films are also used for team education and training. The patients can contact the team through tablet videoconferences, as well. They make the calls through the secure health network using WebRTC technology.

IV. METHODS AND RESEARCH APPROACH

We used a multi-method research approach to evaluate the first 98 patients referred to the team. The patients included were over 60 years of age with multiple conditions, in need of coordinated care and considered to be at risk of experiencing adverse health outcomes. Data were gathered in a prospective registration study and in focus group discussions. For each patient referred to the team, changes in treatment plans and services offered were logged. Data were also collected from the electronic patient records. Group discussions with team members were conducted to support and deepen the understanding of the data and how the team works.

A. *Registration of the descriptive data*

The descriptive data used in this study were registered in two ways:

1. Each team member consecutively logged data by entering information about patients into a pre-defined registration form. Background variables (such as age, sex, total diagnoses, and number of hospital admission in the last 12 months), date of referral to the team and discharge from the team, services provided, and estimated benefits were registered in the case form.
2. The patient records, both from the hospital and the community service records, were then used to verify the registered data and to include additional missing data. Data on re-hospitalisations within 30 days were also included. Two of the authors extracted data from the two electronic record systems simultaneously, while a third author updated and included new data in the registration form.

B. *Focus group discussions*

Focus group discussions can be a useful way to learn about the team's everyday practices. They are helpful in reviewing the team's role as well as technological and organisational barriers. Lastly, they can reveal work practice dynamics and pinpoint common and different experiences among members [15]. Two focus group discussions were conducted including all team members. At the first meeting, eight team members were present. Nine attended the second meeting. Seven attended both meetings. Themes discussed included patient integration, case histories, tasks, perceived benefits, and the main challenges the team experienced. The first discussion lasted approximately one hour; the second ran about two hours. The discussions were audio recorded and then transcribed.

C. *Ethics*

The implementation of the new team model was defined and registered as a quality improvement project at the UNN and falls outside the Health Research Act. No approval from a regional ethics committee is therefore required. Evaluating resources used in such projects do not require patient consent. However, the local data protection supervisor did approve the project. All data are temporarily stored on a secure research server at the UNN.

TABLE I PATIENTS LOCALISATION AT REFERAL DATE

Where the patients are located when referred to the team		
Description	***No***	***%***
Hospital	62	63.3
Home	26	26.5
Nursing home	8	8.2
Rehabilitation centre	2	2.0

TABLE II CHANGES MADE IN TREATMENT AND PLANS

Changes and plans		
Description	***No***	***%***
Increased community services	70	71.4
Offered rehabilitation	18	18.4
Clarified home situation/nursing home	55	56.1
Rapid and complete hospital review	16	16.3
Written follow-up plan completed	53	54.1

V. PRELIMINARY RESULTS

In total, 101 patients were included in this study. Three did not want to receive services from the team, which left 98 for analysis.

The mean age was 80 years, ranging from 54 to 95. The patients had 3.5 diagnoses (ranging 0-7 per patients), and they had been hospitalised 0 to 9 times (mean 2.5) during the last 12 months. Of the patients, 41 were male. Hospital staff referred most of the patients to the team (62%). GPs and healthcare personnel in the municipality referred the remaining patients. Most were inpatients at the time of referral (see Table I).

The team initiated an increase in community services for more than 70 percent of the patients. Of these patients, 25% were offered rehabilitation. For more than half of the patients, the team clarified the home situation and made the transition process from the hospital feel safer for the patient. Table II shows the changes the team made.

The team estimated that some of the patients who received services avoided a hospital admission (15%) or shortened their lengths of stay (15%). Data from the electronic patient records at the hospital showed that only 3% of the patients handled by the team were readmitted within 30 days.

The team members agree that coordination is the most important part of their work, which is a service that did not exist prior to the team's establishment. This involves engaging the appropriate professionals as soon as possible, initiating and arranging meetings, and creating communication channels that include personnel from community care services and the hospitals. The first couple of days after discharge are critical for positive patient outcomes. It is also important that the patients take part in the planning process and are informed about available follow-up services.

Some patients are anxious about how they will manage on their own after a hospital stay. In situations where they have simple and specific needs, one home visit and one phone call to the nursing services can be enough to reassure the patients. The team emphasised the importance of being "interdisciplinary." The combined expertise that the team possesses enables a broad approach to the geriatric field. This was essential to ensuring high quality and effective integrated and patient-centred services.

Organisational barriers and the lack of information systems between the two levels of care were reported as the main challenges of working in a mobile patient-centred team. The team worked across the primary and secondary health systems, and this caused some challenges due to the different organisational cultures. Existing guidelines and procedures within each unit and defined meeting structures did not match well with the mobile patient-centred approach. Furthermore, two separate EPR systems limited the information exchange within the team and between the different units and departments. Another challenge was that the team could not access the patient records outside the office. The team members agreed that resolving some of these technical issues would make everyday teamwork more effective.

The smartphones and tablets were in regular use to contact the office for information during home visits and meetings. Videoconferencing was also used to communicate directly with patients. In one situation, videoconferencing was used to assess a patient whose condition had deteriorated. The team medical doctor spoke directly to the patient, and based on this exchange, pursued further action. Videoconferencing was also used in leg ulcer treatment, where the patient was at home with one of the team nurses and consulted an ulcer specialist at the hospital via video link. However, most videoconferences have been to coordinate health personnel across the different sectors.

VI. Discussion

In this paper, we described an integrated care model that includes mobile and proactive patient-centred interdisciplinary teamwork. We focused on the team's work practice as well as roles, experiences, and challenges during the first year. Furthermore, we analysed changes in treatment plans and reported some preliminary results for the first 98 patients.

These results show that the team initiated an increase in community services for more than 70 percent of the patients. Of these patients, 25% were offered rehabilitation. For more than half of the patients, the team clarified the home situation and thereby made the transition process from hospital to home feel safer. Furthermore, we found that only 3% of the patients handled by the team were readmitted within 30 days. Official statistics from 2013 report that 12.7% of the patients over 67 years old were readmitted after a hospital stay in Tromsø [16].

These positive results correspond to other findings reported in the literature. One study, for instance, reported that multidisciplinary patient-centred intervention offered to all patients 60 years or older reduces hospital readmissions for the patients who received the service [17]. Another study from Sweden reported that patient-centred care is associated with reduced the length of hospital stay [18] [19], reduced patient anxiety and uncertainty [20], reduced medical complications following surgery [21], and cost savings [22].

To understand the mobile patient-centred team described in this paper, it is important to assess how organisational issues and communication patterns in two different organisational cultures affect the team's work practice. The team realised that working across two different healthcare organisations made effective collaboration difficult. To facilitate such a change, a new culture must be created that supports an integrated patient-centred approach to patient care [23].

The strengths of the present study are its real-world setting and the involvement of designated health professionals in designing and developing a patient-centred team model. The model was initiated and is solely driven by clinical personnel at the hospital and the community services. The structure and organization of the team is continuously adapted to the clinical and everyday routines at the hospitals and in the home care settings. This might increase the probability of sustaining the mobile teamwork after the project has ended.

Furthermore, the project boasts the committed engagement of leaders and managers from all three healthcare organisations in the area. The community services, the GPs, and the hospital are all involved and collaborating to improve service integration. This might also increase the likelihood of sustaining the team, to expand the service to include more municipalities in the region and to adapt the service to other patient groups.

VII. Conclusion and Future Work

In this paper, we described an integrated patient-centred model that aims to improve the continuity and quality of care for multimorbid older patients. The practice model consists of a proactive interdisciplinary outreach team, which includes personnel from the Community Nursing Service and the hospital. The team's main task is to follow patients through the system and help them receive appropriate, patient-centred, and timely health services. Preliminary results indicate that the team can improve care, shorten hospital stays, and reduce hospital admissions.

At the time of writing this paper, more than 300 patients received services from the team. The next step is to further develop and adapt the integrated care model and evaluate work practices. Specifically, we will evaluate the use of tablets in a home setting, both from the patients' and the providers' perspective. Future work will also include analysing effectiveness and cost effectiveness using matched controls from two separate hospitals in the region.

Acknowledgement

We thank the management at the UNN and the administration in the Municipality of Tromsø for funding this project. The EU-project RemoAge funded the tablets. We also thank Markus Rumpsfeld, Gro Berntsen, and Johan Gustav Bellika for their contribution to realising the patient-centred model.

References

[1] T. S. Bergmo, G. K. Berntsen, M. Dalbakk, and M. Rumpsfeld, "The effectiveness and cost effectiveness of the PAtient-Centred Team (PACT) model: study protocol of a prospective matched control before-and-after study," BMC Geriatr, vol. 15, p. 133, 2015.

[2] K. K. Andreassen, "Befolkningens størrelse og aldersfordeling," in Seniorer i Norge, E. Mørk, Ed., ed Oslo: Statistics Norway 2010.

[3] C. Hoffman, D. Rice, and H.-Y. Sung, "Persons with chronic conditions: their prevalence and costs," Jama, vol. 276, pp. 1473-1479, 1996.

[4] J. L. Wolff, B. Starfield, and G. Anderson, "Prevalence, expenditures, and complications of multiple chronic conditions in the elderly," Archives of internal medicine, vol. 162, pp. 2269-2276, 2002.

[5] A. Mudge, S. Laracy, K. Richter, and C. Denaro, "Controlled trial of multidisciplinary care teams for acutely ill medical inpatients: enhanced multidisciplinary care," Intern Med J, vol. 36, pp. 558-63, Sep 2006.

[6] L. G. Glynn, J. M. Valderas, P. Healy, E. Burke, J. Newell, P. Gillespie, et al., "The prevalence of multimorbidity in primary care and its effect on health care utilization and cost," Family Practice, vol. 28, pp. 516-523, Oct 2011.

[7] T. Richards, A. Coulter, and P. Wicks, "Time to deliver patient centred care," BMJ, vol. 350, p. 530, 2015.

[8] D. P. Baker, R. Day, and E. Salas, "Teamwork as an essential component of high-reliability organizations," Health Serv Res, vol. 41, pp. 1576-98, Aug 2006.

[9] D. Trivedi, C. Goodman, H. Gage, N. Baron, F. Scheibl, S. Iliffe, et al., "The effectiveness of inter-professional working for older people living in the community: a systematic review," Health Soc Care Community, vol. 21, pp. 113-28, Mar 2013.

[10] S. M. Smith, H. Soubhi, M. Fortin, C. Hudon, and T. O'Dowd, "Interventions for improving outcomes in patients with multimorbidity in primary care and community settings," Cochrane Database Syst Rev, vol. 4, p. Cd006560, 2012.

[11] K. M. Ke, J. M. Blazeby, S. Strong, F. E. Carroll, A. R. Ness, and W. Hollingworth, "Are multidisciplinary teams in secondary care cost-effective? A systematic review of the literature," Cost Eff Resour Alloc, vol. 11, p. 7, 2013.

[12] S. Conroy, C. Ferguson, J. Woodard, and J. Banerjee, "Interface geriatrics: evidence-based care for frail older people with medical crises," British journal of hospital medicine (London, England: 2005), vol. 71, pp. 98-101, 2010.

[13] F. C. Bakker, S. H. Robben, and M. G. Olde Rikkert, "Effects of hospital-wide interventions to improve care for frail older inpatients: a systematic review," BMJ Qual Saf, vol. 20, pp. 680-91, Aug 2011.

[14] A. Coulter and J. Ellins, "Effectiveness of strategies for informing, educating, and involving patients," British Medical Journal, vol. 335, p. 24, 2007.

[15] A. Bowling, Research methods in health: investigating health and health services: McGraw-Hill Education (UK), 2014.

[16]"Kvalitetsindikatorrapport,Helsenorge.no.https://helsenorge.no/kvalitet-seksjon/Sider/Kvalitetsindikatorer-rapporter.aspx?kiid=sannsynlighet_for_reinnleggelse_av_eldre. Accessed January 2016. "

[17] P. M. Stranges, V. D. Marshall, P. C. Walker, K. E. Hall, D. K. Griffith, and T. Remington, "A multidisciplinary intervention for reducing readmissions among older adults in a patient-centered medical home," Am J Manag Care, vol. 21, pp. 106-13, Feb 2015.

[18] L. E. Olsson, J. Karlsson, and I. Ekman, "The integrated care pathway reduced the number of hospital days by half: a prospective comparative study of patients with acute hip fracture," J Orthop Surg Res, vol. 1, p. 3, 2006.

[19] I. Ekman, A. Wolf, L. E. Olsson, C. Taft, K. Dudas, M. Schaufelberger, et al., "Effects of person-centred care in patients with chronic heart failure: the PCC-HF study," Eur Heart J, vol. 33, pp. 1112-9, May 2012.

[20] K. Dudas, L. E. Olsson, A. Wolf, K. Swedberg, C. Taft, M. Schaufelberger, et al., "Uncertainty in illness among patients with chronic heart failure is less in person-centred care than in usual care," Eur J Cardiovasc Nurs, vol. 12, pp. 521-8, Dec 2013.

[21] L. E. Olsson, J. Karlsson, and I. Ekman, "Effects of nursing interventions within an integrated care pathway for patients with hip fracture," J Adv Nurs, vol. 58, pp. 116-25, Apr 2007.

[22] L. E. Olsson, E. Hansson, I. Ekman, and J. Karlsson, "A cost-effectiveness study of a patient-centred integrated care pathway," J Adv Nurs, vol. 65, pp. 1626-35, Aug 2009.

[23] C. A. Orchard, V. Curran, and S. Kabene, "Creating a culture for interdisciplinary collaborative professional practice," Medical Education Online, vol. 10, 2009.

Interaction Space: Older Adults and in-Home Systems

Exploring technology with and for older people

Jeroen Blom, Tara French, Gemma Teal
Institute of Design Innovation
The Glasgow School of Art
Forres, UK
Email: j.blom@gsa.ac.uk, t.french@gsa.ac.uk, g.teal@gsa.ac.uk

Abstract— **This paper discusses important considerations when developing assistive technology with and for older adults. Two case studies demonstrate the use of Experience Labs to engage older adults in a participatory design process in the early stages of development of novel sensor systems. Firstly, we present the 'Interaction Space' as a holistic way to model and understand interaction between people, products, technology and environments when developing complex systems. It is argued that looking at the interaction between older adults and technology benefits from a holistic view of the Interaction Space in which primary and secondary users are continuously acting together with the technology. We then highlight considerations when developing technology systems with participants who are not confident technology users, in order to design meaningful spaces for critical reflection and creative collaboration. We conclude that the Experience Lab approach enables the complex Interaction Space of sensor systems to be dissected into comprehensible elements leading to a better understanding of the impact of the proposed technologies.**

Keywords- older people; participatory design; interaction design; technology; sensor system.

I. INTRODUCTION

The health and care environment is fertile ground for innovation to support patients, families, and health professionals to achieve the Scottish Government '2020 Vision' that "everyone is able to live longer healthier lives at home, or in a homely setting" [1]. Digital technology has the potential to make an important contribution to realising this vision. In this respect, it is becoming increasingly important to involve those who will be using the technology, whether in receipt of or in the delivery of care, at a much earlier stage of the design process [2] to ensure that the proposed technology solution meets their needs and is adopted in use. The field of participatory design offers a wealth of approaches to meaningfully involve end users in the design process [3]. When designing digital technologies with and for patients, families and health professionals, it can be challenging to separate out the different responses and reactions to the proposed technology within a complex interplay of people, objects and spaces with or within which users interact. The concept of an Interaction Space to describe human-computer interaction has been explored in a variety of ways. The Mixed Interaction Space is used to distinguish camera-based interaction from other types of sensor-based interaction in terms of the ability to track a fixed point in relation to the mobile device [4]. The Multimodal Interaction Space presents a framework to explore the different levels, modes and modalities of interaction [5]. Finally, the Continuous Interaction Space describes the physical space above a traditional touch surface to form part of the interaction experience [6]. These interpretations all focus on the physical environment around a device at the moment of active interaction. We propose the Interaction Space as a means of exploring the impact of individual and cumulative interactions with the different parts of the system, as distinct from the physical interaction individuals engage in. The role and position of new technologies within the Interaction Space between humans and systems is of particular interest when considering technological developments in the area of health and care.

In the following part of the introduction we will explore existing knowledge on older adult's perception of technology and the principles of experience design. In Section 2 we will outline the methodology of an Experience Lab and highlight previous work exploring the Interaction Space in Experience Lab projects. In Section 3 we present two case studies describing the context, Interaction Space and findings for each project. Finally, in Section 4 we will discuss how the methodology of an Experience Lab leads to a holistic understanding of the Interaction Space for older adults and in-home systems.

A. Older adults and technology

A key factor contributing to the increased demand on health care services is the global issue of an aging population and shifting demographic [7]. Older adults are therefore a key group to involve in discussions about technology, in order to determine the ways in which technology can benefit everyday lives. General experience with technology is lower for older adults than younger generations, but in health and social care they often form the target users that proposed technologies aim to support. Research on the interaction between older adults and technology is primarily focussed on using computers and accessing the Internet [8]-[11]. Barriers to using technology as expressed by older adults are often related to a lack of confidence leading to a dismissive attitude [8], low awareness of the benefits of technology [9], and the fear of 'breaking something', strengthened by overestimating the price of technological devices [10]. Incentives to start using technology and particularly the Internet relate to a change in health condition, mobility or social contact (e.g., to communicate with family who live at a distance) [11].

Previous research has identified that older adults do show a willingness to adopt new technologies. Particularly in the health and care domain it is recognised that there are several benefits, e.g., to support disease management, medication observation, remote patient monitoring, cognitive fitness and assessment, and social networking [12]. The social aspect of taking part in a training session and subsequently sharing the knowledge gained with peers was also appreciated [10].

Everyday technology is widely perceived to be designed for a younger target group, but this does not mean that supportive technology for older adults needs to be designed specifically for them as this can feel stigmatizing. Rather there is a need to understand the wider context of how older adults live and the potential role for technology. Therefore, when designing a new technology or application of technology, it is important to involve older adults at the beginning of the design process and to employ a participatory approach. This ensures that the ideas taken forward meet the needs and expectations of older adults.

B. *Understanding and designing for experience*

The introduction of technological products or services for older adults is the means to an end and not the end in itself [9]. This is particularly true in the health and care environment where the goal is to enhance the wellbeing of the individual. Wellbeing is rarely achieved by the simple act of using a technology, but rather it depends on the support being provided by other professionals and individuals. This kind of interaction is a continuous interplay between the older adult, other individuals and the technology. Edmund Husserl and Martin Heidegger first described the experience of this interaction in the early 20th century in their philosophy of phenomenology [13]. Phenomenology opposed the Cartesian view that there is a truth in looking at things abstracted from reality, and that the ability to think and analyse determined our place in this world [14]. Phenomenology described the fact that we *have* a body and our presence in this world at this very moment determined how we perceive and experience things. French philosopher and phenomenologist Maurice Merleau-Ponty later argued that perception is inherently interactive: it is an interplay between perceiver and perceived [15]. Interaction does not happen because the person has an exact cognitive model of steps, be it person-to-person, person-to-technology, or person-through-technology-to-person. Rather it is an event that continuously unfolds depending on how things are perceived and responded to. The study of design is therefore focused on designing products and systems aimed at intuitive and engaging interaction that do not require a cognitive model of use [16]. Experiential approaches to design put "experience before functionality… leaving behind oversimplified calls for ease, efficiency, and automation or shallow beautification" [17]. This requires an understanding of what really matters to the desired end users and how their experiences of using technology can be more meaningful.

II. The Interaction Space within an Experience Lab

The Experience Lab methodology provides a way to understand the needs and experiences of users within the design process. Led by the Glasgow School of Art, Experience Labs form a core part of the Digital Health and Care Institute, an innovation centre based in Scotland. There is a clear need to involve users at an early stage in the design process [2] and methods of addressing that need are well documented [18]. The Experience Labs build on human-centred design knowledge and existing research methods, e.g. Living Labs, and aim to provide a safe space for collaboration and creativity, affording the opportunity for experiential learning [19].

In terms of enhancing or achieving wellbeing for older adults through the use of technology, we define the Interaction Space as a conceptual space encompassing all individuals and the technology being proposed. The different streams of interaction between persons or person and technology all contribute to the perceived benefits of the system. We propose that Experience Labs offer a method for exploring the perceived benefits of new technology systems with end users. The Experience Lab aims to create a holistic understanding of the Interaction Space, and enable participants to perceive how different streams of interaction all lead to a continuous and holistic experience. This experiential understanding enables participants to provide insight and feedback to inform the development of the technology.

Exploring the Interaction Space with participants in the early stages of concept development is a challenge, as the proposed technology may be intentionally undefined. To address this, Experience Labs create pop-up spaces for multiple stakeholders (e.g., patients and primary users, academic, business and civic partners) to engage in iterative exploration and early trialling of ideas, concepts and prototypes in a safe and collaborative environment. The pop-up spaces chosen for the Labs are based on real intended use environments. Each Experience Lab, or set of Labs, is bespoke to a project. Typically, projects selected for a Lab propose a conceptual solution, often with limited consideration of the needs of the potential user. The Labs provide the opportunity to step back and (re)discover the needs of the different users and reflect on how the proposed solution can be strengthened in response to the needs and experiences expressed [20].

The goal of an Experience Lab is to enable participants to be at the centre of a participatory design process in which their lived experiences form a rich contribution to ideas and concepts. The sessions in a Lab break down the complex reality of a new supportive technology and provide participants with the tools to experience elements that make up the Interaction Space. Using generative tools and bespoke artefacts, the Labs allow participants to experience a concept that is unknown to them, and reflect and provide feedback as well as share and build on each other's ideas.

Previous work reviewed the Interaction Space of six Experience Lab projects, highlighting three potential roles for technology within the Interaction Space, with different levels of complexity as shown in Figure 1. The Venn diagram shows the potential cross-overs, particularly when technologies are intentionally undefined.

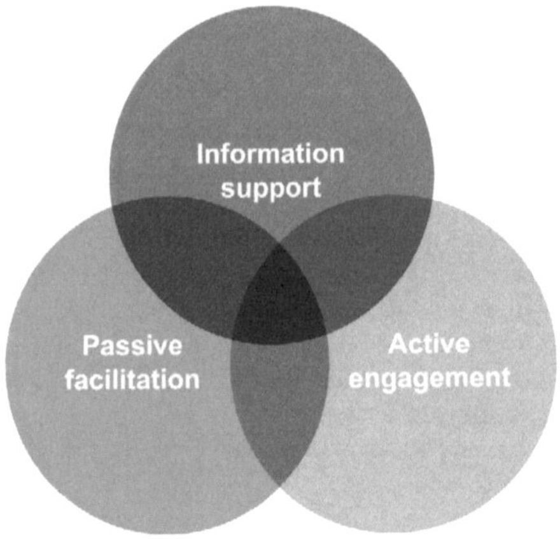

Figure 1. Different roles of technology in the Interaction Space.

The first area involved understanding the way technology provided an information source to the user. Experience Labs for projects such as these focussed on what information needed to be available, when, and how that information was presented (e.g., a directory of services for ambulance clinicians [20]). For these projects, the technology needed to be available when required but did not have a role in facilitating communication with other people.

The second area involved understanding the role of technology in facilitating interaction between people through an online platform. The Interaction Space in this area included a minimum of two people communicating through a digital medium (e.g., a community platform to broker receiving and delivering small services).

The third and most complex Interaction Space involved understanding technology taking an active role by responding to behaviour of people and the animate world. In this space, it was possible for the technology to trigger actions (e.g., sending alerts to carers in response to abnormal behaviour) that influenced how users perceived the environment.

III. CASE STUDIES

The two projects that reflect the complex Interaction Space are further elaborated on in this paper, in particular how the Experience Labs support older adults to understand the potential of the new technology and how it might influence existing relations with other people.

A. *Case study 1: Assisted living for older adults*

1) Context: The project described in this case study explored and developed a new concept for assisted living which aimed to support and empower older adults to live independently at home for longer [21]. The project involved three sequential Experience Labs, which were designed to explore the full potential of the proposed system and user-test the initial hypotheses behind the concept by developing and validating a refined solution with users.

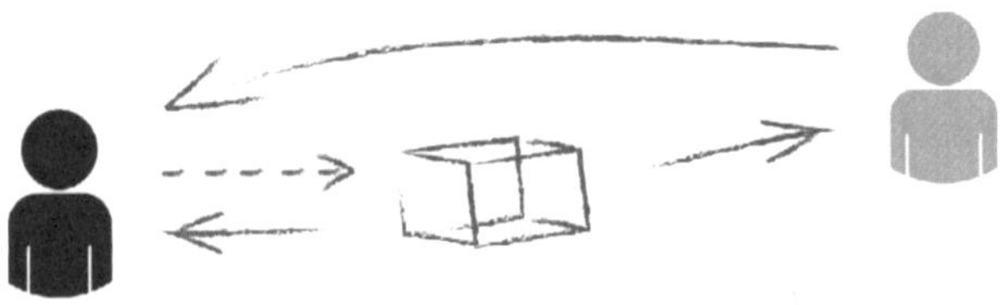

Figure 2. Diagram of the Interaction Space for case study 1, showing potential streams of interaction.

2) Interaction Space: The Interaction Space in this project (see Figure 2) involved the proposed observing system shown as the box, participants who were older adults living at home (blue person) and a family member or friend from the older adult's support network (green person). The streams of interaction, as visualized by the arrows, happen between different people and the system. The system can observe the environment of the older adult (dashed arrow) and potentially trigger a response to the older adult. Another stream of interaction could also be between the system and a carer who may look after the wellbeing of the primary user remotely. The first Experience Lab involved visiting older adults in their homes to understand their home life, routines, networks of support and current use of technology. The second Lab involved a field trip to a department store to allow participants to interact with technology and supportive home products, with the aim of understanding their perceptions of and preferences for technology and the types of scenarios-of-use. These initial Experience Labs provided a rich understanding of the participants' current awareness, use and perceptions of technology and informed the design of the final Lab.

The challenge was to create a realistic environment in which older adults could experience and interact with the proposed assisted living system. The final Experience Lab involved creating a non-functioning prototype, which was operated by a member of the team to give the impression that the system was fully functional. Participants were guided through a role-play scenario using the prototype, allowing them to experience the Interaction Space, physically and interactively, and to discuss the potential use of the system in their own home. The Lab team helped users explore the concept and share their feedback and ideas for improvement. The Lab also provided the opportunity for the family member or friend to observe the system in use. As an extension of the experience, participants were then invited to a workshop environment where they were given the opportunity to explore and visualise ideas using playful materials (e.g., clay, puppets, craft materials and an electronic sensor/actuator kit) to allow them to construct scenarios where they felt the proposed system might provide support.

3) Findings: The sequential nature of the Experience Labs allowed participants to gradually build an understanding of the system. The first Lab took place in the home environment and provided the opportunity to gain perspective on the current role of technology in the everyday lives of older adults: their perceptions and usage. Building on this experience, the field trip to a department store enabled insight to be gathered on the types of technology and products that are preferred by older adults, and provided insight into their buying decisions. Finally, by creating an opportunity to explore the system in a realistic environment, participants were able to experience the concept in a non threatening way and the use of an unfinished aesthetic, together with playful materials, encouraged participants to become 'makers and designers' themselves by making their ideas tangible.

The findings of the Labs indicated that personalisation was of key importance given the differing circumstances of older adults in terms of home environment, networks of support and confidence with technology. The findings revealed that participants were already making adaptations using both low and hi-tech solutions to make everyday life easier and address personal challenges. Through visiting the participants in their homes, it was possible to see these types of solutions already in use. Progressing the Labs sequentially also made it possible to ideate solutions together with reference to existing technology (e.g., products available in the department store) and to generate ideas for the system through scenarios and experiential learning in the final Lab.

B. *Case study 2: A novel indoor tracking system for people living with dementia*

1) Context: The project, initiated by Glasgow Caledonian University, focussed on developing a novel unobtrusive indoor tracking system for people living with dementia. The aim of the system was to recognise abnormal behaviour (e.g., wandering in the middle of the night) and alert the carer for a timely response, thereby positively affecting the safety and wellbeing of individuals. The project aimed to understand the experience of living with dementia, and the perceptions of the indoor behaviour tracking system from the perspectives of people living with dementia and their professional and personal carers.

2) Interaction Space: The Interaction Space in this project (see Figure 3) includes the proposed sensor system shown as the box, the person living with dementia (blue person), a professional carer (red person) and a personal carer or relative (green person). The system is tracking the

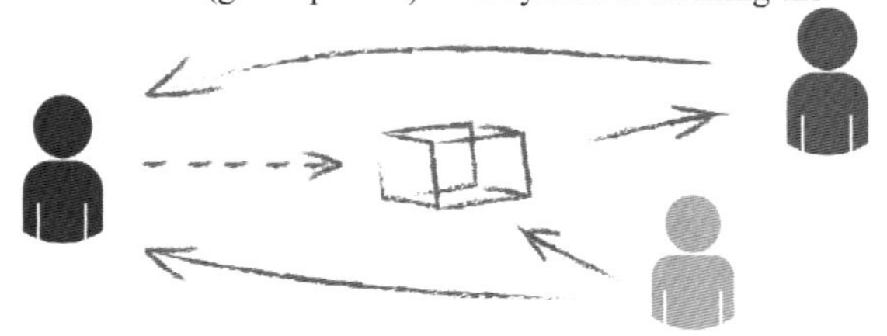

Figure 3. Diagram of the Interaction Space for case study 2, showing potential streams of interaction.

behaviour of the person with dementia (dashed line). It could potentially send continuous data or alerts to a professional carer. A personal carer has the opportunity to log into a platform, allowing them to observe registered behaviour. Both carers can then take action as required, responding to the person living with dementia. The proposed system was at proof of concept stage and therefore it was not possible for participants to engage with the system in the Experience Lab. The challenge was to create an environment that allowed participants, in particular people living with dementia, to understand the proposed system and express lived experiences relevant to the system's potential use, reflecting on the impact the system might have on their lives and the behaviour of others.

The first Experience Lab explored the different elements of the Interaction Space for the tracking system in a series of bespoke activities and design tools. Scenario cards as shown in Figure 4 were used to focus the conversation on experiences of living with dementia, of large indoor environments and the support gained from other people. A second session introduced a small wearable device to present the potential sensor, which participants were asked to wear during lunch. Researchers 'tracked' participants with pen and paper as they moved between rooms, as well as their behaviour within the room. This tracking information was used to discuss potential insights a system might get on behaviour during everyday events. Finally, the appearance of a wearable sensor itself was discussed by relating it to accessories worn by participants on the day and the associated experience.

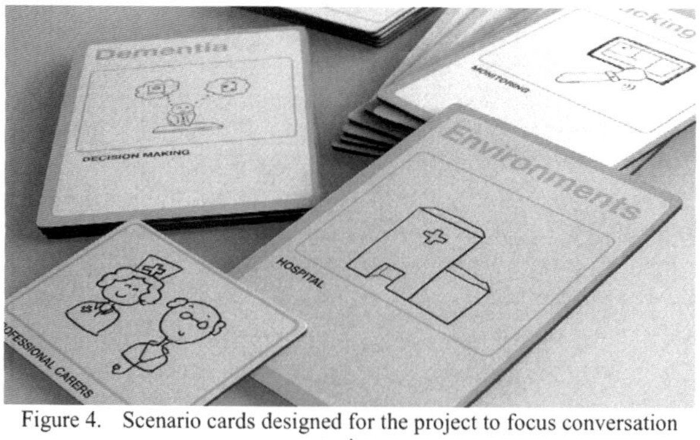

Figure 4. Scenario cards designed for the project to focus conversation topics.

Insights gained from the first Lab led to the development of three scenarios-of-use to discuss potential Interaction Spaces representing shared experiences, concerns and ideas. The scenarios particularly focussed on expressing how the system was introduced, what behavioural aspects were captured by the system, and how and to whom this information was shared. The scenarios were acted out by researchers and filmed, each resulting in a two-minute video. During the second Experience Lab the videos were shown one at a time and participants were invited to reflect on what was shown and how it would affect their individual situation.

3) Findings: The first Lab provided conversation topics including the experience of indoor environments and wearable devices. The sessions relied on the participants to imagine the implementation, opportunities and implications of the indoor tracking system. This led to the identification of a number of concerns in relation to the anticipated support that would be provided and the sharing of positioning data. Participants phrased this as a desire to "know who is at the other end". This phrase embodied concerns about the human support they expected, the reliability of support and the potential privacy issues of tracking in their homes. It is suggested that these concerns emerged because there was no concrete solution participants could either agree or disagree with. The presented Interaction Space was open and relied on input from participants to express their preferences for the tracking system. An aspect of the Interaction Space that participants could conceive very well was the appearance of the sensor. The sensor was required to be in direct or indirect line of sight with the light source, meaning that it would have to be visible and not hidden. Relating it to visible accessories participants felt strongly about not wanting to wear a necklace, preferring an arm-worn device or something you could pin onto clothes. They also wanted the aesthetic design of the sensor to empower individuals rather than show vulnerability.

The second Lab presented videos that showed a 'closed' potential Interaction Space in which it was made clear: the people involved, information shared and support offered by the system and/or carer. Although real life was of course seen to be more complex than the scenarios presented, participants no longer felt as strongly about the privacy concerns. When issues from the first Lab were revisited, participants unanimously responded by acknowledging that the support they were provided with, e.g., by a relative, professional or informal carer, was more important than the privacy concerns they had previously expressed. The involvement of carers was seen as the key value in the system, and the technology was merely seen as a means of initiating timely contact.

Both Experience Labs focussed on creating an environment for participants to understand the proposed technology and the support it could deliver. The technology was well received in the first Lab, yet with some concerns about data sharing. The second Lab addressed these concerns by proposing realistic scenarios showing what information would be shared when and with whom. In the overall project the Interaction Space was described by participants' own experience in combination with the proposed technology in the first Lab and then concretised for further feedback in the second Lab.

IV. DISCUSSION

A. Breaking down complex interactions

The systems presented in both case studies represent two highly complex systems which have the potential to positively impact the lives of older adults. At the initial stages of participatory design in the Experience Labs it can be difficult for participants to imagine the potential use of any complex technological solution. The use depends on a variety of factors such as what the system or sensors are capable of measuring and registering (e.g., movement, light, sound), in what environment they are deployed, the types of events that would trigger a response and how this could possibly lead to false triggers, and what the result of a trigger will be.

The progressive and sequential nature of the Experience Labs described within the case studies provided a way to break down the complexity of the system and introduce participants to the new technology. By exploring the everyday experiences of participants and their current relationship with technology, the Labs helped participants to recognise the ways in which these technologies could impact on their everyday lives. Focussing on separate elements of the system enabled participants to realise what criteria were important to them. For example, case study two overcame challenges in communicating a complex indoor tracking system by breaking the system down into individual elements, with activities in the first Lab separately exploring: experiences of indoor environments, tracking daily activities, support provided by other people and a wearable sensor device. These sessions covered different streams making up the Interaction Space between individuals and the system. Subsequently the final Experience Lab could create an experience prototype embodying these criteria, enabling the participants to reflect on each element as part of a holistic experience and give meaningful feedback.

The older adults involved in the Labs were reasonably familiar with technology through computer and occasional mobile phone use. However, many of the barriers highlighted in the introduction were also supported in the case studies such as lack of confidence with technology and fear of breaking something [9]. Considering this somewhat limited experience with technology, it is essential to incrementally build understanding by relating it to participants' existing technology references, environments and daily life. In this way it becomes easier for participants to imagine how the proposed technology fits with their everyday life and home environment. There is a need to create a balance between an open and suggestive approach when ideating with older adults in order to ensure that the proposed technology is comprehensible to participants. Failure to do so may lead to limited ideas and feedback.

While some might argue that this limits the possibility for radical innovation [22], in practice it uncovers insight into the real needs and aspirations of older people and highlights places where technology is not acceptable, opening up new opportunities to innovate services to support older people. It also offers insight into how new technology can be marketed to support adoption [21]. The use of bespoke design tools as described in the case studies demonstrates a promising method of introducing complex technology to older adults in a meaningful way.

B. Older adults and technology over time

Older adults see technology as a means to an end. It is important to consider the individual streams making up the Interaction Space to enable a valuable discussion with participants regarding the effect on achieving this 'end', particularly when multiple people are involved. For example, in relation to case study two, information picked up by the sensors can be shared with carers to alert them to a safety issue or reassure them that the older person stable. In the case of safety issues, the exchange of potentially private information is less of a concern for the older adults as long as they know it will achieve the end result of getting them human support at a time of need.

The next generation of older adults are often described as digital natives, or at least familiar with using smart devices. However, the development of technological devices and support systems will not slow down, rather it is likely to diversify, for example looking at developments such as the Quantified Self, Internet of Things, virtual immersive environments and autonomous systems. This diversification indicates that the barriers to using technology today as experienced by older adults are not likely to dissolve with the next generation. Alongside the increasing complexity of technology, the ageing process of older adults can amplify barriers to usage, particularly where older adults are living with dementia. Discussing the potential benefit of a supportive sensor system in case study two highlighted an important aspect for participants, namely to introduce a system at an early enough stage when people, particularly when living with dementia, still have the capability to comprehend the changes and benefits. This highlights an additional requirement for the technology, namely ensuring it is not stigmatising and offers progressive functionality that can adapt to the changing needs of the owner. Participants were aware of the progressive nature of their condition and appreciated being involved in the participatory design process at an early stage when they could share what aspects were experienced as empowering or as stigmatising. The progressive nature of long term conditions suggests that this increase in support over time can be seen to be beneficial. The Interaction Space then becomes not only a model to describe how different streams of interaction between person and technology influence other streams, it can also be seen to include an element of time; describing how interactions could be adapted to lead to different results at a later phase of life. Implementations of the support systems in the case studies are recommended to adapt or present opportunities for personalisation to reflect changing needs. The changing needs are expressed as an incentive to start using technological support [11]. The Experience Lab can provide the opportunity to consider the anticipated changing needs and how these can enhance the experienced benefits of a system.

V. CONCLUSION

The Experience Labs provided a platform to involve older adults at an early stage in the design process. The Labs encouraged sharing of experiences and reflecting on proposed concepts of technology to address potential barriers for older adults to accept and use a new system. Through an iterative approach the complexity of support systems was broken down into elements of the Interaction Space that participants could relate to and understand. The Interaction Space model was used to reflect on the case studies in this paper. This presents an opportunity to reflect on and understand the different modes of interaction within a complex system to identify challenges, opportunities and ideas within participatory design research. Further research is required to explore if and how the Interaction Space model could be communicated to facilitate an active dialogue with participants to further support the understanding of a complex system.

ACKNOWLEDGMENT

We would like to thank all our participants who took part in the Experience Labs described in the case studies. We would also like to thank our project partners. We would like to acknowledge the support from our Experience Lab colleagues throughout the projects.

REFERENCES

[1] Scottish Government. Achieving sustainable quality in Scotland's healthcare a '20:20' vision. 2011. Retrieved: January 2016
http://www.gov.scot/resource/0039/00398668.doc

[2] E. B. N. Sanders and P. J. Stappers, "Co-creation and the new landscapes of design," Co-Design: International Journal of CoCreation in Design and the Arts, vol. 4, 2008, pp. 5-18.

[3] M.J. Muller, "Participatory design: the third space in HCI", Human-computer interaction: Development process, 4235. 2003.

[4] T.R. Hansen, E. Eriksson, A. Lykke-Olesen, "Mixed interaction space: designing for camera based interaction with mobile devices". CHI '05 Extended Abstracts on Human Factors in Computing Systems (CHI EA '05). ACM, New York, NY, USA, 2005. 1933-1936. doi:10.1145/1056808.1057060

[5] B. Bongers and G.C. van der Veer, "Towards a multimodel interaction space: categorisation and applications". Personal and Ubiquitous Computing, 11(8), 2007, pp. 609-619, doi:10.1007/s00779-006-0138-8

[6] N. Marquardt, R. Jota, S. Greenberg, J.A. Jorge, "The continuous interaction space: interaction techniques unifying touch and gesture on an above digital surface". Human-

Computer Interaction – INTERACT, 6948, 2011, pp. 461-476, doi:10.1007/978-3-642-23765-2_32

[7] Scottish Government. Demographic change in Scotland report. 2010. Retrieved: March 2015 http://www.gov.scot/Publications/2010/11/24111237/1

[8] Age Concern. Introducing another world: older people and digital inclusion. 2009. Retrieved: March 2016 http://www.ageuk.org.uk/documents/en-gb/for-professionals/computers-and-technology/140_0809_introducing_another_world_older_people_and_digital_inclusion_2009_pro.pdf?dtrk=true

[9] J. Morrison and A. Barnett. Older people, technology and community: the potential of technology to help older people renew or develop social contacts and to actively engage in their communities. Independent Age, 2010.

[10] Sheridan Elder Research Centre (SERC). Introducing older persons to benefits of technology. September, 2005.

[11] T.L. Mitzner, et al. "Older adults talk technology: technology usage and attitudes" Comput Human Behav. November, 2010 1; 26(6): 1710–1721. doi:10.1016/j.chb.2010.06.020.

[12] D. Lindeman, "Briefing Paper: technologies to help older adults maintain independence: advancing technology adoption". Public Health Institute, 2009.

[13] T. Ruspoli, Being in the world [video]. 2010.

[14] C.C.M. Hummels. Matter of transformation: Sculpting a valuable tomorrow. Eindhoven University of Technology September 2012.

[15] D. Abram. The Spell of the Sensuous: Perception and Language in a More-than-Human World. Vintage Books, New York. 1996.

[16] C.C.M. Hummels and P.D. Levy. "Matter of transformation: Designing an alternative tomorrow inspired by phenomenology". Interactions 20, 6, 2013, pp. 42–49.

[17] M. Hassenzahl, "Experience design: technology for all the right reasons", Synthesis Lectures on Human-Centered Informatics 2010 3:1, 1-95.

[18] M. Pallot, B. Trousse, B. Senach, D. Scapin. Living Lab Research Landscape: From User Centred Design and User Experience towards User Cocreation. First European Summer School "Living Labs", Paris, France, August, 2010.

[19] P. Reason and H. Bradbury. The SAGE handbook of action research. London: Sage Publication Inc. 2008.

[20] T. French and G. Teal, "Co-designing a digital directory of services," Procedia Computer Science, vol. 63, September, 2015, pp. 445-450, doi:10.1016/j.procs.2015.08.366.

[21] T. French and G. Teal, " Transforming healthcare through design-led innovation". Design 4 Health European Conference, 2015, pp. 1-11.

[22] D. A. Norman and R. Verganti, "Incremental and radical innovation: Design research vs. technology and meaning change," Design Issues, 30(1), 2014, 78-96.

Co-creation of an Innovation Network: Engagement and User Involvement in Digital Care Services

Torbjørg Træland Meum
Department of Information Systems
University of Agder
Campus Kristiansand, Norway
torbjorg.t.meum@uia.no

Etty Ragnhild Nilsen
Department of Strategy and Finance
University College of Southeast Norway
Campus Ringerike, Norway
etty.nilsen@hbv.no

Abstract— **This work in progress paper presents a study of digital innovation in municipal care services, and we highlight collaboration and communication among various stakeholders in the process of implementing welfare technologies. We characterize innovation in municipal care services as socio-technical networks of organizations that enable innovation and we emphasize mechanisms that enable engagement and mutual learning across professional, organizational and geographical boundaries. In particular, we focus on how knowledge is mediated and how actors and resources are mobilized in the network. The study is further inspired by Action Design Research and we present ongoing activities of the creation of an innovation network.**

Keywords- ***Welfare technology; Innovation; Information Infrastructure; Socio-technical systems; Action Design Research.***

I. INTRODUCTION

Designing information systems is one of the core areas in the interdisciplinary community of health informatics, emphasizing technological, sociological and organizational challenges as key issues in developing and maintaining effective health information systems [1]. In particular, studies of the sociotechnical tradition has highlighted the contextual nature of health information and consider design as the synergy between the specific particularities of health care work, and the informating properties of information and communication technologies (ICT) [2]. Accordingly, new methods and techniques for user involvement in the design process have emerged to bridge the design-reality gap. Several studies have applied user-centric design methods in healthcare and demonstrate the capability to translate user needs into technical requirements [3]. However, the changing nature of health care combined with evolving technological capabilities leads to new challenges to user involvement in the design process.

First and foremost, the evolving use of mobile technologies changes the information pathway in society in general and the healthcare sector in particular. The demographic change in society has led to increased pressure on the organization and performance of health care services. In particular there is a need for new models and technologies to support long-term care of the increasing population of elderly as well as people with chronic illnesses and disabilities. Accordingly, various research fields, such as telecare, assistive technologies and Ambient Assisted Living (AAL) systems have taken part in international research efforts and have led to increased knowledge and insight into currently available solutions and enabling technologies [4]. The term welfare technology is used in Scandinavia and national initiatives in Norway highlight technological solutions that promote safety and enable people to better manage their own health [5]. Despite the promising impacts and opportunities with the use of ICT in elderly care, there is limited and inconsistent evidence about the effects of assistive technologies [6] as well as limited use beyond the pilot-study level [4]. A recent systematic review of AAL systems also argues for more participatory approaches, user feedback and collaborative efforts in the development process [4]. Innovation in mobile technologies has also increased the complexity and introduced new challenges to existing "IT silos" in the e-health field [7]. In line with the increased use of computers as an embedded part of everyday practices, the scope of healthcare technologies has moved from singular tools to networks of systems, practices, and people, i.e., digital infrastructures [7]. As ICT systems have become deeply socially embedded, we need to take into account the dynamic interplay between planned design and context of use as well as integration of new technologies in the socio-technical network. The notion of users in socio-technical networks are often characterized by diversity and the challenge is to manage heterogeneous knowledge resources that span across professional and organizational units [8]. Based on these insights, we focus on user involvement in the adoption and use of welfare technology in municipal care services, and we emphasize the innovation process in particular, i.e., to do something new in order to create value.

The study is based on a research and innovation project on adoption and use of digital surveillance in municipal care services. The opportunity for mutual learning and knowledge sharing across professional and organizational boundaries is one of the research areas in the project and in this paper we focus on user engagement and collaboration in the innovation process. In particular we are interesting in *how actors involved in the project share knowledge and how actors and resources are mobilized during the innovation process*. Furthermore, our study is inspired by user-oriented methods such as Participatory Design [9] and Action Design

Research (ADR) [10]. These methods emphasize design and development as an iterative process of mutual learning and thus enable co-design with community members in the context of their daily lives.

The rest of the paper is organized as follow: Section II provides a brief introduction to the theoretical foundation of the study. The research setting and method is presented in Section III and provides insight into the objectives of the overall research project in order to illustrate how engagement and learning are part of the innovation process. Section IV and V provides insight to the case study, as well as ongoing activities of design and development of a communication platform. Finally, in Section VI we discuss the need to establish a shared information place in the project and further design and development of a digital platform.

II. Digital Innovation and Action Design Research

An official Norwegian report on innovation in the care services has made recommendation for new innovative solutions in order to meet future challenges [11]. In particular, the report highlights the use of new technology as a resource for value creation and emphasize user influence, participation and co-creation in the development of future care services.

Innovation in health care involves interdisciplinary collaboration and is typically distributed across various organizations. Yoo, Lyytinnen and Boland [8] have used the term innovation network to refer to socio-technical networks of organizations that enable innovation [8, p.1]. Moreover, they conceptualize innovation as a series of translations between ideas (either in form of physical products or services), mediated through technology artifacts [8, p. 2]). Of particular interest in our project is what they refer to as "social translation" that takes place at the boundaries of communities where individual actors negotiate and mutually adjust to other's perspectives. Similar studies have emphasized different mechanisms for how information systems evolve as well as the capabilities to generate new services [12] [13]. Overall, these studies have illustrated the generative power of information infrastructures and emphasized knowledge sharing and learning as capabilities during innovation process.

Based on these insights, we have used an ADR approach for the formation of an innovation network across professional and organizational boundaries in our project. ADR provides an opportunity to combine basic principles of traditional Design Research (building and evaluating innovative IT artifacts) [14], while also emphasizing participation and cooperative change [15]. In contrast to traditional design that separates building from evaluating, ADR is characterized as an iterative process that addresses a problem situation encountered in a specific organizational setting by intervening and evaluating [10]. The notion of information technology in ADR is based on the ensemble view of IT artifacts and involves the interaction of design efforts and contextual factors throughout the design process. The design process consists of four stages that include; 1) Problem formulation; 2) Building, Intervention, and Evaluation; 3) Reflection and learning; 4) Formalization and learning. Traditional design is often described as a step-by-step process in which problem formulation (user requirements) is followed by development of the artifact, which in turn is followed by an evaluation. In contrast, the ADR process is a highly iterative process and thus highlights the emergent nature of the ensemble artifact. There are some main features that we consider highly relevant for our project. Firstly, it takes into consideration the dynamic and emergent nature of socio-technical systems and the interplay between planned design and the context of use. Second, the ADR process focuses on participation among researchers, practitioners and end-users throughout the design process. Thirdly, it is strongly oriented toward collaboration and change involving both researchers and subjects [15, p.330]. Eventually the outcome of the design process is formalized and shared with practitioners and generalized as design principles for a particular type of information systems.

In addition, our research is based on previous studies on design and use of digital platforms in healthcare [16] as well as studies on knowledge management [17] and trans-situated learning [18].

III. Research Setting and Metod

The study is based on an ongoing research project in the municipal care services involving partners from different professional communities and organizational units. The project originates from a health innovation cluster (Arena Health Innovation) that was formed in 2009 as a partnership between academia, public sector and academia. A key objective among partners in the innovation cluster was to identify capabilities to improve services and provide technologies to support the needs of municipal care. Based on ongoing activities in the health innovation cluster, a pilot project was initiated and carried out between June 2013 and May 2014. Moreover, experiences and findings from the pilot project have been prolonged to a larger research project lasting from 2014 to the end of 2017.

The primary research objective of the overall project is to identify factors for successful implementation of welfare technologies in municipal care services. Communication and collaboration among various stakeholders is a challenge and one of the research areas in the project is how knowledge and skills are shared among stakeholders during the innovation process. Stakeholders involved in the project include eight municipalities, two research institutions, as well as two providers who are partners in the innovation cluster. All municipalities started with the same basic technology that includes a web-based portal and sensors embedded in security blankets and door handles. In an iterative process of design and use, new features will be added and the number of service receivers will increase in line with the innovation process. Accordingly, research and innovation goes "hand in hand" and this involves adjustment of technology, identification of new services and technological capabilities. Thus, the project requires a high level of engagement from all stakeholders who have signed a consortium agreement that involves commitment to participate in all planned activities.

Data collection has been a combination of participant observations, semi-structured interviews and archival documents. A main source of data collection has been participant observation at workshops. In total, we have participated at five workshops that have gathered all the key people involved in the project, that is, nurses, nursing assistants, vendors, employees at the IT departments, and researchers. As researchers, we have had a dual role during these meetings. First, we have acted as facilitators by participating in the preparation as well as practical support during the workshops. Secondly, we have acted as observers and followed closely emerging discussions, reflections and interaction between various actors. In addition, we participated in local project meetings in the municipalities as well as informal meetings in the project. In order to get more in-depth understanding of the social context of interaction and engagement in the project, we have carried out 5 interviews with project managers in municipalities. Finally, various documents such as strategy documents, minutes of meetings, evaluations, reports and social media posts were collected. Analysis of data was based on an interpretative approach to qualitative research [19] and we have used NVivo to identify and categorize topics related to communication and interaction among stakeholders in our study.

IV. Organizing for User Engagement

As mentioned, one of the research areas in the project is communication and collaboration, and one of the aims was to create an environment for learning and knowledge sharing. Organization of workshops has thus been an effort to strengthen collaboration and five workshops have been arranged so far in the project. On an average, 30 – 40 people from different professional fields and organizational units attended the meetings. A manager in Arena Health Innovation has played a key role in the planning and organizing the workshops and the main theme has been service innovation. Moreover, the agendas for the meetings have been based on activities described in the project plan as well as emerging issues during the meetings. The structure of the workshops has been a combination of presentations and group work. Some key people have been hired to lecture on the various topics, as well as facilitate the group work. For example, service designers have attended several workshops and provided methods and tools for the preparation of scenarios and user stories during group work. The purpose of the group work was to reflect on practice; identify actors involved, how the use of technology affects work processes as well as new service areas. Several of the methods and tools used during the workshop sessions have contributed to the identification and visualization of user needs, demands and expectations of the project and thus provided valuable knowledge for further progress. Some of the issues that have been discussed are related to communication, collaboration and organizational aspects. For example, reliable technological solutions are a primary goal and the need for guidelines and procedures have been identified. Guidelines for the assessment and mapping of the use of digital surveillance, risk and vulnerability analysis and user support has been highlighted, and the ongoing work on these issues has been the topic of the last two workshops. Discussion of these issues has facilitated reflection and mutual learning and thereby raised awareness of the use of digital surveillance in care services. Nevertheless, several challenges have also been identified and have affected further work in the project. For example, it was pointed out that healthcare professionals and technologists speak different languages and thus makes it difficult to obtain a shared understanding of issues in everyday practice. Another challenge has been the need to co-develop procedures and guidelines and to maintain involvement in the time between workshop sessions. As mentioned, project members in the study are located in eight municipalities distributed in four different counties in southern Norway. This involves long traveling distances and limited opportunities for face-to-face meetings between the workshops. To deal with these issues we have looked at online communication as a resource for sharing knowledge and experience in the project.

V. Co-creation of a Innovation Network

To promote engagement and interaction among participant in the project, we have started the design and development of a digital platform. Design activities in the first phase have been inspired by similar studies in the healthcare domain [16] as well as studies on digital platforms and innovation networks in general [8] [18]. Furthermore, the identified needs to share experiences, co-development of procedures and guidelines, and interdisciplinary interaction has formed the basis of the initial user requirements. In addition, we searched for available solutions to build on what already exists. We did not find, however, any available solutions that met the requirements for safety and accessibility for all stakeholders in our study. Although several municipalities use platforms such as SharePoint, these are not available to all the users in our project and were thus not relevant solutions. We have also considered social networking sites such as Facebook as a resource in the project. Several of the project members use Facebook and we have created an account for the project. At the moment there are 33 members in the group, all participants in the project. In addition, some of the project members also participate in several other Facebook groups. Most of the posts in this and similar Facebook groups deal with general news on welfare technology, links to public documents or policies, as well as news and pictures from workshops. However, Facebook is an open network and does not meet the requirements for safety when sharing internal documents. In order to move forward, we made contact with a local partner and made an agreement with a provider who has developed similar solutions. The local provider, who has previously been involved in activities in the Arena Health Innovation cluster, wanted to participate in the project and is now involved in the ongoing work to develop a solution.

In the first stage, we (the researchers) acted as mediators between the vendors and users. Empirical data from the workshop sessions formed the basis for user needs, and we had several meetings with the vendor in order to provide input to the development of the first version of a platform solution. As mentioned, we have identified the need to co-develop guidelines and procedures, as well as a shared awareness among stakeholders involved in the project. Thus, the main features of this first solution are the ability to share documents, coordinate activities, video meetings, messages and posts. Further development depends on feedback from users and we have invited one of the project members to a video meeting to make a brief test of basic features. In particular, the ability to video meetings was considered an important feature to maintain interaction between project managers in the municipalities. Thus it was decided to test the use of the system in a real-world environment and all project leaders in the eight municipalities were invited to participate. These ongoing activities and future development are planned as an iterative process of evaluation, re-design and intervention in the organizational context. So far, in the process we have focused on interactions between project members in the municipalities. Further in the process we will also include technologists and providers, and we will be open to emerging needs and capabilities in order to expand the innovation network.

VI. Discussion and Further Work

In this research in progress paper we have focused on engagement and mutual learning in digital innovation in the municipal care services. In particular, we highlighted how the project participants in our study play a key role in the innovation process that is characterized by reflection in practice and mutual learning [10]. Adoption and adjustments of the socio-technical system depends on ongoing shaping by organizational use, perspectives, and participants (ibid. p. 44). Skills and knowledge that emerge in the daily use of digital surveillance are valuable contributions to the innovation process. However, articulating this kind of knowledge (situated learning) is not a straightforward matter but requires translation and transformation across domain-specific contexts [17]. The interdisciplinary environment that characterizes our study indicates the need to create a "shared design space" that enables translation of different meanings as well as negotiation of interests and making trade-offs between actors [17]. In addition, the project members are dispersed across several municipalities and this limits the face-to-face interaction. Vaast and Walsham [18] have argued that the ability of sharing resources and experiences in such an environment depends on a supportive information infrastructure. They have also proposed a model for "trans-situated" learning supported by an information infrastructure.

The project is still at an early stage and several ongoing activities will have an impact on further progression. A primary goal is to create a digital innovation network that provides user utility in the organizational context. The iterative process means that the technological solution and the organizational setting will continuously be subject to reinterpretation, reformulation and redesign [16]. Moreover, reflection and learning, and formalization and learning are important stages in the ADR method. The research process involves more than simply solving a problem. It must also contribute to knowledge that can be applied to a wider class of problems [10]. This means that we strive to develop design principles that can be generalized to the design of innovation networks as well as insight into collective actions that shape new processes and services in innovation networks.

References

[1] J. Aarts, J. Callen, E. Coiera, and J. Westbrook, "Information technology in health care: socio-technical approaches.," International journal of medical informatics, vol. 79, no. 6, Jun-2010, pp. 389–90,.

[2] M. Berg, J. Aarts, and J. Van der Lei, "ICT in Health Care : Sociotechnical Approaches What Are Sociotechnical," Methods Inf. Med., vol. 4, 2003, pp. 297–301.

[3] R. Heeks, "Health information systems: failure, success and improvisation.," Int. J. Med. Inform., vol. 75, no. 2, Feb. 2006, pp. 125–37.

[4] M. Memon, S. R. Wagner, C. F. Pedersen, F. H. A. Beevi, and F. O. Hansen, "Ambient assisted living healthcare frameworks, platforms, standards, and quality attributes.," Sensors, vol. 14, Mar. 2014, pp. 4312–41,

[5] Norwegian Directorate of Health, "Helsedirektoratets anbefalinger på det velferdsteknologiske området," Oct. 2014.

[6] J. Barlow, D. Singh, S. Bayer, and R. Curry, "A systematic review of the benefits of home telecare for frail elderly people and those with long-term conditions.," J. Telemed. Telecare, vol. 13, no. 4, Jan. 2007, pp. 172–9,

[7] B. Bygstad, O. Hanseth, and D. Truong Le, "From IT Silos to Integrated Solutions . A Study in E- Health Complexity," Proc. ECIS, May 2015, pp. 0–15.

[8] Y. Yoo, K. Lyytinen, and R. J. Boland, Jr., "Distributed Innovation in Classes of Networks," Proc. 41st Annu. Hawaii Int. Conf. Syst. Sci. (HICSS 2008), Jan. 2008, pp. 1–9.

[9] T. Robertson and J. Simonsen, "Challenges and Opportunities in Contemporary Participatory Design," Design Issues, vol. 28, no. 3, Jul. 2012, pp. 3–9.

[10] M. K. Sein, O. Henfridsson, S. Purao, M. Rossi, and R. Lindgren, "Action Design Research," MIS Q., vol. 35, no. 1, Mar. 2011, pp. 37–56.

[11] The Ministry Health and Care Services, "Innovation in the Care Services, NOU2011: 11," 2011, pp. 1–40.

[12] B. Bygstad, "Generative mechanisms for innovation in information infrastructures," Inf. Organ., vol. 20, no. 3–4, Jul. 2010, pp. 156–168.

[13] C. U. Ciborra and G. F. Lanzara, "Formative contexts and information technology: Understanding the dynamics of innovation in organizations," Accounting, Manag. Inf. Technol., vol. 4, no. 2, 1994, pp. 61–86.

[14] A. R. Hevner, S. T. March, J. Park, and S. Ram, "Design Science in Information System Research," MIS Q., vol. 28, no. 1, Mar. 2004, pp. 75–105.

[15] R. Baskerville and M. D. Myers, "Special Issue on Action Research in Information Systems: Making IS Research relevant to Practice-Foreword," MIS Q., vol. 28, no. 3, Sept. 2004, pp. 329–335.

[16] P. Spagnoletti, A. Resca, and Ø. Sæbø, "Design for social media engagement: Insights from elderly care assistance," J. Strateg. Inf. Syst., vol. 24, no. 2, May 2015, pp. 128–145.

[17] P. R. Carlile, "Transferring, Translating, and Transforming: An Integrative Framework for Managing Knowledge Across Boundaries," Organ. Sci., vol. 15, no. 5, Oct. 2004, pp. 555–568.

[18] E. Vaast and G. Walsham, "Trans-Situated Learning: Supporting a Network of Practice with an Information Infrastructure," Inf. Syst. Res., vol. 20, no. 4, Dec. 2009, pp. 547–564.

[19] G. Walsham, "Interpretive case studies in IS research : nature and method," Eur. J. Inf. Syst. No. 4, May 1995, pp. 74–81.

Assessing Electronic Health Records: Are Basic Assumptions in "Health Technology Assessment" Useful?

Anne Granstrøm Ekeland
University Hospital of North Norway
Norwegian Centre for e-health research
Tromsø, Norway
E-mail: anne.granstrom.ekeland@telemed.no

Abstract—**The use of Health Technology Assessment (HTA) in the field of information and communication technologies (ICT) is limited. The Norwegian health authorities and international networks call for steps to strengthen its use to support the development of good e-health services. Randomized controlled trials (RCT) are the gold standard approach in case studies in HTA and scholars have raised questions concerning their relevance in e-health assessments. Failure of basic philosophical assumptions inherent in RCT to reflect empirical features of e-health is one explanation. In a sociotechnical perspective, this paper explores empirical features of "Common Implementation of Clinical Systems" (FIKS), a large-scale electronic health record program in North Norway. Drawing on documents, information and presentations over a 4 year period, it discusses how empirical features correspond to assumptions of RCT. Also considering scientific literature from assessments of electronic records, complimentary assumptions are presented. The objective is to contribute to a knowledge base for improving HTA of ICT. Results show that RCT assumptions of a stable world, fixed interventions and controlled implementation processes differed from empirical processes. Hence, RCT approaches fail to address important features of the program and produce knowledge that fully demonstrate (causes of) empirical benefits or pitfalls. The paper briefly considers complementary assessment perspectives. Embedding assumptions of a world in flux where social, technical and clinical entities influence each other in dynamic processes should increase the relevance of HTA of ICT, and affect real time developments. Further exploration of assumptions that encourage participatory and process assessment approaches is timely.**

Keywords- Health Technology Assessment (HTA); challenging assumptions, approaches and methods; programs of electronic health records; assumptions in constructive assessments

I. Introduction

A. Background

The paper is positioned within a sociotechnical perspective and seeks to produce insights into the coherence between assumptions in health technology assessment (HTA) and the practices that are assessed. The need for assessments of information and communication technology (ICT) programs have been strongly expressed for instance by The parliament in United Kingdom (UK) in a summary of the National Health Service (NHS) information technology (IT) program. "The original objective was to ensure every NHS patient had an individual electronic care record which could be rapidly transmitted between different parts of the NHS, in order to make accurate patient records available to NHS staff at all times. This intention has proved beyond the capacity of the department to deliver and the department is no longer delivering a universal system. Implementation of alternative up-to-date IT systems has fallen significantly behind schedule and costs have escalated" [1].

Health Technology Assessment is expected to produce knowledge to help decide on and procure technology and services that are accurate, cost effective and with expected value and quality [2]. For those purposes, the Norwegian health authorities and international scientific networks for the conduct of HTA call for steps to strengthen its use in the field of ICT. As part of this, the Northern Norwegian Health Authorities (NNHA) in 2016 funded a three-year project for developing and adapting HTA approaches and tools. The project builds upon the White paper to the Storting: one patient – one record [3]. This paper is part of the project.

The need for adapting and developing assessments for the field of e-health is expressed in several scientific publications, some of them referred to in section C below. A common idea is that established assessments have weaknesses in that they produce less relevant and timely knowledge. In this paper, weaknesses connected to basic philosophical assumptions in HTA are addressed. More specifically, those expressed in the gold standard approach of Randomised Controlled Trials (RCT) and related to development of electronic health records (EHR) in North Norway.

The research question is how assumptions of RCT are amenable to empirical features of "Common Implementation of Clinical Systems" (in Norwegian an acronym, FIKS), a large-scale program for developing and implementing a new EHR [4, 5]. The paper also briefly comments on approaches and methods of RCT, relying on equal assumptions. FIKS started in 2012 and is scheduled to last through 2016. It lies within the jurisdiction of the NNHA in North Norway. One common electronic patient record for all hospitals in the northern region of Norway is a goal. No pre-implementation evaluation has been carried out.

The objective of the paper is to contribute to a knowledge base for addressing weaknesses of HTA of ICT and to briefly point to alternative assumptions which might strengthen its use for the benefit of patients, health professionals, policy makers, leaders and industry.

In sections B and C to follow, HTA, its assumptions and weaknesses are presented. In section II an account of FIKS is given, followed by methods and materials of the investigation. The results and discussion section is divided in three sections each addressing different assumptions in RCT: a singular reality (context), a clear definition of the intervention and a controlled implementation process. Section IV points to approaches and methods accommodating different assumptions: a reality in flux, and interventions and implementation as ongoing socio/technical/medical achievements. In conclusion, the paper argues that exploring such assumptions should strengthen the relevance of HTA methodology for e-health.

B. Health Technology Assessment (HTA) and RCT

HTA is a research field defined as "the systematic evaluation of the properties and effects of a health technology, addressing the direct and intended effects of this technology, as well as its indirect and unintended consequences, and aimed mainly at informing decision making regarding health technologies. HTA is conducted by interdisciplinary groups that use explicit analytical frameworks drawing on a variety of methods [6]."

The purpose of HTA is to establish a decision basis for the procurement of the right health technologies [7]. The European network for HTA, EUnetHTA justifies the research field as follows: "Health care decision making requires the right evidence at the right time. Every day there are new health technologies available that can improve patient outcomes and refine health system efficiency. HTA is a tool to review technologies and provide evidence of the value these technologies can deliver to patients and their families, health system stakeholders, and to society more broadly [6]."

Health technologies comprise "Diagnostic and treatment methods, medical equipment, pharmaceuticals, rehabilitation and prevention methods, but also organisational and support systems used to deliver healthcare" [8]. The electronic patient record is a type of system that not only comprises of technology, but also involves vulnerable information, social interactions, relationships, and competencies among users (e.g. doctors, nurses, patients) as well as organizational structures, routines and coordination components.

In HTA, different products have been developed to support knowledge-based decisions in health care, such as systematic reviews, meta-analyses, modelling and assessments of new medical methods. All tools draw on basic philosophical assumptions and form a coherent approach. The gold standard tool in case assessments are Randomised Controlled Trials (RCT). RCT is a type of scientific (often medical) experiment, where the people being studied are randomly allocated to one or other of the different treatments or interventions under study. RCTs are often used to test the efficacy or effectiveness of various types of medical interventions. The interventions are assumed to be clearly defined and demarcated and may provide evidence for adverse effects, such as drug reactions.

C. Assumptions and Weaknesses

Accurate assumptions to guide approaches is imperative for the production of useful knowledge to the confidence of different stakeholder groups.

A basic assumption underlying an RCT is that of a singular reality amenable to objective scientific measurement to provide universal evidence for outcome of specified interventions. A relatively stable situation and causal variables and linkages has to be identified in order to be able to generalize and repeat outcome. One challenge for applying RCT for ICT and e-health programs is that empirical situations in general are more messy and in flux [9].

In their assessment of electronic health records in the National Health Services (NHS) IT program in United Kingdom, Greenhalgh and colleagues addressed this challenge. They asserted that e-health initiatives occur in complex and fast moving socio-political arenas. Evidence is produced by, and fed back into a political process of deciding priorities and allocate resources to pursue them [10]. Interpreting practice in context can therefore be an alternative to the production of evidence for universal truths in controlled experiments as recommended in RCT.

A second assumption underlying RCT is that of a clear demarcation and definition of the intervention, including a fixed start and endpoint. In ICT programs, this can be difficult to achieve given the fast-paced technological development and the seemingly endless range of possibilities for novel service delivery platforms. It normally takes years to accomplish an RCT and this is described as the most formidable challenge threatening to upset the very promise of potential solutions: The rate of emerging technologies and services is far outpacing the field's capacity to demonstrate the conceptual or empirical benefits [11].

A third challenge described is pressure to roll out new ICT services before pilots are fully evaluated. Implementation is hence assumed to be a linear operation where readymade technological applications are "rolled out" to an organization and can be objectively assessed. Human interaction might be considered as an obstacle. The alternative is proposals to address person-to-person models to understand how collegiate and interpersonal elements of care delivery can be better embodied in assessments and as such brought to consciousness for influencing development [12]. In design, the influential openEHR standard represents a model driven approach allowing clinical personnel to be involved in development processes [4]. OpenEHR is a virtual community working on interoperability and computability. Its main focus is electronic patient records (EHRs) and systems [13]. Other innovation and design philosophies/practices such as "Design thinking" involve future thinking and creativity as main assumptions behind innovation [14].

The three challenges and assumptions behind them are interconnected. It seems that basic assumptions and subsequent approaches of RCTs could fail as guiding principles for addressing all important aspects that affect the relevance of ICT. Evidence for positive or negative effects based upon erroneous assumptions might support both over optimist and over pessimist expectations for future development.

In the rest of the paper, steps to address these challenges connected to the FIKS program are discussed. The research question is specified and discussed in three parts: How are assumptions about the reality, the intervention, the process of implementation and subsequent approaches and methods of RCT, amenable to empirical features of FIKS? Based upon scientific literature, complementary assumptions capable of improving HTA for ICT are presented.

II. THE FIKS PROGRAM, METHODS AND MATIERALS

A. FIKS

FIKS is a large-scale program for developing and implementing a new electronic health record system, running from 2012 through 2016. The costs are estimated to EUR 90 million and the vendor (DIPS) is the largest EHR vendor in Norway [4]. The aim is to introduce a single electronic patient record at the eleven North Norwegian hospitals, including radiology, lab, pathology and electronic requisition of laboratory services for general practices in the region [15]. An official information sheet of FIKS is published in English [5]. Additional description of the program and implementation is provided in the results and discussion sections.

B. Methods and Materials

The paper is based upon a mixed data material consisting of documents, web sites, information from advisors and presentations of the FIKS program to different actors in the hospitals in North Norway over a period of 4 years, from 2012 to February 2016. In addition, document studies of papers and reports from two large scale evaluation and assessment projects in UK connected to the National Health Services (NHS) ICT program were studied: "The UK Summary Care Record Programme" [10] and "Healthcare Electronic Records in Organisations" [16]. A number of scientific papers were recommended in publications from the two programs with a focus on assessment traditions. These are reflected both in the background and discussion sections.

Triangulation is a social science technique that facilitates validation of data through cross verification from two or more sources [17]. In particular, it refers to the application and combination of several research methods in the study of the same phenomenon. Such techniques were applied to combine information from the multiple sources refined into useable assemblages. These culminated to form recognizable examples for the discussion of assumptions and approaches. The discussion sections also draw on arguments developed with the support of the MethoTelemed team, whose contribution is acknowledged [18].

III. RESULTS AND DISCUSSION

A. Assumption one: A Singular Reality Amenable to Scientific Measurement and Control

This section will substantiate that the context of FIKS and the program itself can be understood in terms of complexity, multiplicity and dynamism. To single out a clear distinction between the context and the program is also not a straightforward task. Relatively stable variables depicting the reality, or context, will, however also be distinguished and commented. This is one premise in RCT for defining external and internal causal variables and linkages in order to be able to repeat outcome in controlled ways.

The context of FIKS is a number of mutually dependent actors, representing a myriad of interests trying to accomplish a unified vision. This is documented in the web page informing that the FIKS project implements the new systems in close cooperation with health authorities, Health Nord ICT (An organization established by the Regional Health Authorities designated to implement ICT services) and the various system suppliers.

In addition, the different hospitals where the implementation occurs, represent different socio-political and institutional contexts. The context is therefore complex, interconnected and politicized, as health political decisions affect resources necessary to add affordances of, and accommodation of the record. Concerning affordances, the web page of FIKS informs that the next generation patient record is under development and is tested in the region. Some upcoming milestones in 2016 on the path to one common medical record are listed:

- One common medical record at the hospitals in Hammerfest and Kirkenes cities
- University Hospital North Norway employs regional radiology solutions including one common radiology archive
- The hospitals in Helgeland, Mo I Rana, Mosjøen and Sandnessjøen employ one common medical record (DIPS) [5]

This information tells us that both the contexts and the intervention consist of multiple, developing and mutually dependent components.

Also the historical process accounts for a dynamic and inter-woven character of the context and intervention.

Already in 2011, necessary contracts for the program were signed, showing the different industrial actors involved. These are some of the milestones presented on the web page:

2011

- Helse Nord Regional Trust signed a contract with Sectra
- Helse Nord Regional Trust signed a contract with Tieto
- Helse Nord Regional Trust signed a contract with DIPS

2013

- Helse Nord Trust signed a contract with CompuGroup Medical Norge (CGM)
- Helse Nord Trust signed a contract with Infodoc
- Merging of pathology systems in two major hospitals

2015

- Helse Nord Trust signed a contract with Hove Medical Systems [5]

The assumptions for the conduct of an RCT, a controlled, measurable and relatively stable reality, are not reflected in the empirical features of the context/intervention. Rather, the multiple and mutually dependent actors and interests, depict a reality and situation under development and flux, depending on negotiations, shifting political conditions and resources.

B. Assumption Two: A Clear Demarcation and Definition of the Intervention

On the web page as well as the Facebook page, the goal of FIKS is described as an ambition that the people of the north will have their clinical history assembled in one patient record and that the practice of sending records between hospitals will end. An ambition refers to a work process, and not to a defined and fixed intervention as assumed in RCT. The notion rather refers to assumptions of a creative process like in "Design Thinking". One of the first features of the intervention was described in 2013: "Moving the databases of the hospitals in Helse Nord to one central common database, is an important condition for the implementation of common patient administration and treatment systems and one common electronic record for the individual hospitals in North Norway."[5]

New ingredients were added to the service. Events were planned as an ongoing deployment process and the program was described in terms of technical/operational events:

- Connection of Narvik medical center against Health Nords regional solution for electronic requisitioning of Laboratory Services 11/25/2015
- Connection of Leirfjord medical office against Health Nord regional solution for electronic requisitioning of Laboratory services 11/25/2015
- Connection of Nordreisa medical office against Health Nords regional solution for electronic requisitioning of Laboratory Services 11/24/2015
- Connection of Træna medical office against Health Nords regional solution for electronic requisitioning of Laboratory Services 11/24/2015
- Connection of Skjervøy medical office against Health Nords regional solution for electronic requisitioning of Laboratory services[19]

By distinguishing events this way, the grounds are laid out for RCT of each part of the process, but the resources needed for accomplishing such an endeavor would be vast. There are also interconnections between the parts and therefore difficult to single out as clearly demarcated interventions.

C. Assumption Three: Implementation as a Linear and Controlled Operation

The described pressure to roll out new ICT services before pilots are fully evaluated involves an assumption that implementation is a linear, top down and controlled process that can be distinguished from context and socio-political or human processes. The description of the implementation of FIKS, however, clearly points to an ongoing and changing process where different components should be aligned. This is how it is expressed on Facebook: The FIKS program in Helse Nord consists of six projects intended to develop and implement joint electronic record systems at the hospitals in Northern Norway:

- One joint electronic record (DIPS)
- New features of the electronic record (DIPS Arena)
- Laboratory Information System
- Radiology Systems ("Sectra", "RIS" and "PACS")
- Joint pathology system in Tromsø and Bodø
- Electronic requisition of laboratory services[19]

In 2011 DIPS decided to use the OpenEHR framework for developing its next generation EHR for the hospital market [4]. This involves negotiations on development directions. The role of interaction between different participants in the process is a collegial and interpersonal process, enacted as different meetings for dialogue and negotiations:

- 11.26.2015: Workshop (EPR Development): Theater nurse meeting - Planning and booking DIPS Arena
- 11.26.2015: Workshop (EPR Development): Theater nurse meeting - meeting with clinicians
- 11/26/2015: Workshop (EPR Development): Decisions in psychiatry - new module in DIPS Arena
- 11/12/2015: Operation Planning (EPR Development): Meeting with clinicians at University Hospital of North Norway [5]

This process adds to the formerly documented process of negotiating contracts with producers and vendors.

FIKS also designates and educates super users, and state that the competencies of employees are crucial for the development of good record systems: "Close to 190 super users at Nordlandssykehuset are ready to be educated on use and routines of the new electronic record and become leading DIPS experts [5].

The description of the process of implementation shows a multitude of inter-related operational, interactional and relational processes. A linear, pre-defined and controlled "roll-out" process is not present, as assumed for RCT.

D. *Summary and discussion*

Among challenges in applying a HTA assessment framework for the study of an electronic patient record, is that HTA tools form a coherent approach and draw on common basic assumptions, which differ from empirical features. The basic assumptions of a stable reality amenable to objective measurement, a defined intervention and an operational and linear implementation process fails to address the empirical features described. This issue has been discussed in HTA, related to innovation research. Two different models or approaches are described in the core model (reference 20, page 15). "The linear diffusion model perceives new technology as an external stable entity that is brought to a (health care) system and induces change. A competing paradigm, the translation model, presumes that technology undergoes change in the environment it is brought into. Hence the final impact will not depend on the original technology only" [20]. In the case of FIKS, both the technology, the health care setting and the implementation process seem to be in a state of mutual translation. The empirical features of the applications and services connected with the record are highly diverse and constantly in flux within shifting social and organisational contexts.

Challenges were connected to discrepancies between assumptions and features of the context or "reality" within which the electronic record is embedded, the intervention itself and the implementation process. In the next section, approaches that build upon other assumptions are addressed.

IV. COMPLIMENTARY ASSUMPTIONS - THE CONSTRUCTIVIST UMBRELLA

Challenges concerning the validity of evidence in the face of involvement of different stakeholders have been articulated within the HTA tradition, which is looking to overcome such challenges. One of the tools of HTA in that respect is consensus conferences with different stakeholders [21]. Such conferences have been subject to investigations and the following assertion strengthen the argument of a shifting social reality and the need to consider social relations as drivers for both intended and unintended outcome:

"Consensus development programs are not immune to the economic, political, and social forces that often serve as barriers or threats to evidence-based processes. Organizations that sponsor consensus development conferences may do so because they have certain expectations for the findings of these processes, and may find themselves at odds with evidence-based findings. Other stakeholders, including from industry, biomedical research institutions, health professions, patient groups, and politicians seeking to align themselves with certain groups, may seek to pressure consensus development panelists or even denounce a panel's findings in order to render desired results, the evidence notwithstanding." (Goodman 2004:49) [22]

The citation points to the importance of addressing social/political interests and processes to understand the way evidence can be produced and affect results of ICT use in health services.

In contrast to assumptions in RCT about a relatively stable and objective reality, a fixed intervention and a linear and controlled implementation process, constructivist traditions assume flux. This implies a reality or context under development, interventions which are also subject to change, and implementation as partly unpredictable and depending on for instance resource allocation. Implementation is considered an on-going process where support or lack of support strongly influence the outcome. Therefore, it is not considered possible to generalise evidence based outcome in order to repeat good results in new or future settings. Assessment results are aiming to be fed back into a pragmatic and political process of deciding priorities and allocate resources to pursue them.

In this perspective, validation is considered to be obtained through negotiations between the context, the researchers, the intervention and other stakeholders. Context is considered by involving different stakeholders' interests and validity is addressed by asking what the study is valid for (Aguinaldo 2004:127) [23].

Such assumptions and resulting approaches may have particular strengths where the goal is to develop good e-health services, to the confidence of users, professionals, policy makers and payers, and as a lead market in Europe. Thus, obtaining balance between different validity claims is a vast challenge.

This paper has focused on the fluctuating character of reality, interventions and implementation. Nevertheless, there are also stable features of the three elements. To bridge the gaps between assumptions of the two traditions, the positivist (as in HTA) and the constructivist should be important for assessments of e-health. This point is also noted by Ammenwerth [24]. One goal should be to open the borders between traditions and identify how evaluators may draw on the benefits the different ones have to offer.

Answers to the question "Does it work?" to produce evidence for universal truths, need to be supplemented by a whole range of answers to questions that reflect the complexity of most e-health interventions. "How does it work?" "What components are vital to success, and which are redundant?" "Why does it work in this context (and equally important not work)?" "Is this an appropriate and acceptable way of tackling the problem?" "How is quality produced and defined within certain innovation processes? Moreover, "who owns the definition of success?"

In process approaches, investigations are directed towards the conditions that are included in development processes, with a purpose to feed results back into the process for dialogue and improvements. The "intervention" is shaped and adjusted in and through practices of professional-social interaction between participants (doctors, nurses, patients) and the organisational, economic, political and ideological settings these practices are embedded in. The intervention also contribute to shape these settings as the approach presupposes that all entities are in mutual shaping. Controlling conditions will be the crucial task for future results of innovations. Process investigations may produce knowledge to this end.

V. CONCLUSION AND FURTHER WORK

The paper has substantiated empirical features of a reality in flux, an intervention under development and implementation processes as ongoing negotiations for the FIKS program. HTA assumptions of a stable reality, a fixed intervention and a controlled implementation process were not present. Steps to strengthen HTA use for ICT are timely. Knowledge about conditions for large processes with escalating costs is important as conditions built into the programs vastly influence the effects that emerge and manifest. Embedding assumptions of a world in flux where social, technical and clinical entities influence each other in dynamic processes should increase the relevance of HTA of ICT, and affect real time developments. Further exploration of assumptions that encourage participatory and process assessment approaches is timely.

ACKNOWLEDGMENT

I am grateful to the North Norwegian Health Authorities who funded the project and to communication advisor Elin Vinje Jenssen, who rapidly translated the fact sheet of FIKS to English.

REFERENCES

[1] 1. The Parliament, U.K. The National Programme for IT in the NHS: an update on the delivery of detailed care records systems, HC (2010-12). [cited 2016 14.03]; Available from: http://www.publications.parliament.uk/pa/cm201012/cmselect/cmpubacc/1070/107003.htm.

[2] HTA Glossary. [cited 2016 14.03.]; Available from: http://htaglossary.net/HomePage.

[3] The Norwegian Government. One inhabitant - one record, Meld. St. 9 (2012–2013). [cited 2016 14.03.]; Available from: https://www.regjeringen.no/no/dokumenter/meld-st-9-20122013/id708609/?q=Meld

[4] B.Christensen and G. Ellingsen. Evaluating Model-Driven Development for large-scale EHRs through the openEHR approach. International Journal of Medical Informatics, 2016. 89: p. 43-54.

[5] Helse Nord. Implementation of common clinical ICT systems (FIKS). [cited 2016 14.03.]; Available from: http://www.helse-nord.no/about-fiks-in-english/implementation-of-common-clinical-ict-systems-fiks-article135571-42080.html.

[6] Eunethta. Mission, Vision & Values. [cited 2016 14.03]; Available from: http://www.eunethta.eu/about-us/mission-vision-values.

[7] B. Jennett. Health technology assessment. BMJ : British Medical Journal, 1992. 305(6845): p. 67-68.

[8] eunethta. HTA Common Questions. [cited 2016 14.03.]; Available from: http://www.eunethta.eu/about-us/faq#t287n73.

[9] T. Scott. Implementing an Electronic Medical Record System: Successes, Failures, Lessons. 2007: Radcliffe.

[10] T.Greenhalgh and J. Russell. Why Do Evaluations of eHealth Programs Fail? An Alternative Set of Guiding Principles. PLoS Med, 2010. 7(11): p. e1000360.

[11] D.J. Jones. Future Directions in the Design, Development, and Investigation of Technology as a Service Delivery Vehicle. J Clin Child Adolesc Psychol., 2014. 43((1)): p. 128–142.

[12] D.Avison and T. Young. Time to rethink health care and ICT? Commun. ACM, 2007. 50(6): p. 69-74.

[13] What is OpenEHR? [cited 2016 14.03.]; Available from: http://www.openehr.org/what_is_openehr.

[14] Cohen, R. Design Thinking: A Unified Framework For Innovation. 2014 [cited 2016 15.03]; Available from: http://www.forbes.com/sites/reuvencohen/2014/03/31/design-thinking-a-unified-framework-for-innovation/#76b8788256fc.

[15] Kallbekken, S. Første steg på veien mot en pasientjournal i Helse Nord. [cited 2016 15.03]; Available from: http://www.unn.no/aktuelt/forste-steg-pa-veien-mot-en-pasientjournal-i-helse-nord-article108606-10106.html.

[16] EPSRC. Healthcare Electronic Records in Organisations (HERO). Available from: http://www.researchperspectives.org/rcuk/DD37EA56-9633-4962-831C-D59E401B2216_HERO--Healthcare-Electronic-Records-In-Organisations-.

[17] A.K. Bekhet and J.A. Zauszniewski. Methodological triangulation: an approach to understanding data. Nurse Res, 2012. 20(2): p. 40-3.

[18] A.G. Ekeland, A. Bowes and S. Flottorp. Methodologies for assessing telemedicine: a systematic review of reviews. Int J Med Inform, 2012. 81(1): p. 1-11.

[19] FIKS. Helse Nord FIKS. [cited 2016 15.03.]; Available from: https://www.facebook.com/HelseNordFIKS/info/?tab=page_info.

[20] eunethta. HTA Core model for medical and surgical interventions. [cited 2016 15.03]; Available from: http://meka.thl.fi/htacore/model/HTA%20Core%20Model%20for%20Medical%20and%20Surgical%20Interventions%201.0r.pdf.

[21] NICHSR. Monitor impact of HTA. [cited 2016 15.03.]; Available from: https://www.nlm.nih.gov/nichsr/hta101/ta101011.html.

[22] C. Goodman. HTA 101: introduction to health technology assessment. 2004, National Information Center on Health Services Research and Health Care Technology (NICHSR): http://aaz.hr/resources/pages/57/2.%20HTA101-%20Introduction%20to%20health%20technology%20assessment.pdf. p. 137.

[23] J. Aguinaldo. Rethinking Validity in Qualitative Research from a Social Constructionist Perspective: From "Is this valid research?" to "What is this research valid for?". The Qualitative Report, 2004. 9(1): p. 127-136.

[24] E. Ammenwerth, J.Brender, P. Nykanen, H. U. Prokosch, M. Rigby and J. Talmon. Visions and strategies to improve evaluation of health information systems. Reflections and lessons based on the HIS-EVAL workshop in Innsbruck. Int J Med Inform, 2004. 73(6): p. 479-91.

Denk je zèlf! Developing a Personalised Virtual Coach for Emotional Eaters using Personas

Aranka Dol[1], Olga Kulyk[2], Hugo Velthuijsen[1], Lisette van Gemert-Pijnen[2], Tatjana van Strien[3]
[1]Institute for Communication, Media & IT, Hanzehogeschool UAS, Groningen
[2]Department of Psychology, Health and Technology, University of Twente, Enschede
[3]Behavioural Science Institute and Institute for Gender Studies, Radboud University, Nijmegen
[3]Department of Health Sciences and the EMGO Institute for Health and Care Research, VU University, Amsterdam
e-mail: [1]{a.dol, h.velthuijsen}@pl.hanze.nl, [2]{o.a.kulyk, j.vangemert-pijnen}@utwente.nl, [3]t.vanstrien@psych.ru.nl

***Abstract* — Obesity is a fast growing societal threat, causing chronic conditions, physical and psychological health problems, as well as absenteeism and large healthcare costs. Despite numerous attempts to promote physical activity and healthy diet, existing interventions do not focus on often occurring emotional causes of obesity. There is a need for self-management support of this vulnerable target group: emotional eaters. This paper presents the results of the design case study focusing on a holistic development of a personalised virtual mHealth coach that provides self-management training *'Denk je zèlf!'* (Dutch for 'Develop a wise mind and counsel yourself'). Target group are young adults with emotional eating disorder and obesity. The contextual inquiry study was conducted to get insights into the needs and experiences of the target users, including interviews and questionnaires with emotional eaters, obesity treatment patients and healthcare practitioners. Personas and user stories were derived from these results and translated into a new *'Denk je zèlf!'* virtual coach, based on Dialectical Behaviour Therapy and experience sampling measures to capture user experience and emotional state. This paper makes two main contributions: (a) combining holistic design with behaviour therapy in one virtual mHealth coaching application for emotional eaters; (b) applying Personas to guide the design. Preliminary results suggest that an online self-management training might be useful for the target group. Future research will be aimed at iterative evaluation and further development of the dialectical dialogues for the virtual coach and content for the education and instruction modules.**

Keywords-obesity; emotional eating; Dialectical Behaviour Therapy; personalised care; virtual coach.

I. INTRODUCTION

The fast growth of obesity is a threat for humanity. *Obesity* is defined as an abnormal or excessive fat accumulation that may impair health and is classified as such with a Body Mass Index (BMI) of 30 kg/m2 or higher [1]. Obese patients often suffer from physical, metabolic, or psychological comorbidities, such as cardiovascular conditions, diabetes II, depression, etc. [2]. The number of years spent in relative unhealthiness (suffering from illnesses, disabilities) has impact on societal costs due to incapacity for work, absenteeism and large costs for healthcare [3]-[5]. Nearly 50% of the Dutch population suffer from being overweight or obese [2].

Obesity is associated with low socioeconomic background, unhealthy family lifestyle, bad eating habits and lack of physical activity. Increasing physical activity and reducing food intake (dieting) are considered cornerstones in the prevention and treatment of obesity. Though many of existing interventions are successful and make patients losing weight in the short run, long term randomised studies demonstrate that diets are not the answer [6]. Existing interventions (the Big Two – eat less, exercise more, such as Weight Watchers [7], My Diet coach [8] and Lose it! [9] (online weight loss programs) do not offer the ultimate solution in the long run, since people are not able to maintain their bodyweight over a longer period of time.

A. Emotional eaters and obesity

Recent studies show that a certain group (40%) of the obese population overeat due to negative emotions [10]. Emotional eating (EE) is an atypical stress reaction. A normal reaction to stress and negative emotions would be loss of appetite. Emotional eaters show this atypical behaviour because they confuse negative emotions with hunger. They have a narrow view on what happens in their body (poor interoceptive awareness) and they are having difficulties identifying and describing emotions and feelings (high alexithymia). Emotional eaters are having problems with emotion regulation - the ability to keep one's emotional system in a healthy condition [11].

Diets and behaviour therapies do not help people with high degrees of emotional eating as they do not treat the emotions resulting in the problem of emotional eating. Most treatment programs for obesity do not focus on emotion regulation [12][13].

B. Emotional eaters and eHealth

The majority of emotional eaters have a long history of dieting, followed by the inevitable overeating and starting dieting again. They gain weight because of poor emotion regulation, not just due to bad eating habits and/or due to an insufficient level of physical activity [14]. For numerous times, they have tried to loose weight and when the emotional eating behaviour kicked in again, they gained more weight than when they started their previous dieting episode. This is an example of the so-called '*yo-yo effect*' in health behaviour [15][16]. It is highly imaginable that this is ground for an accumulation of disappointments and a growing disbelief one will ever succeed.

Evidence indicates that eHealth therapy can be just as effective as the face-to-face treatments. Evidence-based therapeutic procedures could be delivered online [17]. eHealth interventions allow for an effective therapeutic relationship. In addition, this special target group needs personalised anonymous support that is always available. Not only emotional eaters need the moral support, a personalised self-management intervention could also clear away obstacles that hold emotional eaters from face-to-face contact with a therapist.

Obese emotional eaters form a very vulnerable group of people. To avoid further setbacks, they need support they can rely on, one that is trustworthy and auspicious, but realistic, in compliance with their needs. Such support needs to be accessible and comfortable so that one feels safe and secure. There is a need for supportive training programs for this specific target group of emotional eaters. This paper presents the design case study aiming at the development of a virtual mHealth coach application for self-management of young emotional eaters with obesity.

The rest of the paper is structured as follows. In section II, related work is discussed on (online) eHealth interventions for emotional eaters and obesity. After that, in section III the approach and methods are presented for developing Personas and applying them to guide the design process. In section IV, the results are presented including the Personas and the first prototype of a virtual coach. Finally, conclusion and discussion are presented in section V.

II. RELATED WORK

A. Dialectical Behaviour Therapy

Dialectical Behaviour Therapy (DBT) is relatively new in treating emotional eating behaviour. DBT was originally designed to help people that are suffering from Borderline Disorder [18]. The therapy focuses on the process of 'reduction of ineffective action tendencies linked with dysregulated emotions' [19]. Recent research on deployment of Dialectical Behaviour Therapy (DBT) shows positive results in weight loss management and weight maintenance in obese emotional eaters [20]-[23]. DBT might be successful in patients where insufficient effect was achieved with Cognitive Behavioural Therapy (CBT) [12] [24][25].

B. The dialectical focus

One of the most powerful 'mechanisms of change' or mediators in DBT is the dialectical focus. Since an invalidating environment plays an important role in the life of emotional eaters, it is important that they are treated with a well-balanced mix of being validated in their perception of negative emotions and being confronted with a practical focus on changing problem behaviour. "Based in the biosocial theory, DBT has a unique approach to targeting behavioural dysfunction that is not typically seen among other cognitive-behavioural treatments; one key difference is the emphasis placed on emotions and emotion dysregulation." [19].

C. DBT and eHealth

There is a broad variety of (blended) eHealth self-management treatments available but the majority of them is focused on weight loss and behaviour change. The discussion on the effectiveness of such interventions inches along [26]-[32]. Little knowledge in the field of eHealth treatment using Dialectical Behaviour Therapy or even emotion regulation is acquired so far, let alone emotion regulation focused on emotional eating behaviour.

Results of one quasi-experimental study on the effectiveness of the mobile "DBT Coach", that focused only on one particular skill in DBT (Opposite Action), showed that emotion intensity decreased within each coaching session in participants with Borderline Personality Disorder [33][34]. The target group uses the DBT Coach when it is needed most for them – after getting engaged in dysfunctional behaviour. One publication discusses the lack of user-friendliness of a DBT self-management mHealth application [35].

A small number of self-management mHealth apps can be found at the Play Store (Android) and at the App Store (iOS, Apple). However, they lack scientific grounding, user involvement in the design process, psychological aspects and personalization.

III. APPROACH AND METHODS

The objective of this research is to develop a personalised self-management intervention based on Dialectical Behaviour Therapy for young emotional eaters with obesity. Development is guided by the CeHRes roadmap (Center for eHealth Research, at the University Twente) – a holistic eHealth framework for developing eHealth interventions based on a participatory design process and persuasive design approach to maximize the impact of the intervention [36]. This study focused on the contextual inquiry and early design phases.

First, the contextual inquiry phase was carried out. In order to get familiar with what kind of support emotional eaters really need, it is important to understand the target group [36]. Even though interventions can be evaluated as positive according to effectiveness, as long as the target group will not be captivated by its design and functionalities, they are not going to use it.

The employment of user profiles and Personas as a tool to inform design is still rare in social sciences [36]. We followed the LeRouge classification model [37] and Van Velsen's additions to it [38] as a guideline to develop two Personas to guide the design of a virtual mHealth coaching intervention. Personas contain information on their Internet skills and smart phone use, demographic facts and healthcare specifics such as current practices in managing own healthcare, support network, information seeking attitude etc.

To gather input for user profiles and Personas, questionnaires were circulated via social media and the network of contacts. The target group to be reached for was "young adults, 18-44 years of age" and self-declared emotional eaters. Examples of questionnaire questions: "*For what purposes do you use your smartphone?* (social media, news gathering, mail, gaming)", "*At what specific moment in time would you like to/are you in need to get in contact with a help system?*", "*What kind of support do you expect from a smartphone application?*" We approached healthcare practitioners for expert interviews. In the next sections, we present the results of this design case study, including Personas, user stories, architecture and design of the new *'Denk je zèlf!'* virtual coach.

IV. RESULTS

A. Questionnaires and interviews

In total, 321 responses were collected with the questionnaires and thirteen interviews with healthcare practitioners (dieticians, physical therapists and psychologists) were conducted. The interviews were transcribed and coded. We used a free coding style not to lose the richness of the data. In addition, six obesity therapy patients were interviewed to get insight into the daily needs and experiences of the emotional eaters.

Data extraction from the questionnaires was processed according to the method of LeRouge [37] for creating user profiles and Personas - (a) personal and demographic information; (b) technical capabilities and limitations; (c) needs and desires concerning support and care. Data derived from the interviews with experts and patients learned about eating styles and the problems emotional eaters run into.

B. Personas and user stories

Two Personas were derived from the questionnaires data: *Lisanne* (25 years) and *Anita* (46 years). Figure 1 shows their personal profiles. Lisanne is a highly educated young woman. She is an obese emotional eater. Her eating behaviour is caused by a negative self image. Eating gives her a feeling of comfort as long as the eating lasts. Afterwards she feels guilty and depressed. Anita is a 46 years old mother of two. She is worrying about the family's financial situation and overeats in stressful situations. Anita left school at an early age. She is from a low socio-economic background.

What we learned by creating the Personas is that we got a better understanding about the specific needs of the target group. We obtained a better comprehension of the moments and situations that cause emotional eating behaviour and the kind of interference that might be helpful to them. While in the design process you can ask questions such as: "*What would Lisanne think of this? Would she like it? Would she consider this as useful?*"

To illustrate how Lisanne will benefit from the application we created a scenario (Figure 2) in which she is about to give in to emotional eating. Use-case scenarios [39][40] were derived from Personas, interviews and questionnaire results by describing the user goals, motivations, actions and reactions while using the new virtual mHealth coaching application.

C. eHealth Intervention Architecture

The first prototype of the smartphone application 'Denk je zèlf!' personalised virtual coach is developed within the pre-design phase.

Personal profiles emotional eaters	
Lisanne, 25 years, highly educated	Anita, 46 years, limited education

"I wish that I was more confident about myself…."	*"When I am stressed out I start snatching the cookie tin"*
Background	
Lives alone, single / High school graduate / Suffers from obesity / Co-morbidity unknown / l. 1.62 mtr., w. 91 kgs.	Lives with husband and two sons / low SES / Suffers from obesity / Co-morbidity unknown / l. 1.65 mtr., w. 85 kgs.
Attributes	
Feels insecure about her body / Is not able to sense the difference between hunger, appetite or emotion / Fears to experience what she really feels.	Is worried about putting on more weight / Suffers from stress and finds comfort in food / Has a sweet tooth / Hates to be patronized (authorities).
User needs	
Education on eating behaviour / Self-confidence / Help to set realistic goals / Support from peers.	Stop gaining more weight / Education on healthy food / Getting support from her family / Stop harassing thoughts.

Figure 1. Key Personas Lisanne and Anita

Scenario Lisanne and the e-DBT 'Denk je zèlf!' virtual coach

Lisanne comes home after a busy and frantic day at the real estate agent's office. It has been a turbulent day and it felt like a lot of her colleagues were hot-tempered, judging by their blunt manners. Due to lack of time Lisanne skipped lunch. She satisfied her appetite with unhealthy snacks and cookies.

At home she should start cooking her dinner but she is too worn out and totally not inspired. Her stomach is rumbling and she takes a quick glance at the store cupboard. She notices butter biscuits and potato chips. She feels tempted to rip open the bag of potato chips and plunge into it, grabbing chips by the handfull at the same time. Binge eating lies in ambush for attack and she senses a crying need for some kind of support, for someone who could provide her with advice to pull her through this situation.

She reaches for her phone and activates the 'Denk je zèlf!' app. The virtual coach welcomes her with: Hi Lisanne, how can I help you?

Lisanne starts typing:

-When I come home from work I start craving for snacks and chocolate… I just can't resist them…

-Hi Lisanne, the greater part of emotional eating occurs at night, due to feelings of loneliness or experiencing stress, but it can also happen because of irregular eating behaviour. Shall we give it a try to investigate this?

The virtual coach refers Lisanne to the behaviour chain analysis.

Lisanne finds comfort in the reassuring words of the virtual coach and starts with the behaviour chain exercise.

Figure 2. Scenario Lisanne.

The training consists of a series of education and instructions modules on emotional eating behaviour and emotion regulation. A personalised virtual coach will guide the user through four different modules. First an 'intake procedure' will take place: the user will be invited to make a commitment never to lose oneself in emotional eating behaviour again, followed by educational modules on mindfulness, emotion regulation and stress tolerance (Figure 4). The training offers exercises based on practical experience in daily life. Modules are replaceable – they can be substituted by modules with content that is focused on users with low socialeconomic status.

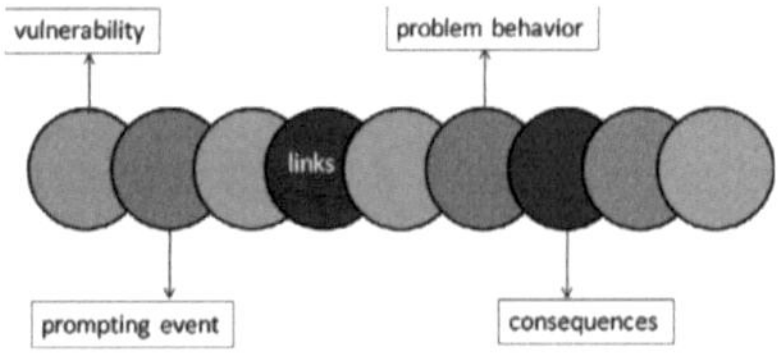

Figure 3. Behaviour chain analysis model by Linehan [17]

Users are invited to fill in their behaviour chain analysis (Figure 3) and emotion diary on a daily basis. Reminders to do so are sent out on fixed moments (by agreement with the user). Both components are considered indispensable in the face-to-face training being daily 'homework' for the participants. The behaviour chain analysis is to be performed at the moment a participant has given into cravings and bad eating behaviour or is just about to do so.

The behavioural chain analysis is utilised to analyse problem behaviour and determine prompting events and vulnerability factors. People can also fill in new skillful solutions and ways to prevent prompting events and think of solutions to reduce vulnerability in the future.

D. *'Denk je zèlf!' Personalised Virtual Coach*

The virtual coach was invented to meet the needs of the user for immediate support. Every time a user is experiencing negative emotions he can connect to the virtual coach and ask questions and start a dialogue. The virtual coach is the very heart of the e-DBT 'Denk je zèlf!' training. It supplies the user with so called dialectical dialogues – providing them with answers to their need for change and to their need for acceptance. According to Lynch et al. [19], dialectical theory is defined as: "The thesis (behaviour change) brought forth the antithesis (the need for acceptance), and both acceptance and change-based strategies were integrated into the treatment package *(synthesis)*. Dialectical theory provides the theoretical undercurrent needed to balance and synthesise these strategies. Core acceptance-based strategies derive from client-centered approaches and Zen practice and these involve mindfulness skills, validation, and radical acceptance." [19, pp.463]. The goal of the training is to teach people developing their own wise mind and to learn making decisions that have consequences for the quality of life. By providing dialectical dialogues the virtual coach helps the user practicing this process of decision-making by identifying the possible consequences of choosing either one of them. The output of the virtual coach is personalised by data derived from the behavioural chain analysis and the emotion diary. The virtual coach is a self-learning system. Ecological momentary assessment (EMA), often called as experience sampling measures (ESM), is applied within virtual mHealth coaching application to assess behavioural aspects [41][42], for instance by assessing subjective momentary states several times a day via a user-experience diary integrated in a virtual mHealth coach application.

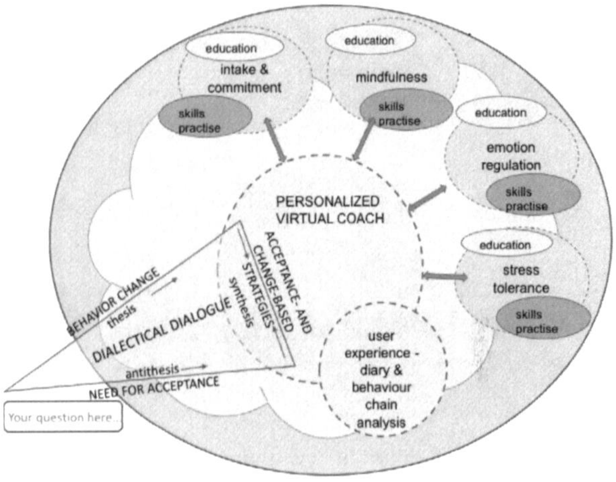

Figure 4.
Schematic drawing of 'Denk je zèlf!' mHealth intervention

In order to develop the virtual coach, including the two vital parts of the coach - the behaviour chain-analysis and the emotion diary - Personas and use-case scenarios were translated into user stories. Next, user stories are translated into a prototype of the virtual coach. Java-based virtual assistant (developed on the Play framework) makes use of an open-source natural language parser named *Alpino*. This Dutch linguistic language analyser [43] is self-learning and produces 'tree diagram' data in XML format. The output is stored in a graph database (NEO4J).

V. DISCUSSION AND CONCLUSION

In the process of developing a self-management intervention for a vulnerable group of obese emotional eaters, it is essential that the design and application meets the needs and expectations of the target users. By questioning the target group and mapping out their needs and wishes, user profiles were defined and two Personas Lisanne and Anita were formulated. Personas provided the needed guidance to define user stories that were translated into a first prototype of the personalized virtual coach *'Denk je zèlf!'*. This new interactive prototype will be iteratively evaluated together with users and healthcare practitioners.

The next step is to conduct a field test with members of the target group on the usability and the content of the virtual coach. The content – dialectical dialogues - is collected from handbooks for therapists and derived from online user groups. The users will be given small assignments such as navigating to the virtual coach and starting a conversation. We ask them to judge the replies of the coach by their faithfulness and truthfulness. The output will not only benefit the quality of the interaction with the virtual coach but the information will also be used to enrich the characters of the two Personas.

Lisanne and Anita served as a starting point for a concept design (Figure 5) and they will guide further development of the user interface and content of *'Denk je zèlf!'* modules intake, mindfulness, emotion regulation and stress tolerance.

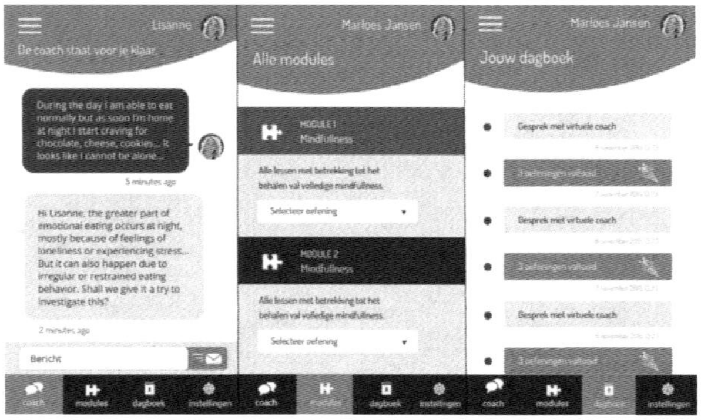

Figure 5. Concept design of the virtual coach

ACKNOWLEDGMENT

We thank the students of the minor Healthy Ageing and Applied Psychology at Hanzehogeschool for collecting data and supporting this study.

REFERENCES

[1] World HO. WHO | Obesity and Overweight. January 2015; Available at: http://www.who.int/mediacentre/factsheets/fs311/en/ [retrieved: March 4, 2016]

[2] Statistics Netherlands (Centraal Bureau voor de Statistiek). Webmagazine, Available at: http://www.cbs.nl/nl-NL/menu/themas/gezondheid-welzijn/publicaties/artikelen/archief/2014/2014-3939-wm.htm?RefererType=RSSItem [retrieved: March 4, 2016]

[3] National Institute for Public Health and the Environment (RIVM). Available at: https://www.volksgezondheidenzorg.info/onderwerp/overgewicht/cijfers-context/gezondheidsgevolgen#node-maatschappelijke-gevolgen 05-01-2016 [retrieved: March 4, 2016]

[4] L. Kristén, A. Ivarsson, J. Parker, and K. Ziegert, "Future challenges for intervention research in health and lifestyle research - A systematic meta-literature review. International," Journal of Qualitative Studies on Health and Well-Being, v. 10, 2015.

[5] T.I. Van den Berg, S.M. Alavinia, F.J. Bredt, D. Lindeboom, L.A. Elders, and A. Burdorf, "The influence of psychosocial factors at work and life style on health and work ability among professional workers," Int Arch Occup Environ Health, 81(8), 2008, pp. 1029-1036.

[6] T. Mann, A.J. Tomiyama, E. Westling, A. Lew, B. Samuels, and J. Chatman, "Medicare's search for effective obesity treatments: Diets are not the answer," Am Psychol, 04;62(3), 2007, pp. 220-233.

[7] Weight Watchers. Available at https://www.weightwatchers.com/us/ [retrieved April 13, 2016].

[8] My Diet Coach. Available at http://www.mydietcoachapp.com [retrieved April 13, 2016].

[9] Lose It! Available at https://www.loseit.com [retrieved April 13, 2016].

[10] T. Van Strien, A. Cebolla, E. Etchemendy, J. Gutiérrez-Maldonado, M. Ferrer-García, C. Botella, and R. Baños, "Emotional eating and food intake after sadness and joy," Appetite, 7/1;66, 2013, pp. 20-25.

[11] T. Van Strien, C.P. Herman, and M.W. Verheijden, "Eating style, overeating and weight gain. A prospective 2-year follow-up study in a representative Dutch sample," Appetite, 12;59(3), 2012, pp. 782-789.

[12] M.A. Roosen, D. Safer, S. Adler, A. Cebolla, and T. van Strien, "Group dialectical behavior therapy adapted for obese emotional eaters; a pilot study," Nutr Hosp, 07/20;27(4), 2012, pp. 1141-1147.

[13] Z. Cooper et al. "Testing a new cognitive behavioural treatment for obesity: A randomized controlled trial with three-year follow-up," Behav Res Ther, 8;48(8), 2010, pp. 706-713.

[14] P.G. Koenders and T. van Strien, "Emotional eating, rather than lifestyle behavior, drives weight gain in a prospective study in 1562 employees," Journal of Occupational and Environmental Medicine, 11;53(11), 2011, pp. 1287-1293.

[15] A.I. Concepción, "Effects of diets and their role in weight control," Psychol Health Med, 05;12(3), 2007, 321-327.

[16] A.J. Blair, V.J. Lewis, and D.A. Booth, "Does emotional eating interfere with success in attempts at weight control?" Appetite, 10;15(2), 1990, 151-157.

[17] J. Ruwaard, "The efficacy and effectiveness of online CBT," Department op Clinical Psychology, University of Amsterdam, 2012.

[18] M.M. Linehan, "Skills training manual for treating borderline personality disorder," New York, NY US: Guilford Press, 1993.

[19] T.R. Lynch, A.L. Chapman, M.Z. Rosenthal, J.R. Kuo, and M.M. Linehan, "Mechanisms of change in dialectical behavior therapy: Theoretical and empirical observations," J Clin Psychol, 04;62(4), 2006, pp. 459-480.

[20] D.L. Safer, C.F. Telch, and E.Y. Chen, "Dialectical Behaviour Therapy for Binge Eating and Bulimia," New York: The Guilford Press, 2009.

[21] D.L. Safer, A.H. Robinson, and B. Jo, "Outcome from a randomized controlled trial of group therapy for binge eating disorder: Comparing dialectical behavior therapy adapted for binge eating to an active comparison group therapy," Behavior Therapy, 03;41(1), 2010, pp. 106-120.

[22] C.F. Telch, W.S. Agras, and M.M. Linehan, "Dialectical behavior therapy for binge eating disorder," J Consult Clin Psychol, 12;69(6), 2001, pp. 1061-1065.

[23] E.Y. Chen, L. Matthews, C. Allen, J.R. Kuo, and M.M. Linehan, "Dialectical behavior therapy for clients with binge-eating disorder or bulimia nervosa and borderline personality disorder," Int J Eat Disord, 41(6), 2008, pp. 505-512.

[24] A.H. Robinson and D.L. Safer, "Moderators of dialectical behavior therapy for binge eating disorder: Results from a randomized controlled trial," Int J Eat Disord, 05;45(4), 2012, pp. 597-602.

[25] P. Daansen, "Eetbuien en affectregulatie: het gebruik van dialectisch-gedragstherapeutische principes bij de behandeling van eetstoornissen," Directieve Therapie, 26, 2006, pp. 79-89.

[26] M. Bardus, J. Smith, and C. Abraham, "Are eHealth interventions for obesity prevention effective? A systematic review of reviews," The European Journal of Public Health, v. 25 (suppl 3), 2015.

[27] B. Khokhar et al. "Effectiveness of mobile electronic devices in weight loss among overweight and obese populations: a systematic review and meta-analysis," BMC Obesity, 2014, 1:22.

[28] M. De Zwaan et al. "INTERBED: internet-based guided self-help for overweight and obese patients with full or subsyndromal binge eating disorder. A multicenter randomized controlled trial," Trials, 11/21;13, 2012, pp. 220-220.

[29] J. Harvey-Berino et al. "Internet delivered behavioral obesity treatment," Prev Med, 08;51(2), 2010, pp. 123-128.

[30] M.J. Hutchesson et al. "eHealth interventions for the prevention and treatment of overweight and obesity in adults: a systematic review with meta-analysis," Obesity Reviews, 16, 5, 2015, pp. 376-392.

[31] G.M. Manzoni, F. Pagnini, S. Corti, E. Molinari, and G. Castelnuovo, "Internet-based behavioral interventions for obesity: An updated systematic review," Clinical Practice and Epidemiology in Mental Health, 2011, 7, pp. 19-28.

[32] M. Svensson et al. "The Change in Eating Behaviors in a Web-Based Weight Loss Program: A Longitudinal Analysis of Study Completers," Journal of Medical Internet Research, 11;16(11), 2014, 1-1.

[33] L.A. Dimeff, S.L. Rizvi, I.S. Contreras, J.M. Skutch, and D. Carroll, "The mobile revolution and the DBT Coach," the Behavior Therapist, 09;34(6), 2011, pp. 104-110.

[34] S.L. Rizvi, L.A. Dimeff, J. Skutch, D. Carroll, and M.M. Linehan, "A Pilot Study of the DBT Coach: An Interactive Mobile Phone Application for Individuals with Borderline Personality Disorder and Substance Use Disorder," Behavior Therapy, 12/01;42(4), 2011, pp. 589-600.

[35] M. Washburn and D.E. Parrish, "DBT Self-Help Application for Mobile Devices," Journal of Technology in Human Services; 31(2), 2013, pp. 175-183.

[36] E.J. Van Gemert-Pijnen et al. "A holistic framework to improve the uptake and impact of ehealth technologies," J Med Internet Res, vol. 13, no. 4, 2011, p. e111.

[37] C. LeRouge, J. Ma, S. Sneha, and K. Tolle, "User profiles and personas in the design and development of consumer health technologies," Int J Med Inf, 82(11), 2013, pp. 251-268.

[38] L. Van Velsen, L. Van Gemert-Pijnen, N. Nijland, D. Beaujean, and J. Van Steenbergen, "Personas: The linking pin in holistic design for eHealth," The Fourth International Conference on eHealth, Telemedicine, and Social Medicine (eTELEMED 2012), IARIA, pp. 128 – 133, ISSN: 2308-4359, ISBN: 978-1-61208-179-3.

[39] J.M. Carroll, "Five reasons for scenario-based design," Interacting with Computers, 13(1), 2000, pp. 43-60.

[40] A. Pommeranz, W-P. Brinkman, P. Wiggers, J. Broekens, and C.M. Jonker, "Design Guidelines for Negotiation Support Systems: An Expert Perspective Using Scenarios," VTT symposium, 258, 2009, pp. 323-330.

[41] S. Shiffman, A.A. Stone, and M.R. Hufford, "Ecological momentary assessment," Annu Rev Clin Psychol, 2008, 4:1–3.6.

[42] I. Myin-Germeys et al. "Experience sampling research in psychopathology: opening the black box of daily life," Psychol. Med. 39, 2009, pp. 1533–1547.

[43] University of Groningen, "PIONIER Project," Available at http://www.let.rug.nl/vannoord/alp/Alpino/ [retrieved: March 4, 2016].

Identifying Important Components for Life Style Changes Using an Online Complex E-health Intervention in General Practice

a qualitative interview study

Carl J. Brandt
Research Unit of General Practice, Dept. of Public Health
University of Southern Denmark
Odense, Denmark
e-mail: cbrandt@health.sdu.dk

Jesper Bo Nielsen
Research Unit of General Practice, Dept. of Public Health
University of Southern Denmark
Odense, Denmark
e-mail: jbnielsen@health.sdu.dk

Jane Clemensen
CIMT- Centre for Innovative Medical Technology
University of Southern Denmark
Odense, Denmark
e-mail: Jane.Clemensen@rsyd.dk

Jens Søndergaard
Research Unit of General Practice, Dept. of Public Health
University of Southern Denmark
Odense, Denmark
e-mail: jsoendergaard@health.sdu.dk

***Abstract*-In a prospective pilot study, we recently found a 7.0 kg mean weight loss within a 20 months' intervention period using an online complex e-health approach. In order to tailor a randomized controlled trial testing an online platform with app technology, accelerometers etc., we conducted a qualitative interview study to identify important determinants for weight loss management using digital solutions. Results showed that the main themes for incitements identified were life events in close family, the establishment of support, a honest and trustworthy relationship with health professionals, and/or supportive peers and ways to monitor the behavioral change with nudging from a reference person. The primary barriers were self-inflicted obstacles, an experience of lack of self-efficacy and ways to keep up appearances when discussing personal health issues with peers. The primary roles of the referents were the experience of honest and trustworthy forums to discuss personal challenges. We can conclude that the important factors are the availability of behavioral change monitoring and empathic relevant feedback. The opinions of referents matter, and long-term success depends on the ability to establish a strong positive support online and/or offline. E-health solutions can support healthy living, but further investigations are needed to establish updated relevant solutions that can be used in daily practice.**

Keywords: obesity; internet community; e-health; self-efficacy; patient empowerment

I. INTRODUCTION

Obesity is an increasing problem, and more than 30% of the European population is obese, with a body mass index (BMI) > 30 kg/m2. It is well documented that obesity increases the risk of type 2 diabetes, cardiovascular diseases, and joint and musculoskeletal diseases, and that it decreases fertility and increases the risk of spontaneous abortion.

In a prospective pilot study, we investigated the effect of online dietician advice combined with access to an Internet community using the existing commercial online Internet weight loss management program not using app technology [1]. We have previously described the usability of this platform and found a mean weight loss from baseline of 7.0 kg. (95 % CI: 4.6 - 9.3 kg) after a mean maintenance period of 20 months among 21 patients with an initial BMI of 36.4 kg/m2 [2].

We now plan to perform a Randomized Controlled Trial (RCT) based on an updated online platform using app technology, accelerometers etc. [3]. Prior to this RCT study, we here report a qualitative interview study designed to identify issues of importance for weight loss management using digital solutions today. We interviewed patients who have previously successfully participated in an online complex e-health intervention by exploring: 1.What is the experience using supportive e-health solutions offered in relation to a healthier life style? 2. What are the incitements and barriers for personal life style changes in general and when using e-health solutions? 3. What is the role of peers and health professionals when changing life style in general and when using e-health tools seen from a patient perspective?

II. METHOD

A. *Design*

Qualitative, semi-structured, individual interviews.

B. *Setting*

General practice in the Southern Region of Denmark.

C. *Participants*

Ten overweight patients who had previously used an online complex Internet e-health intervention successfully. The intervention tool was used as a supplement to a dietician working in the general practice while making sure that the relationship was established in person before beginning the online communication. The same dietician met the patient in the clinic and handled the online communication with the patient. Apart from the dietician, the patients could meet and share health data/information with other patients in online forums. This developed to become a strong community, where it was possible to discuss many personal issues as well [2].

D. *Analysis*

Interviews were transcribed as soon as possible after each interview and uploaded to a common database. Anonymous transcripts were analyzed by the researchers (CJB, JC, JBN and JS) using thematic analysis. The identified themes were compared between the different researchers. Overlap and consistency were reached. The findings were then related to the Social Action Theory (SAT)[4], the Theory of Triatic Influence (TTI) [5] and the Social Cognitive Theory (SCT) [6].

III. RESULTS

All but one of the participants used smartphones; the last one used a tablet. They all used apps and the Internet, but had no longer access to the Internet support system [1]. Five years after the initial intervention, they all still used Internet or apps to benefit their health. Everyone looked up recipes, some had joined weight loss Facebook groups, and some used an app service monitoring their physical activity. They were not aware of all the functionality that could support them for free using their smartphones, but the majority acknowledged that the provided e-health solution had helped them. Following the two years intervention study five years ago, they have all gained weight, but not to the level before the intervention.

A. *The main themes for incitements identified were:*

1) The establishment of an honest and trustworthy relationship with a health professional with whom it was possible to talk about everything and to be heard:

a) Offline: "I think it was really nice to see CA (the dietician), also because she could see it was my head that controlled me a lot"

b) Online: "I don't know if I would always use it (report my diet intake) because sometimes I would just not report anything… to help me not being confronted with the fact that I had eaten more than I should"

c) Miscellaneas: Some of the patients revealed that they did not trust the dietician who had been assigned to them, and they stated the lack of trust as the main reason for them to discontinue the intervention. One of the participants then established a contact to another health coach that helped him along with significant weight loss without a single face to face meeting.

2) Find ways to monitor the behavioral change with nudging from a referent person:

a) Offline: "One who had to see that I kept my weight. If not, I would have said I had done it without having done it….. to satisfy others".

b) Online: "and I can't cheat it (An objective activity monitoring fitness app: "Endomondo"). It is not possible for me to pretend I have gone for a walk when I haven't".

3) Goal setting:

a) Online: "Today, when I have to go to work at noon, then I go for a walk during the morning. Sometimes I walk four times a week, but it should be at least 15 km per week."

4) Support from partner:

a) Offline: "I do have support from home – from my husband. Many times he has encouraged me to stop smoking".. "This thing with saying: "hey we have to remember vegetables""…"He may also say: "Do you really have to eat that?".. He is very observant to what I eat. A little judgmental if I eat something that is not healthy"

5) Life events in close family have a strong impact for a majority of the participants:

a) Offline: "My husband got type 2diabetes, and then I also lost 10-12 kg in the spring…."

B. *For barriers the main themes were:*

1) Self-inflicted obstacles:

a) Offline: "I had this knot under my left foot removed in surgery two years ago. Three months ago, I had the same operation in the other foot. So I'm not able to do it (exercise), and I still can't use regular foot wear. I still use sandals".

2) Experience of lack of self-efficacy:

a) Offline: "It is probably because I can eat a little candy again or buy a chocolate bar when I'm out shopping…. And I have problems saying no thanks".

3) Keeping up appearances when discussing personal health issues and about lying to referent others to avoid judgmental behavior:

a) Offline: "Yes, if someone says: "Well, wouldn't you like to lose weight?", then I say: "Yes, but it is very hard because bla, bla, bla, and then the same story goes. So it is an excuse for not losing weight"

b) Online: "If it is someone I would never meet, then I could be honest, but as soon I know that it is someone I am going to meet, then I add to the story".

C. *The main themes for the role of peers were:*

1) The experience of honest and trustworthy forums to discuss personal challenges, in relation to peers:

a) Offline: " Then I have my friend, who I met when I joined weight watchers. She has all these ideas as to how to combine different food ingredients….. and that has my other friend as well, but she has now joined this low fat high something……".

b) Online: "…I'm more in need for a closed group of people who have the same problems as me or at least are in the same position as me" and "I could do it (be honest), if it was online, but not if I had to look them in the eyes."

2) *The need for acknowledgement from referent others:*

a) *Offline:* "The worst thing was that nobody called me up" and said: "Why is it you have not scheduled a new appointment?"

b) *Online:* "It is the same with Endomondo. I have friends there. And then I have to show I walk. Not to have them believe I'm not active"

IV. Discussion

To our knowledge, this is the first qualitative interview study among the users of a successful weight loss Internet platform showing what elements are important for success in an updated e-health solution. Establishing an honest and trustworthy space is important. This matches the social ecological point of view that establishing a good relationship to the patient is of paramount importance for coaching patients to make difficult decisions [6]. To establish an honest space with your health professional can be difficult in a 100% online solution. The provided solution delivered a combination that seemed to strengthen the relationship even though not all patients seemed to have that need. Goal setting and monitoring the behavioral change are important for many patients. Goal-directed action is the basis of the social action theory [4] and, by using e-health solutions, this can be done in real time both objectively (Endomondo/accelerometers) and subjectively through reporting. The spouse plays an important role for all our participants. The participants all experienced support from their spouse during their weight loss, which is in accordance with the social cognitive theory underlying the importance of the opinion of others [6]. This was also the reason why life events in close family could move the patients to decide that weight loss or other life style changes were important enough to do something about. Barriers centered around self-inflicted obstacles, lack of self-efficacy and keeping up appearances, which very much related to the intrapersonal stream described in the Theory of Triatic Influence (TTI) [5]. To close the gap patients could describe how the supportive role they experienced both offline and online from referent others made a difference. This interpersonal aspect is the second stream in TTI and confirms that meeting others in the same situation is important. The third stream in TTI is the sociocultural environment that changes to some extent when a person meets up online with new people who also want to loose weight or stop smoking. Basically, you can change your referent others online, something that might be of paramount importance, especially for people with little support from home [7]. Offline, many of the same psychological elements are found in anonymous forums like Alcoholics Anonymous. The findings can be included in the complex e-health solutions used in the future RCT study [3]. The study will be conducted in a general practice setting using local dieticians out of respect for the importance of a good relation to the patients found in this article. Patients will monitor activity, diet and sleep through simple monitoring features in combination with an online community [3].

V. Conclusion

The major findings were that many of the important issues addressed by the participants can be supported online. Patients need to have a personal and trustworthy relationship to at least one health professional who can boost their self-efficacy when needed. Another important factor is the availability of behavioral change monitoring and empathic relevant feedback. The opinions of referents others matter, and long-term success depends on the ability to establish a strong positive support. E-health solutions can support healthy living, but further investigations are needed to establish updated, relevant, and complementing solutions that can be used in daily practice.

VI. Acknowledgement

Thanks to Public Health students Betina Bruun and Merete Moll for editorial contributions to this paper. Thanks to the staff at Stenstrup Lægehus for making this study possible. Thanks to Region Syddanmark.

References

[1] www.SlankeDoktor.dk [accessed March 2016]

[2] Brandt, V., Brandt, C. J., Glintborg, D., Arendal, C., Toubro, S., & Brandt, K. (2011). Sustained weight loss during 20 months using a personalized interactive internet based dietician advice program in a general practice setting. International Journal on Advances in Life Sciences, 3(1&2), 23-28.

[3] Brandt, C. J., Brandt, V., Pedersen, M., Glintborg, D., Toubro, S., Nielsen, J. B., Eysenbach, G., Brandt, K., Søndergaard, J. (2014). Long-term effect of interactive online dietician weight loss advice in general practice (LIVA) protocol for a randomized controlled trial. International journal of family medicine, 2014.

[4] Ewart, C. K. (2009). Changing our unhealthy ways: Emerging perspectives from social action theory. Emerging theories in health promotion practice and research, 2, 359-389.

[5] Flay, B. R., Snyder, F., & Petraitis, J. (2009). The theory of triadic influence. Emerging theories in health promotion practice and research, 2, 451-510.

[6] Bandura, A. (2004). Health promotion by social cognitive means. Health education & behavior, 31(2), 143-164.

[7] Winzelberg, A. J., Classen, C., Alpers, G. W., Roberts, H., Koopman, C., Adams, R. E., Heidemarie, E., Parvati, D., Taylor, C. B. (2003). Evaluation of an internet support group for women with primary breast cancer. Cancer, 97(5), 1164-1173

Feasibility of a Second Iteration Wrist and Hand Supported Training System for Self-administered Training at Home in Chronic Stroke

Sharon M Nijenhuis[1,2], Gerdienke B Prange-Lasonder[1,2], Farshid Amirabdollahian[3], Francesco Infarinato[4], Jaap H Buurke[1,2], Johan S Rietman[1,2]

[1]Roessingh Research and Development
Enschede, the Netherlands

[2]University of Twente
Enschede, the Netherlands

[3]Adaptive Systems Research Group
University of Hertfordshire
Hatfield, United Kingdom

[4]Rehabilitation Bioengineering Laboratory
IRCCS San Raffaele Pisana
Rome, Italy

Email: sharonnijenhuis86@gmail.com, g.prange@rrd.nl, f.amirabdollahian2@herts.ac.uk, francesco.infarinato@sanraffaele.it, j.buurke@rrd.nl, j.s.rietman@rrd.nl

***Abstract*—Telerehabilitation allows continued rehabilitation at home after discharge. The use of rehabilitation technology supporting wrist and hand movements within a motivational gaming environment could enable patients to train independently and ultimately serve as a way to increase the dosage of practice. This has been previously examined in the European Supervised Care & Rehabilitation Involving Personal Telerobotics (SCRIPT) project using a first prototype, showing potential feasibility, although several usability issues needed further attention. The current study examined feasibility and clinical changes of a second iteration training system, involving an updated wrist and hand supporting orthosis and larger variety of games with respect to the first iteration. The paper is relevant for the conference, reporting a new telemedicine service, combining physical orthotic support with remote offline supervision, for telerehabilitation at home after stroke. Nine chronic stroke patients with impaired arm and hand function were recruited to use the training system at home for six weeks. Evaluation of feasibility and arm and hand function were assessed before and after training. Median weekly training duration was 113 minutes. Participants accepted the six weeks of training (median Intrinsic Motivation Inventory = 4.4 points and median System Usability Scale = 73%). After training, significant improvements were found for the Fugl-Meyer assessment, Action Research Arm Test and self-perceived amount of arm and hand use in daily life. These findings indicate that technology-supported arm and hand training can be a promising tool for self-administered practice at home after stroke.**

Keywords-stroke; upper extremity; telerehabilitation; dynamic orthotic device; rehabilitation games; home training.

I. INTRODUCTION

Stroke is one of the most common causes of adult disability. Even in the chronic phase, motor problems still persist in the majority of stroke patients [1], leading to difficulties in performing activities of daily living independently. Motor problems that persist in the chronic phase may be partly due to learned nonuse of the affected upper limb [2]. Stimulating the use of the affected upper extremity has shown to overcome the learned nonuse [3].

Technology has the potential to support rehabilitation since it can provide high-intensive, repetitive, task-specific, interactive treatment of the impaired upper extremity. Besides, it has the potential to accurately quantify therapy and monitor patients' progress, while also providing immediate feedback to patients, as well as therapists. Rehabilitation robotics has been shown to be as effective as conventional rehabilitation for the hemiparetic arm [4]-[7]. Most research so far has shown significant improvements in upper limb motor function, although evidence of the transfer of robotic training effects to activities in daily life remains limited, as is observed for most interventions in stroke rehabilitation, including conventional therapy [8]. To maximize independent use of the upper extremity in daily life, it is important to include functional movements of both the proximal and distal arm and hand into post-stroke training, since a generalization effect to improvements on the entire upper extremity was found [9][10].

Most robotic devices are mainly suitable for the clinical setting with direct supervision of a therapist [5]. A next step would be to provide such systems at home, to enable self-administered practice of the arm and hand after stroke [11]. This is especially interesting since an increasing number of stroke survivors is expected, which will result in increased demands on healthcare systems [12]. New ways of providing healthcare services, such as teleconsultation and remote monitoring and treatment in the patient's home are therefore needed.

Telemedicine systems for upper extremity exercise showed promising results in improving health of stroke patients [13]. In addition, healthcare professionals and participants reported good levels of satisfaction and acceptance of telerehabilitation interventions [13][14]. This is in line with the precursor of the current study, using a passive device with three motivational rehabilitation games for arm and hand training at home [15][16], showing that the training was motivational which was underlined by an average weekly training duration of 105 ± 66 minutes. Usability showed potential, although several usability issues needed further attention. Clinical evaluations showed modest changes in arm and hand function [15].

In the current study, we expanded this research by using a next iteration of the developed training system. Lessons learned regarding usability findings, therapeutic benefit and practical issues which were obtained in the first iteration of patient measurements with the SCRIPT system [15][16] have been taken into account for the design of the second prototype. An updated passive wrist and hand orthosis with improved user interface including nine exercise games was provided. The new games further focused on functional exercises incorporating versatile grasping gestures. The second prototype was evaluated with a new group of chronic stroke patients. The objective of the current study was to examine feasibility (user acceptance and adherence) and clinical changes in arm and hand function of a second generation technology-supported arm and hand training system at home in chronic stroke. The paper is structured as follows: the methods of the study are presented in Section 2, the results are shown in Section 3, followed by the discussion in Section 4 and conclusion in Section 5.

II. METHODS

A. *Participants*

Participants were recruited from the Roessingh Rehabilitation center, Enschede, the Netherlands and IRCCS San Raffaele Pisana, Rome, Italy. Participants were eligible for inclusion if they (1) had a stroke between 6 months and 5 years ago; (2) were between 18 and 80 years of age; (3) had limited arm and hand function because of the stroke, but having at least active control of 15° elbow flexion and having active finger flexion of at least a quarter of the passive range of motion; (4) were living at home and having internet access; (5) were able to understand and follow instructions; (6) had no additional orthopedic, neurological, or rheumatologic disease of the upper extremity; and (7) no severe neglect or uncorrected visual impairments. All participants provided written informed consent before participation. The study protocol was approved by the local medical ethics committees (Medisch Spectrum Twente, Enschede, the Netherlands and the IRCCS San Raffaele Pisana ethics committee, Rome, Italy).

B. *Study design*

This feasibility study has a longitudinal design. The participants received six weeks of self-administered technology-supported training for the arm and hand at home. Evaluation of arm and hand function was based on a baseline measurement pre-training and an evaluation measurement within one week post-training, performed at the research lab of the rehabilitation center.

C. *Training intervention*

Participants used a technology-supported training system [16], which consisted of a slightly adapted version of the SCRIPT dynamic wrist and hand orthosis [17], a mobile arm support (SaeboMAS, Saebo Inc., Charlotte NC, USA) and a computer with webcam and touchscreen displaying exercise games (Figure 1). The mobile arm support was used to support the weight of the proximal arm. The wrist and hand orthosis is a passive exoskeleton worn on the forearm and hand, customized to the hand size of each participant. It provides extension forces to the wrist and fingers via passive leaf springs and elastic tension cords [17]. The orthosis was equipped with sensors to measure joint excursions of the wrist and hand, which allowed control of nine exercise games. A green marker placed on the hand plate of the orthosis was used to track the location of the orthosis by means of a camera placed on top of the screen to incorporate translational movements of the arm.

The exercise games consisted of various difficulty categories, to match the progress of individual participants. The categories were classified in a game difficulty schedule, ranked according to increasing complexity. Complexity was higher when a game required multiple movement planes (from 1D to 3D), involved a higher number of gestures to control the game, movements with progression from proximal to distal movements or gross to fine manipulation. The gestures needed to control the games were hand opening and closing, wrist flexion and extension, forearm pronation and supination, and reaching forwards, backwards and moving left or right. For hand opening and closing we could distinguish a general grasp, cylinder grasp, tripod grasp and lateral grasp, which have been shown to be reliable hand postures which could be recognized during performance of rehabilitation games [18]. Translational movements of the hand were integrated with wrist and hand movements (e.g., moving the hand to a target, grasping, transferring to a different target, releasing) to emphasize functional, task-specific movements.

The training environment was available within a motivational user interface including feedback on performance, which was displayed on the touchscreen. The general recommendation for training was about 30 minutes per day, six days a week. Participants could train at the time

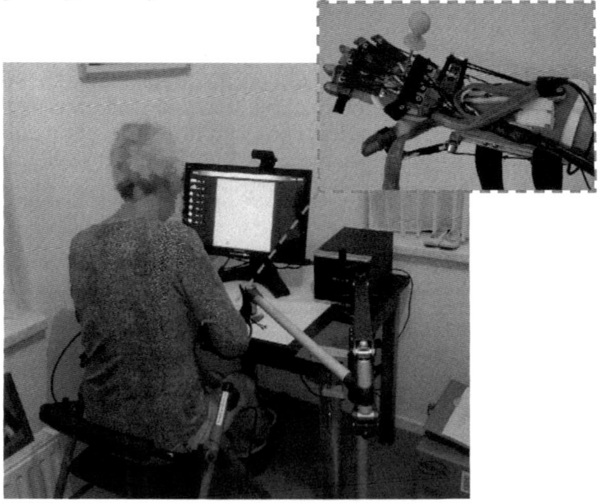

Figure 1. Training system at home.

of the day they preferred and were allowed to practice additionally if they wished to. A trained healthcare professional followed the participants' training progress remotely, offline, via another user interface available on a secured website and provided feedback by means of sending motivational messages when training duration was low. In addition, the healthcare professional visited each participant once per week to ensure competence with the training system, informally monitor progress, and to answer potential questions. Based on performance of the exercise games, in addition to ranking on the game difficulty schedule, the healthcare professional decided if a participant could move up to the next category of games. The professional adjusted the training program for the participants remotely.

D. *Evaluation*

1) *User experience*

The frequency and duration of training were automatically stored within the system and displayed in the user interface. The total minutes of training per week were counted to provide the total weekly training duration. These weekly training durations of all six weeks of training were used to calculate the average amount of practice over six weeks.

Motivation during training was measured using the Intrinsic Motivation Inventory (IMI) questionnaire [19]. It provides qualitative information about the content and level of motivation that a participant experienced during the training period (maximum score = 7). A higher score represents higher motivation during training, with a neutral score of four.

The System Usability Scale (SUS) is a 10-item scale providing a global view of subjective usability [20][21]. The questions were scored on a 5-point Likert scale ranging from 'strongly agree' to 'strongly disagree'. Scores are translated to 0–100%, with a higher score representing better usability. Interventions with scores in the 90s are exceptional, scores in the 70s and 80s are promising, and with SUS scores below 50 one can be almost certain that the intervention will have usability difficulties in the field. The SUS and IMI were completed during the post-evaluation measurement only.

2) *Arm and hand function tests*

Clinical tests were used to quantify general arm function before and after the training. The scales used are valid, standardized assessments, which were performed according to their specific test protocols.

The upper extremity part of the Brunnstrom Fugl-Meyer assessment (FM) evaluates motor status and the degree of synergy-development in the arm (maximum score = 66) [22]. Separate scores were calculated for proximal (maximum = 42) and distal components of the FM (maximum = 24).

The Action Research Arm Test (ARAT) evaluates coordination, dexterity and upper extremity function on the subtests grasp, grip, pinch, and gross arm movement (maximum score = 57) [23].

The Motor Activity Log (MAL) is a semi-structured interview for hemiparetic stroke patients to assess the perceived use of their paretic arm and hand (amount of use (AOU) and quality of movement (QOM)) during activities of daily living (maximum score = 5 per subsection) [24].

The Stroke Impact Scale (SIS) was used to assess changes in function, activity and participation following stroke. The questionnaire assesses eight domains related to function, activities and participation. Each domain score has a range of zero to hundred percent, with a higher score indicating better quality of life [25].

E. *Statistical analyses*

Statistical analyses were performed with IBM SPSS Statistics 19 for Windows. Outcomes were non-parametrically tested for statistical significance due to the small sample size. Descriptive statistical methods (median with interquartile range (IQR)) were used to describe the participant characteristics and all outcome measures. Outcomes of each clinical scale were compared between both evaluation measurements using the Wilcoxon signed rank test. The level for significance was set at $\alpha \leq .05$.

III. RESULTS

A. *Participants*

Nine participants (six in the Netherlands, three in Italy) were included in the study. Two participants were lost to the study. One because of personal problems not related to this study and one having recurrent technical problems with the system. The characteristics of the remaining seven participants are shown in Table 1.

B. *User experience*

The participants actually used the system, but with a large amount of variation between and within individuals (Figure 2). Median weekly training duration for the group, averaged over six weeks, was 113 (IQR 69 – 158) minutes. One participant (E01) exceeded the advised training duration of 180 minutes per week once, and two other participants (D08 and E03) exceeded the advised duration several times.

The median score on the SUS was 73% (IQR 60% – 83%). On individual level, four participants rated usability over 70% and three participants between 50 and 70%.

Overall, the participants enjoyed the six weeks of training, as reflected in the overall median score on the IMI of 4.4 points (IQR 3.9 – 6.0 points). Table 2 shows individual participant results and group medians on all outcomes.

TABLE I. PARTICIPANT CHARACTERISTICS

	N = 7[a]
Sex	5 male / 2 female
Age	57 (44 – 67) years
Months post stroke	21 (9 – 33)
Type of stroke	5 infarction / 2 hemorrhage
Affected body side	4 left / 3 right
Dominant arm	0 left / 7 right
Fugl-Meyer score (maximal 66 points)	37 (30 – 45)
Action Research Arm Test score (maximal 57 points)	26 (21 – 28)
Stroke severity	6 moderate / 1 severe

a. Results are shown as absolute numbers or median (interquartile range)

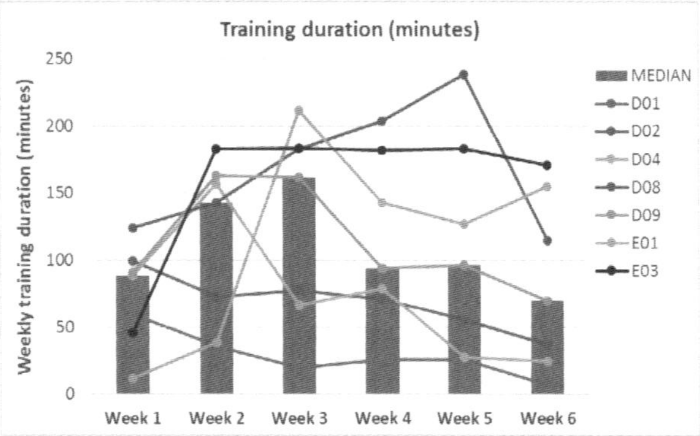

Figure 2. Individual (colored lines) and group median (grey bars) weekly training duration.

C. *Arm and hand function tests*

On group level, the Wilcoxon signed rank test showed a significant improvement after training for the FM total, FM proximal part, ARAT, and MAL AOU (Table 3). The FM showed a median improvement of 4.0 points ($P = 0.034$) for the total scale, and median improvement of 3.0 points ($P = 0.027$) for the proximal part only. The ARAT improved with median 2.0 points ($P = 0.045$) over training, and the MAL AOU with median 0.2 points ($P = 0.046$).

Examination of individual scores (Table 2) shows quite substantial improvements for one participant (E01), grossly exceeding the minimal clinically important difference (MID) for FM and ARAT [26], constituting changes greater than 10% of the total score. Participant E03 approaches MCID for FM with six points improvement. In addition, two participants (D09 and E03) exceeded MCID for the MAL QOM, of which E03 also exceeded MCID on MAL AOU [27], and participant E01 approaches MCID for both MAL AOU and QOM.

IV. DISCUSSION

A second generation technology-supported arm and hand training system was evaluated in patients with chronic stroke in their own home. The present findings show that training at home using the training system was feasible, since patients accepted the training well (median SUS = 73%) and were motivated (median IMI = 4.4). This was reflected in a median weekly training duration of 113 minutes (i.e., approximately 15 minutes per day). Participants showed improvements in arm and hand function, dexterity and self-perceived amount of arm and hand use in daily life.

TABLE III. ARM AND HAND FUNCTION TESTS (MEDIAN (IQR))

Outcome	Pre measurement	Post measurement	*P*-value
FM	37 (30 – 45)	41 (33 – 49)	**0.034**
FM Proximal	21 (21 – 31)	25.0 (23 – 32)	**0.027**
FM Distal	14 (9 – 17)	16 (10 – 18)	0.131
ARAT	26 (21 – 28)	28 (23 – 31)	**0.045**
MAL AOU	0.8 (0.4 – 1.5)	0.9 (0.8 – 1.4)	**0.046**
MAL QOM	0.8 (0.4 – 1.2)	0.8 (0.7 – 1.3)	0.249
SIS	61.4 (50.6 – 68.6)	66.2 (50.8 – 72.1)	0.128

Abbreviations: FM = Fugl-Meyer assessment, ARAT = Action Research Arm Test, MAL AOU = Motor Activity Log Amount of Use, MAL QOM = Motor Activity Log Quality of Movement, SIS = Stroke Impact Scale.

The motivation outcomes of the current study indicate that patients perceived the training as motivating, to a similar extent as interventions applying rehabilitation technology in a clinical setting [28][29]. With a median SUS score of 73% the training system was rated as promising. This might be related to a large variation of games available in the current study, which was much appreciated by the participants. Although participants positively valued the training system, several usability issues were identified and should be considered when implementing further design adaptations.

In particular, some games caused errors after leaving the pause screen resulting in incorrect saving of the data, controlling one game was not clear for some participants, and another game had limited fluent game control in poor day light. The game utilized the position of a marker on the orthosis to determine hand position in space, however poor day light impacted on capturing this position accurately. Although these issues were not major, if repeated or cumulative, they are likely to result in frustration and might influence the motivation and attitude towards use of the system, which can negatively affect the adherence to training over time [30].

In the current study, participants were able to make their own decisions about their training schedule, without direct, real-time supervision of a therapist. The rationale for this was to remove the training constraints and increase therapy availability. Compared to previous telerehabilitation studies in which training sessions are often scheduled beforehand and with direct supervision [13][31], the achieved training duration of median 113 minutes per week was substantial,

TABLE II. INDIVIDUAL PARTICIPANT RESULTS AND GROUP MEDIANS ON ALL OUTCOME MEASURES

Participant	Average weekly training duration (minutes)	IMI (1-7)	SUS (%)	FM change (max = 66)	ARAT change (max = 57)	MAL AOU change (0-5)	MAL QOM change (0-5)	SIS change (%)
D01	29	4.0	60	-1	2	-0.04	-0.26	1.8
D02	69	3.4	73	1	-1	0.00	0.00	-0.6
D04	74	3.9	68	3	2	0.28	-0.18	-3.7
D08	168	6.0	83	4	0	0.05	0.05	10.6
D09	113	6.7	95	4	3	0.18	0.67	3.2
E01	115	4.4	58	15	13	0.44	0.44	11.0
E03	158	4.7	73	6	3	0.68	0.56	9.3
Median (IQR)	**113 (69 – 158)**	**4.4 (3.9 – 6.0)**	**73 (60 – 83)**	**4 (1 – 6)**	**2 (0 –3)**	**0.2 (0.0 – 0.4)**	**0.1 (-0.2 – 0.6)**	**3.2 (-0.6 – 10.6)**

Abbreviations: IMI = Intrinsic Motivation Inventory, SUS = System Usability Scale, FM = Fugl-Meyer assessment, ARAT = Action Research Arm Test, MAL AOU = Motor Activity Log Amount of Use, MAL QOM = Motor Activity Log Quality of Movement, SIS = Stroke Impact Scale, IQR = Interquartile range.

suggesting that stroke patients do have the incentive to train at home and were able to use the system. This adherence falls within the range reported in other recent studies into technology-supported home-based self-administered upper limb therapy programs after stroke [30][32][33]. On the other hand, most participants did not reach the advised training duration, which is also comparable to these previous home-based studies [15][30][32][33].

One of the major assumptions concerning tele-rehabilitation using technologies is that when patients accept the technology and clinically benefit from it, they will actually use such a system when provided. However in practice, several factors, such as low motivation, fatigue and musculoskeletal issues can result in limited adherence [34], while training dose is an important factor in rehabilitation outcome [35]. When considering individual results, participants who had a rather high amount of training per week (>100 minutes), showed substantial improvements on arm and hand function. This suggests that actual adherence during self-administered practice is a highly relevant outcome. Moreover, stimulation of adherence should receive wide attention when designing and implementing home-based training interventions.

Several strategies can be considered for stimulation of adherence. Research has shown that regular patient-therapist contact during treatment has a motivational effect and can increase therapy adherence [36]. In our study, additional motivational strategies were implemented in a subset of our games [37], to increase participants' independent training time at home. This comprised more direct feedback about training duration, such as showing motivational messages about the duration after completion of each game, or continuously showing a timer during game play. We incorporated approaches from the field of psychology and education theory to further overcome this barrier, such as setting the correct balance between supporting and challenging, to maximize adherence to therapy [37]. However, our adaptive game-difficulty setting was not available in all games. In addition, patients' self-discipline might also play a role in this kind of unsupervised home training. So in future, it is also valuable to look closer at other characteristics such as patients' attitudes, personality traits, coping skills and commitments in daily life [30][32] to enable understanding of the most suitable patients for this kind of self-administered training.

The extent of improvement in motor function of the arm in the present study corresponded with those found in other robot-aided studies in chronic stroke in a clinical setting [4]-[7], and with therapy programs for the upper limb performed at home [13]. Perceived use of the affected arm in daily life as assessed by the MAL did significantly change on group level after six weeks training, which was also reflected in an improved dexterity capacity as measured by ARAT. Three of the games contained functional movements: integration of reaching, grasping and transportation simultaneously, and the inclusion of specific grasps, such as cylinder grasp, lateral grasp and palmar prehension grasp, to represent handling of different objects. This might have played a role in the improvements on activity level in the current study, since task-specificity is an important factor in restoration of arm and hand function after stroke [9][38]. However, these improvements are still modest on group level and not clinically relevant in terms of functional improvements. On the other hand, games with more complex gestures were only made available to patients after some progress was made in simpler games which could impact on the extent of these modest improvements. When these aspects are incorporated more prominently and in more games, exercises become even more functional and task-specific, which is expected to further enhance the clinical impact.

This study was performed with chronic stroke patients, limiting bias from spontaneous recovery and simultaneous other treatments. Home-based training could be considered at an earlier stage, where larger treatment effects would be expected. Further, only data of seven chronic stroke patients with mainly moderate stroke are available. Findings of the present study can only be partly generalized to other stroke survivors because of the small number of participants in this study. Future research should consider implementing a large randomized controlled trial, with sufficient statistical power to compare the effects with a control group.

V. CONCLUSION

The positive results for motivation, usability and actual training duration in this study indicate that home-based technology-supported arm and hand training is a feasible tool to enable self-administered practice at home. The improved arm and hand function and increased performance on activity level (both actual as well as perceived) indicates that self-administered home-based training can have a clinical value. Such an application has the potential to allow a higher dose of treatment than would be possible when depending on therapist availability in a conventional setting, if adherence can be stimulated further. Future research regarding telerehabilitation should therefore pay attention to adherence (stimulation) and the functional nature of exercises. Furthermore, identification of factors associated with better treatment outcomes (e.g., time post stroke, stroke severity and personal characteristics) is needed in order to understand who would benefit most from this technology-supported training at home.

ACKNOWLEDGMENT

We are grateful to the SCRIPT consortium for their contributions and dedication to the project: University of Hertfordshire Higher Education Corporation (United Kingdom), R.U.Robots Limited (United Kingdom), The University of Sheffield (United Kingdom), University of Twente (the Netherlands), Roessingh Research and Development BV (the Netherlands), MOOG BV (the Netherlands), IRCCS San Raffaele Pisana (Italy), User Interface Design GmbH (Germany).

REFERENCES

[1] Hendricks, H.T., van Limbeek, J., Geurts, A.C., and Zwarts, M.J.: 'Motor recovery after stroke: a systematic review of the literature', Arch Phys Med Rehabil, 2002, 83, (11), pp. 1629-1637

[2] Taub, E., Uswatte, G., Mark, V., and Morris, D.: 'The learned nonuse phenomenon: implications for rehabilitation', Eura Medicophys, 2006, 42, pp. 241-255

[3] Taub, E., and Morris, D.M.: 'Constraint-induced movement therapy to enhance recovery after stroke', Current atherosclerosis reports, 2001, 3, (4), pp. 279-286

[4] Mehrholz, J., Pohl, M., Platz, T., Kugler, J., and Elsner, B.: 'Electromechanical and robot-assisted arm training for improving activities of daily living, arm function, and arm muscle strength after stroke', Cochrane Database Syst Rev, 2015, 11, pp. CD006876

[5] Maciejasz, P., Eschweiler, J., Gerlach-Hahn, K., Jansen-Troy, A., and Leonhardt, S.: 'A survey on robotic devices for upper limb rehabilitation', Journal of Neuroengineering and Rehabilitation, 2014, 11, pp. 3

[6] Prange, G.B., Jannink, M.J., Groothuis-Oudshoorn, C.G., Hermens, H.J., and Ijzerman, M.J.: 'Systematic review of the effect of robot-aided therapy on recovery of the hemiparetic arm after stroke', J Rehabil Res Dev, 2006, 43, (2), pp. 171-184

[7] Basteris, A., et al.: 'Training modalities in robot-mediated upper limb rehabilitation in stroke: a framework for classification based on a systematic review', J Neuroeng Rehabil, 2014, 11, (1), pp. 111

[8] Wagenaar, R.C., and Meijer, O.G.: 'Effects of stroke rehabilitation: a critical review of the literature ', Journal of Rehabilitation Sciences, 1991, 4, (3), pp. 61-73

[9] Timmermans, A.A., Seelen, H.A., Willmann, R.D., and Kingma, H.: 'Technology-assisted training of arm-hand skills in stroke: concepts on reacquisition of motor control and therapist guidelines for rehabilitation technology design', J Neuroeng Rehabil, 2009, 6, pp. 1

[10] Balasubramanian, S., Klein, J., and Burdet, E.: 'Robot-assisted rehabilitation of hand function', Curr Opin Neurol, 2010, 23, (6), pp. 661-670

[11] Brennan, D.M., Mawson, S., and Brownsell, S.: 'Telerehabilitation: enabling the remote delivery of healthcare, rehabilitation, and self management', Stud Health Technol Inform, 2009, 145, pp. 231-248

[12] Howard, G., and Goff, D.C.: 'Population shifts and the future of stroke: forecasts of the future burden of stroke', Ann N Y Acad Sci, 2012, 1268, pp. 14-20

[13] Johansson, T., and Wild, C.: 'Telerehabilitation in stroke care--a systematic review', J Telemed Telecare, 2011, 17, (1), pp. 1-6

[14] Piron, L., et al.: 'Satisfaction with care in post-stroke patients undergoing a telerehabilitation programme at home', Journal of telemedicine and telecare, 2008, 14, (5), pp. 257-260

[15] Nijenhuis, S.M., et al.: 'Feasibility study into self-administered training at home using an arm and hand device with motivational gaming environment in chronic stroke', Journal of Neuroengineering and Rehabilitation, 2015, 12, (1), pp. 89

[16] Amirabdollahian, F., et al.: 'Design, development and deployment of a hand/wrist exoskeleton for home-based rehabilitation after stroke - SCRIPT project', Robotica, 2014, 32, (Special Issue 08), pp. 1331-1346

[17] Ates, S., et al., "Technical evaluation of and clinical experiences with the SCRIPT passive wrist and hand orthosis," in 7th International Conference on Human System Interactions (HSI), 2014, pp. 188-193.

[18] Leon, B., et al.: 'Grasps recognition and evaluation of stroke patients for supporting rehabilitation therapy', Biomed Res Int, 2014, 2014, pp. 318016

[19] McAuley, E., Duncan, T., and Tammen, V.V.: 'Psychometric properties of the Intrinsic Motivation Inventory in a competitive sport setting: a confirmatory factor analysis', Res Q Exerc Sport, 1989, 60, (1), pp. 48-58

[20] Brooke, J.: 'SUS: A Quick and Dirty Usability Scale', in Jordan, P.W., Thomas, B., Weerdmeester, B.A., and McClelland, I.L. (Eds.): 'Usability Evaluation in Industry' (Taylor & Francis., 1996)

[21] Bangor, A., Kortum, P., and Miller, J.: 'Determining what individual SUS scores mean: Adding an adjective rating scale', Journal of usability studies, 2009, 4, (3), pp. 114-123

[22] Fugl-Meyer, A.R., Jaasko, L., Leyman, I., Olsson, S., and Steglind, S.: 'The post-stroke hemiplegic patient. 1. A method for evaluation of physical performance', Scand J Rehabil Med, 1975, 7, (1), pp. 13-31

[23] Lyle, R.C.: 'A performance test for assessment of upper limb function in physical rehabilitation treatment and research', Int J Rehabil Res, 1981, 4, (4), pp. 483-492

[24] van der Lee, J.H., Beckerman, H., Knol, D.L., de Vet, H.C., and Bouter, L.M.: 'Clinimetric properties of the motor activity log for the assessment of arm use in hemiparetic patients', Stroke, 2004, 35, (6), pp. 1410-1414

[25] Duncan, P.W., et al.: 'The stroke impact scale version 2.0. Evaluation of reliability, validity, and sensitivity to change', Stroke, 1999, 30, (10), pp. 2131-2140

[26] van der Lee, J.H., Beckerman, H., Lankhorst, G.J., and Bouter, L.M.: 'The responsiveness of the Action Research Arm test and the Fugl-Meyer Assessment scale in chronic stroke patients', J Rehabil Med, 2001, 33, (3), pp. 110-113

[27] van der Lee, J.H., et al.: 'Forced use of the upper extremity in chronic stroke patients: results from a single-blind randomized clinical trial', Stroke, 1999, 30, (11), pp. 2369-2375

[28] Prange, G.B., et al.: 'The effect of arm support combined with rehabilitation games on upper-extremity function in subacute stroke: a randomized controlled trial', Neurorehabil Neural Repair, 2015, 29, (2), pp. 174-182

[29] Colombo, R., et al.: 'Design strategies to improve patient motivation during robot-aided rehabilitation', J Neuroeng Rehabil, 2007, 4, pp. 3

[30] Standen, P.J., et al.: 'Patients' Use of a Home-Based Virtual Reality System to Provide Rehabilitation of the Upper Limb Following Stroke', Phys Ther, 2014, 11, pp. 11

[31] Laver, K.E., et al.: 'Telerehabilitation services for stroke', Cochrane Database Syst Rev, 2013, 12, pp. CD010255

[32] Wolf, S.L., et al.: 'The HAAPI (Home Arm Assistance Progression Initiative) Trial: A Novel Robotics Delivery Approach in Stroke Rehabilitation', Neurorehabil Neural Repair, 2015, 29, (10), pp. 958-968

[33] Donoso Brown, E.V., et al.: 'Preliminary investigation of an electromyography-controlled video game as a home program for persons in the chronic phase of stroke recovery', Arch Phys Med Rehabil, 2014, 95, (8), pp. 1461-1469

[34] Jurkiewicz, M.T., Marzolini, S., and Oh, P.: 'Adherence to a home-based exercise program for individuals after stroke', Topics in Stroke Rehabilitation, 2011, 18, (3), pp. 277-284

[35] Kwakkel, G.: 'Intensity of practice after stroke: More is better', Schweizer Archiv für Neurologie und Psychiatrie, 2009, 7, pp. 295-298

[36] Jette, A.M., et al.: 'Home-based resistance training: predictors of participation and adherence', Gerontologist, 1998, 38, (4), pp. 412-421

[37] Basteris, A., Nijenhuis, S.M., Buurke, J.H., Prange, G.B., and Amirabdollahian, F.: 'Lag–lead based assessment and adaptation of exercise speed for stroke survivors', Robotics and Autonomous Systems, 2015, 73, pp. 144-154

[38] Kitago, T., and Krakauer, J.W.: 'Chapter 8 - Motor learning principles for neurorehabilitation', in Michael, P.B., and David, C.G. (Eds.): 'Handb Clin Neurol' (Elsevier, 2013), pp. 93-103

A Renewed Framework for the Evaluation of Telemedicine

Stephanie Jansen-Kosterink
Cluster Telemedicine
Roessingh Research and Development
Enschede, the Netherlands
e-mail: s.jansen@rrd.nl

Miriam Vollenbroek-Hutten & Hermie Hermens
Telemedine Group
University of Twente
Enschede, the Netherlands
e-mail: m.vollenbroek@rrd.nl; h.hermens@rrd.nl

Abstract— **The aim of this paper is to present a renewed framework for the evaluation of telemedicine that provides better insight into the real potential of telemedicine and as such fosters implementation in daily clinical practice. This study first evaluates the current literature on the use of the framework proposed by Dechant et al., 1996. Physical rehabilitation is used as casus. After screening, 40 relevant papers were included. Results show that the technology used and the clinical purposes are diverse and that the majority of the technology used was not implemented in daily clinical practice. The staged approach to the evaluation of telemedicine proposed by Dechant et al., 1996 was rarely applied. From the papers included it becomes clear that the following aspects are important to consider in the evaluation of telemedicine: (1) the type of telemedicine in terms of technology used, its level of maturity and its clinical purpose and (2) the way the telemedicine is implemented in daily clinical practice (service configuration).**

Keywords-Telemedicine; evaluation; framework.

I. Introduction

It is widely acknowledged that telemedicine has great potential in healthcare to overcome the problems related to our ageing community, to increase the quality and accessibility of care, and to restrain the rise of imperative healthcare costs. The current state is that the amount of evidence regarding the effectiveness of telemedicine is growing [1][2]. However, even proven effective telemedicine services often fade away and are not implemented into healthcare [3][4][5]. It deserves a further analysis to what factors impede the uptake of these services and what is needed to speed up its implementation [6][7]. One of the questions directly related to this, is whether the evaluation studies currently being performed provide sufficient evidence to convince healthcare professionals, policy makers and insurance companies.

An evaluation framework is the first step to secure a proper evaluation. Currently, only a few evaluation frameworks are available. The most common evaluation framework is the stage model of drug evaluation [8]. This model has been developed by the Food and Drug Administration and provides guidelines for demonstrating the safety and efficacy of new drugs as a prerequisite for marketing. In 1996, an analogous model for evaluation of new technologies was proposed by Dechant et al.[9]. In this framework, the type of assessment is tailored to the development life cycle of the technology. This so-called staged approach differentiates between telemedicine evaluation at application (stage 1-2) and global level (stage 3-4). Evaluation of a telemedicine service starts with an evaluation of the technical efficacy (accuracy and reliability) of the application and evaluation of the primary objective of the service in terms of access, quality or cost (stage 1-2). During the subsequent deployment a comprehensive evaluation is necessary, using multiple endpoints such as accessibility, quality and cost of care (stage 3). The last step of evaluating a telemedicine service is to examine whether the overall evaluation of a technology in one system, applies in other settings (stage 4). An advantage of this evaluation framework is that it takes into account the iterative process of the development of the technology. However, considering the fast development of new technology the obsolete of this evaluation framework could be a disadvantage.

Proper evaluation is essential to convince the various stakeholders of the added value of telemedicine and to come to sustainable implementation in daily clinical practice. Therefore, the aim of this paper is to create and present a renewed framework for evaluation of telemedicine starting from the framework proposed by Dechant et al. [9] that provides better insight in the real potential of telemedicine and as such fosters implementation in daily clinical practice. In section II the methods of this paper are described. Section III addresses the results in four topics; telemedicine service, added value, use of an evaluation framework and refinement of the evaluation framework. Section IV describes the discussion. The acknowledgement and references close the paper.

II. Methods

To present the current state of the evaluation of telemedicine for physical rehabilitation, a computerized literature search of the Medline and Scopus databases were conducted in January 2014. The search strategy and keywords used for both databases are shown in Table 1. In addition to this search, the online versions of three journals in telemedicine (Journal of Telemedicine and Telecare, Journal of Telemedicine and e-Health and International Journal of Telemedicine and Applications) were manually searched for additional relevant references.

TABLE I. SEARCH STRATEGY

For Medline database	
Step 1	teletreatment OR telerehabilitation OR telehealth OR telecare OR ehealth OR telemedicine [MeSH Terms] OR therapy, computer assisted [MeSH Terms] OR ambulatory monitoring [MeSH Terms] OR computer [MeSH Terms] OR Technology [MeSH Terms] OR Internet [MeSH Terms] OR telecommunication [MeSH Terms]
Step 2	physical therapy* OR physiotherapy OR exercise [MeSH Terms] OR physical therapy modalities [MeSH Terms]
Step 3	home OR home based* OR outpatient OR home care services [MeSH Terms]
Step 4	1 AND 2 AND 3
For Scopus database	
Step 1	teletreatment OR telerehabilitation OR telehealth OR telecare OR ehealth OR telemedicine OR computer assisted therapy OR ambulatory monitoring OR telecommunication
Step 2	Physical therapy OR physiotherapy OR exercise
Step 3	home OR outpatient
Step 4	1 AND 2 AND 3

Papers were included when: (1) they were designed as an evaluation study; (2) they concerned patients and not healthy subjects; (3) the telemedicine intervention utilized remote treatment by means of ICT; (4) the treatment focused on physical rehabilitation or exercising and (5) they were written in English, German or Dutch. Papers were excluded when: (1) no results of the evaluation were provided; (2) they only gave a description of the telemedicine service or the proposed evaluation; (3) no healthcare professionals were involved in the service delivery; (4) they concerned patients with mental illnesses; (5) they were duplicates of other already included paper and (6) they were published before 01-01-2000.

Potential eligibility of the papers was first identified from the titles and abstracts identified during the searches. Two reviewers (CSvdV and SMJK) read all titles and/or abstracts independently. If an abstract did not give sufficient information about the study, the full-text paper was obtained for further review. Then the reviewers evaluated full-text papers independently and reached consensus about whether or not the papers should be included. Papers were not blinded for authors and journals.

To gain insight into the evaluations performed in the studies, a data extraction form was developed to systematically describe:

- the technology used in the telemedicine service;
- the clinical aim for which the telemedicine service is used;
- the way the telemedicine service was implemented in daily clinical practice i.e. service configuration;
- the outcome of the evaluation study on the domains accessibility, quality of care and cost of care as suggested by Dechant et al.[9],
- and whether or not the author refers explicitly to an evaluation framework as a starting point.

After assessing all full-text papers, the reviewers reached consensus and completed the data extraction form. The outcome of the data extraction form will be presented in the results section. Based on these results, the evaluation framework proposed by Dechant et al. [9] is refined, to involve all aspects and to increase the use of it as the standard framework for evaluation of telemedicine services.

III. RESULTS

Based on our literature search, we started with a set of 1511 citations. These were analyzed and 1413 citations were excluded following screening. We retrieved 98 potentially relevant papers in full text. We excluded 62 of these based on the pre-specified inclusion and exclusion criteria. Main reasons for exclusion were that technology used did not utilize remote treatment and the participants of the evaluation study were healthy subjects. The literature search provided us with 36 papers. The manual search of the online version of the journals in telemedicine by screening of titles, abstracts and full-texts left us with 4 relevant papers in full text. In total, we retrieved 40 relevant papers.

A. *Telemedicine service*

Technology used: Various technologies are described in the 40 papers included. In 24 (60%) papers, a videoconference system (synchronous communication technologies) was used to enable contact between the patient and healthcare professional. This was used to have remote face-to-face contact during exercising [10-25] or a scheduled face-to-face contact [26][27][[28][29][30][31][32][33]. In six (15%) papers, patient and professional had contact by an asynchronous communication technology, such as email on a weekly basis [29][34][35][36] or as short messaging technology after an exercise session [37][38].

In 26 (65%) papers, sensor-based technologies were used for a variety of reasons. In more detail: in eight papers to guarantee secure exercising [25][32][33][38][39][40][41][42]; in seven papers to monitor patient's progression or adherence [27][34][35][36][38][46][47]; in three papers to deliver automatic and professional feedback to the patients [43][44][45] and in nine papers to detect the motions of a patient [11][12][16][26][28][29][30][31][48]. Exercise-application are used in 18 (45%) papers to activate patients to perform exercises and to rehabilitate in their own environment [22][26][27][28][29][30][31][33][35][36][37][41][43][44][45][47][48][49] and in four (10%) papers, virtual reality or game technologies are used to stimulate the patient to execute the requested exercises [11][12][16][22]. In 72.5% of the included papers the telemedicine service used two or more of above mentioned technologies.

Clinical purpose: Clinical purpose is an important characteristic to describe a telemedicine application and was hardly addressed in the included papers. Based on the technology used three different clinical purposes can be identified:

- Consultation (27.5%): to enable a real-time one-to-one or group based contact between patient and healthcare professional during the rehabilitation session [10][13][14][15]17][18][19][20][21][23][24];
- Safety (20%): to enable a safe environment to rehabilitate independently. In these cases, during a remote rehabilitation session, ECG or saturation level was monitored [25][32][34][38][39][40][42][46];

- Remote supervision and exercising (52.5%): to remotely supervise the patient using sensor-based technology and to enable the patient to exercise by means of a technology supported exercise-application [11][12][16][22][26][27][28][29][30][31][33][35][36][37][41][43][44][45][47][48][49].

Service configurations: This characteristic of the telemedicine application was in most included papers not addressed. In 15 papers telemedicine was delivered to the patients as a follow-up treatment [10][12][14][15][19][20][23][24][26][32][34][37][38][40][41] after a period of conventional rehabilitation patients prolonged their rehabilitation at home by means of telemedicine. In the remaining 22 papers, the telemedicine technology was evaluated as being an autonomous treatment. In none of the included papers telemedicine was delivered as addition or (partially) replacement of the conventional treatment.

B. Added Value

Telemedicine has the potential to increase the accessibility of care, to increase the quality of care and to decrease the costs of care. This added value of telemedicine is widely accepted and determined by the characteristics of telemedicine: technology used, clinical purpose and service configuration. To evaluate the true potential of telemedicine it is important to relate the outcome of the evaluation to the hypothesized added value of the telemedicine services beforehand.

Accessibility: All telemedicine services have the potential added value to increase the accessibility of healthcare, because technology used allows remote contact among patient and healthcare professional. From a patient point of view increase accessibility means no geographical obstacles or absence of work [33][39]. Accessibility was not directly parameterized as outcome in the evaluation of the telemedicine intervention. However, 25 of the included papers assessed the patients' experience in terms of satisfaction and usability. Overall it can be stated that patients are satisfied with the telemedicine interventions and the interventions evaluated are "easy to use". Next to the accessibility for the patient, there is also accessibility from the healthcare professional's point of view what can be defined as the ability to treat more patients simultaneously or to treat patients from a larger geographical area. In none of the included papers these potential added value was addressed.

Quality of care: Telemedicine services that support remote supervision and actuate patients to exercise in their own environment have the potential added value to increase the quality of healthcare as these telemedicine services give patients the ability to excise more often, independently from the availability of a healthcare professional or treatment facilities. In 21 of the included papers the evaluated telemedicine services gave patients the ability to rehabilitate independently. Quality of care was assessed in nineteen of these papers. Eleven studies used a prognostic cohort and concluded that telemedicine services induced positive changes [28][29][31][33][34][38][40][42][45][48][49]. The other 8 studies were randomized controlled trials (RCT). Seven of these RCT-studies found telemedicine services at least as effective as conventional care [26][35][41][43][44][46][47]. Only 1 of these RCT-studies concluded that telemedicine was more effective as conventional care [30]

Costs of care: Telemedicine services delivered as (partial) replacement of the conventional treatment have the potential added value to reduce costs. From a healthcare professional point of view cost can be reduced when the technology used give the professional the ability to increase the efficiency of the treatment. Only four of the included papers investigated the costs relating to the evaluated telemedicine service. One service was implemented as follow-up treatment [14] and the other three as autonomous treatments [33][43][44]. Given the results of these 4 papers it can be stated that the efficiency of the treatment can be increased by a decrease in preparation and consultation time [43][44] or by lowering travel costs for professionals [14] and patients [33][44].

C. Use of an evaluation framework

The evaluation framework proposed by Dechant et al. [9] was only used in two of the included papers [43][44]. Both papers were stage 4 evaluation studies. Applying the four stages of the evaluation framework proposed by Dechant et al. [9] to categories, the included papers, show that most papers (55%) present the results of a stage 1-2 evaluation. The included papers focused mainly on clinical effectiveness (45%), feasibility (42,5%), user-experience (7,5%) and adherence (5%).

D. Refinement of the evaluation framework

The staged approach to the evaluation of telemedicine purpose by Dechant et al.[9] is rarely applied in the included papers. From the reviewed papers, it becomes clear that the following aspects are important to consider in evaluation:

- The type of telemedicine application in terms of which technology used and its level of maturity and clinical purpose for which it is being used.
- The context in which the telemedicine application is being used such as the service configuration

Once having these defined the main outcome criteria and the design of the evaluation can be defined. Taking this into account and looking at the framework proposed by Dechant et al. [9], the stages of evaluation are well defined but their content can be further refined in the following way:

Stage I: The first stage of telemedicine evaluation focuses on the feasibility and usability of the technology used in an experimental design with a small number of subjects or even case studies. This type of evaluation design allows researchers to gain detailed information which can be used for further improvement of the telemedicine service. The telemedicine service is evaluated as a standalone service and evaluation endpoints focus on feasibility and usability of the technology used.

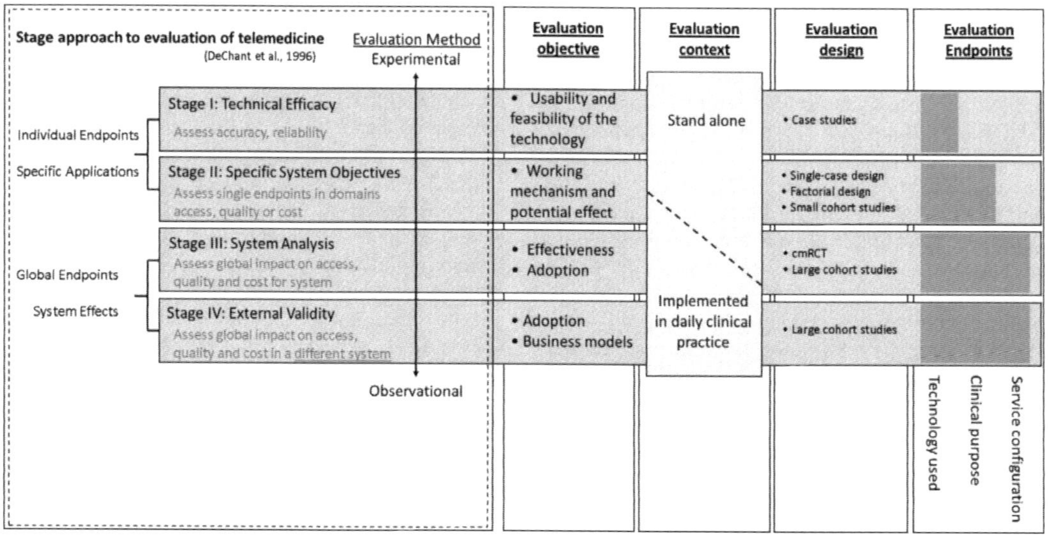

Figure 1. The refinement of the staged approach to evaluation of telemedicine.

Stage II: The technology used in the second stage is stable and evaluation is focused on gaining an initial idea about the potential added value for clinical practice and possible working mechanisms. For this, evaluation can be performed using the telemedicine service as a standalone service. Designs that can be used focus on studying processes in often small groups of subjects rather than on examining the effectiveness. Suitable designs are cohort studies with a small sample size (n<50) or single-case design (or N = 1 designs) [50]. The evaluation endpoints within this stage should focus on the potential added value of the telemedicine service mapped on both the technology used and the clinical purpose that is supported.

Stage III: This stage starts when earlier studies indicate that the telemedicine service has potential and focuses on showing the effectiveness of the telemedicine service and/or adoption of the service by its end-users. In order to identify these aspects, it is important that the telemedicine services are evaluated in the way they will be implemented in daily clinical practice. Although, randomized controlled trials (RCTs) are considered the gold standard for evaluating the safety and effectiveness of medical interventions their characteristics do not fit well with the evaluation of telemedicine services [50]. An alternative for a conventional RCT might be the "cohort multiple randomized controlled trial" (cmRCT) being introduced by Relton et al.[51]. The evaluation endpoints at this stage should not only focus on a previously defined value expected for each technology used and the clinical purpose that is supported but also take into account the way the telemedicine service is being implemented in daily clinical practice.

Stage IV: The fourth stage evaluation elaborates the adoption as addressed in stage III. To ensure further implementation, involvement of every stakeholder (healthcare professionals, patients, technology providers, insurance companies and policy makers on a local and national level) is important. This means that evaluation here should focus on the business models and concrete business cases. Without information on the cost and effectiveness of telemedicine services, decision makers run the risk of introducing services that are not cost-effective for society [53]. This evaluation can only be performed in an adequate way when the service is implemented in daily clinical practice as only in this case the true added value can be evaluated. The studies performed in this stage are large-scale cohort studies (n≥50) [54]. As addressed in stage III the evaluation endpoints in this stage should focus on the expected value of the telemedicine service depending on the application that is being used (technology used and clinical purpose) but also on the way it has been implemented in daily clinical practice (service configuration).

The refinement of the staged approach.to evaluation of telemedicine are presented in figure 1.

IV. DISCUSSION

The aim of this paper was to create and present a renewed framework for telemedicine evaluation that provides better insight in the real potential of telemedicine services and as such fosters implementation in daily practice. For this the use of the evaluation framework proposed by Dechant et al. [9] was analyzed using the current state of the evaluation of telemedicine service for physical rehabilitation as casus. Focusing on the characteristics of telemedicine for physical rehabilitation it can be concluded that the technology used and the clinical purpose were diverse and the majority of the telemedicine was not implemented in daily clinical practice. The level of maturity of the evaluated telemedicine was low and therefore most evaluations focused on feasibility, user-experience and adherence (stage 1-2 evaluation). In the following years, the level of maturity of telemedicine will increase and it is expected that more stage 3-4 evaluations will be published.

It is desirable that these evaluation studies relate the outcome of the evaluation to the hypothesized added value of the telemedicine beforehand to evaluate the true potential, focusing on accessibility of care, quality of care and costs of care [51, 55-59].

Based on the results, a refined version of the staged approach to the evaluation of telemedicine [9] for physical rehabilitations were presented and created, which of course need to be further validated in other cases to see whether this framework is useful and is generalizable for telemedicine evaluation in general.

ACKNOWLEDGMENT

We thank Karlijn Cranen and Charlotte van der Vos for their contribution.

REFERENCES

[1] A.G. Ekeland, A. Bowes, and S. Flottorp, "Methodologies for assessing telemedicine: a systematic review of reviews", Int J Med Inform, 2012. 81(1): p. 1-11.

[2] A.G. Ekeland, A. Bowes, and S. Flottorp, "Effectiveness of telemedicine: a systematic review of reviews", Int J Med Inform, 2010. 79(11): p. 736-71.

[3] J. Hendy, et al., "An organisational analysis of the implementation of telecare and telehealth: the whole systems demonstrator", BMC Health Services Research, 2012. 12(1): p. 1-10.

[4] T.H. Broens, et al., "Determinants of successful telemedicine implementations: a literature study", J Telemed Telecare, 2007. 13(6): p. 303-9.

[5] G. Pare, M. Jaana, and C. Sicotte, "Systematic review of home telemonitoring for chronic diseases: the evidence base", J Am Med Inform Assoc, 2007. 14(3): p. 269-77.

[6] A. Obstfelder, K.H. Engeseth, and R. Wynn, "Characteristics of successfully implemented telemedical applications", Implement Sci, 2007. 2: p. 25.

[7] R. Harrison, A. MacFarlane, and P. Wallace,"Implementation of telemedicine: the problem of evaluation", J Telemed Telecare, 2002. 8 Suppl 2: p. 39-40.

[8] Fundamentals of Clinical Trials. Fifth Edition. Anticancer research, 2016. 36(1): p. 480.

[9] H.K. DeChant, W.G. Tohme, S.K. Mun, W.S. Hayes, and K.A. Schulman, "Health systems evaluation of telemedicine: a staged approach", Telemed J, 1996. 2(4): p. 303-12.

[10] J.C. Lai, J. Woo, E. Hui, and W.M. Chan, "Telerehabilitation - a new model for community-based stroke rehabilitation", J Telemed Telecare, 2004. 10(4): p. 199-205.

[11] L. Piron, P. Tonin, E. Trivello, L. Battistin, and M. Dam, "Motor tele-rehabilitation in post-stroke patients", Med Inform Internet Med, 2004. 29(2): p. 119-25.

[12] L. Piron, et al.,"Satisfaction with care in post-stroke patients undergoing a telerehabilitation programme at home", J Telemed Telecare, 2008. 14(5): p. 257-60.

[13] Y.K. Wong, E. Hui, and J. Woo, "A community-based exercise programme for older persons with knee pain using telemedicine", J Telemed Telecare, 2005. 11(6): p. 310-5.

[14] M. Tousignant, P. Boissy, H. Corriveau, and H. Moffet, "In home telerehabilitation for older adults after discharge from an acute hospital or rehabilitation unit: A proof-of-concept study and costs estimation", Disabil Rehabil Assist Technol, 2006. 1(4): p. 209-16.

[15] M. Tousignant, et al., "A randomized controlled trial of home telerehabilitation for post-knee arthroplasty", J Telemed Telecare, 2011. 17(4): p. 195-8.

[16] M.K. Holden, T.A. Dyar, and L. Dayan-Cimadoro, "Telerehabilitation using a virtual environment improves upper extremity function in patients with stroke", IEEE Trans Neural Syst Rehabil Eng, 2007. 15(1): p. 36-42.

[17] S.J. Page, and P. Levine, "Modified constraint-induced therapy extension: using remote technologies to improve function", Arch Phys Med Rehabil, 2007. 88(7): p. 922-7.

[18] T.C. Pickett, et al., "Telehealth and constraint-induced movement therapy (CIMT)", Clinical Gerontologist, 2007. 31(1): p. 5-20.

[19] L. Eriksson, B. Lindstrom, and L. Ekenberg, "Patients' experiences of telerehabilitation at home after shoulder joint replacement", J Telemed Telecare, 2011. 17(1): p. 25-30.

[20] L. Eriksson, B. Lindstrom, G. Gard, and J. Lysholm, "Physiotherapy at a distance: a controlled study of rehabilitation at home after a shoulder joint operation", J Telemed Telecare, 2009. 15(5): p. 215-20.

[21] T.J. Kelechi, A. Green, B. Dumas, and S.S. Brotherton, "Online coaching for a lower limb physical activity program for individuals at home with a history of venous ulcers", Home Healthc Nurse, 2010. 28(10): p. 596-605.

[22] J. Kowalczewski, S.L. Chong, M. Galea, and A. Prochazka, "In-home tele-rehabilitation improves tetraplegic hand function", Neurorehabil Neural Repair, 2011. 25(5): p. 412-22.

[23] H. Corriveau, M. Tousignant, S. Gosselin, and P. Boissy, "The use of telerehabilitation to provide an exercise program to improve balance in a post-stroke population: preliminary results", in: Impact Analysis of Solutions for Chronic Disease Prevention and Management, M. Donnelly, et al., Editors. 2012, Springer Berlin Heidelberg. p. 58-65.

[24] H. Corriveau, M. Tousignant, S. Gosselin, and P. Boissy, "Patients satisfaction with an in-home telerehabilitation exercise program and physiotherapists' satisfaction toward technology for an acute stroke population: A pilot study", 2013. p. 753-757.

[25] A.E. Holland, et al., "Telerehabilitation for people with chronic obstructive pulmonary disease: feasibility of a simple, real time model of supervised exercise training", Journal of Telemedicine and Telecare, 2013. 19(4): p. 222-226.

[26] H. Hermens, et al.,"Clinical assessment of the HELLODOC tele-rehabilitation service", Ann Ist Super Sanita, 2008. 44(2): p. 154-63.

[27] T.W. Burkow, et al., "An easy to use and affordable home-based personal eHealth system for chronic disease management based on free open source software. Stud Health Technol Inform, 2008. 136: p. 83-8.

[28] S.H. Brown, C.A. Lewis, J.M. McCarthy, S.T. Doyle, and E.A. Hurvitz, "The effects of Internet-based home training on upper limb function in adults with cerebral palsy", Neurorehabil Neural Repair, 2010. 24(6): p. 575-83.

[29] P.E. Bilde, et al., "Individualized, home-based interactive training of cerebral palsy children delivered through the Internet", BMC Neurol, 2011. 11: p. 32.

[30] H. Deng, et al., "Complex versus simple ankle movement training in stroke using telerehabilitation: a randomized controlled trial", Phys Ther, 2012. 92(2): p. 197-209.

[31] J. Langan, K. DeLave, L. Phillips, P. Pangilinan, and S.H. Brown, "Home-based telerehabilitation shows improved upper limb function in adults with chronic stroke: A pilot study", Journal of Rehabilitation Medicine, 2013. 45(2): p. 217-220.

[32] S. Scalvini, et al., "Home-based versus in-hospital cardiac rehabilitation after cardiac surgery: A nonrandomized controlled study", Physical Therapy, 2013. 93(8): p. 1073-1083.

[33] P. Zanaboni, L.A. Lien, A. Hjalmarsen, and R. Wootton, "Long-term telerehabilitation of COPD patients in their homes: Interim results from a pilot study in Northern Norway", Journal of Telemedicine and Telecare, 2013. 19(7): p. 425-429.

[34] N. Smart, B. Haluska, L. Jeffriess, and T.H. Marwick, "Predictors of a sustained response to exercise training in patients with chronic heart failure: a telemonitoring study", Am Heart J, 2005. 150(6): p. 1240-7.

[35] M.H. van den Berg, et al., "Using internet technology to deliver a home-based physical activity intervention for patients with rheumatoid arthritis: A randomized controlled trial", Arthritis Rheum, 2006. 55(6): p. 935-45.

[36] M.H. van den Berg, et al., "Engagement and satisfaction with an internet-based physical activity intervention in patients with rheumatoid arthritis", Rheumatology (Oxford), 2007. 46(3): p. 545-52.

[37] C. Lau, R.S. Churchill, J. Kim, F.A. Matsen, 3rd, and Y. Kim, "Asynchronous web-based patient-centered home telemedicine system", IEEE Trans Biomed Eng, 2002. 49(12): p. 1452-62.

[38] K. Ueshima, H. Kamata, N. Kobayashi, J. Kamata, and K. Hiramori, "Medically directed home-based exercise using a stepping device with ECG telemetry monitoring in patients with previous myocardial infarction", J Cardiopulm Rehabil, 2002. 22(2): p. 105-8.

[39] P.A. Ades, et al., "A controlled trial of cardiac rehabilitation in the home setting using electrocardiographic and voice transtelephonic monitoring", Am Heart J, 2000. 139(3): p. 543-8.

[40] S. Scalvini, et al., "Home-based exercise rehabilitation with telemedicine following cardiac surgery", J Telemed Telecare, 2009. 15(6): p. 297-301.

[41] E. Piotrowicz, et al., "A new model of home-based telemonitored cardiac rehabilitation in patients with heart failure: effectiveness, quality of life, and adherence", Eur J Heart Fail, 2010. 12(2): p. 164-71.

[42] P. Bernocchi, et al., "Six-month programme on lifestyle changes in primary cardiovascular prevention: a telemedicine pilot study", Eur J Cardiovasc Prev Rehabil, 2011. 18(3): p. 481-7.

[43] L. Sandsjo, P. Larsman, R.M. Huis in 't Veld, and M.M. Vollenbroek-Hutten, "Clinical evaluation of a myofeedback-based teletreatment service applied in the workplace: a randomized controlled trial", J Telemed Telecare, 2010. 16(6): p. 329-35.

[44] S.M. Kosterink, R.M. Huis in 't Veld, B. Cagnie, M. Hasenbring, and M.M. Vollenbroek-Hutten, "The clinical effectiveness of a myofeedback-based teletreatment service in patients with non-specific neck and shoulder pain: a randomized controlled trial", J Telemed Telecare, 2010. 16(6): p. 316-21.

[45] R.M. Huis in 't Veld, et al., "Relation between patient satisfaction, compliance and the clinical benefit of a teletreatment application for chronic pain", J Telemed Telecare, 2010. 16(6): p. 322-8.

[46] T. Marios, S. Dalton, and N.A. Smart, "The effect of tele-monitoring on exercise training adherence, functional capacity, quality of life and glycemic control in patients with type II diabetes", Journal of Sports Science and Medicine, 2012. 11: p. 51-56.

[47] A. Zutz, A. Ignaszewski, J. Bates, and S.A. Lear, "Utilization of the internet to deliver cardiac rehabilitation at a distance: a pilot study", Telemed J E Health, 2007. 13(3): p. 323-30.

[48] W.K. Durfee, S.A. Weinstein, E. Bhatt, A. Nagpal, and J.R. Carey, "Design and Usability of a Home Telerehabilitation System to Train Hand Recovery Following Stroke", Journal of Medical Devices, 2009. 3(4): p. 041003-8.

[49] J. Finkelstein, O. Lapshin, H. Castro, E. Cha, and P.G. Provance, "Home-based physical telerehabilitation in patients with multiple sclerosis: a pilot study", J Rehabil Res Dev, 2008. 45(9): p. 1361-73.

[50] N. Hobbs, D. Dixon, M. Johnston, and K. Howie, "Can the theory of planned behaviour predict the physical activity behaviour of individuals?", Psychol Health, 2013. 28(3): p. 234-49.

[51] D. Kairy, P. Lehoux, C. Vincent, and M. Visintin, "A systematic review of clinical outcomes, clinical process, healthcare utilization and costs associated with telerehabilitation", Disabil Rehabil, 2009. 31(6): p. 427-47.

[52] C. Relton, D. Torgerson, A. O'Cathain, and J. Nicholl, "Rethinking pragmatic randomised controlled trials: introducing the "cohort multiple randomised controlled trial" design", BMJ, 2010. 340: p. c1066.

[53] D. Hailey, "The need for cost-effectiveness studies in telemedicine", J Telemed Telecare, 2005. 11(8): p. 379-83.

[54] P. Zanaboni and R. Wootton, "Adoption of telemedicine: from pilot stage to routine delivery", BMC Med Inform Decis Mak, 2012. 12: p. 1.

[55] M. Rogante, M. Grigioni, D. Cordella, and C. Giacomozzi, "Ten years of telerehabilitation: A literature overview of technologies and clinical applications", NeuroRehabilitation, 2010. 27(4): p. 287-304.

[56] R. Wootton, "Twenty years of telemedicine in chronic disease management - an evidence synthesis", J Telemed Telecare, 2012. 18(4): p. 211-20.

[57] C. Laplante and W. Peng, "A systematic review of e-health interventions for physical activity: an analysis of study design, intervention characteristics, and outcomes", Telemed J E Health, 2011. 17(7): p. 509-23.

[58] J. Munro, N. Angus, and S.J. Leslie, "Patient focused internet-based approaches to cardiovascular rehabilitation-a systematic review", J Telemed Telecare, 2013. 19(6): p. 347-53.

[59] M.H. van den Berg, J.W. Schoones, and T.P. Vliet Vlieland, "Internet-based physical activity interventions: a systematic review of the literature", J Med Internet Res, 2007. 9(3): p. e26.

The Analysis of Youngster with Fever by Using Instantaneous Pulse Rate Variability

Po-Hsun Huang
Institute of Biomedical Engineering
National Chiao Tung University
Hsinchu, Taiwan R.O.C.
e-mail: pohsun.iie03g@nctu.edu.tw

Chia-Chi Chang
Department of Electronics Engineering and Institute of Electronics, Institute of Biomedical Engineering,
Biomedical Electronics Translational Research Center
National Chiao Tung University
Hsinchu, Taiwan R.O.C.
e-mail: ccchang.cs@gmail.com

Tzu-Chien Hsiao
Department of Computer Science,
Institute of Biomedical Engineering,
Biomedical Electronics Translational Research Center
National Chiao Tung University
Hsinchu, Taiwan R.O.C.
e-mail: labview@cs.nctu.edu.tw

Abstract—**Body temperature is an important homeothermic autoregulation of humans. Body temperature is usually controlled by autonomic nervous system (ANS). When a person has suffered a virus or bacterial infection, it causes fever. The ANS rises the body temperature by tachycardia and cutaneous vasomotion. The common non-invasive indication for ANS is based on heart rate variability (HRV) and pulse rate variability (PRV). However, these two methods are limited by the timescale of time series. Instantaneous PRV (iPRV) was proposed as a surrogate of HRV in frequency domain and breaks the restriction of timescale. In this paper, thirty subjects were recruited for the study; the body temperature and the photoplethysmography (PPG) signal were acquired in supine position for ten minutes. Fifteen fever subjects, whose body temperature was higher than 37.9 °C, are in Group Fev., the others are in Group Ctr. The results show that the power of iPRV in normalized low frequency (nLF) in Group Fev.(0.63±0.13, $p<0.05$) was significantly higher than that in Group Ctr.(0.47±0.16). And the power of iPRV in normalized high frequency (nHF) in Group Fev.(0.38 ±0.13, $p<0.05$) was significantly lower than that in Group Ctr.(0.54±0.16). Besides, the nVHF might have a potential for indicating the peripheral circulation and providing more information about body regulation.**

Keywords-photoplethysmography; instantaneous pulse rate variability; body temperature; fever; autonomic nervous system; peripheral circulation.

I. INTRODUCTION

Body temperature, which usually refers to the rectal temperature in animals, is an important homeothermic autoregulation of humans. In many mammals, the preoptic area (POA), a region of hypothalamus, is considered as the thermoregulatory center. The normal body temperature is about 36.5°C~37.5 °C and it is affected by the environmental temperature. When the receptor receives an environment temperature change, POA controls some organs activities or reaction to regulate the body temperature, and then the body temperature also sends feedback to POA for maintaining the body temperature. Sometimes, an abnormal rise in body temperature occurs, which is higher than normal body temperature, and we call it fever. It is usually caused by infection of bacterium or virus. When a person gets fever, the hypothalamus changes the set point to the higher temperature and it makes the body feel cold because the body temperature is lower than the set point. Therefore, POA tries to rise the body temperature [1].

There are many mechanisms to regulate the body temperature like cutaneous vasomotion, tachycardia, etc. Most of them are controlled by the autonomic nervous system (ANS), which is dominated by POA. So, the body temperature regulation can be observed by ANS activities [2].

Heart rate variability (HRV) [3], a common assessment for cardiovascular circulation in clinic, has been widely used as an ANS activities observation. It is measured by a non-invasion instrument, electrocardiogram (ECG), and calculated by beat-to-beat interval (RRI, which is the interval between R peaks in ECG.) series to get time domain or frequency domain index. However, the study of HRV is limited by timescale because of RRI (Fig 1(a)).

Pulse rate variability (PRV), another non-invasive method, was proposed for replacing HRV to provide not only ANS activities but also the information of peripheral circulation (Fig. 1(b)). PRV is measured by photoplethysmography (PPG) and calculated by peak-to-peak interval (PPI) instead of ECG and RRI. Previous studies have determined that PRV acts as a surrogate of HRV during non-stationary conditions [4][5]. Nevertheless, timescale of PPI still restricted the study of PRV.

Because both HRV and PRV have timescale restrictions, a new measurement named instantaneous pulse rate variability (iPRV) was proposed [6]. iPRV adopts the PRV technique and the Hilbert-Huang transform (HHT) [7] for breaking the timescale limitation. However, there is a problem called mode-mixing problem that occurs during the process of empirical mode decomposition (EMD) in the HHT. For solving this problem, ensemble EMD, a noise-assisted technique, was proposed [8]. Previous studies have determined that iPRV is reliable by using PPG during non-stationary condition [9][10], and iPRV can not only observe ANS activities like HRV and PRV, but also provide higher frequency band information [6]. Nevertheless, the meaning of this frequency band still is undetermined.

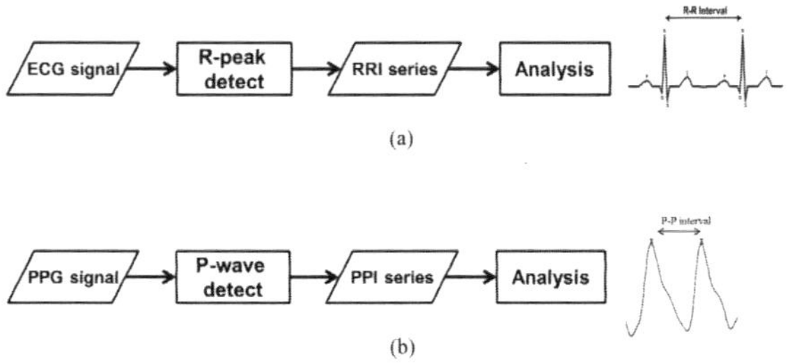

Figure 1. The steps of HRV and PRV: (a) HRV analysis and R-R interval. (b) PRV analysis and P-P interval

Body temperature is part of a mechanism of body regulation. We can assess the health status of a person by observing the variety of body reactions, such as ANS activities and peripheral circulation, when body temperature change. Peripheral blood vessels constrict when a person gets fever. It is possible to observe the regulation of peripheral blood vessels in VHF. This paper applied the iPRV method to compare the difference between fever patients and others and to find out the meaning of the index calculated by iPRV.

II. METHOD

A. *Subjects and data collection*

The experiment was carried out in Yo-Yo Clinic, Kaohsiung, Taiwan. The body temperature was measured by ear thermometer (Radiant TH889, Radiant Innovation Inc.) and the signal was acquired by PPG (Nonin 8500, Nonin Medical Inc.) with 200 Hz sampling rate.

Thirty subjects (19 males; age 10.5±2.9), whose ages were 7 to 18 years old, were recruited in this study. Fifteen subjects (10 males; age 10.1±2.4) with fever, whose body temperature was higher than 37.9 °C, served as Group Fev. and the others served as Group Ctr. We measured the body temperature of all recruited subjects before the experiment. Then participants wore the PPG sensor on the right index finger and rested quietly in supine position for 10 minutes. This experiment was approved by the institutional review board of the National Chiao Tung University. Informed consent was obtained from all subjects before the experiment.

B. *iPRV procedure*

iPRV applies the frequency extension method based on HHT [7] to PRV for the continuous-time heart rate rhythm. There is an important process during HHT called EMD. The steps of EMD are shown in Fig. 2. First, find the local maximum and local minimum, compute upper envelope and lower envelope by cubic spline and calculate the mean of these two envelopes. Then use the input data to subtract the average envelope. The above steps make up what is called the sifting process. After the sifting process, use the input data to subtract the average envelope and determine if the result is intrinsic mode functions (IMF) or not. It is IMF if the data satisfies by following two requirements: one is that the number of local extrema must be equal to the number of zero-crossings or differ at most by one, and two, is that, at any point, the value of the average envelope must approximately equal to zero. If it is not IMF, replace the input data by the result and redo the sifting process. Else, output the IMF and subtract the input data by this IMF to get the residue. Determine if the residue is a monotonic function or not. If it is not monotonic function, let residue be the input

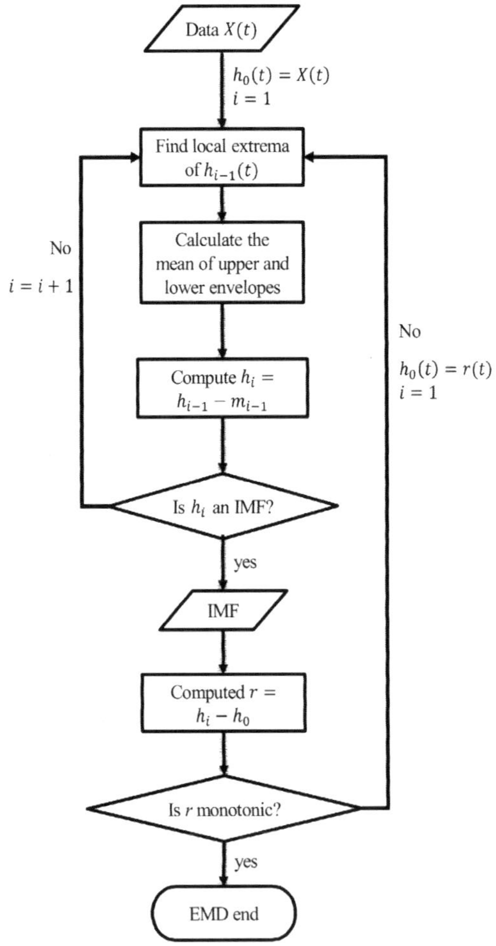

Figure 2. The EMD algorithm.

data and redo the above steps. Otherwise, the EMD finishes, and there are many IMFs decomposed by EMD.

However, there is a problem called the mode-mixing problem that may occur during the EMD process. For solving it, the ensemble EMD method (EEMD) was proposed [8]. EEMD is similar to EMD, but it adds white noise before the sifting process to solve the mode-mixing problem. In addition, EEMD does many times the EMD and averages the output IMFs and average the output IMFs to reduce the effect of white noise (Fig. 3).

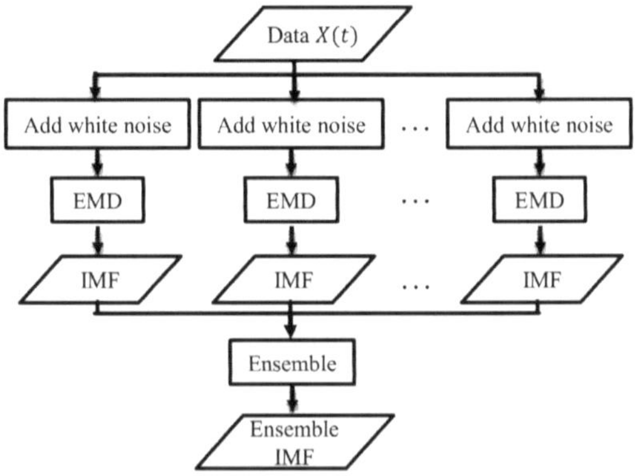

Figure 3. The EEMD algorithm.

After the EEMD process, one of the IMFs which is sinusoid-like is considered as the heartbeats component and is used for calculating the instantaneous frequency (IF) by normalized direct quadrature (NDQ) [11].

The steps of iPRV are shown in Fig 4. One of the IMFs extracted from the input PPG data by using EEMD is used as continuous-time heart rate rhythm. The IF is computed by NDQ and gets the instantaneous period (iPeriod) estimated by the inversion of IF of IMF. The Fast Fourier transform (FFT) is applied for power spectral analysis to estimate each frequency band including low frequency (LF, 0.04 to 0.15 Hz), high frequency (HF, 0.15 to 0.4 Hz) and very high frequency (VHF, 0.4 to 0.9 Hz) [3][6]. Besides, the iPRV spectrum can find the heartbeats peak in much higher frequency band (Fig. 5). The spectral analysis programs in this study were developed by using a commercial software platform (LabVIEW version 2013, National Instruments Corp., Austin, USA).

The normalized LF power (nLF) and normalized HF power (nHF) without VHF is calculated as in

$$nLF = LF / (TP - VHF) \quad (1)$$

$$nHF = HF / (TP - VHF) \quad (2)$$

where TP means total power and is computed as

$$TP = LF + HF + VHF \quad (3)$$

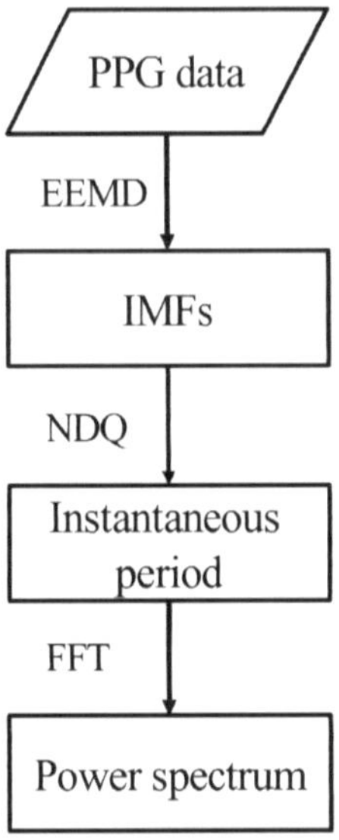

Figure 4. The iPRV procedure.

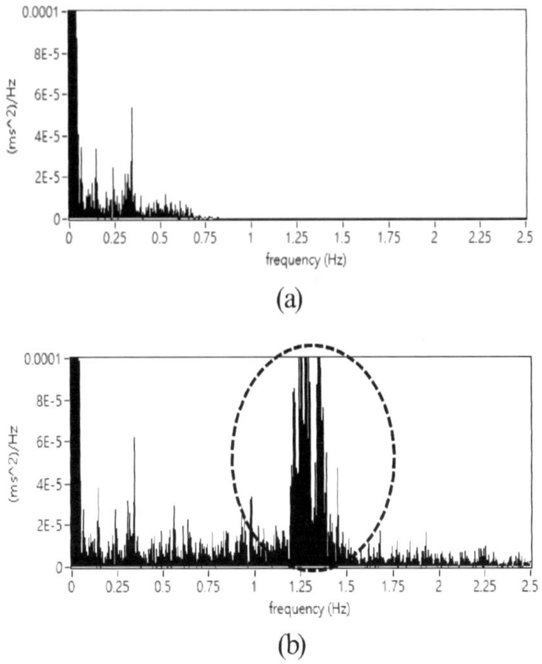

Figure 5. The comparision between PRV and iPRV spectrum: (a) PRV spectrum. (b) iPRV spectrum, which can observe the heartbeats peak on spectrum.

And, if we calculate the normalized power including the VHF, the representation is as follow

$$nLF \text{ (including VHF)} = LF / TP \quad (4)$$

$$\text{nHF (including VHF)} = \text{HF} / \text{TP} \quad (5)$$

$$\text{nVHF (including VHF)} = \text{VHF} / \text{TP} \quad (6)$$

LF to HF ratio is computed as in

$$\text{LF-HF ratio} = \text{LF} / \text{HF} \quad (7)$$

III. RESULTS

The results of the body temperature, heartbeats per second and power spectral analysis between the two groups are shown in Table 1. The processing of the iPRV analysis between Group Fev. and Group Ctr. is illustrated in Fig. 6.

TABLE I. THE RESULT BETWEEN TWO GROUPS

	Group Fev.(15)	*Group Ctr.(15)*
Body temperature (°C)	38.41±0.46	36.33±0.32*
Heartbeats (Hz)	1.92±0.18	1.5±0.22*
nLF	0.63±0.13	0.47±0.16*
nHF	0.38 ±0.13	0.54±0.16*
nLF (including VHF)	0.33 ±0.12	0.26±0.11
nHF (including VHF)	0.19 ±0.07	0.31±0.12*
nVHF (including VHF)	0.49 ±0.14	0.43 ±0.1
LF-HF ratio	2.08 ±1.36	1.1±0.87*

The form is (mean ± standard deviation), * means p-value<0.05 compared with Group Fev.

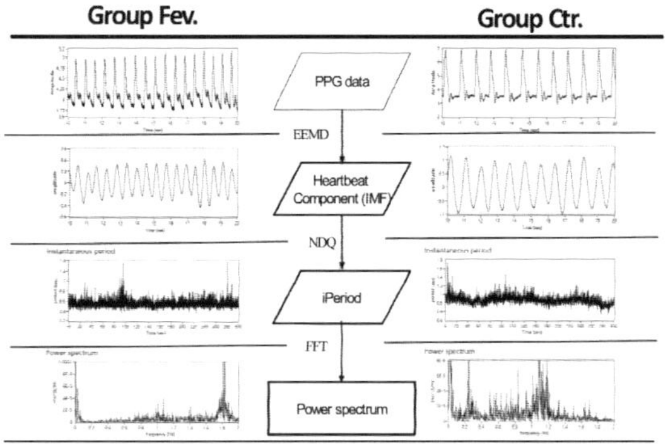

Figure 6. The iPRV analysis between two groups

The body temperatures in Group Fev.(38.41±0.46, p<0.05) are significantly higher than those in Group Ctr.(36.33±0.32). The heart rates in Group Fev.(1.92±0.18, p<0.05) are also significantly higher than those in Group Ctr. (1.5±0.22).

In frequency domain analysis, the nLF values in Group Fev.(0.63±0.13, p<0.05) are significantly higher than those in Group Ctr.(0.47±0.16) . The LF-HF ratios in Group Fev.(2.08 ±1.36, p<0.05) are also significantly higher than those in Group Ctr.(1.1±0.87). On the other hand, nHF values in Group Fev.(0.38 ±0.13, p<0.05) are significantly lower than those in Group Ctr.(0.54±0.16). If we calculate the normalized power including VHF, nHF values (including VHF) in Group Fev.(0.19±0.07, p<0.05) are significantly lower than those in Group Ctr.(0.31±0.12). In the nLF values (including VHF) and nVHF values (including VHF) can be observed a trend, and that trend is that the mean values of nLF (including VHF) in Group Fev.(0.33±0.12) are higher than those in Group Ctr.(0.26±0.11) and the mean of nVHF values (including VHF) in Group Fev.(0.49±0.14) are lower than those in Group Ctr.(0.43 ±0.1).

IV. DISCUSSION

This study compares the difference between Group Fev. and Group Ctr. by iPRV analysis. The result shows that the person whose body temperature is higher usually had higher number of heartbeats. The relationship between body temperature and heartbeats per second is drawn in Fig 7. However, it is not absolute because the heart rate is not only influenced by body temperature. It is also influenced by the environment temperature, others diseases, age and gender. Nonetheless, the body temperature is still an important factor to affect the heart rate.

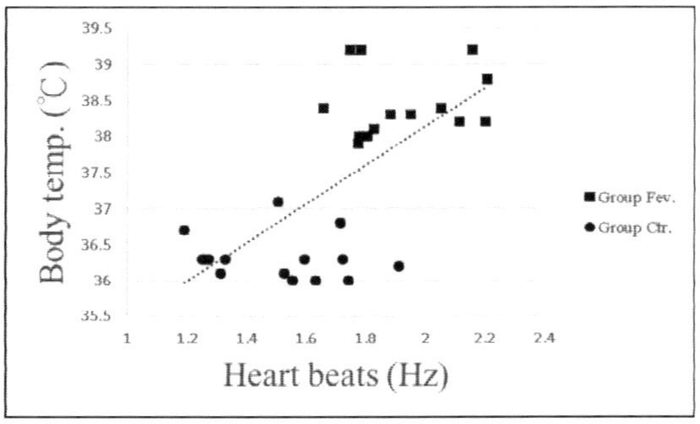

Figure 7. The relationship between body temperatur and heartbeats per second in all subjects.

Heart activities are usually dominated by ANS. It is considered that the heart rate increases because the sympathetic system is more excited than the parasympathetic system, and, on the contrary, the heart rate decrease because the parasympathetic system is more excited than the sympathetic system. The conventional HRV analysis in frequency domain is usually used for the ANS activity observation. The nLF is the percentage of LF in the sum of LF and HF is usually thought as the activities of sympathetic, and the nHF is the percentage of HF in the sum of LF and HF is usually considered as the activities of parasympathetic. The ratio of LF to HF is considered as the regulation of sympathetic [3]. The correlation of iPRV and HRV in the frequency domain analysis during non-stationary conditions has been examined in LF and HF [6][9]. HRV analysis in frequency domain rarely uses VHF because of the timescale problem, but iPRV breaks the restriction and provides more information in VHF.

The ANS rises the body temperature when someone has fever [1][2]. Therefore, the sympathetic is more excited than the parasympathetic. Regarding the results in Table 1., the value of nLF and LF-HF ratio in Group Fev. is significantly higher than Group Ctr. This means that the sympathetic is actually more excited in fever patients. This result is the same as the hypothesis.

If calculating the normalized power including VHF, only the nHF values (including VHF) in Group Fev. are significantly lower than those in Group Ctr. There is a trend, and that trend is that the mean value of the nLF (including VHF) and nVHF (including VHF) are larger in Group Fev., though this is not significant. Maybe the nLF values (including VHF) and the nHF values (including VHF) do not only represent the activities of sympathetic and parasympathetic, but have more regulation mixed in them. For VHF, maybe there is too much information to indicate the difference between the two groups. The meaning of these normalized powers need further research to examine.

For the new frequency band, VHF proposed by iPRV, previous studies have assumed that VHF has possible meaning of cardiac output or peripheral circulation [6][12][13]. In this study, the fever patients have the constriction of peripheral blood vessels, and it increases blood pressure. These situations cannot be observed in LF and HF by conventional HRV analysis. So, the VHF band might also contain information of regulation between the peripheral blood vessels and blood pressure, and the normalized power calculated including VHF might contain the regulation of ANS and peripheral circulation. The VHF bandwidth is very wide (0.4 to 0.9 Hz), it is possible to find more useful information by separating the band into more numbers of narrow bandwidth. It needs further research and investigation for verification.

V. Conclusion

This study compares the difference between the fever patients and people without fever by iPRV analysis and the results showed that iPRV can explore ANS function as conventional HRV analysis and, furthermore, provides more information owing to the new frequency band (VHF). The VHF has a possibility to indicate the regulation between the peripheral blood vessel and blood pressure or more regulation among ANS activities, heart activities and peripheral circulation. Nevertheless, it still needs further work to verify its feasibility. Even so, it still has the potential to be a useful indicator for healthcare by iPRV analysis.

Acknowledgment

This work was fully supported by the Taiwan Ministry of Science and Technology under grant numbers MOST-103-2218-E-009-016, MOST-103-2221-E-009-139, and MOST 104-2220-E-009-003, and in part by 'Aim for the Top University Plan' of the National Chiao Tung University and Ministry of Education, Taiwan, R.O.C. This work was also supported in part by the UST-UCSD International Center of Excellence in Advanced Bioengineering sponsored by the Taiwan National Science Council I-RiCE Program under grant number NSC-101-2911-I-009-101.

References

[1] K. Nakamura, "Central circuitries for body temperature regulation and fever," American Journal of Physiology-Regulatory, Integrative and Comparative Physiology, Vol. 301, Sep. 2011, pp. 1207–1228.

[2] L. K. McCorry, "Physiology of the autonomic nervous system," American journal of pharmaceutical education, Vol. 71:78, Aug. 2007, pp. 1–11.

[3] M. Malik, "Heart rate variability," Annals of Noninvasive Electrocardiology, Vol. 1, Apr. 1996, pp. 151–181.

[4] S. Lu et al, "Can photoplethysmography variability serve as an alternative approach to obtain heart rate variability information?" Journal of clinical monitoring and computing, Vol. 22, Jan. 2008, pp. 23–29.

[5] E. Gil, M. Orini, R. Bailón, J. M. Vergara, L. Mainardi, and P. Laguna, "Photoplethysmography pulse rate variability as a surrogate measurement of heart rate variability during non-stationary conditions," Physiological Measurement, vol. 31, Aug. 2010, pp. 1271-1290.

[6] C. C. Chang, T. C. Hsiao, and H. Y. Hsu, "Frequency range extension of spectral analysis of pulse rate variability based on Hilbert–Huang transform," Medical & Biological Engineering & Computing, vol. 52, 2014, pp. 343-351.

[7] N. E. Huang, et al., "The empirical mode decomposition and the Hilbert spectrum for nonlinear and non-stationary time series analysis," Proceedings of the Royal Society of London. Series A: Mathematical, Physical and Engineering Sciences, vol. 454, Mar. 1998, pp.903-995.

[8] Z. Wu and N. E. Huang, "Ensemble empirical mode decomposition: a noise-assisted data analysis method," Advances in Adaptive Data Analysis, vol. 1, Jan. 2009, pp. 1-41.

[9] P. C. Lin, K. C. Hsu, C. C. Chang, and T. C. Hsiao, "Reliability of instantaneous pulse rate variability by using photoplethysmography," Workshop on Biomedical Microelectronic Translational Systems Research (WBMTSR 2014) EPFL, Aug. 2014.

[10] P. C. Lin, P. H. Huang,, C. C. Chang, H. Y. Hsu, and T. C. Hsiao, "A novel index of photoplethysmography by using instantaneous pulse rate variability during non-stationary condition," Consumer Electronics - Taiwan (ICCE-TW), 2015 IEEE International Conference on, Jun. 2015, pp. 100 –101, doi: 10.1109/ICCE-TW.2015.7216799

[11] N. E. Huang, Z. Wu, S. R. Long, K. C. Arnold, X. Chen, and K. Blank, "On instantaneous frequency," Advances in Adaptive Data Analysis, vol. 1, Apr. 2009, pp. 177–229.

[12] C. C. Chang, H. Y. Hsu, and T. C. Hsiao, "The interpretation of very high frequency band of instantaneous pulse rate variability during paced respiration," Biomedical Engineering Online, vol. 13, Apr. 2014, pp. 46-56.

[13] P. C. Lin, C. C. Chang, T. C. Hsiao and H. Y. Hsu, "The Circulation Assessment of Daily E-health by Using Instantaneous Pulse Rate Variability during Nonstationary Conditions," The Seventh International Conference on eHealth, Telemedicine and Social Medicine, (eTELEMED 2015) IARIA, Feb, 2015, pp. 209–212, ISBN: 978-1-61208-384-1.

Planning for Sustainability:

Multiple Technologies for Different Care Models

Anthony P. Glascock
Department of Anthropology
Drexel University
Philadelphia, PA 19104, USA
email: glascock@drexel.edu

Rene A. Burke, Sherri T. Portnoy, Shaleea Shields
NHS Human Services
Lafayette Hill, PA 19444, USA
email: rburke1@nhsonline.org, SPortnoy@nhsonline.org, ssheilds@nhsonline.org

Abstract—**The integration of eHealth into the routine delivery of care is an important element in the maintenance of health care costs. This paper reports on the findings from a series of projects assessing the effectiveness of several eHealth technologies to both contain costs and aid in the provision of care to individuals with Intellectual and Developmental Disabilities or Severe and Persistent Mental Illness. Data were collected from 11 different locations in order to determine if the technologies resulted in an improvement in living conditions and to judge the acceptance of the technologies by both the staff and individuals. Findings indicate that the technologies have exceeded expectations resulting in plans to expand to other facilities and to build a new residence in which all of the technologies will be installed.**

Keywords—telecare; eHealth, multiple technologies; demonstration project; intellectual and developmental disabilities; severe and persistent mental illness

I. INTRODUCTION

The need to contain health care costs has seen the development of new technologies, as well as the innovative use of existing technologies, in an ever enlarging number of care models. Thus, it is not surprising that new and reconfigured eHealth technologies are increasingly being used to provide care and services to individuals with Intellectual and Developmental Disabilities (IDD) and those with Severe and Persistent Mental Illness (SMI) [1][2]. As in the use of new technologies, in each and every care model there are challenges to using innovative technologies in addressing the needs of these two populations, but the necessity to contain, and if possible, reduce the cost of providing care to these populations, makes the use of technology, in some form, inevitable. However, three different trends intersect in such a way that result in the urgency to develop more efficient care models for the IDD and SMI populations: the cost of care; the aging of the two populations and a reduction in the number of qualified staff providing care to the IDD and SMI populations.

First, deinstitutionalization, in both the United States and Europe took place from the mid-1960's through the 1970's, had a dramatic impact on the care models used for both populations [3][4]. Prior to the deinstitutionalization, the majority of care was provided for individuals with IDD and SMI in large institutions which were often dehumanizing. The impact of deinstitutionalization on the cost of care varied significantly, but the impact on the care model was significant as most individuals with IDD and SMI moved into the community, many living in some form of group home. The number of individuals living in any particular group home varies based upon the needs of the individuals. The most common number of residents is four with some group homes having as many as eight to ten residents. In the most severe cases, an individual lives alone in a residence with 24 hour supervision. Although costs of providing care to the residents in group homes vary based upon the needs and location, the average cost is between $40-50,000 per year per resident and if an individual needs to live alone, the cost can top $150,000 per year [5][6].

Second, the rapidly increasing number of such individuals brought about by the same demographic factors as for the general population is adding even greater cost to the care of IDD and SMI populations [7]. As individuals with IDD or SMI age, they are as susceptible to chronic illnesses as the general population, but the cost of caring for them is much greater. For example, care for an individual with IDD who has congestive heart failure costs approximately eight times more than for a person without IDD [5]. The ratios for other chronic diseases and individuals with IDD versus SMI vary somewhat, but the reasons are consistent. Many individuals with IDD and SMI make poor lifestyle decisions—use tobacco products and abuse alcohol and drugs. In addition, many are unable to self-manage disease, e.g., adhere to complicated medication regimes and follow complex health care instructions. Thus, greater cost of care for individuals with IDD and SMI when combined with the cost of residential care in general, results in a compelling reason for attempting to use technology to contain costs.

A third reason is that there is an increasing imbalance between the growing needs of the IDD and SMI populations and the number of qualified staff available to provide care. Projections from the federal government suggest that the need for trained staff will increase by over 30% in the next decade, while the supply of individuals who traditionally have filled these jobs is expected to increase only by 7% [8]. In addition, the high turnover rate for individuals caring for these populations adds another dimension to the staffing challenges. It is estimated that turnover for direct support professionals (DSPs) ranges between 50-70% depending on the specific jobs undertaken, e.g., residential care versus in-home care. This high turnover rate adds at least $2,500 in direct expenses plus a minimum of an additional $1,000 of indirect expenditures for an organization to replace a single DSP, thus adding to the ever increasing cost of providing care to these populations [9].

In the next section, a brief discussion of attempts at using technology to provide care for these populations is presented, while in the third section the overall Project is described including a short description of the organization undertaking the Projects, as well as the overall goals and objectives to be achieved by the introduction of the new technologies. In Section IV, the different Projects undertaken are described along with the care models employed and the technologies introduced. The next section discusses the methodologies employed in gathering data on the individuals with disabilities and mental illness and staff in order to assess the effectiveness of the technologies in care delivery, while Section VI offers a discussion of preliminary findings. What has been learned from the analysis is summarized in the Discussion Section, while plans for the roll out of the technologies to other facilities and the financial model to pay for this roll out comprise the concluding section.

II. RELATED WORK

It became apparent from the earliest discussions about the use of technology in the provision of care to individuals with disabilities, that there had been few other attempts to utilize technologies in a similar matter. This was the case, even though what little research was available that the increase in the use of technology benefits the quality of life for people with disabilities and that, "(m)ore often than not, people with intellectual and developmental disabilities (IDD) end up on the side of the divide with others who do not have access to or use technology" [10]. The vast majority of studies that exist focus on children within the context of school [11][12], rather than adults living in the community Those studies that do focus on similar populations are either, not helpful because of advances in technology since being undertaken, or are not residential based [13][14]. Thus, the effort to use a series of technologies within the NHS care system became more than just a localized demonstration; it became a test of the resolve to make these technologies widely available within the larger IDD community [15].

III. PROJECT

NHS Human Services, through its subsidiaries, is one of the United States' leading non-profit providers of community-based human and special education services. With nationally recognized programs in multiple states, NHS offers a full range of integrated services to children and adults in the areas of mental health, addictive diseases, autism, intellectual and developmental disabilities, juvenile justice, foster care treatment, education and other specialized programs. In particular, NHS is a leader in developing treatment plans for people with dual diagnoses and other multiple challenges and is in the forefront of the use of innovative technologies.

A. The Strategic Plan

In 2012, in order to accomplish the objectives outlined in the NHS Mission and Strategic Plan, the organization embarked on an assistive technology Project. This was partly in response to the growing IDD and SMI populations, partly based on the financial realities faced by the organization in providing care to these populations and partly in response to The Rights of People with Cognitive Disabilities to Technology and Information Access [16].

The planning process was inclusive and there was a recognition that in order to "do it right" it would take time to put everything in place. The first step was to form a committee led by an upper administrator, and employees were encouraged to review the current state of technology use in the delivery of care to the IDD and SMI populations and to propose sites at which new technologies could be used. Members of the committee attended several professional meetings in the United States and met with representatives of companies and organizations that used and produced various technologies in order to determine what technologies were available for use.

During the remainder of 2012, proposals were received and evaluated based upon specific criteria: administrative and staff buy-in; existence of suitable technology; evidence that technology would enhance care provision; evidence that, if successful, the technology could be used at a large number of other care facilities within the organization; and a financial model showing that the technology was sustainable—the organization would be reimbursed for its use. Finally, there was an attempt to achieve a rough balance among the different care models employed throughout the organization. This process took over a year which afforded a thorough evaluation of the resources available at each of the selected sites. The final decision was confirmed at an all-day meeting of administrators and representatives from each of the chosen sites in the fall of 2013.

B. The Project Goals

As planning progressed, three main goals emerged: 1) to determine which, if any, of the technologies being tested can allow for an improvement in living conditions and the care being delivered in the selected facilities; 2) to judge the acceptance of the technologies by both the staff and individuals with disabilities and mental illness; and 3) to assess whether the technologies should be rolled out to other facilities with similar care models. In order for any of the technology Projects to be deemed successful, it was necessary to determine if the new technologies allowed for an improvement in living conditions and the care being delivered in that the individuals with disabilities and mental illness express that their lives are better after the introduction of the technologies than before. It was also necessary to determine if care delivered is more timely, efficient and cost effective than the care delivered without the technologies.

It also became apparent that it was essential to ascertain if staff could properly use the new technology, that they believed in its effectiveness and accepted that the technology would require that they did their jobs differently. If staff did not accept the changes that would be brought about by the introduction of the technology, there was no reason to go forward with a roll out to additional facilities. Likewise, it was necessary to determine if the individuals with disabilities and mental illness accepted the use of the new technologies in the care that they received, if they were intimidated or not by the technologies and if they would willingly comply with requirements for the use of the technologies. Finally, even if it was determined that the new technologies provided improved care, were accepted by staff and individuals with disabilities and mental illness, it was still vital to find out if the care provided with the new technology was reimbursable as a billable expense. Consequently, the locations, the care models and the technologies had to be carefully selected in order to ensure that all three goals were achievable.

IV. THE TECHNOLOGIES AND LOCATIONS

During the planning process, the Assistive Technology Executive Steering Committee received proposals from local management teams which had assessed individual needs, staff interest and available dollars. The Executive Committee narrowed the proposals to three that were to be part of the first phase of the Project: Communication Technologies Project (CTB); the Smarthome Project (SHP); and the Biometrics Project (BMP). Although the initial goal was to have all three of the Projects begin simultaneously, issues, e.g., renovation delays, equipment problems, made this impossible. As a result, the CTP has advanced at a faster rate than the other two Projects. Also, as each of the Projects has progressed, sites have been added and subtracted, once again resulting in different timelines for the three Projects.

A. The Communication Technologies Project (CTP)

The CTP began in the spring of 2014 with the selection of sites and upgrading of wireless routers. Work continued during that summer with the focus of training staff in the use of AbleLink software [17] that had been selected for use and during the fall of 2014, staff and individuals with disabilities were surveyed and the Glasgow Depression Scale (GDS) administered to all individuals with disabilities participating in the Project [18] [19]. The main goals of the CTP was fourfold: to enable individuals with disabilities to stay in touch with family and friends; to allow a greater ability for them to communicate with members of the support services team; to encourage them to acquire basic computer skills; and to permit safe and secure access to the internet in order for them to pursue their particular interests. Seven sites were selected for inclusion in the CTP. Five of the sites are group homes and two program centers, all in Western Pennsylvania. The group homes are single sex residences for between three and six individuals. In contrast, the two program centers serve between 90 and 130 individuals on any given day.

The hardware introduced into the five group homes was iPads and laptops while at the day programs all-in-one desktop computers, laptops, tablets and iPads were made available. The hardware was customized to meet the needs of the IDD population, e.g., large keyboards, headphones. After much research a software package designed specially for individuals with cognitive disabilities—AbleLink—was purchased and installed. AbleLink allowed individuals to experience a more self-determined and fulfilled life through an empowering technology characterized by a person-centered design philosophy. Several AbleLink applications were installed that allowed individuals to use email (voice activated), Skype and webcam broadcasts, along with providing prompts for tasks that increased independence. The personalized laptops in the homes allowed the individuals to utilize the skills learned at the day program in their own homes.

B. Smarthome Project (SHP)

The SHP required a remodeling of a residential unit which faced construction problems delaying the start of the Project several months. However, by the summer of 2014, the four residents were able to move into the remodeled facility and be administered surveys and the GDS. The main objectives of the SHP were to increase the independence of the four IDD residents and to conserve energy through the use of "green" appliances and more efficient heating and air conditioning systems (HVAC). To achieve the goal of increasing the independence of residents, a Smart TV was installed, iPads and remote controls for lighting and window blinds were made available to the residents. Additionally, motorized cabinetry and cook tops and sinks were installed in a lowered position to allow wheelchair access. Finally, to reduce the amount of energy consumed, remote control HVAC systems and smaller and more easily accessible dishwashers and refrigerators were installed.

C Biometric Project (BMP)

The BMP began in the summer of 2014 with the development of protocols, the installation of the technology, training of staff, retrospective data collection, the creation of event and error forms and administering surveys to residents and staff and the GDS to the residents. The main goal of the Project was to use technology to reduce the number of emergency room visits and hospitalizations and thus, by doing so curtail costs by delivering care in a more timely manner and at a lower level of care [20]. Two group homes with four residents each were selected for inclusion along with a Long Term Structured Residence (LTSR), a locked facility that served eight male individuals with SMI with serious and persistent mental illness. The technology installed was a basic vital signs monitoring system including a digital scale, blood pressure cuff and pulse oximeters. The software included with the system allowed data to be sent to an external location and was configured to send alerts when the data collected went outside preconfigured parameters. Planning to expand this

Project to include the large Delaware County Adult Behavioral Health Outpatient Clinic is currently underway.

V. METHODS

There were several challenges to the selection of the methods to use to collect data on the three Projects. First, the Projects were not a test of the technologies themselves, as it was already known that they worked. Instead, the objective of the Projects was to determine how the selected technologies could be used to enhance the provision of care, while at the same time curtailing the cost of that care. Thus, the methods had to capture specific data on various components of care delivery. This entailed collecting data on the staff at each of the sites, both the way they used the technologies and their level of acceptance and willingness to change how they did their jobs. Data also had to be collected on the level of acceptance of the technologies by individuals at each of the sites. If individuals were uncomfortable with the use of the new technologies it would not be possible to roll out the technologies to other facilities. Secondly, although ideally the same methods of data collection would be used at each of the sites, this proved impossible because of the differences in the nature of the sites and the care models employed.

A. Communication Technologies Project

The main challenge was to develop questions that could be answered by individuals with disabilities and would, at the same time, provide the data necessary on which to make future decisions [21] [22]. Achieving these twin goals necessitated the development of a project-specific questionnaire for the CTP, which included simple straightforward questions and took no more than 15 minutes to administer. The questions asked included:

- Which of the following electronic devices do you use to communicate with friends, family or other people?
- How much help do you need to use these devices?
- When you want to communicate with friends, family or other people, how often is the device available?
- What devices do you use to play games or watch movies?
- How much help do you need to use these devices to play games or watch movies?
- When you want to play games or watch movies, how often is the device available?

The main goal of asking these questions was to determine the amount of change that took place in both device use and amount of help needed by individuals with disabilities during the length of the Project. Staff was trained to administer the questionnaire to individuals with disabilities at each of the sites with the goal being that the same staff member at each of the sites would administer the questionnaire at initiation of the Project and at three, six and 12 month intervals. However, this proved not to be possible because of the high rate of staff turnover. The staff questionnaire was self-administered and, similar to the questionnaire for individuals with disabilities, was repeated at three, six and 12 month intervals. The GDS was administered by staff members at the inception of the Project and six and 12 month intervals. Once again, staff turnover prevented the same staff member from administering the Scale at each of the intervals.

B. Smarthome Project

The methods used for the SHP were closely matched to those used for the CTP: Project specific questionnaires were given to the residents at the initiation of the Project and three, six and 12 month intervals; likewise the GDS was administered at the initiation of the Project and six and 12 month intervals. Questions focused on the ability of the residents to undertake basic tasks within the home, e.g., meal preparation, putting away groceries, controlling the lighting and blinds in their rooms, using computers and other electronic devices, using email to communicate with family and friends. Thus, it was possible to determine changes in both the residents' ability to use the new technologies and the impact on the technology of residents' well-being.

C. Biometric Project

The methods employed for the BMP, to a large extent, mirrored those for the other two Projects with a couple of exceptions. Staff was surveyed at the beginning of the Project and after six and 12 months. Questions for the staff focused on:

- The comfort level of staff members in the use of the biometric devices;
- The reliability of the devices;
- The acceptance of the devices by individuals with disabilities; and
- The perceived change in the quality of care with the use of the biometric devices;

Similarly residents at the three facilities were administered questionnaires and the GDS at the inception of the Project and at six and 12 month intervals. In addition to these instruments, event and error forms were developed for use. The event forms were used to record each event triggered by a biomedical alert, the actions taken by staff in response to the event and the outcome, e.g., a visit by a nurse, emergency room visit or hospitalization. The error forms were used to record problems with the various devices comprising the vital signs array, steps taken to correct the problem, the potential risk to the health/safety of the residents and how the problem was resolved.

D. Limitations of the Methods

There were several factors which limited the effectiveness of the data collection and the quality of the data. First, as stated previously, the fact that individuals at all the sites had either developmental and intellectual disabilities or were diagnosed with severe mental illness limited, to a certain extent, the type of questions that could be asked and often required prompting by the staff member administering the instrument Secondly, although not optimal from a research perspective, given the scope of the three Projects, it was necessary for staff to administer the questionnaires and GDS. These staff members were para-professional whose main responsibility was not research, but instead, was the delivery of care. In addition, staff turnover was also an issue because in many cases the same staff member was not available to re-administer the questionnaires and scales at the designated intervals. Finally, and perhaps most importantly, the Projects were not research per se, but a real world evaluation of the effectiveness of technology within challenging care models. In other words, the information collected was that which could help NHS determine whether the technology installed in the sites should be rolled out to other facilities, rather than what would necessarily be collected in a controlled research project.

VI. PRELIMINARY FINDINGS

At this stage, the findings from all three Projects are preliminary and some are more preliminary than others. The CTP started before the other two Projects and as a result the data collected is more complete and comprehensive. However, there is sufficient data from the other two to allow for initial analysis that can provide useful information to NHS on what works and what doesn't and to allow informed decisions about the future use of technology within the organization. Data collection will continue at all three Projects until, at least, September 2016. CTP eHealth technologies were installed at three additional sites in August 2015 and data will be collected at these three sites, at least, through September 2016. At this time, a comparative analysis of data from all sites will be undertaken and a decision whether to continue data collection will be made. There are preliminary plans to include an outpatient clinic as part of the BMP, but delays in finding an eHealth technology that would allow the collection and integration of vital signs information has put the project on hold.

A. Community Technology Project

As stated in the previous section, data were collected on both individuals with disabilities and staff at the five group homes and two day programs: questionnaires and the GDS for the individuals with disabilities; and questionnaires for the staff. In order to track changes in the activities and well-being of the individuals with disabilities, questionnaires were administered at the initiation of the Project and at three, six and 12 month intervals. The findings from a comparison of the data collected from these four sets of questionnaires are, from an organization perspective, very encouraging. Questions were asked about the use of electronic devices, both the number of devices used and the purpose for the use of the device. Answers to these questions showed a distinct pattern of the increase in both device use and the number and type of applications used. Sixteen of the 35 individuals (44%) for which data on all four sets of questionnaires are available were using more devices after 12 months than at the initiation of the Project, while fifteen (43%) were using more applications than in the prior 12 months. For the majority of individuals, the added device was a laptop that was made available in their residences. The pattern that emerged was quite clear. Individuals learned to use new applications on the desk-top computers at the day programs and then used the applications on the laptop when they returned to their residences.

The findings for the amount of help that individuals required to use the new devices and applications are a bit more complicated to interpret. The raw findings are: 2 (6%) of the individuals did not need help throughout the twelve months; 11 (31%) of them had no change in the level of help needed to access the devices and applications; 12 (34%) increased the level of help needed to access the devices and applications; and 10 (29%) decreased the level of help needed to access the devices and applications. These data are confusing enough, but in addition, there is no distinct relationship between the individuals who increased their use of devices and applications and the need for help. The amount of staff time required to train staff in the use of the technologies and to help individuals with new devices and application is a key factor in the decision to expand this Project to other facilities and therefore, having more usable findings is extremely important. As a result, a better way of measuring the level of help needed by individuals is being considered for use in the newer sites that have been added to the CTP.

Similarly, the findings from the GDS are ambivalent. Although there is a slight overall decrease in the number of answers that reflect a depressive state for over one-third of the individuals with disabilities, there is no apparent relationship between an increase in the use of devices and applications and a decrease in a depressive state. Once again, a better way of measuring the change in wellbeing is being sought.

Although the main conclusion that can be drawn from the three staff surveys is that the staff believes strongly that the technologies introduced have been greatly beneficial, the findings did expose some problems. A full quarter of staff believed that the technology was not useful for all individuals with disabilities. In particular, those individuals who had problems with reading grew frustrated when attempting to use the various applications. Secondly, almost half of staff reported that there were problems with the applications periodically crashing and/or having difficulty in getting the applications to work properly. However, the data did indicate that over time, the technological problems decreased significantly. Finally, the data confirmed the high rate of staff turnover, as only 12 of the 50 staff who completed at least one survey completed all three. In fact, an equal number—12—of staff completed only the last survey as those who had completed all three. Although staff turnover is a major issue with providing care for IDD and SMI populations (see the Introduction), such turnover will increase the costs of rolling out the technologies to other facilities.

B. Smarthome Project

There were no problems with data collection for the SHP. All four individuals with disabilities completed the three questionnaires and GDS administered upon initiation and three, six and 12 months into the Project. Nevertheless, the simple fact that there were only four residents in the study does limit the ability to generalize and reach firm conclusions about the wisdom of expanding the Project to other facilities.

The findings are largely positive as three out of the four residents expressed that over the twelve months of the Project their level of independence had increased: three out of the four residents expressed an increase in the ability to operate blinds and lights without help; two out of the four residents expressed an increase in the ability to undertake chores in the kitchen without help; and one out of the four residents recorded a greater ability to communicate with family. Answers on the GDS indicated that two of the four residents experienced a slight decrease in their level of depression. Staff also filled out the GDS for the residents and once again, it appeared that the same two residents experienced a decline in their level of depression.

In addition to the quantitative data collected, more informal interviews with both residents and staff revealed a very high level of satisfaction with the modifications made in the residence and the addition of the Smart TV, iPads and remote controls for blinds and lights. In particular, staff indicated that the mood of the residents had become more positive and that residents are much more active in the kitchen and taking pride in their increased independence.

C. Biometric Project

The findings for the BMP are the most preliminary of the three Projects, primarily because of equipment issues that delayed its start. Nevertheless, there are sufficient data to draw some conclusions that can be used as the Project is expanded to other facilities. For this analysis the two residential facilities serving IDD residents will be lumped together, while the findings for the LTSR are presented separately.

Staff surveys at the two IDD facilities showed that at the beginning of the Project over one-half of the staff did not know how to use at least one of the devices that was being installed. However, by the six month mark, all but one staff member, not only could use all of the devices, but were comfortable using them. The six month survey also indicated that, overall, staff were very positive about the use of the biometric equipment: a clear majority believed that care had improved with the use of the equipment; and all staff believed that residents had accepted the use of the equipment and were comfortable with its use. Over the first six months of the Project, there were 10 instances when one or more vital sign reading was beyond the safe range. In seven cases, a physician was contacted and in three cases, a nurse was contacted. Although in none of these cases was hospitalization necessary, four residents were put on outpatient observation in order to more carefully track their vital signs.

The only negative finding was the number of problems with the equipment recorded in 32 error logs. Just over 50% of the errors were a failure of the data to upload from the device to the iPad, which was used to record and forward the data to the nursing staff. In one-third of the error logs, the problem was with the devices not actually recording any data, e.g., the blood pressure cuff not indicating a reading. These findings have led to a reevaluation of the vital signs system being used.

All staff at the LTSR, when surveyed, expressed a high level of familiarity with all equipment used in the Project, both at the inception and six months later. Eight of the nine staff reported that the residents were comfortable with the use of the vital signs array, but one-third reported that the equipment was not as reliable as they would have liked. This unreliability was reflected in the nine error reports that indicated both problems with uploading data and the blood pressure and oximeter cuffs not generating a reading. Finally, there were eight events when one or more vital sign reading was beyond the safe range. In four of the cases, the nurse was contacted and the resident more closely monitored for the next 24 hours.

VII. DISCUSSION

As stated at the beginning of the previous section, the three Projects have been on different timelines with the result being that the extent of data collection and analysis varies. However, there are sufficient findings to reach preliminary conclusions for each of the Projects separately and when combined. In particular, it is possible to determine what has exceeded expectations, what has worked as hoped and what has not worked as well as hoped. Additionally, valuable lessons have been learned that can be used to help move the Projects to the next stage.

A. Successful Implementation

The CTP has been a tremendous success and has far exceeded expectations. The vast majority of individuals embraced the new technology and the applications made available through the Project. These individuals were able, in a relatively short period of time, to use the technology to communicate with family and friends, safely surf the internet in order to pursue their individual interests and to play games and watch movies—none of which they could do on their own before the Project. Although not all individuals at the seven facilities were able to utilize the technology, the vast majority could and they were able, over time, to do so with less and less staff help. Staff was equally pleased with the introduction of the technology and consistently reported that individuals were happy with their increased independence.

The SHP also exceeded expectations. Even though the numbers of residents impacted by the introduction was small, four, the findings clearly show that they benefitted from having the new technologies. Their independence increased over the duration of the Project because they were able to undertake tasks that they could not accomplish prior to the introduction of the new technology. The simple ability to control the lights and blinds in their own rooms, not only increased their level of independence, but staff reported that the residents' mood became increasingly positive over time.

B. Meeting Implementation Objectives

Although the BMP did not exceed expectations, it certainly succeeded in meeting the objectives set out at the beginning of the Project. The vital signs system is able to record, upload and send data to an external location as was hoped. In addition, the system was able to determine when readings are outside established norms and this information was used at all locations to take action, e.g., notify nurses, inform physicians. It is too early to determine if the use of the technology has reduced emergency room visits and hospitalizations, but staff believe that the system is able to allow more timely care and, as a result, the well-being of individuals has increased.

C. Hardware and Software Problems

Even though the CTP far exceeded expectation, there have been some issues surrounding the reliability of both the hardware and applications used. Most of these problems have been resolved during the course of the Project, but some problems still linger. The biggest issue with the technology occurred in the BMP. The number of error reports filed at the three facilities, confirm the overall impression that a different vital signs system needs to be used as the Project moves forward. The encouraging conclusion is that, even with the problems with the technology, the results were sufficiently encouraging to plan for the Project's expansion.

D. What Has Been Learned

In the last three and one-half years, many lessons have been learned about the process of incorporating new technologies into NHS's various care models. Some are more important than others, but some are absolutely essential to rolling out the technologies to additional facilities. Among the most important are: detailed planning is indispensable; there must be buy-in at all levels—board, ceo, upper administration, management and line personnel; one technology must be working before the next one is introduced; and everything takes longer than originally thought.

In the early stages of the Project, many individuals at NHS believed that things were moving too slowly; they were anxious to "get-on-with-it". This urge to move quickly is natural, especially from individuals who have been recruited because they are enthusiastic about the introduction of new technologies. Nevertheless, taking the time to plan every step of the Project was vital to success. Even with careful planning, mistakes were made and problems encountered. Likewise, there must be buy-in at every level of the organization and this also takes time. Without buy-in and commitment, there is the tendency to "cut the losses" when problems arise. The buy-in of the NHS Board, CEO and upper administrators was key to the continuation of the Project when things went wrong.

The incremental approach to the introduction of the technologies also proved to be a wise decision. Once again, there was a push to introduce "everything" at once, but the plan to make sure that one technology worked before installing a second, allowed staff and individuals with disabilities and severe mental illness to adjust to the first change before a second was introduced. Finally, although initially people involved with the Project were confident that the timeline for their slow and cautious approach was realistic, as the three Projects got underway there was a realization that the amount of time necessary to get 11 different sites up and running was going to take longer than anyone had anticipated. Fortunately, the fact that there was total buy-in at all levels of the organization allowed the Projects to proceed at the slower pace required.

VIII. CONCLUSION

When the Assistive Technology Project was being planned in 2012 there were doubts whether, because of its scale and complexity, it could achieve its objectives. Even as the first technologies were being installed and problems emerged, there were concerns that trying to evaluate three distinct technologies in 11 locations was just too ambitious. However, the Project leadership persevered and by the end of 2015, it is impossible to conclude otherwise than that the Project has been a success.

A. Improvements for Next Stage

Based on an analysis of the results from the three Projects, several adjustments have been made to both the technologies employed and the implementation protocols. First, continual problems with the biometric instruments have led to a search for more robust products from companies that offer greater technical support. Similarly, dissatisfaction on the part of staff and individuals with disabilities with several products in the SHP has resulted in the selection of more sophisticated equipment that incorporates more sensor technology. Third, results have indicated the greatest degree of success at the day centers. As a result, the plans are to include a greater number of these centers and fewer group homes in the next phase of the project. Finally, because the use of the GDS did not produce the level of meaningful information as anticipated, it's been replaced by satisfaction scales.

B. Expansion

The best measure of this success is that NHS has made the decision to extend all three technologies to additional locations. The CTP has already been expanded to three new locations bringing the total number of participants to just under 200 and there are plans to include more facilities during 2016. NHS is also working with AbleLink to develop new applications specifically targeting the IDD population, as well as working to refine and enhance existing applications. As indicated previously, the Delaware County Adult Behavioral Health Outpatient Clinic will be added to the BMP in the summer of 2016. The complexity of incorporating vital signs monitoring into an outpatient clinic is, from NHS's perspective, outweighed by the opportunity to effectively offer psychiatric services, primary care physicians and pharmacy services in one location along with an integrated medical record. If successful, there are plans to extend the BMP to other outpatient locations.

Perhaps the most significant indication of the success of the overall Project is that there are plans to build, ground up, a new facility in Western Pennsylvania which will include technologies from all three Projects. This $650,000 residence will house six individuals and incorporate the green appliances, design features—lower sinks and cabinets—and remote control features from the SHP, with the AbleLink applications from the CTP and the vital signs array from the BMP. It is hoped that this new facility will be the model for the necessary replenishment of the aging housing stock that is currently in use.

C. Financial Model

From the inception, one of the key components of the Project was to construct a financial model that would allow NHS to be reimbursed for the care delivered by the use of the new technologies. The Project itself, costing over $200,000 in real money and much more when the amount of staff time expended is included, has been financed by grants. Although grant funding is absolutely satisfactory for a project whose goal is to evaluate the appropriateness of new technologies in the delivery of care, it is not a satisfactory means for developing a sustainable financial model. A sustainable financial model can only exist if the care delivered with the use of the new technologies is reimburseable by Medicare (the health insurance program for people in the United States who are 65 or older. Medicare Part B covers certain doctors' services, outpatient care, medical supplies, and preventive services.) and, especially, Medicaid (the U.S. government program, financed by federal, state, and local funds, of hospitalization and medical insurance for low income persons of all ages.) as billable services. The problem is that, currently, most of the care delivered in the three Projects is not billable and thus, not reimburseable, but this is changing. Virginia and Pennsylvania have granted Medicaid waivers that could allow reimbursement for care delivered with the new technologies as soon as mid-2016. Individuals from NHS are presently developing proposals for both states that would allow the services to be reimburseable.

If the three Projects have done nothing else, they have confirmed that the use of technology to aid in the delivery of care to individuals with IDD and SMI is inevitable. The increasing numbers, along with the aging of the IDD and SMI populations, is increasing the cost of care exponentially at the same time as the number of people available to deliver the care is stagnating. The only way to maintain, let alone enhance, the level of care to these populations is through the innovative use of technology and the only way to make this happen is to develop a means of reimbursing this care. This must and will occur; the only question remaining is when?

REFERENCES

[1] G. Demiris, B. Hensel, M. Skubic, and M. Rantz, "Senior residents' perceived need of and preferences 'smart home' sensor technologies," Inter J of Tech Ass in Health Care, vol. 24, no. 1, pp. 120-124, 2008.

[2] E. Stip and V. Rialle, "Environmental cognitive remediation in schizophrenia: Ethical implications of 'smart home' technology," Canadian J of Psychiatry, vol. 50, no. 5, pp. 281-291, 2005.

[3] E. Novella, "Mental health care and the politics of inclusion: a social systems account of psychiatric deinstitutionalization," Theor Med Bioeth, vol. 31, pp. 411-427, 2010.

[4] M. Knapp, J. Beecham, D. McDaid, T. Matosevic, and M. Smith, "The economic consequences of deinstitutionalization of mental health services: lessons from a systematic review of European experience," Health and Social Care in the Community, vol. 19, no. 2, pp. 113-125, 2011.

[5] National Council on Disability. [Online]. Available from: http://www.ncd.gov/publications/2012/DIToolkit/Costs/inDetail/ retrieved March 2016.

[6] National Institute of Mental Health. [Online]. Available from: http://www.nimh.nih.gov/statistics/index.shtml, retrieved March 2016.

[7] Administration on Aging—Projected Future Growth of Older Population. [Online]. Available from: http://www.aoa.gov/ Aging_Statistics/future_ growth/future_growth.aspx, retrieved March 2016.

[8] U.S. Department of Health and Human Services Office of the Assistant Secretary for Planning and Evaluation. [Online]. Available from: aspe.hhs.gov/.../supply-direct-support-professionals-serving-individuals- intellectual-disabilities-and-other-developmental-disabilities-report-c, retrieved March 2016.

[9] The New Jersey Council on Developmental Disabilities Recommendations on the Direct Support Professional Workforce. [Online]. Available from: https://www.njcdd.org/ 2012-04-23.../direct-support-professionals retrieved March 2016.

[10] E. Tanis, S. Palmer, M. Wehmeyer, D. Davies, S. Stock, K. Lobb, and B. Bishop, "Self-Report computer-based survey of technology use by people with intellectual and developmental disabilities," Intellectual and Developmental Disabilities, vol. 50, no. 1, pp. 53-68, 2012.

[11] D. Hammond, D. Whatley, K. Ayres, and D. Gast, "Effectiveness of video modelling to teach iPad use to students with moderate intellectual disabilities", Education and Training in Autism and Intellectual Disabilities, vol. 45, no. 4, pp. 525-538, 2010.

[12] B. Bryant, R. Byant, M. Shih, and S. Seok, "The role of assistive technology in support of needs assessment for children with disabilities", Exceptionality, vol. 18, no. 4, pp. 203-213, 2010.

[13] A. Carey, M. Friedman, and D. Byrant, "Use of electronic devices by people with intellectual disabilities", Mental Retardation, vol. 43, no. 5, pp. 322-333, 2005.

[14] B. Weinberg, and A. Kjellberg, "Participation when using cognitive assistive devices from the perspective of people with disbilities", Occupational Therapy Instruction, vol. 17, no. 4, pp. 168-176, 2010.

[15] B. Bryant, S. Seok, M. Ok, and D. Bryant, "Individuals with intellectual and/or developmental disabilities use of assistive technology devices in support provision", J Spec Educ Technol, vol. 27, no. 2, pp. 41-57, 2012.

[16] The Rights of People with Cognitive Disabilities to Technology and Information Access. [Online]. Available from: http://www. colemaninstitute.org/declaration, retrieved March 2016.

[17] Ablelink Technologies Home Page. [Online]. Available at http://www. ablelinktech.com, retrieved March 2016.

[18] J. Mindham and C. A. Espie, "Glasgow Anxiety Scale for people with an Intellectual Disability (GAS-ID): development and psychometric properties of a new measure for use with people with mild intellectual disability", J of Intellectual Disability Research, vol. 47, no. 1, pp. 22-30, 2003.

[19] H. Hermans and H. M. Evenhuis, "Characteristics of instruments screening for depression in adults with intellectual disabilities: Systematic review", Research in Developmental Disabilities, vol. 31, no. 6, pp. 1109-1120, 2010.

[20] Centers for Disease Control, Adults with Disabilities and Vital Signs. [Online]. Available from: http://www.cdc.gov /vitalsigns/disabilities, retrieved March 2016.

[21] M. Tasse, R. Schalock, J. Thompson, and M. Wehmeyer, Guidelines for interviewing people with disabilities. American Association on Intellectual and Developmental Disabilities. Washington, DC, 2005.

[22] K. Caldwell, "Dyadic interviewing: A technique valuing interdependence in interviews with individuals with intellectual disabilities", Qualitative Research, vol. 14, no. 4, pp. 488-507, 2013.

Automated Nursing Agent: A Software Agent for At-Home Elderly Care

Amir Hossein Faghih Dinevari, Osmar Zaiane
Department of Computing Science
University of Alberta
Edmonton, Alberta, Canada
e-mail: {a.dinevari,zaiane}@ualberta.ca

Abstract—**The global increase in elderly population is a concern for many developed countries. The concern has to do with the financial and infrastructure challenges to assist, take care, and manage the aging population that typically requires more care, specialized personnel, and expensive facilities. While the population is indeed aging, particularly in developed countries, many of the elderly prefer remaining at home instead of moving to nursing homes. It has been demonstrated that companionship, such as other residents or pets, can significantly help preserve mental and physical health of the elderly. There has been considerable interest in building computer programs that work as an intelligent personal assistant such as Siri and Google Now, but these typically serve as knowledge navigators to search for information or do simple tasks on a smartphone. We propose an intelligent software agent that runs on tablet devices and operates as a companion to the elderly, assisting them in information retrieval, reminding them of relevant events, and more importantly, conversing with them as a human would do. In this paper, we introduce the Automated Nursing Agent (ANA) and present the technical challenges developing such a software agent.**

Keywords-Software Agent; Intelligent Agents; Automated Conversation; Natural Language Processing.

I. MOTIVATION

The need for companionship has been identified by many people and experts in psychology and geriatrics in different countries and cultures. Every human being feels the necessity of having a companion to some extent with whom to talk and share life events, and other important personal issues [1]. This need for having a companion is even more important for older people. Indeed, loneliness and social isolation can predict, for the elderly, declining health and poor quality of life [2]. Social interactions can reduce the mentioned effects of loneliness and social isolation. This is demonstrated to be true for the working people, let alone the elderly living alone [3][4].

Nowadays, thanks to advancements in healthcare and medicine, the life expectancy in developed countries has increased significantly [5][6]. Another noteworthy issue is the population decline in most of these countries [7]. These changes resulted in a higher percentage of elderly population compared to last decades. Many live assisted but many choose to live at home [8][9]. This trend will accentuate. According to the U.S. Census Bureau projections, by 2050, one-in-five Americans will be 65 or older, and at least 400,000 will be 100 years or older [10]. Today, the 65 or older represent 13% of the US population denoting around 41 million people. With a growing number of elderly in developed countries, there are not enough caregivers or nurses to take care of this aging population [11]. In Canada, 92% of the 5 million seniors aged 65 and over live in private homes [12]. Remaining at home will be encouraged for financial, social and health reasons. Services and facilities for the elderly, like nursing homes, will not always be available and the services cannot scale with the predicted demand of the senior population. Having a software agent at home could be a solution to help attenuate the need for companionship if this software agent running on a tablet-like device can indeed converse in a fluent manner and provide the required access to information and remote assistance.

The need for companionship has also been acknowledged for patients in hospitals [13][14]. Hospitals can be lonely places for patients. This is particularly more accentuated for older adults who may have few, if any, family members or acquaintances living close enough to visit them. Some hospitals even created Companion Care Programs where volunteers come to interact with patients, sharing small talk and activities.

Aside from companionship, the elderly have various needs and their living conditions can vary from one individual to another. Thus, personalized care and attention are required. The elderly who can take care of themselves often choose and remain at home instead of joining care facilities. More often than not, they opt for pets as companions. Pets have proven to be beneficial for mental health in the case of lonely seniors [15][16], but their entertaining interaction falls short from intelligent and knowledgeable conversation. In many cases, while choosing to remain at home, some seniors are not totally autonomous and capable of looking after themselves, therefore, there is a need to have a family member or a caregiver to live with them or check on them frequently. When it is not possible, a software agent could communicate with caregivers in cases of emergency. Moreover, doctors, caregivers, or family members could remotely inquire about the state of the senior, communicating directly with the agent.

We present here the architecture of a software agent, ANA, to attempt to remediate the current shortcomings. ANA, for Automated Nursing Agent, is a software agent running on Android tablet devices, specifically designed for the companionship of the elderly. ANA verbally tends to a person, at home, in need. It looks after an elderly at home so as to promote health through companionship. Nursing is about providing care and is an integral part of the health care system. It encompasses the promotion of health, prevention of illness, and care of someone in physical or mental need.

Unlike intelligent personal assistant software agents such as Apple Siri, Google Now, Microsoft Cortana, and many others, which are designed to answer information requests or to help owners of handheld devices to verbally express commands on their devices, ANA piggybacks on Google APIs for information requests but extends this capability by building a personalized knowledge base to provide personalized conversations with answers but also questions and general statements. We are currently focusing on conversations related to family, weather, health, and cooking. Moreover, ANA also reminds the senior about events and facts recorded in previous conversations, documents health-related events, and is accessible remotely by authorized caregivers and family members.

In Section 2, we first present some scenarios illustrating the usefulness of the agent and in Section 3, we discuss our methodology to address those scenarios. In Section 4, we conclude the paper and talk about the current state of our project.

II. SCENARIOS

ANA is not competing with intelligent personal assistants. In fact, it uses Google APIs to access information and search the Web knowledge. However, it adds new capabilities to make it more personal. For instance, it uses a database of jokes and learns to choose jokes to tell based on the appreciations of the previous jokes by the senior using reinforcement learning. Other activities provided by ANA include reading, playing games, and providing means to tell their life story. By encouraging the elderly to share life experiences, it improves the connection between the agent and the user. ANA stores personal information in a personalized knowledge base in addition to using a general knowledge base like Freebase [17] or Google Knowledge Graph [18]. In the following, we present some example scenarios using ANA.

A. Scenario 1. Conversations initiated by the elderly

One of the main characteristics that an intelligent system needs to have to be considered as a companion is the ability to have a normal, fluent conversation similar to human verbal interactions. There are three types of interactions in conversations: making a statement (declarative), asking a question (Interrogative), and giving an order (Imperative).

1) Declarative statements

There are different types of declarative statements. All the statements in talking about a past or future event or telling a story are considered declarative. People talk about how they are feeling, who they met, family members and friends, and also, the events that may happen during that day or in the near future. A person who is being talked to usually acknowledges the understanding of the story by saying some other declarative statements. This acknowledgment also shows that the person is interested in the story. Also, there can be some questions that may come up to the listener's mind, in case of having some vague details in the story. It is very common to ask about some details when you are listening to someone. Also, people expect that the others that they are talking to remember some details of the previous conversations and when they are referring to those told stories later, that person remembers some of it and is not completely clueless.

All these details should be implemented in our system to make elderly feel comfortable talking to a machine because without having these details, the conversation will seem unnatural and they would rather not use the system. ANA stores such information in a personalized knowledge base as well as a calendar for events and medical conditions.

2) Interrogative statements

Questions are considered as interrogative statements. There can be different types of questions that need different responses. In a normal conversation, it is usual to ask something personal from the person that we are talking to. Although the system that we are talking about is not a person, it should give some kind of a response that makes sense and keeps the conversation going.

The other form of questions can be related to the person who is talking. They can ask about how they look or want to know your opinion about some related issues and problems. This kind of question is very hard to handle by a machine since it needs to have opinions about things that are happening around it, but like the previous type of question that we talked about, it needs to give answers that make sense.

The third type of question is related to the subjects that have been talked about before and need remembering. To answer these questions, we should refer to the previous conversations that we had with the person. The system should gather information from conversations that it had with the person in order to answer these questions. Remembering details will make the conversations more natural and closer to what happens in real life. Even missing some details by the system is not a big deal since it can happen with real people too.

Questions can be about general knowledge. They are easier to handle compared to the others that we have talked about. To answer these questions, we can check different online knowledge bases that are available widely. If the answer is not found, the machine says it does not know it. It is very unlikely that someone knows answers to every question hence our system is not an exception. This is similar to what current intelligent personal assistants do. Also, the goal is not answering questions but being a companion, so missing answers to some questions in this context is fine.

There also can be other questions that are imperative in essence. These statements should be handled like imperative statements.

3) Imperative statements

All the commands and requests are considered imperative. Human beings are usually able to help each other with different tasks but it does not necessarily mean that they can help with all the tasks. Our system also does not need to address all the commands, but it is favorable if it can do as many as possible. Right now, having physical abilities in the system is not one of our goals. Controlling physical objects can be done by a computerized system, but there is a need for additional equipment to add this capability, which is against our current objective of having a cheap system for end users. Even if the system is not going to do anything about physical commands, it will try to give some sort of suggestions or try to come up with a witty response. For other types of commands, the system should be able to do them easily. These commands can be things like calling, texting, sending an email, checking up some information online, finding a recipe, or reminding the person of something. After doing any of these commands, there should be some sort of acknowledgment for the action in a way that the person knows that the command is done.

B. Scenario 2. Conversations initiated by system

In Scenario 1 we considered responses that are necessary for a natural conversation, however, in a normal conversation, each subtopic can be initiated by both sides of the conversation. If someone is always asking questions or making statements while the other person just responds to each one of them, the conversation will be one-sided and boring.

In order to avoid situations like this, we need to have functionalities in our system to sometimes initiate a conversation. One of the things that should be considered is that the subtopics in a discussion are usually related. Going completely off-topic can make the conversation feel unnatural. This is the case with chatterbots, software agents that are mostly built for fun.

To find related topics, we can refer to previous conversations between person and system. To Start a new topic, the system searches for empty facts or attributes in its personalized knowledge base and creates a statement to initiate a conversation that leads to filling the missing facts. These facts will be used later for other conversations.

Initiating conversation can be used for continuing current conversation after a long pause by the person. These long pauses are known as awkward silences and people try to fill them with starting a conversation. This human-like behavior can make the system very similar to a human companion.

C. Scenario 3. Caregiving functionality and remote access

In previous scenarios, we considered all the different characteristics that can make our system act like a human. While it is very interesting to mimic human behavior, the main purpose of this system is to ensure or improve the welfare of elderly when they live on their own and without direct human supervision.

Keeping track of medication for elderly can be challenging. Caregivers usually remind them of taking their medicines and they make sure that they took them. They also keep track of what medicine they took and when they took it. This information can be helpful for doctors. If we implement all the necessary components to handle scenario 1 and 2, it is easy to add these functionalities since they can be accommodated in the form of statements.

Caregivers also ask elderly about how they feel. Their responses in a period of time can be helpful in identifying unusual symptoms and patterns that can be caused by a disease. Like the previous case, it is also a very small extension to other scenarios that we have described.

One of the vital functionalities that is easy to spot by a human is seeing something unusual in elderly. Based on the severity of the situation, caregiver contacts the appropriate person. This can be a little tricky to implement but it is necessary to have in our system. A red flag for the system can be unresponsiveness of the person which can be a trigger for contacting a family member, doctor, or an ambulance. The system should be able to deal with this kind of situations in real-time.

Having a system keeping track of every aspect of elderly's life can be very helpful. This information can be monitored regularly by a doctor to see how well the person is doing and check on some unusual behavior patterns that can be a sign of a particular discomfort, disease, etc. Also, if it is necessary, this information can be viewed by family members to check on their loved ones and see how well they are doing. Because we have a computerized system, it can be possible to add remote access to this information via Internet to authorized people.

III. METHODOLOGY

The system design is the first concern that we need to deal with. The system should be able to the cases that we have described in the previous section. In order to cover those functionalities, a server-client model is necessary. The server side architecture is not a concern since we do not need separate servers for each client. Each server can support multiple clients but for the client side, the device price and availability are important. Having a device that is manufactured in large scale and support many functionalities, while keeping the price low for end users can be an ideal choice. Tablet devices satisfy our criteria. They have a large screen which makes reading easier for elderly that have vision impairment. They are affordable and they support many functionalities for keeping information like task management, contacts, emails, calling, and much more. Because of the affordability of Android devices compared to others, we consider Android tablets for our current prototype.

Most of the libraries that are dealing with natural language processing are designed to deal with text while input in Scenarios 1 and 2 is speech. To convert speech into text Android has built-in libraries for Speech to Text and Text to Speech conversion. We have used these libraries for current prototype.

In Scenario 1, we have talked about different types of statements that our system may encounter. Classification is necessary for categorizing statements. We have used a classifier developed by Quinn and Zaiane [17] for this purpose.

For all the statements, natural language processing is necessary to extract information that we need. We have talked about remembering information in the conversation. After processing of the statements, we store them in a database that we call Personalized Knowledge Base (PKB). This information can be used as described in the scenarios.

For answering questions, we first look into PKB. If we could not find the answer, we look in a global knowledge base. We are using Google Knowledge Graph as our general knowledge base. If look-up in general knowledge base was unsuccessful, the system is going to acknowledge that by saying it could not find the answer but it will show web search results for that question. For the imperative statements, all the necessary functionalities are included in the operating system.

In Scenario 2, we talked about starting conversations by the system. It is achievable by going through some templates and PKB to find an appropriate conversation opener in both cases of starting a conversation from the beginning or continuing the conversation that is started by elderly. Also, there can be different pointers for knowing when the system should start a conversation or continue it. A long pause while the conversation is not ended or not interacting with the system for a long time can be a sign to start a conversation.

For Scenario 3 all the cases except remote access can be considered an extension to Scenario 2. For calling people Android has communication abilities like phone calls that can take care of emergencies. The remote access part will be implemented by a Web-based user interface that can be accessed with all the internet enabled devices. To prevent unauthorized access, there will an authentication mechanism with unique identifiers. Figure 1 shows a simplified flowchart of the system.

It may seem that implementing such a system is a trivial task but it is not. Natural language processing and information extraction from plain text or voice are very complicated tasks. They are research topics that are actively being worked on in universities and research centers. By checking projects like Siri, Google Now, and Cortana, which are being supported by huge technology companies, it can be concluded that the conversing ability for machines is in its primary stages.

Companionship is important for the mental and physical wellbeing of an elderly. Seeking companionship can be disconcerting for seniors and often not possible. Building an intelligent automated software agent to converse with older people and provide information and verbal assistance is a goal of ours and would beneficial to society. We have started building a prototype but the undertaking is challenging and provides many interesting open research problems.

IV. CONCLUSION

In this paper, we talked about benefits of an automated system for elderly care, when they can live by themselves in their own places and. We talked about our proposed system, its specifications, and how we can implement it. Currently, we have a proof of concept prototype with some of these functionalities. We are working on improvements to the current system to be used in a field study. We will use the improved version to gather results about the effectiveness of the system and how people feel about conversations with the agent.

If a system with these specifications can be accepted by people, it can reduce imposed costs to national health care system by increasing the percentage of elderly population.

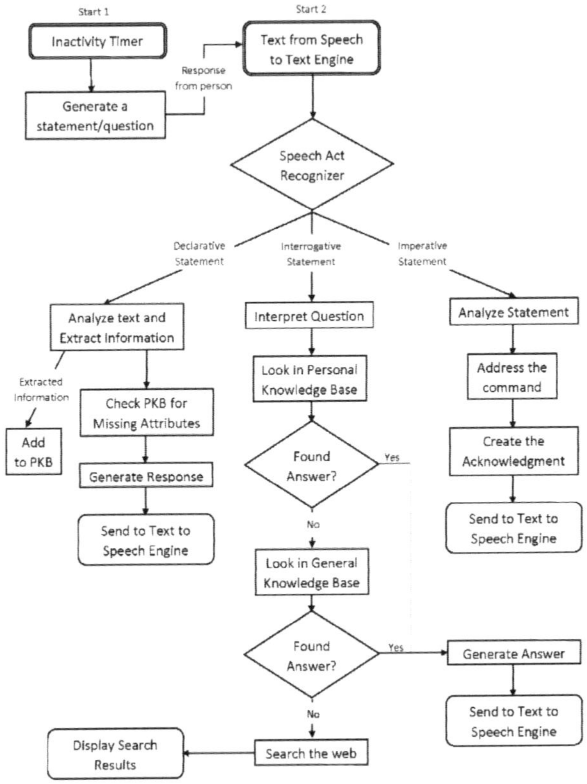

Figure 1. Flowchart of the system.

Also, it can reduce the number of caregivers or at least reduce their responsibilities and their working hours.

REFERENCES

[1] R. F. Baumeister and M. R. Leary, "The need to belong: desire for interpersonal attachments as a fundamental human motivation," Psychological bulletin, vol. 117, no. 3, 1995, p. 497.

[2] R. P. Meyer and D. Schuyler, "Old Age and Loneliness," Prim Care Companion CNS Disord., vol. 13, no. 2, 2011, pp. e1-e2.

[3] B. Buunk and M. Peeters, "Stress at work, social support and companionship: Towards an event-contingent recording approach," Work and Stress, vol. 8, no. 2, 1994, pp. 177-190.

[4] J. B. Unger, A. C. Johnson and G. Marks, "Functional decline in the elderly: Evidence for direct and stress-buffering protective effects of social interactions and physical activity," Annals of Behavioral Medicine, vol. 19, no. 2, 1997, pp. 152-160.

[5] C. D. Mathers, G. A. Stevens, T. Boerma, R. A. White and M. I. Tobias, "Causes of international increases in older age life expectancy," The Lancet, vol. 9967, 2015, pp. 540-548.

[6] S. Katz, L. G. Branch, M. H. Branson, J. A. Papsidero, J. C. Beck and D. S. Greer, "Active Life Expectancy," N Engl J Med., vol. 309, 1983, pp. 1218-1224.

[7] A. H. Gauthier, The state and the family: A comparative analysis of family policies in industrialized countries, Oxford University Press Catalogue, 1998.

[8] P. Kemper, "The use of formal and informal home care by the disabled elderly," Health services research, vol. 27, no. 4, 1992, p. 421.

[9] R. Bernabei, F. Landi, G. Gambassi, A. Sgadari, G. Zuccala, V. Mor, L. Z. Rubenstein and P. Carbonin, "Randomised trial of impact of model of integrated care and case management for older people living in the community," BMJ, vol. 316, no. 7141, 1998, p. 1348.

[10] P. D. Census Bureau, "Projections of the Population by Selected Age Groups and Sex for the United States," 2012. [Online]. Available: http://www.census.gov/population/projections/data/national/2012/summarytables.html. [Accessed 2016.03.18].

[11] W. Lutz, W. Sanderson and S. Scherbov, "The coming acceleration of global population ageing," Nature, vol. 451, no. 7179, 2008, pp. 716-719.

[12] Canada Statistics, "Living Arrangements of Seniors," 2011. [Online]. Available: http://www12.statcan.ca/census-recensement/2011/as-sa/98-312-x/98-312-x2011003_4-eng.cfm. [Accessed 2016.03.18].

[13] J. J. Essen, F. Cardiello and M. M. Baun, "Avian Companionship in Alleviation of Depression Loneliness, and Low Morale of Older Adults in Skilled Rehabilitation Units," Psychological Reports, vol. 78, no. 1, 1996, pp. 339-348.

[14] J. A. Carletti, "Volunteers provide companionship therapy under social service supervision," Psychiatric Services, vol. 15, no. 12, 1964, pp. 691-693.

[15] F. Walsh, "Human-Animal Bonds I: The Relational Significance of Companion Animals," Family Process, vol. 48, no. 4, 2009, pp. 462-480.

[16] T. F. Garrity, L. F. Stallones, M. B. Marx and T. P. Johnson, "Pet ownership and attachment as supportive factors in the health of the elderly," Anthrozoös, vol. 3, no. 1, 1989, pp. 35-44.

[17] "Freebase," [Online]. Available: https://www.freebase.com/. [Accessed 2016.04.15].

[18] "Google Knowledge Graph Search API," [Online]. Available: https://developers.google.com/knowledge-graph/. [Accessed 2016.04.15].

[19] K. Quinn and O. Zaiane, "Identifying Questions & Requests in Conversation," in Proceedings of the 2014 International C* Conference on Computer Science & Software Engineering, 2014.

A Compact Dual-Band Antenna for Wearable e-Health Devices

Haider Khaleel Raad[1*], Casey White[2], Hussain Al-Rizzo[3], Ayman Isaac[3], and Ali Hammoodi[3]
[*1]Department of Physics, Engineering Physics Program
Xavier University
3800 Victory Pkwy, Cincinnati, OH 45207, USA
[2]Deposition Sciences Inc., Lockheed Martin
Santa Rosa, California 95403, USA
[3] Dept. of Systems Engineering
University of Arkansas at Little Rock
2801 S. University Ave., Little Rock, Arkansas 72204, USA
raadh@xavier.edu, whiteca@seawolf.sonoma.edu, hmalrizzo@ualr.edu, axabbosh@ualr.edu, aihammoodi@ualr.edu

Abstract- **In this paper, a flexible, compact, ultra-low profile dual band antenna is presented. The proposed design is suitable for wearable and flexible Wireless Body Area Network (WBAN) devices geared for tele-health technology. The antenna has a 50.8 μm thickness and fed by a Coplanar Waveguide (CPW) structure. The proposed design is based on a planar monopole and resonates at 2.45 GHz and 5.2 GHz which makes it suitable for Wireless Local Area Network (WLAN) technologies. Design and analysis of the proposed printed monopole antenna have been carried out using the full wave simulation software CST Microwave Studio, which is based on the Finite Integration Technique (FIT). The proposed design has the merits of compactness, light weight, wide bandwidth, and high efficiency which suggest that the proposed design is a reasonable candidate for integration within wearable and flexible telemedicine systems.**

Keywords- e-Health; Telemedicine; WBAN; Wearable Antennas; WLAN.

I. INTRODUCTION

The light weight, efficient energy consumption, low manufacturing cost, fabrication simplicity, convenience, in addition to the abundance of inexpensive synthesized films and substrates, make wearable and flexible electronics an appealing candidate for the modern electronics market. Moreover, recent developments in miniaturized energy storage, flexible photo voltaics, and auto-powered electronic components have paved the road for the commercial success to such devices [1] [2].

Wearable electronics, which can be bent, reconfigured, and flexed, would substantially expand the applications of conventional electronic devices.

Generally, wearable and flexible electronics would often require the integration of antennas operating at specific frequency bands to provide wireless connectivity which is considered an essential requirement and greatly demanded by today's information oriented consumers [3]. Applications of wearable wireless systems include but not limited to the fields of personal communication, medical devices, warfare, aeronautics, and entertainment [4]-[6]. Obviously, the efficiency of flexible wireless systems directly relies on the characteristics of the integrated antenna unit, which must be light weight, conformal, low profile, electrically and physically small, and robust. At the same time, these antennas have to exhibit high efficiency and good radiation characteristics [7]-[11].

On the other hand, Wireless Local Area Network (WLAN) is recognized as the most reliable standard for short range wireless connectivity. The continuous developments in WLAN technologies require the integration of all IEEE 802.11a/b/g/n standards of the 2.4 GHz (2400 –2484 MHz), and 5.2 GHz (5150–5350 MHz) bands into a single radiating element. In the past decade, a plethora of flexible and wearable antenna designs have been reported in the literature, which are based on different topologies and substrate types. Electro textile based antennas seem to be a flexible, low profile solution for wearable applications; however, their substrate materials are likely to introduce discontinuities [12]. Technically, conventional microstrip (patch) antennas are not preferred in wearable electronics since the bandwidth is a function of the substrate's thickness. In [13], a flexible inverted-F antenna printed on a paper-based film was proposed for integration within flexible displays. Although paper based substrates are cost effective, they are found to lack mechanical robustness when used in applications that require high extents of flexing and bending. Moreover, they tend to have a high loss factor which compromises the efficiency of the antenna.

In [14], a wearable and flexible aperture coupled antenna is reported. This technique significantly enhances the impedance bandwidth, however, it leads to a serious increase in the antenna's overall thickness; moreover, it involves a multi-layer fabrication process.

In this paper, we propose a Co-Planar Waveguide (CPW) fed dual band antenna deposited on a 50.8 μm Kapton Polyimide substrate. In addition to its ultra-low thickness, compactness, and flexibility, the design has a more reasonable feed than the abovementioned feeding techniques since it offers several advantageous characteristics such as: reduced radiation losses, improved impedance bandwidth, and more importantly, lower fabrication cost and reduced complexity since both radiating element and ground plane are laid on the same side of the substrate, which enables roll to roll production and a consistent thickness.

In Section II, the design of the proposed wearable dual band antenna is presented. In Section III, simulation results of return loss and radiation patterns of the proposed design are discussed. Finally, conclusions are given in Section IV.

II. ANTENNA DESIGN

The topology of planar monopole and dipole antennas have received significant attention over other antenna types especially in wearable applications. For WLAN technologies, printed monopoles are preferred over other antenna topologies due to their relatively large impedance bandwidth, low thickness, fabrication simplicity, and omni-directional radiation pattern, which is highly desired in WLAN schemes.

As shown in Figure 1, the antenna consists of a winding branched radiating element fed by a CPW (please note that the grey colored area represents the metallization of ground plane and the radiating element). This miniaturization technique lengthens the current path, which consequently gives rise to a lower resonance and in turn reduces the structure size without significant degradation to the radiation characteristic of the antenna. The antenna structure is positioned on a 31 mm × 34 mm Kapton Polyimide substrate with a dielectric constant of 3.4 and a loss tangent of 0.002. To feed the radiating element, a CPW transmission line, which consists of a central strip with a width of 4 mm and a 1 mm gap between the central strip and the coplanar ground plane is used.

To achieve the desired dual-band behavior, U-shaped structures, which comprise both horizontal and vertical sections are introduced. The introduction of these sections is to produce two distinct current paths and thus dual resonant modes are enabled. Obviously, the longer arm gives rise to the lower resonance while the short arm is responsible for the upper resonance. It is worth mentioning that a parametric study was conducted to investigate the gaps and ground plane size effects on the resonant frequencies and return loss of the antenna. As mentioned previously, the branched radiating element is fed by a CPW feed reduces the fabrication complexity as both the radiating element and ground plane are deposited on the same side of the substrate. The dimensions of the antenna are depicted in Table I.

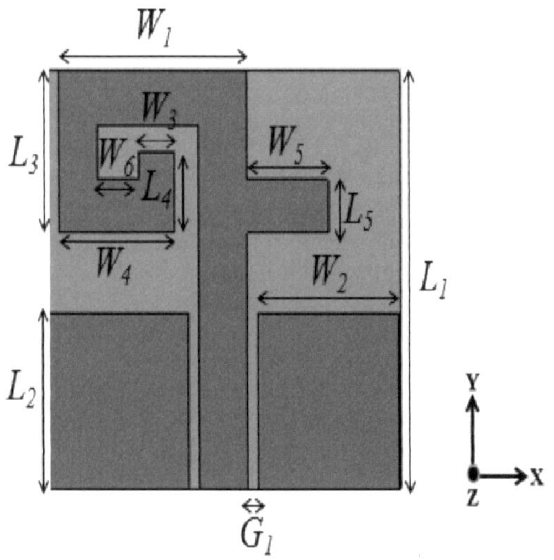

Figure 1. Geometry and dimensions of the proposed dual band printed monopole antenna.

TABLE I. DUAL BAND PRINTED MONOPOLE ANTENNA (DIMENSIONS IN MILLIMETER)

L1	31	***W2***	14
L2	13	***W3***	3.5
L3	12	***W4***	11.5
L4	6	***W5***	8
L5	4	***W6***	4
W1	19	***G1***	1

III. RESULTS

Design and analysis of the proposed printed monopole antenna have been carried out using the full wave simulation software CST Microwave Studio which is based on the Finite Integration Technique (FIT) [18].

As can be seen in Figure 2, the simulated return loss for the antenna is 22.8 dB at 2.5 GHz, with a -10 dB bandwidth of 662 MHz (27%), while the simulated return loss is 25 dB at 5.35 GHz with a bandwidth of 2125 MHz (40%). The principal planes (E and H) were also obtained using CST microwave studio. E-plane (*YZ* cut) and H-plane (*XZ* cut) far-field radiation patterns for the resonances under consideration are depicted in Figure 3.

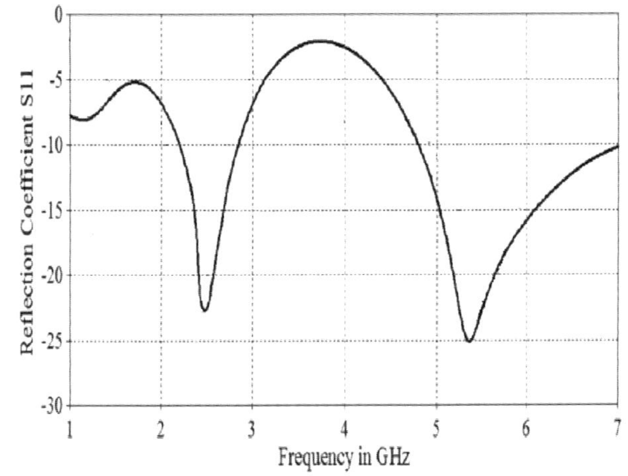

Figure 2. Simulated reflection coefficient S_{11} for the dual band printed monopole.

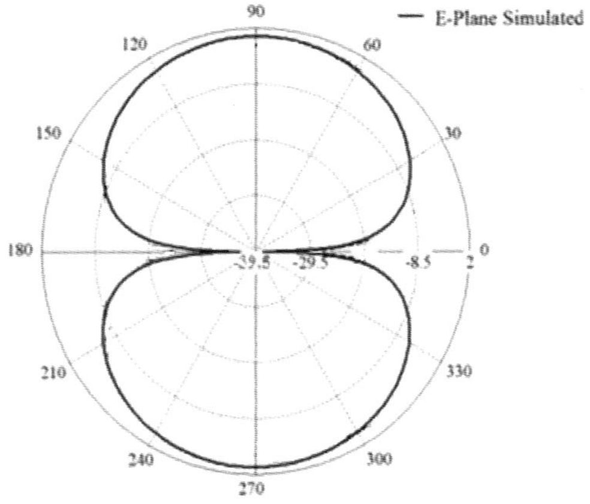

(a)

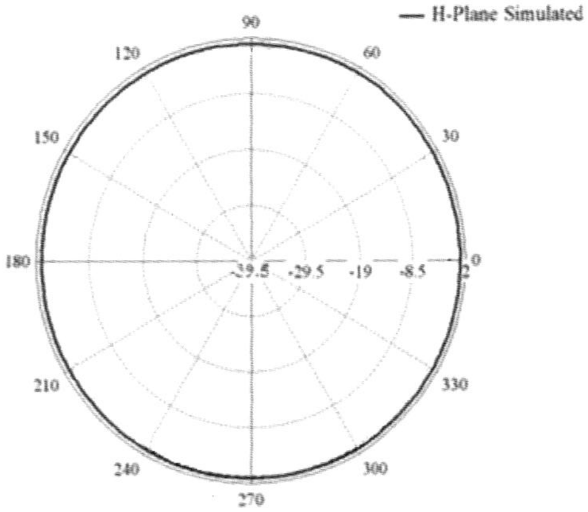

(b)

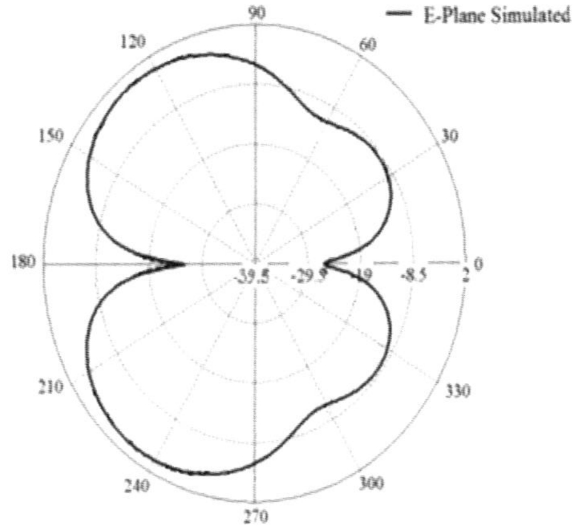

(c)

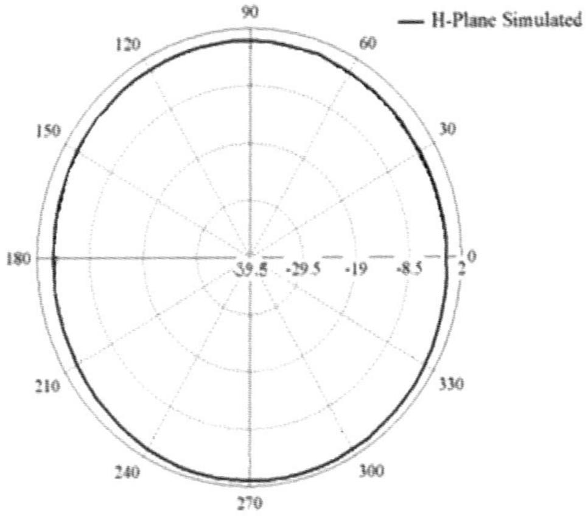

(d)

Figure 3. Simulated E-plane (*YZ*) and H-plane (*XZ*) radiation patterns for the dual band printed monopole at 2.45 GHz ((a) and (b)); and at 5.35 GHz ((c) and (d)).

It can be seen that the radiation power is omni-directional at both resonant frequencies. The antenna achieved gains of 1.68 dBi and 1.64 dBi at 2.45 GHz and 5.35 GHz, respectively, which are typical values for omni-directional antennas. It is worth mentioning that the proposed antenna is suitable for low-power applications (less than 10 mW), hence, it is not expected to expose the user to any significant hyper-thermal effect due to electromagnetic radiation, and thus, Specific Absorption Rate (SAR) analysis would be redundant in this case [15]-[17].

IV. CONCLUSION

In this paper, the design, and simulation of wearable, flexible and compact printed dual band antenna are discussed in details. The reported design is based on a Kapton Polyimide substrate, which is known for its flexibility, mechanical robustness and low dielectric losses. Ultra flexibility, compactness, mechanical robustness, along with excellent radiation characteristics and efficiency suggest that the reported design is a feasible candidate for integration within wearable and flexible WBAN systems.

REFERENCES

[1] H. Bahrami, S. A. Mirbozorgi, R. Ameli, L. A. Rusch and B. Gosselin, "Flexible, Polarization-Diverse UWB Antennas for Implantable Neural Recording Systems," in *IEEE Transactions on Biomedical Circuits and Systems*, vol. 10, no. 1, pp. 38-48, Feb. 2016. [accessed May 2015]

[2] P. Vanveerdeghem, P. Van Torre, C. Stevens, J. Knockaert and H. Rogier, "Flexible dual-diversity wearable wireless node integrated on a dual-polarised textile patch antenna," in *IET Science, Measurement & Technology*, vol. 8, no. 6, pp. 452-458, 11 2014. [accessed January 2016]

[3] J. Hu, "Overview of flexible electronics from ITRI's viewpoint," VLSI Test Symposium (VTS), 2010 28th , vol., no., pp.84, April, 2010. [accessed March 2016]

[4] H. R. Khaleel, H. Al-Rizzo, D. Rucker, and Y. Al-Naiemy, "Flexible printed monopole antennas for WLAN applications," Antennas and Propagation (APSURSI), 2011 IEEE International Symposium on , vol., no., pp.1334-1337, 3-8 July, 2011. [accessed January 2016]

[5] M. N. Srifi, S. K. Podilchak, M. Essaaidi, and Y. M. M. Antar, "Compact disc monopole antennas for current and future ultrawideband (UWB) applications," IEEE Trans. Antennas Propag., vol. 59, no. 12, pp. 4470–4480, Dec. 2011. [accessed September 2015]

[6] M. Hack, R. Hewitt, K. Urbanik, A. Chwang, and J. J. Brown, "Full color top emission AMOLED displays on flexible metal foil," in IMID/IDMC'06 Dig., vol. 16, no. 1, p. 305, 2006. [accessed July 2015]

[7] C. Hertleer, H. Rogier, L. Vallozzi, and L. Van Langenhove, "A Textile Antenna for Off-Body Communication Integrated Into Protective Clothing for Firefighters," Antennas and Propagation, IEEE Transactions on , vol.57, no.4, pp.919-925, April 2009. [accessed March 2016]

[8] R. K. Raj, M. Joseph, C. K. Aanandan, K. Vasudevan, and P. Mohanan, "A new compact microstrip-fed dual-band coplanar antenna for WLAN applications," IEEE Trans. Antennas Propag., vol. 54, no. 12, pp. 3755–3762, Dec. 2006. [accessed November 2015]

[9] H.-D. Chen, J.-S. Chen, and Y.-T. Cheng, "Modified invertedL monopole antenna for 2.4/5 GHz dual-band operations," IEE Electron. Lett., vol. 39, no. 22, Oct. 2003. [accessed September 2015]

[10] J. W. Wu, H. M. Hsiao, J. H. Lu, and S. H. Chang, "Dual broadband design of rectangular slot antenna for 2.4 and 5 GHz wireless," IEE Electron. Lett., vol. 40, no. 23, Nov. 2004. [accessed January 2016]

[11] P. Salonen, J. Kim, and Y. Rahmat-Samii, "Dual-band E-shaped patch wearable textile antenna," IEEE Antennas and Propagation Society Symposium, vol. 1 , pp. 466– 469, 2004. [accessed December 2015]

[12] D. Anagnostou, A. Gheethan, A. Amert, and K. Whites, "A Direct-Write Printed Antenna on Paper-Based Organic Substrate for Flexible Displays and WLAN Applications," Display Technology, Journal of , vol.6, no.11, pp.558-564, Nov. 2010. [accessed June 2015]

[13] H.R. Khaleel, H. Al-Rizzo, and D. Rucker, "Compact Polyimide-Based Antennas for Flexible Displays," IEEE Display Technology, Journal of , vol.8, no.2, pp.91-97, Feb. 2012. [accessed January 2016]

[14] Hertleer, C., Tronquo, A.; Rogier, H., Vallozzi, L., and Van Langenhove, L., "Aperture-Coupled Patch Antenna for Integration Into Wearable Textile Systems," Antennas and Wireless Propagation Letters, IEEE , vol.6, no., pp.392-395, 2007. [accessed August 2015]

[15] H. Ko, R. Kapadia, K. Takei, T. Takahashi, X. Zhang, and A. Javey, "Multifunctional, flexible electronic systems based on engineered nanostructured materials", Nanotechnology, Vol. 23, pp.11, 2012. [accessed April 2015]

[16] A. Rida, S. Nikolaou and M. M. Tentzeris, "Broadband UHF RFID/sensor modules for pervasive cognition applications," IEEE Conferences Ant. & Prop., pp. 2344-2347, Nov., 2009. [accessed January 2016]

[17] H. R. Khaleel, A. Abbosh, H. Al-Rizzo, and D. Rucker, "Flexible and Compact AMC Based Antenna for Telemedicine Applications," Antennas and Propagation, IEEE Transactions on , vol. 61, no. 2, pp. 524-531, Feb. 2013. [accessed February 2015]

[18] Software information can be accessed online at: http/www.cst.com. [accessed January 2016]

Proposal for A KINECT-Based Auscultation Practice System

Yoshitoshi Murata, Kazuhiro Yoshida
Faculty of Software and Information Science
Iwate Prefectural University
Takizawa, Japan
e-mail: y-murata@iwate-pu.ac.jp, kyoshida@ipu-office.iwate-pu.ac.jp

Natsuko Miura, Yoshihito Endo
Faculty of Nursing
Iwate Prefectural University
Takizawa, Japan
e-mail: natsuko@iwate-pu.ac.jp, y-endo@iwate-pu.ac.jp

Abstract—**Students in medical and nursing schools have to practice auscultation. A humanoid simulator is effective for learning disease sounds and correct stethoscope location. Such humanoid simulators, however, are too expensive for most nursing schools to buy. In this paper, we propose a low-cost system for the practice of auscultation. In this system, students themselves play the role of a patient, instead of a humanoid, and stethoscope locations on the body are measured with KINECT. Practicing students hear disease sounds, synchronized with the movement of breathing, through earphones. Movements of the upper body from breathing are also detected by KINECT. Experimental results with a prototype showed that our system could perfectly detect stethoscope locations on a body, except for a few lower points, and it could also detect respiratory changes. There are a few challenges, however, to be solved in future work.**

Keywords-simulator; auscultation; nursing; KINECT.

I. INTRODUCTION

Generally, practicing auscultation is a required subject for students in medical and nursing schools. Humanoid-type simulators have been developed [1][2][3][4][5], and there are several reports that such simulators improve auscultation skills [6]. Unfortunately, these simulators are generally too expensive for most nursing schools to buy.

We are developing a new, low-cost auscultation simulator whose concept is different from that of the existing humanoid simulators. In this simulator, students themselves are the practice subjects instead of a humanoid model, and it is possible to detect the location of a stethoscope with KINECT that is a line of motion sensing input devices by Microsoft [7].

In the case of existing humanoid simulators, it is possible to know correct stethoscope locations by marking these points on a mannequin, but it is impossible to detect whether a stethoscope is actually placed correctly on a mannequin. Moreover, correct locations vary among patients according to body size. Our proposed simulator can both show correct locations on a body and detect whether a stethoscope is placed on correct points. The correct locations are normalized with respect to the positions of both shoulder joints and both hip joints.

In addition, most humanoid simulators cannot simulate the timing of breathing or the synchronized forward and backward movements of the upper body. Our simulator, however, can detect these forward and backward movements and provide exhalation and inhalation sounds synchronized with those movements.

We have developed a prototype system and evaluated it experimentally. The results showed that our system could perfectly detect stethoscope placement on a body at eight of ten points. Because two lower points were easily shadowed from KINECT by the T-shirt worn by a student acting as patient, our system could not always detect a stethoscope. Moreover, our system could detect changes in breathing, but it sometimes made a mistake in counting the number of them and the detection delay for respiratory changes was slightly larger than expected.

After introducing related works in Section II, we describe the concepts and features of our system in Section III. The key technologies of our simulator and the evaluation results are described in Section IV. The key points are summarized in Section V.

II. RELATED WORKS

Many kinds of patient simulators have been developed and provided as medical and nursing training tools [1][2][3][4][5]. Since we propose a new type of simulator for practicing auscultation, we first discuss existing auscultation simulators, which are divided into two groups: the humanoid model type, and the virtual reality type.

A. *Humanoid model type*

Kyoto Kagaku Co., Ltd. provides the Lung Sound Auscultation Trainer (LSAT) [2], shown in Figure 1, for respiratory auscultation. There are several small speakers inside a mannequin. Disease sounds are recorded from real patients. This simulator also works for cardiac auscultation by changing from respiratory sounds to cardiac sounds. Sakamoto Model Corporation provides the Sakamoto auscultation simulator [3]. This simulator also works for both respiratory and cardiac auscultation. Sakamoto provides a transparent cover for this simulator, as shown in Figure 2, to illustrate correct stethoscope locations.

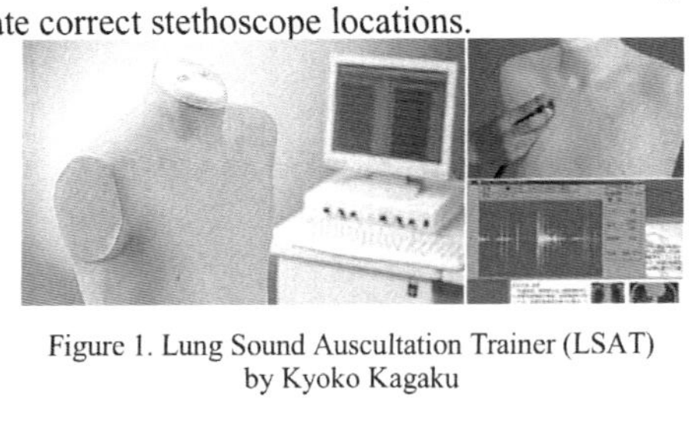

Figure 1. Lung Sound Auscultation Trainer (LSAT) by Kyoko Kagaku

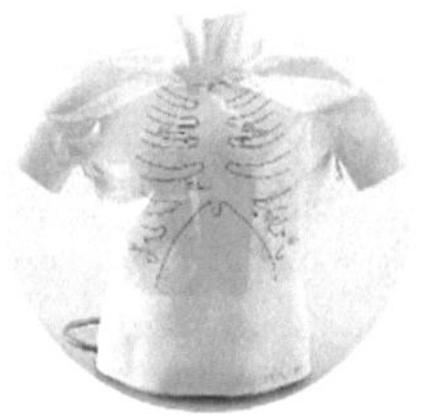

Figure 2. Transparent chest cover, by Sakamoto Model, to illustrate correct stethoscope locations

Although the above two simulators are focused on the upper body, they simulate disease sounds, not the motion of the upper body. On the other hand, the SimMan® 3G [4] by Laerdal is an advanced patient simulator that can simulate the characteristics of a real patient, including the blood pressure, heart beat, chest motion, and so on. It is too expensive, however, for a general nursing school to buy.

B. Virtual reality type

Zadow, et al. developed the SimMed system for medical education [5]. By using an interactive multi-touch tabletop to display a simulated patient, as shown in Figure 3, they have created an immersive environment that supports a large variety of learning scenarios. The simulated patient can show skin changes and be animated to show realistic bodily and facial movements. By its nature, the setup allows scenarios to be repeated easily and to be changed and configured dynamically.

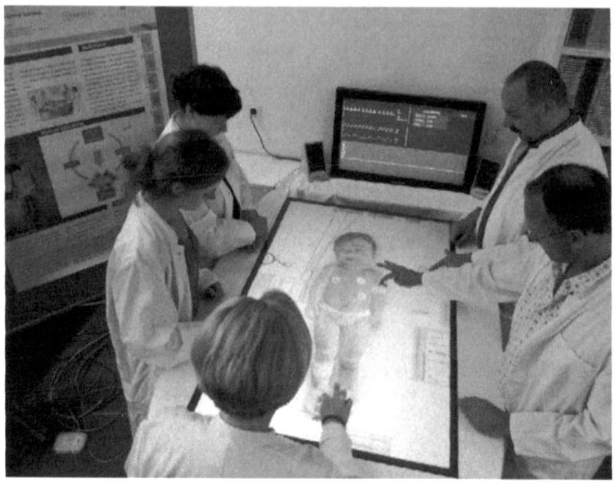

Figure 3. The SimMed system

SimMed is substantially lower in cost than a full-scale humanoid simulator. It has many functions, however, and is still too expensive for most nursing schools. Moreover, while students can touch the virtual patient on a display, they cannot physically feel the motion of the virtual patient.

III. CONCEPT OF KINECT-BASED SIMULATOR

Among the nursing skills that students have to learn, are: the recognition of different sounds between different kinds of disease and the knowledge about placing correct points for locating a stethoscope on a body. Moreover, in the case of respiratory auscultation, students have to listen to respiratory sounds for more than one cycle. Therefore, an auscultation practice system requires the following functions:

- Simulating real disease sounds at different points on the body.
- Showing correct points for locating a stethoscope on an operation display.
- Judging whether a stethoscope is located on showed points.
- Judging whether a stethoscope is fixed on a body for more than one cycle in respiratory.

As introduced in the above section, existing auscultation simulators represent patients with mannequins or virtual reality technology, and disease sounds played through speakers mounted on a humanoid model or an external speaker. Most disease sounds are recorded from real patients. Students learn differences in disease sounds and correct stethoscope locations on the body by marking them on a humanoid model or hearing lectures by a teacher. Since most such models, however, do not have functions for detecting stethoscope locations, they cannot show whether these locations are correct. Furthermore, since the humanoid model has only a single size, it is difficult to learn differences depending on body size.

With our practice system, on the other hand, students themselves act as patients, instead of mannequins. The stethoscope locations and forward and backward movement of a body from breathing are measured with KINECT, as shown in Figure 4. Students hear disease sounds, generated by a PC, through earphones. The sound volume for each point is different for locating a stethoscope as with a real patient.

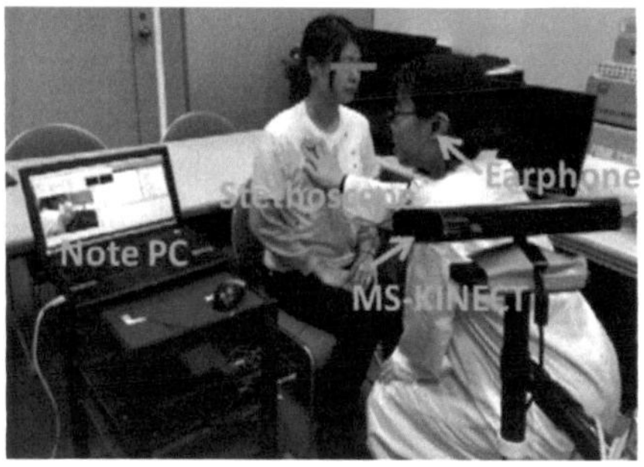

Figure 4. Auscultation practice with our proposed system

As introduced in the next section, color tracing technology is used to trace a stethoscope, and select "yellow green" for tracing. Therefore, we decided a white T-shirt for a student acting as a patient and a white coat for a student acting as a nurse wearing to remove "yellow green" and likeness in experiments.

IV. DETECTION TECHNOLOGIES

The following capabilities are required to implement our proposed auscultation practice tool:

- Tracing a stethoscope.
- Detecting a stethoscope's location.
- Detecting a stethoscope on a body.

- Detecting forward and backward movements of the upper body from breathing.
- Automatically adjusting correct points for locating a stethoscope on a body according to body size.

Since KINECT automatically generates position data for a stethoscope while it is traced, we examine each of the above four issues except the second one.

A. Tracing a stethoscope

Two candidate tracing methods are shape tracing and color tracing. Since a stethoscope is held by hand, the shape of a stethoscope as viewed through a video camera changes over time. Therefore, we chose color tracing, rather than shape tracing. The process of color tracing is shown schematically in Figure 5. Video data from the BGR (Blue, Green, and Red) 32 output of KINECT is converted to HSV (Hue, Saturation, and Value) color data. First, the traced object, i.e., a stethoscope, is pointed, and its hue histogram is generated and stored. Then, masking data are generated for each frame from the HSV data, the minimum saturation, and the maximum and minimum brightness.

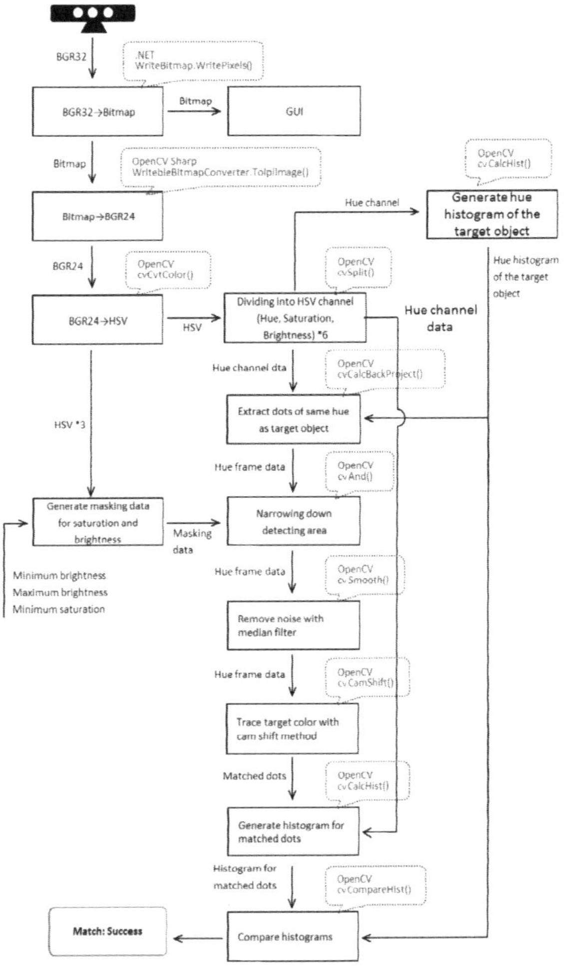

Figure 5. Process of color tracing

The video data for practice is also converted to HSV data, and target areas are separated with the masking data. Noise, including the hue data of the target area, is reduced by a median filter. The output data from the median filter is then traced by the cvCamShift function of Open CV [8]. Since cvCamShift sometimes outputs incorrect data, the hue histogram for the output data is repeatedly compared with the pre-stored hue histogram. When these histograms are equivalent, the color tracing is successful.

We used an experiment to select a target color from among seven choices: "red", "green", "light blue", "yellow", "yellow-green", "pink", and "orange". We used KINECT v1 and examined whether it could detect only a target color. We show the resulting data for the top three colors in Figure 6. "Yellow-green" had the best performance, with no portions having the target color except the target. "Light blue" also performed well, but since the color of a stethoscope tube is "light blue", that was detected. In the case of "yellow", dots of the same color appeared in the bottom-left region of the image. The other colors exhibited more dots of the same color or had a smaller target size. Hence, we chose "yellow-green" as the target color for our experiments.

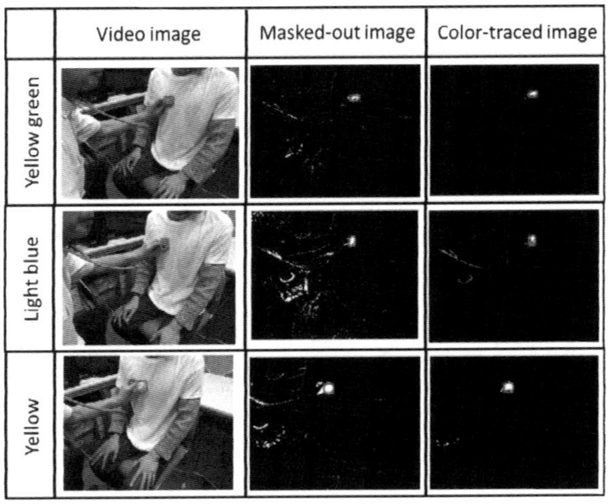

Figure 6. Experiment for deciding a target color

B. Detecting a stethoscope on a body

At this time, we think it is unnecessary to determine whether a stethoscope is exactly placed on a body, so we do not use any sensors on the stethoscope. Instead, we estimate a stethoscope is located on a body, where a stethoscope is placed and fixed within the distance S from a body surface for T seconds. For now, we use S = 10 cm, and T = 0.3 second to achieve balance between certainty and fast recognition. If stricter detection is required, we can change these parameters. Before measuring length L_{st} between a stethoscope and a KINECT, we determine whether the stethoscope is within the outline of a body, by using a pre-installed program on a KINECT to get an outline.

We experimentally examined whether this method was useful. A student acting as patient wore a white T-shirt with dots marking correct stethoscope locations; and a student acting as nurse placed a stethoscope on these marked dots.

We marked 10 dots on the T-shirt, as shown in Figure 7. The participants were five male students and five female students. The experimental results are listed in Table I.

Figure 7. T-shirt with dots marking stethoscope locations for a female participant

TABLE I. COUNT OF DETECTING A STETHOSCOPE PLACED ON A BODY

Point #	1	2	3	4	5
Male	5	5	5	5	5
Female	5	5	5	5	5
Point #	6	7	8	9	10
Male	5	5	5	5	3
Female	5	5	4	3	2

The proposed system sometimes missed when the stethoscope was placed at certain points (points 8, 9, 10). As seen from Figure 4, a stethoscope placed on one of these points would sometimes be shadowed from KINECT, especially for women, since these points are below the breast. Possible solutions for this problem are the follows:

- Switch from a T-shirt to clothing more fitted to the body.
- Attach some dimensional marker to the stethoscope.

C. *Detecting upper body motion*

Burba et al. used chest motion to detect breathing [9], since there are two main types of breathing: chest respiration and abdominal respiration. Therefore, we select six points for measuring movement of the chest and abdomen in this experiment, as shown in Figure 8. Since the stethoscope location is not always fixed, we do not consider it in this experiment. The upper three measuring points are the inner junctions of five vertical lines equally dividing the space between both shoulders into four regions and a horizontal line halfway between the height of the center of the spine and the average height of both shoulders. The lower three measuring points are the inner junctions of the above-mentioned vertical lines and a horizontal line through the hip center. We adopt the same direction of more than 4 points as the resultant direction.

We designed our system to detect the changes from expiration to inhalation and vice versa as quickly as possible. Since there were small but rapid changes in the output data, we used a moving average of 30 samples, with a sampling period of 10 ms. We then have specified that when the sampling data continued to increase 3 times, the breathing mode was expiration; and when the sampling data continued to decrease 3 times, the breathing mode was inhalation.

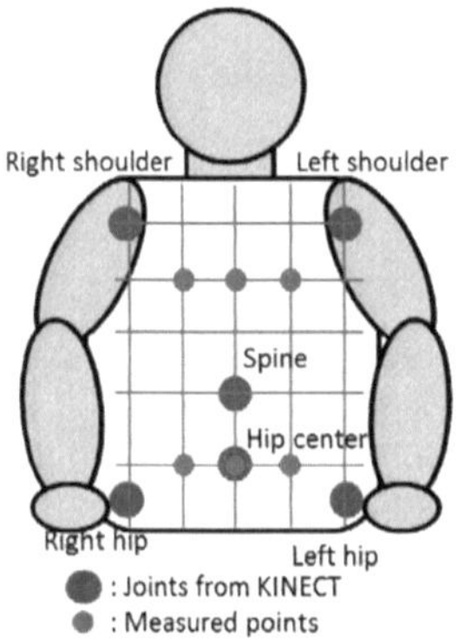

Figure 8. Measuring points for breathing motion

We measured the number of breathings and their periods with our proposed system and compared the results with other data obtained by participants keying the up and down arrow keys on a keyboard. The experimental results are shown in Table II. We measured them for two KINECTs that are K-1 and K-2. Both of them were the same model. At first, we measured for many participants with K-1. Our system counted more and more breaths than did keying for most participants. Data shown in Table II are some of them. In addition, the output data for some measuring points had extraordinary values, as shown in Figure 9 (2). We changed K-1 to K-2 to clear the reason of this problem. Our system with K-2 could count more accurately than that with K-1. However, our system counted fewer breaths than did keying for participant D. And, the extraordinary values were sometimes measured with K-2, too.

As shown in Figure 9 (1), our system detected changes in breathing with a delay of about 1 s relative to keying when breathings were correctly detected. The measured delay is bigger than we expected.

We think these problems can be derived from the above extraordinary output data, sagging T-shirt, and the above algorithm. The detecting algorithm would detect fluctuation of sagging T-shirt as movement from breathing. A sagging T-shirt would sometimes hide upper body motion from the detecting algorithm. We are re-programming the proposed system to KINECT v2 [7] and we are adding an algorithm to remove extraordinary data. We plan to switch from a T-shirt to clothing more fitted to the body; and optimize parameters of the detecting algorithm.

TABLE II. NUMBER OF BREATHS

	Participant	Keying		Proposed system	
		Inhalation	Expiration	Inhalation	Expiration
K-1	A	12	12	16	16
	B	10	10	39	39
	C	14	14	15	15
K-2	C	11	11	11	11
	D	10	10	7	7
	E	15	15	15	15

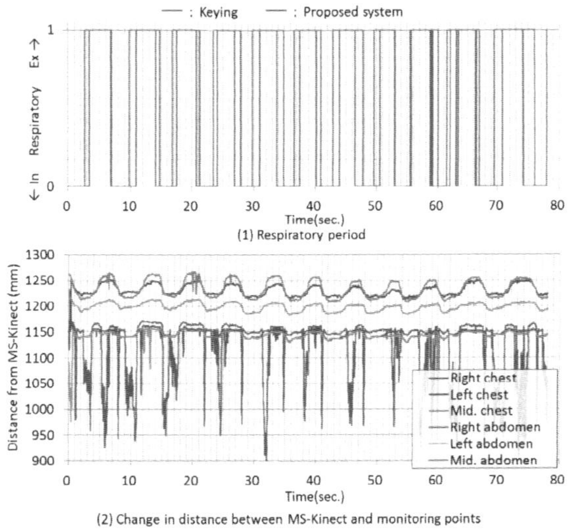

Figure 9. (1) Respiratory period, and (2) change in distance between KINECT and measuring points

D. Automatically adjusting according to body size

The correct points for placing a stethoscope depend on the size of the body; they differ a little between men and women. Also, their X and Y values vary according to the distance between a student playing a patient and a KINECT. Therefore, we estimate correct positions for placing a stethoscope with respect to the positions of both shoulder joints and both hip joints.

Two sets of correct position data and the positions of both shoulders and both hips are measured and stored as standard man and woman data. Since the origin of the output of KINECT's depth camera is at the upper left, it is difficult to compare body sizes among people. Therefore, for each person we reset the origin to the junction of a horizontal line connecting the average right and left hip heights and a vertical line passing through the midpoint between the right and left hip heights. Here, we assume that a person is sitting upright, and that the human body is a little unsymmetrical. The result of resetting the origin for each person is illustrated in Figure 10. The estimated left-side (X_{LE}, Y_{LE}) and right-side (X_{RE}, Y_{RE}) locations are calculated with the following equations by using the above stored standard locations and the measured positions of both shoulders and both hips:

$$X_{LE} = X_{LS} * X_{mls}/X_{sls} \quad , \qquad (1)$$

$$Y_{LE} = Y_{LS} * Y_{mls}/Y_{sls} \quad , \qquad (2)$$

$$X_{RE} = - X_{RS} * X_{mrs}/X_{srs} \quad , \qquad (3)$$

$$Y_{RE} = Y_{RS} * Y_{mrs}/Y_{srs} \quad , \qquad (4)$$

where

(X_{LS}, Y_{LS}) is the left-side standard location,

(X_{sls}, Y_{sls}) is the standard left shoulder position,

(X_{srs}, Y_{srs}) is the standard right shoulder positon,

(X_{pls}, Y_{pls}) is the measured left shoulder position of the patient, and

(X_{prs}, Y_{prs}) is the measured right shoulder positon of the patient.

We measured the ten points marked on a T-shirt and skeleton data for three men and three women to validate the proposed automatically adjusting algorithm. Each man wore the same T-shirt like that in Figure 8. Since the T-shirt was relatively small, it should have closely fit each person, and we think the marked points should have adjusted to each person correctly. After selecting one man and one woman each as the standard, we estimated correct points a stethoscope placing by using the above equations and the data for the standard person.

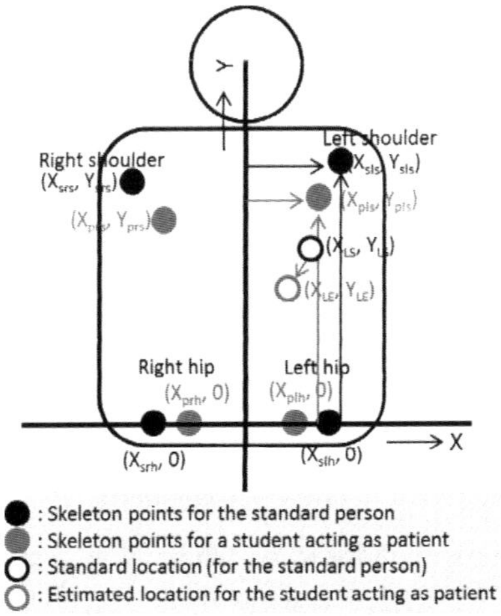

Figure 10. Illustration of adjusting locations for body size

Since there was not a big difference between the left-side data and the right-side data, we showed only the left-side data in Table III. The location relationships between measured and estimated correct points for a female and male participant are shown in Figure 11. In this figure, shoulder points and correct points a stethoscope placing for a standard person, and shoulder points and estimated and measured correct points a stethoscope placing for other participant are presented.

The data unit in Figure 11 is not the millimeter, but the pixel. In these figures, 10 pixels correspond roughly to 4 cm. Differences between estimated and measured positions depend on participants and measuring points. Maximum difference is about 8 cm. In case of Figure 11 (1), shoulder and hip positions of Female-1 are the same as those of a standard participant. Therefore, estimated correct positions of Female-1 are the same as those of a standard participant. However, measured points are different from them. The reason for this difference must be that each person did not wear the T-shirt symmetrically between right and left. In case of Figure 11 (2), differences between estimated and measured Y values were bigger for lower points, because the T-shirt was less elastic in the vertical direction.

We think the above experiment is not appropriate to evaluate the proposed estimation scheme for adjusting locations to different body sizes after experimenting. If participants wear a T-shirt in bilaterally symmetric and pull it down the hip position, the estimated positions must be approximately equal to measured positons. However, it is difficult for participants to wear the T-shirt correctly. Therefore, we have to validate the proposed method automatically adjusting algorithm by some other experiment.

TABLE III. COMPARISON BETWEEN MEASURED AND ESTIMATED LOCATIONS

Location		Male-1		Male-2		Female-1		Female-2	
		X	Y	X	Y	X	Y	X	Y
2	measured	28	136	46	157	53	167	52	157
	estimated	33	142	42	154	35	164	36	155
4	measured	28	118	43	140	55	144	53	133
	estimated	32	127	40	137	45	140	46	131
6	measured	27	94	44	118	52	123	50	107
	estimated	31	107	39	116	44	120	45	113
8	measured	30	51	43	78	55	68	50	59
	estimated	33	72	41	77	56	68	57	64
10	measured	40	36	49	59	62	52	65	54
	estimated	38	57	48	62	63	55	65	52

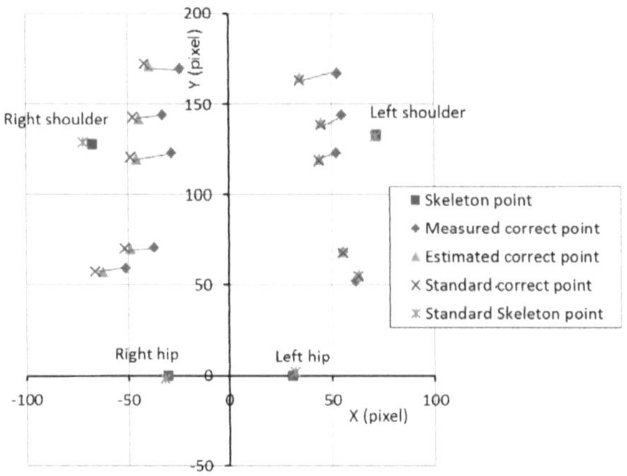

(1) Female-1

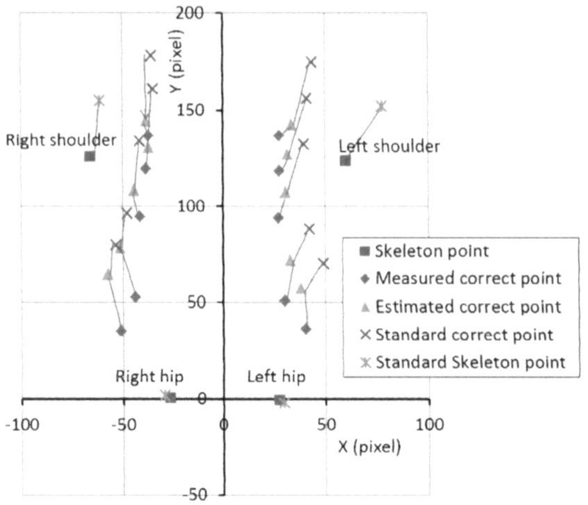

(2) Male-1

Figure 11. Example of measured and estimated correct points

V. CONCLUSION

We have proposed a new auscultation practice system for medical and nursing students. In this system, students themselves play the role of a patient instead of a humanoid model, and the locations for stethoscope placement on the body are measured with KINECT. Therefore, this practice system would have low cost.

In addition, the system can judge whether stethoscope locations are correct. Practicing students hear disease sounds, synchronized with the movement of breathing, through earphones or a speaker.

We developed a prototype system and evaluated experimentally. The results showed that our system could perfectly detect stethoscope placement on a body, except for two lower points, and it could detect respiratory changes. However, it sometimes made a mistake to count the number of them and the detection delay for respiratory changes was slightly larger than expected. We have to solve these problems for detecting respiration. After solving them, we plan to develop a real learning system, consisting of a learning unit for teaching correct locations and disease sounds and an evaluation unit for testing.

W used a white-T shirt and white coat for practicing students in experiments to remove "yellow green" and likeness. However, our system works well for usual clothes which have several colors except "yellow green".

REFERENCES

[1] Patient simulators for nurse & nursing care training, Kyoto Kagaku Co., Ltd., https://www.kyotokagaku.com/products/list02.html#cate_head01 [retrieved: March, 2016]

[2] Lung Sound Auscultation Trainer "LSAT", Kyoto Kagaku Co., Ltd., https://www.kyotokagaku.com/products/detail02/m81-s.html [retrieved: March, 2016]

[3] Sakamoto Model Corporation, Sakamoto auscultation simulator, http://www.sakamoto-model.com/product/emergency/m164/ [retrieved: March, 2016]

[4] Laerdal, SimMan 3, http://www.laerdal.com/us/SimMan3G [retrieved: March, 2016].

[5] Ulrich von Zadow, L andet al. , "SimMed: Combining Simulation and Interactive Tabletops for Medical Education," CHI '13, Proceedings of the SIGCHI Conference on Human Factors in Computing Systems, pp. 1409-1478, April 2013.

[6] John Butter, William C. McGaghie, Elaine R. Cohen, Marsha E. Kaye, and Daiane B. Wayne, "Simulation-Based Mastery Learning Improves Cardiac Auscultation Skills in Medical Students," Springer, Journal of General Internal Medicine, Volume 25, Issue 8, pp. 780-785, August 2010.

[7] Meet Kinect for Windows, https://developer.microsoft.com/en-us/windows/kinect, [retrieved: March, 2016]

[8] Open CV, http://opencv.org/ [retrieved: March, 2016]

[9] Nathan Burba, Mark Bolas, David M. Krum, and Evan A. Suma, "Unobtrusive Measurement of Subtle Nonverbal Behaviors with the Microsoft Kinect," IEEE, Virtual Reality Short Papers and Posters (VRW), pp. 1-4, 2012.

Co-creating with Stakeholders: Ideating eHealth Applications to Support Antibiotic Stewardship in Hospitals

M. van Limburg, H.C. Ossebaard, M.J. Wentzel, J.E.W.C. van Gemert-Pijnen
Center for eHealth Research, Department of Psychology, Health and Technology, Faculty of Behavioral, Management and Social sciences, University of Twente, Enschede, the Netherlands
a.h.m.vanlimburg@utwente.nl; h.c.ossebaard@utwente.nl; m.j.wentzel@utwente.nl; j.e.w.c.vangemert-pijnen@utwente.nl

Abstract— **Inappropriate prescription of antibiotics can lead to complications with hospital infections and increased antimicrobial resistance. Antibiotic stewardship programs have been developed to influence antibiotic prescription behavior in hospitals at a multidisciplinary level. Infectionmanager.com is part of such a program. It is an online platform for stakeholders in the antibiotic therapy process that provides eHealth applications tailored to their roles and needs. This paper focuses on the process of stakeholder involvement to design an implementation strategy for our infectionmanager.com. We used business modeling tools in a focus group and in individual interviews. By analyzing the antibiotic therapy process with a value-driven dialogue with all stakeholders, we ideated possible eHealth technologies based on what stakeholders find valuable in their daily practices. We also conclude an implementation strategy as a basis for human-centered design to develop these eHealth technologies with end-users that can be used as a basis to sustainably implement the platform in its intended setting.**

eHealth technology; implementation; stakeholder co-creation; business modeling; antibiotic prescription.

I. Introduction

Increasingly antimicrobial resistance threatens the health and safety of patients and citizens and is, therefore, a major concern for public health authorities [1]. Hospital-acquired infections resistant to antibiotics yearly cause an estimated 4 million infections in Europe, resulting in around 37,000 deaths per year [2]. A change in prescription behavior is imperative since up to 50% of all prescribed antibiotics is reported inappropriately prescribed, avoidable or affects the effectiveness of these antibiotics [3]. Infection control experts have developed antibiotic stewardship programs (ASP) or antimicrobial stewardship interventions (AMS) to ensure prudent use of antibiotics for better patient outcomes, lower risk of adverse effects, promotion of cost-effectiveness and reduction of resistance levels [4].

As part of EurSafety Health-net project we co-created with relevant stakeholders, an online infection control platform called infectionmanager.com. Stakeholders are defined as all people or organizations that are influenced by or influence the eHealth technology used to reinforce antibiotic stewardship [5]. Its purpose is to provide a platform with several eHealth technologies that support ASP and implementation advice for all stakeholders involved with infection prevention and control and antibiotic therapy. We focus on a bottom-up approach, where stakeholders are actively involved to determine what eHealth technologies are deemed valuable for ASP.

Many eHealth technologies floundered because they failed to involve the intended users [6]. To be successful, implementation research for eHealth should start involving stakeholders [7]. They need to be involved ab initio in designing eHealth technology and participate in its implementation [6, 7]. Stakeholder interaction, stakeholder relationships and added value(s) offered through the eHealth technology need to be understood for determining a fitting implementation.

We had the following research objectives: 1) Understanding the antibiotic therapy process from stakeholder-perspective and identifying the stakeholders; 2) Understanding problems that stakeholders encounter in the antibiotic therapy process; 3) Identifying which improvements and opportunities stakeholders see in this process; 4) Ideating what eHealth applications are required; 5) Designing an implementation strategy for a platform with eHealth applications to support ASP.

Section II introduces the methods used to involve stakeholders in our business modeling research, using focus groups and interviews. In section III, we provide the business modeling results of our process and problem analysis (objectives 1 to 3) as well as ideated eHealth opportunities and implementation strategy inferred from values found in co-creation research with stakeholders (objectives 4 and 5) [8]. In sections IV and V, we discuss and conclude that stakeholder involvement and business modeling adds to the existing ASP guidelines, as it helps to understand problems with antibiotic therapy from a bottom-up perspective.

II. Methods

A. CeHRes Roadmap

Central to our research is the Center for eHealth research (CeHRes) roadmap. It supports researchers to develop, design and implement eHealth technologies using a holistic approach that combines human-centered design principles and business modeling principles [6]. In this paper, we focus primarily on its business modeling research activities. Our method of business modeling is all about determining what the added-value of the eHealth technology should be together with the stakeholders. Subsequently, this co-created added-value is the foundation for an implementation strategy for the technology. By arranging value-driven dialogues with stakeholders, researchers can discuss and understand value needs [8]. We define a value as any ideal or interest a stakeholder aspires with regard to an eHealth technology [9]. In this paper, we used business modeling tools that are part of the CeHRes roadmap, to co-create and ideate possible interventions based on value needs of stakeholders.

In short, a problem analysis (applied in the form of a focus group) was used to understand the antibiotic therapy process from a stakeholder's perspective and to understand which stakeholders are most influential in the antibiotic therapy process. This problem analysis was also used to analyze problems that stakeholders face in the antibiotic therapy process. These problems and bottlenecks were the starting point for value-driven dialogues. We used follow-up interviews to discuss improvements and opportunities and possible eHealth applications to support ASP and the value needs and to organize these needs in an implementation strategy.

B. Focus group with scenarios

We organized a focus group with stakeholders in a pilot hospital participating in the EurSafety Health-net project [10]. Based on a literature scan and expert recommendations, we selected healthcare professionals from the pulmonary ward: clinical microbiologist, clinical pharmacist (2x), chest physician (2x), residents (2x), nurse, ward manager, nurse manager and quality manager. We prepared a complex scenario, known as scenario-based testing [10], with a fictive patient to invoke discussion over the exact choice of antibiotic therapy. With the scenario, we discussed the roles of stakeholders, which communication and information needs are present and critical issues. The focus group was recorded and transcribed in Excel for analysis.

C. In-depth interviews

We organized semi-structured one-on-one interviews of one hour with a resident, a clinical microbiologist, a nurse manager and two ASP experts to further specify their views on the added-value of ASP with eHealth. As we focus on implementation, we addressed the following topics to discuss possible value needs: what should the added value of the interventions be? Who need ASP and how can they be involved? What is required in terms of infrastructure and resources? Who pay for ASP and what are its benefits? We discussed value needs and organized these in a possible business model for ASP with the interviewed stakeholders. The interviews were recorded and transcribed in Excel for analysis.

III. Results

A. Understanding the antibiotic therapy process and identifying the stakeholders

The focus group concluded the primary stakeholders in every form of therapy are physician, patient and nurse. A physician prescribes medication and a nurse usually administers all medication. In the case of prescribing antibiotics, a clinical microbiologist may be consulted by a physician or resident for interpreting laboratory results or advice with non-routine antibiotic therapy. Clinical pharmacists check all prescribed medications, including antibiotics. They also occasionally give extra advice to physicians or nurses or provide background information on medication. Nurses use this information regularly when administering antibiotics. This list of key stakeholders is comparable with the ASP stakeholders suggested by experts in available literature [4, 11] Other stakeholders in the focus group stated they play a more facilitating role and do not directly influence antibiotic therapy. They, however, facilitate the other stakeholders in terms of organizational aspects, resources and support with protocols.

B. Understanding problems that stakeholders face in the antibiotic therapy process

We report the results of the problem analysis in three topics: communication, information/documentation and critical moments/bottlenecks in the process:

Communication: Stakeholders expressed two important moments of communication that are vital in antibiotic therapy: 1) Physicians need information from the clinical microbiologist or clinical pharmacist usually communicated over phone. Contact with a microbiologist or pharmacist should be 24/7 possible; 2) Nurses take daily care of patients and frequently need patient-specific instructions from a physician and occasionally need additional information concerning intravenous delivery from protocols, a physician or a resident (face-to-face), or a clinical pharmacist (phone). A common communication problem expressed by the focus group is that communication by phone that information gets 'lost in translation' as the clinical view is explained from the perspective of the physician.

Information/documentation: All stakeholders required information sources and documents. They reported information comes in mixed forms, either digitally in intranet information systems or hardcopy in folders or pocket cards. A few common complications were mentioned: 1) Finding the right information can be problematic due to multiple and different information sources and require research; 2) The patient information system is not directly accessible by the microbiologists thus they rely on patient information shared verbally by the physician; 3) Culture exchange and laboratory results on the cultures is prone to delays. The faster this information is available; the faster antibiotics can be adjusted.

Critical moments/bottlenecks: The most important critical moments regarding patients' health are reported to be in the therapy process itself (diagnostics, prescription, administration), in the communication of laboratory results (correctness and timeliness) and the physical responses of the patient on the therapy (timely adjustment of medication). Stakeholders also expressed critical issues apart from patient-related ones: 1) Antibiotics should be stocked; 2) (Too) many diverse information sources for protocols, information, etc.; 3) Delays in sending cultures and communication of laboratory results cause prolonged antibiotic therapy that may be unnecessary; 4) Laboratory results are sent to the physician/resident who requested them, which is not ideal when physicians/residents share the care of patients and the one who requested the results is not available or present. 5) There is a focus on efficiency; which means available resources, time and personnel are limited and a balance has to be found in improving quality with ASP and efficiency; 6) Communication per phone can have a negative effect on quality of antibiotic therapy if not all information is shared. The more informed the decision making for therapy is, the better.

C. *Identifying which improvements and opportunities stakeholders see in this process*

Stakeholders mentioned the following improvements and opportunities how the antibiotic therapy process can be improved: 1) A bed-side audit consisting of a physician and microbiologist would improve clinical assessment for appropriate antibiotic therapy (apps #4, #5, #6, see D). Clinical pharmacists would also benefit from joining this bed-side audit to assist with prescription details. A bed-side audit is not needed at every prescription, yet with complex treatments it might be beneficial to visit the patient as a team; 2) Looking for uniformity in protocols and provided information can also be beneficial for antibiotic therapy (apps #1, #2, #3, see D). In fact, a resident stated many available protocols could be replaced with a solid uniform one; 3) Education to disseminate new protocols, changes in protocols or generally news regarding antibiotics could improve knowledge and awareness of ASP (app #7, see D).

Two desired improvements were also mentioned that are outside the scope of an antibiotic stewardship program as they are hospital-wide issues that would be beyond our ability to change: 1) Improvement in timely logistics with the microbiology laboratory, by adequately sending cultures to the laboratory and getting timely results might be needed; 2) Current IT systems require better data connectivity and information sharing.

D. *Ideating what eHealth applications are required*

Based on the antibiotic therapy process and problem analysis, we ideated the following eHealth opportunities for the infectionmanager.com with stakeholders, available guidelines and literature and expert opinions. The following paragraphs provide an overview of opportunities for eHealth applications based on what stakeholders deem helpful in the antibiotic therapy process translated into a short, general description of each eHealth application.

1) *App #1: Antibiotic prescription information*

An important step is the start of this therapy, known as empiric therapy. For common infections or antibiotics, the physician can rely on his/her experience, yet for certain infections or antibiotics, extra information is needed to verify the right antibiotic, dose and duration. Physicians and residents usually use an antibiotic formulary that contains most of this information. Additionally, there are national guidelines and local guidelines and protocols that can be used as well. An antibiotic prescription information application can bring all these sources of information together in one place so that physicians have to search less. Another strong point of this application is that it can fulfill the need for uniformity in protocols and information if content of the application has a consistent presentation.

2) *App #2: Antibiotic prescription decision support*

In the pilot hospital, another intervention was recently implemented called 'Surviving sepsis'. This is a little pocket card to help physicians and residents signal life-threatening infections by scoring a few parameters in a checklist. A possibility for an ASP application is to replicate this pocket card with an antibiotic prescription decision support that can make this checklist go more in-depth towards, e.g., suggesting possible infections and possible therapies. This application can be an expansion of the antibiotic prescription information application.

3) *App #3: Antibiotic administration information*

Nurses administer antibiotics to patients. With common antibiotics, the delivery is done on experience, yet for uncommon antibiotics, information is needed regarding the delivery. For example, the exact flow rate of an intravenous antibiotic or if an antibiotic has to be given before or after dinner. Nurses - and occasionally also physicians and residents - check an information system of the pharmacy or national guidelines and local guidelines and protocols. Digital sources can be accessed via the Computer-on-Wheels but some protocols are available as printed copies. The antibiotic delivery/administration information application can bring all these sources of information together in one place. That way nurses, physicians and residents do not have to search in multiple sources. Also, this application can fulfill the need for uniformity in protocols and information, if content of the application has a consistent presentation.

4) App #4: Information patient care transfer

The care of patients is transferred between multiple disciplines: Physicians transfer the care to nurses or physicians can transfer among each other or to residents. An opportunity for an application can be to provide an infection-specific or antibiotic-specific checklist of important therapy details that need to be shared during transfer. A nurse gave as an example that he sometimes had to verify with the attending physician or resident when to stop therapy or that therapy was continued longer than officially stated in protocols.

5) App #5: Facilitating the team/audit

A possible ASP application is a bed-side assistance application to quickly access information: 1) Patient information (primary parameters (age/weight), history, allergies); 2) Antibiotic medication information (antibiotic formulary, guidelines); 3) Laboratory results.

This bed-side assistance application can be useful for microbiologists, infectious disease physicians, pharmacists and physicians (depending on the ASP team formation) to obtain information while doing audit rounds. Providing antibiotic medication information is easiest to manage as the technology and content will have much in common with the information applications described above. Patient information and laboratory results, however, require connectivity with existing IT systems or some manual preparation beforehand.

6) App #6: Alerts/notifications

Within the antibiotic therapy process there are a few critical moments where an alert or notification application can be helpful: 1) In the reviewing process, done by clinical pharmacists, an alert or messaging system for important messages per receipt can be used to notify physicians to re-evaluate therapy or to provide patient-specific information such as conflicting medications. This can be combined with a restriction-approval strategy as suggested as a possible ASP strategy in IDSA/SHEA guidelines for implementing antibiotic stewardship programs in hospitals [11]; 2) When laboratory results are available for a certain patient, the currently attending physician or resident can be notified in addition to sending results directly to the physician or resident who requested them; 3) When an antibiotic is prescribed and administered, after a certain time a re-evaluation is in order. This is described as a day-3 bundle that after two or three days the effectiveness of antibiotic therapy can be assessed [12]. At that point, the prescribed antibiotic can be continued, adjusted or even stopped. An eHealth opportunity here is to give an alert or notification when the re-evaluation should take place or co-create an application providing a daily list of antibiotic therapies to re-evaluate.

7) App #7: E-Learning

In the focus group, it became apparent that education is important. Also in ASP guidelines, education is an important supplementary strategy in implementing ASP interventions [11]. For every educational element of ASP there is an eHealth possibility to provide that education using E-Learning applications. The following educational activities were deemed interesting when implementing ASP applications: 1) All personnel needs to be informed about the importance of ASP in general to gain awareness and understanding; 2) All personnel needs to be informed why a team is performing audits; 3) Physicians, residents, nurses, microbiologists, infectious disease physicians and pharmacists should stay up-to-date with new information, guidelines, protocols, etc.; 4) Training in using other eHealth applications implemented for ASP.

E. Ideating an implementation strategy for a platform with ASP applications

In the follow-up interviews, we discussed with stakeholders and two ASP experts what the expected-added value of the platform and eHealth applications should be. All stakeholders unanimously agreed that the most important benefit from optimized antibiotic therapy will be a reduction in length-of-stay. The length-of-stay of patients will reduce when they have optimal therapy, can go home sooner with oral antibiotics, and have less risk of complications with infections. This has beneficial consequences for the quality of care as well as less antibiotic use and thus less antibiotic costs.

All stakeholders also agreed that hospital management needs to be convinced that ASP and subsequently using the infectionmanager.com platform and its eHealth applications is beneficial. There is supportive evidence in literature of already existing ASPs that they are beneficial [13], however, the role of technology in ASP is rather limited. Using eHealth applications as part of an ASP in the hospital may lead to improvements when integrated with ASP initiatives within the hospital. Proving the beneficial effects of individual parts of a program is difficult as results are always reported over a program consisting of multiple interventions. Therefore, the platform can be implemented as part of (starting) ASP initiatives, but requires these existing ASP initiatives as a prerequisite. It can be part of a program and the team can choose which eHealth applications they want to embed in the ASP in their hospital. Nonetheless, the platform can also improve its own value by providing information how to set ASP initiatives up using the platform and its eHealth applications to facilitate teams with the introduction of the eHealth applications in their program.

Based on the above, the strongest business case for the hospital management is that this reduced length-of-stay that will reduce costs and improve patient safety. One interviewee said the hospital management needs to understand that "investing at the beginning will pay off at the end". The investment mostly requires FTEs and resources for stakeholders to prepare and perform ASP activities. Also for the platform and eHealth applications, minimal costs are necessary for general maintenance. The revenue is in cost reductions: all costs related to length-of-stay can be reduced but also when speedier interventions in antibiotic therapy can be made, complications can also be reduced and thus reduce costs and negative effects for the patient.

IV. Discussion

We involved stakeholders in our research to ideate eHealth applications and a possible implementation strategy for improving the antibiotic therapy process in hospitals. Stakeholders expressed complications mostly with information sharing and, related, communication. With a focus group and interviews, we had value-driven dialogues with stakeholders to assess opportunities for eHealth. In this article, we conclude seven applications that are worthwhile to implement to support antibiotic therapy processes in a hospital. In sum, these applications mostly need to provide therapy or patient specific information to assure optimal therapy.

With a growing body of literature of expert recommendations on antibiotic stewardship [3, 4, 11, 14], the available ideas for interventions for ASP are expanding. Nonetheless, these expert recommendations still need to be implemented locally in every hospital. Despite guidelines, these local implementations are still diverse and Patel et al pose that little guidance is offered on the practical aspects of implementing ASPs and that e.g., non-academic hospitals in the US need to overcome implementation issues by accounting for unique characteristics of their institutions [15]. The strong point of our bottom-up focus is that by combining value-driven dialogues with available guidelines, we assessed which interventions are required and supported by the stakeholders themselves. The stakeholders, therefore, added a local relevance to the possible eHealth technologies.

ASPs can be very comprehensive programs and contain multiple interventions for basically any process where antibiotics are involved. We decided to focus on primary care processes in the hospital and specifically the antibiotic therapy process as a first focus to implement possible interventions for ASP as one of the key pillars of antibiotic stewardship is a more optimal prescription of antibiotics.

When we looked in a review for the use of eHealth in ASP in literature, we noticed eHealth technology is rarely mentioned or used. Some ASPs make use of existing software systems like electronic prescribing, electronic patient records, but few technological tools specifically designed for ASP were present. eHealth is attributed to helping efficiency and thus, can be helpful to optimize efforts and resources for ASP. Especially as manpower and (financial) resources are most attributed barriers that hinder ASP implementation.

The first step in our ASP research was to prepare a plan of action for possibilities to implement eHealth applications in ASP. By putting the antibiotic therapy process central, we identified stakeholders that fit the local, Dutch context and daily processes in the piloting ward. Also, the seven possible eHealth applications to support ASP were ideated based on this process. These ideated applications are input for further research and development to design eHealth applications that fit the needs of end-users and that fits the overall goals of an ASP. Further research is also needed in the conditions necessary to implement ASP and further analysis and co-creation with stakeholders is necessary to determine a fitting implementation strategy for ASP.

Currently, there are three information applications in development. An antibiotic delivery/administration information application for nurses (app #3) finished its pilot with nurses at the pulmonary ward and is implemented in other participating hospitals [16]. Residents saw this application and asked whether they could have a spin-off information application. We also developed a similar information application called the antimicrobial therapy information application (app #1), containing German antibiotic therapy information tested in three German hospitals. Preparatory research for an antibiotic prescription decision support application (app #2) is in progress. The research and development of the other remaining possible ASP applications are still to be planned.

eHealth applications available via infectionmanager.com need to be part of a larger, more comprehensive ASP. Commonly these ASPs require an ASP team, surveillance of antibiotic use and infections, local guidelines, education and audits) [4, 11]. An implementation strategy for the infectionmanager.com therefore has to align with starting ASP initiatives. Helpful would be to add implementation advice inside the infectionmanager.com how to embed eHealth applications to support starting ASP initiatives. We are also working on an additional implementation tool to help to implement these apps in a wholesome ASP.

Limitations in this research are that we held a focus group in one ward in one hospital, who offered to participate in our research. Although some problems may be local and hospital-specific, we decided to provide generic tools that focus on the structure of presenting information and knowledge. The content of these eHealth applications can be hospital-specific and altered as hospital professionals see fit. Later, we also expanded our research to multiple hospitals to implement the applications elsewhere as there was interest in trying the applications in other ASPs. Based on this expansion we can test whether the structures are robust and how the content changes.

When we started our ASP research, very few hospitals were active with ASP. That was also the main reason why we collaborated with the pulmonary ward as they offered to be part of our pilot. Later when interest in ASP arose at other hospitals, mainly due to new guidelines in The Netherlands, we got more interest in our eHealth applications as well.

In future research, we will expand our research to other hospitals and research the necessary components for ASP to help hospitals with their implementation of ASP and the eHealth applications available via infectionmanager.com. For designing these applications, we used the ideas of the ideated eHealth technologies and we use the principles of the CeHRes roadmap, human-centered design for requirements engineering and designing the technology for end-users and business modeling for stakeholder-based value-creation to embed the eHealth applications in ASP [6-9].

V. Conclusion

This example case of infectionmanager.com demonstrates the use of business modeling methods to assess possible eHealth technologies for ASP. We argue for more awareness among researchers who design and develop technologies targeted for healthcare that they should not only focus on designing great eHealth technology but also focus on the implementation of these technologies. This infectionmanager.com example case demonstrated that analyzing the implementation of technology early on the in any eHealth technology development process contributes to the understanding of what the added-value of this technology should be. In terms of our example case, it adds up to the available expert guidelines (that also hardly deal with eHealth) for antibiotic stewardship. There is a role for eHealth in ASP when one looks at the opportunities how technology can improve processes. These applications are worthwhile to develop for ASP and a basis for an implementation strategy for eHealth applications within ASP.

References

[1] World Health Organisation. "Antimicrobial resistance - Fact sheet N°194" http://www.who.int/mediacentre/factsheets/fs194/en/ [retrieved 10-12-2015].

[2] European Centre for Disease Prevention and Control. "Healthcare-associated infections." http://ecdc.europa.eu/en/healthtopics/Healthcare-associated_infections/pages/index.aspx. [retrieved 10-12-2015].

[3] Owens Jr, RC, Ambrose, PG. "Antimicrobial stewardship and the role of pharmacokinetics–pharmacodynamics in the modern antibiotic era". Diagnostic microbiology and infectious disease, 2007. 57(3): p. S77-S83.

[4] MacDougall, C, Polk, RE. "Antimicrobial stewardship programs in health care systems". Clinical microbiology reviews, 2005. 18(4): p. 638-656.

[5] Freeman, RE. "The stakeholder approach revisited". Zeitschrift für Wirtschafts-und Unternehmensethik, 2004. 5(3): p. 228-241.

[6] van Gemert-Pijnen, JEWC, Nijland, N, Ossebaard, HC, van Limburg, AHM, Kelders, S, Eysenbach, G. "A holistic framework to improve the uptake and impact of eHealth technologies". Journal of medical Internet research, 2011. 13(4).

[7] van Limburg, M, van Gemert-Pijnen, JEWC, Nijland, N, Ossebaard, HC, Hendrix, RMG, Seydel, ER. "Why business modeling is crucial in the development of eHealth technologies". Journal of medical Internet research, 2011. 13(4).

[8] van Limburg, M, Wentzel, MJ, Sanderman, R, van Gemert-Pijnen, JEWC. "Business modeling to implement an infection control eHealth portal: a reflection on co-creation with stakeholders". JMIR research protocols 2015;4(3):e104

[9] Van Velsen, L, Wentzel, J, van Gemert-Pijnen, JEWC. "Designing eHealth that matters via a multidisciplinary requirements development approach." JMIR research protocols, 2013. 2(1).

[10] Wentzel, J, van Limburg, M, Karreman, J, Hendrix, R, van Gemert-Pijnen, L. "Co-creation with stakeholders: A web 2.0 antibiotic stewardship program." The Fourth International Conference on eHealth, Telemedicine, and Social Medicine. 2012.

[11] Dellit TH, Owens RC, McGowan JE, Gerding DN, Weinstein RA, Burke JP, Huskins WC, Paterson DL, Fishman NO, Carpenter CF, Brennan PJ. "Infectious Diseases Society of America and the Society for Healthcare Epidemiology of America guidelines for developing an institutional program to enhance antimicrobial stewardship." Clinical infectious diseases. 2007. 44(2):159-77.

[12] Pulcini C, Defres S, Aggarwal I, Nathwani D, Davey P. "Design of a 'day 3 bundle' to improve the reassessment of inpatient empirical antibiotic prescriptions". Journal of antimicrobial chemotherapy, 2008. 61(6): p. 1384-1388.

[13] Davey P, Brown E, Charani E, Fenelon L, Gould IM, Holmes A, Ramsay CR, Wiffen PJ, Wilcox M. "Interventions to improve antibiotic prescribing practices for hospital inpatients". Cochrane Database Syst Rev, 2013. 4.

[14] Centre for Disease Control. "Core elements of Hospital Antibiotic Stewardship Programs" http://www.cdc.gov/getsmart/healthcare/implementation/core-elements.html [retrieved 10-12-2015].

[15] Patel, D, MacDougall, C. "How to make antimicrobial stewardship work: practical considerations for hospitals of all sizes". Hospital Pharmacy, 2010. 45(1): p. S10-S18.

[16] Wentzel, J, van Velsen, L, van Limburg, M, de Jong, N, Karreman, J, Hendrix, R, van Gemert, JEWC. "Participatory eHealth development to support nurses in antimicrobial stewardship". BMC medical informatics and decision making, 2014. 14(1): p. 45.

The Grid, Classification of eHealth Applications Towards a Better (re)Design and Evaluation

Saskia M. Akkersdijk*, Saskia M. Kelders†, Julia E. W. C. van Gemert - Pijnen‡ and, Louise M. A. Braakman - Jansen §
Center for eHealth Research & Disease Management
University of Twente,
Enschede, the Netherlands
Email: {s.m.akkersdijk*, s.m.kelders†, j.vangemert-pijnen‡, l.m.a.braakman-jansen§}@utwente.nl

***Abstract*—eHealth holds a diverse range of successful and unsuccessful applications. It is often unclear why this difference in success appears because eHealth is usually approached as a black box. Additionally, we tend to evaluate non-treatment-like and treatment-like applications the same way. Both approaches focus on searching for effects and outcomes. This results in applications that are wrongfully put away. Placing an application on the continuous dimensions of use-structure and caregiver involvement on the grid presented in this paper helps to make conscious and better decisions when (re)designing and evaluating eHealth applications. Positioning on the grid influences the terms 'user' and 'usage' but also has implications for the way we can evaluate and (re)design eHealth applications. The grid is a tool to gain insight, facilitate thought processes, and start discussions, and is not meant to be a formal and rigid model. This tool helps making conscious choices in (re)design and evaluation of applications.**

***Keywords*—eHealth; classification; evaluation; design.**

I. Introduction

In eHealth, there is a diverse range of successful and unsuccessful applications and interventions. When observing these applications we observe that applications that represent a form of treatment tend to be more successful than those that are more supportive of nature. Often it is unclear why one application is successful, when the other is not. A reason why we cannot always explain the difference in success is because eHealth is often approached as a black box. With a black box we search for its effects and focus on outcomes. It would be better to examine eHealth technology from a holistic perspective, in which the technology itself also has value, and we focus on the mechanisms behind the success. To find these mechanisms it is necessary to open the black box. An important reason why we observe differences in success is that we evaluate non-treatment-like applications the same way we evaluate the treatment-like interventions, focused on outcome measures or usage numbers. This results in applications that are maybe wrongfully put away because they do not measure up to the measurements they are wrongfully compared to. In this paper we search for a tool to give a better inside in the application, which helps with a (re)design and evaluation of that application.

One of the ways applications are often evaluated is measuring to which extent therapies are followed as intended. This measurement of adherence is one of the primary determinants of success in treatments [1], and overall effectiveness of health systems decreases by poor adherence [1]. Although adherence might be one of the primary determinants of success in therapy there are also examples of applications with a low adherence that are successful. An example of this is QuitNet; a program for smoking cessation [2][3][4]. Adherence to this program is low (23%) [5], but the program can be successful in promoting cessation and preventing relapse [2]. These studies show that it is possible for an eHealth application to have a low adherence but still be successful for a certain group of users.

Studies with eMentalHealth interventions often find a high dose - response relationship (also called a usage - outcome relation). An example of this is the study of Bolier et al. [6]. Donkin et al [7] further explored the usage - outcome relation. The study of Donkin investigates which usage metrics are important in predicting and explaining outcomes for an internet-delivered trail targeting depressive symptoms for those with risk factors for or diagnosis of cardiovascular disease (Cardiovascular Risk E-couch Depression Outcome (CREDO)). Their study shows that there is not always a linear dose-response relation, but could be curvilinear (e.g., reaches a saturation point where no further benefit is obtained), or even more complex.

There is a broad range of different eHealth applications and variety in how these applications should be used. These variations can be put on a continuous scale. At one end of the continuum (see Figure 1) we see applications forcing the user to use the intervention in a specified way, for example a fixed use of the modules of an intervention. An example of such an application is the Web-based 'Living to the full' intervention. 'Living to the full' consists of nine lessons which have to be completed in a specific order in a 12-week period [8]. These applications are often a (web-based) program of a method, course or intervention.

At the other end of the continuum we see applications that leave the usage free, without a strict protocol for each user. QuitNet, the application that was mentioned before for cessation treatment, is such an application. This website offers advice and assistance to quit smoking. Usage frequency of the program and how the program is used is left to the user.

Another important factor that varies among different eHealth applications is the involvement of a caregiver. Some eHealth applications are used in close collaboration between patient and caregiver, others with no involvement of a caregiver at all and all variations in between. Caregiver involvement is often found to be necessary to ensure adherence and increase effects for web-based interventions for people with depression symptoms, or chronic conditions [9][10][11][12][13][14].

Knowing where your application is positioned on these two dimensions can help with (re)designing and evaluating your application. These two dimensions form a grid, and applications can be put somewhere on this grid depending on its usage-structure and caregiver involvement. The positioning of an application on the two dimensions influences the term 'user' and 'usage' but also has implications for the way we can (re)design and evaluate the application. The aim of this paper is to present a tool to give a better inside in the application, which helps with a (re)design and evaluation of that application.

In Section 2 (The Grid), we will take a closer look at the grid, after which we will discuss implications based on the different positions an application can take on the grid in Section 3 (Implications). We will end this paper with a discussion and conclusion in Section 4 (Discussion and Conclusion).

II. The Grid

Based on our observations of eHealth applications and the extremes we see, we could classify each application on the following grid based on their characteristics:

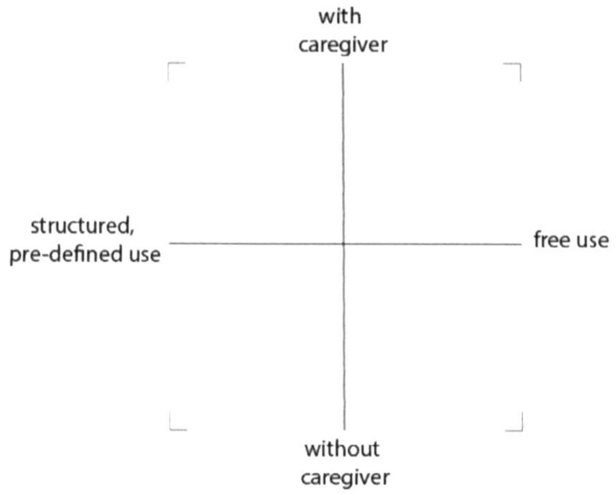

Fig. 1. The grid

In this section, we will take a closer look at the two dimensions of the grid. We will first look at the dimension of use-structure, after which we will look at the dimension of caregiver involvement. Finally, we will describe some eHealth applications and their position on the grid.

The dimension of use-structure has at one end of the continuum applications that force the user to use the intervention in a specific way(railroad them). This can be in a specific order, for a specific number of times/lessons, or for a specific duration. These interventions often have a specific end that is known beforehand and are often based on theories about mental health behavior like acceptance and commitment therapy (ACT) or cognitive behavioral therapy (CBT). Because they often find their origin in know theories and therapies they are often more 'treatment' like and help deliver a form of short term care. As discussed in the introduction, 'Living to the full' is a good example for this end of the continuum. 'Living to the full' has been published as a self-help book [15] and is based on mindfulness [16][17] and ACT [18]. The intervention consists of nine lessons which have to be completed in a specific order in a 12-week period. Whether participants worked through a lesson in one session or in multiple sessions was up to them [8].

At the other end applications that leave the usage free without a strict protocol for each user. There is no specific order or duration for which this application should be used, therefore they have no specific end. These free-to-use applications often focus more on support and long term care. As discussed in the introduction, 'QuitNet', the application for cessation treatment, is a good example for this end of the continuum. This website offers advice to quit smoking, assistance in setting a quit date, tailored information, assessment of motivation and nicotine dependence, practical counseling (skills training and problem solving), tailored assistants in selecting pharmacotherapies and intra- and extra-treatment social support. How QuitNet is used is completely up to the user.

The vertical dimension represents a form of caregiver involvement, which varies among eHealth applications. For example treatment-driven applications involve caregivers usually, while lifestyle interventions can be used autonomously. Research finds caregiver involvement important, but it is not clear what the dosage and frequency of involvement should be [9][10][11][12][13]. With applications that target people with chronic conditions, there often is some form of caregiver involvement. However, these applications often struggle to find their fit into daily life, and adherence is often low [19]. Users find it difficult to embed these application in their own life, while caregivers struggle to embed them into their daily practice [20]. Nonetheless, caregiver involvement is often found to be necessary to ensure adherence and increase effects for web-based interventions for people with depression symptoms, or chronic conditions [14].

To illustrate the positioning of an eHealth application on the grid, we will now position 'Living to the full' on the grid, after which we will describe another application ('My Health Platform') and its position on the grid.

As discussed in the introduction,'Living to the full' (LttF) consists of nine lessons which have to be completed in a specific order in a 12-week period. The intervention is used without involvement of a caregiver.

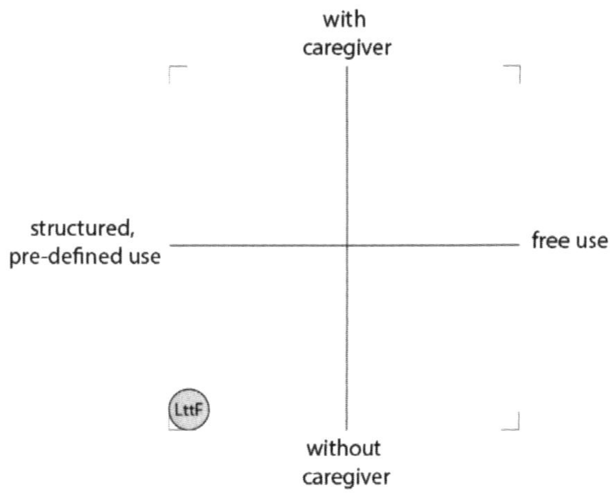

Fig. 2. Positioning of 'Living to the full' on the grid

We would place 'Living to the Full' at the bottom of the left corner on the grid for the following reasons:

- horizontal dimension: Usage of 'Living to the Full' (such as how it is used, how often) is pre-defined. Exactly when (time) a lesson is completed is left to the user. Therefore we would place 'Living to the Full' just a small bit to the right out on the horizontal dimension.
- vertical dimension: 'Living to the full' is a standalone program without caregiver support, usage is completely left to the user. Therefore we would place 'Living to the Full' completely at the bottom of the vertical dimension.

'My HealthPlatform' (MHP) is an online platform to support self-care and self-management for people with a chronic illness (e.g. increased cardiovascular risk, COPD, Diabetes mellitus type 2). It is designed to help users keep an overview of and be a director of their own health and lifestyle, alone or in cooperation with a caregiver or expert. In MHP they can monitor their health, find information about their condition, but also use one of the lifestyle coaches. While the usage of most of the platform is unstructured, the coaches follow a 12 week schedule.

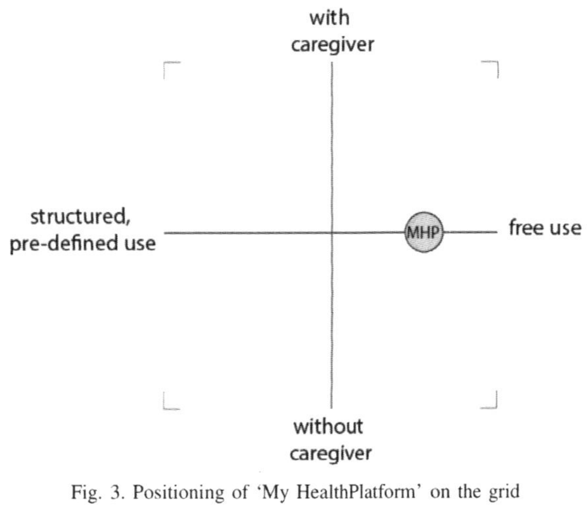

Fig. 3. Positioning of 'My HealthPlatform' on the grid

We would place 'My HealthPlatform' at the right-hand side, in the middle of the vertical dimension on the grid for the following reasons:

- horizontal dimension: Usage of MHP (such as how it is used, how often, and whether or not a coach is used) is left to the user. We would not place MHP completely at the right side, because the coaches do require the user to use them in a specific way and for a predetermined number of weeks.
- vertical dimension: Usage of MHP is mostly left to the user. When MHP is used in cooperation with a caregiver, the caregiver is able to see at home measurements of the user, which provides more insight in the health status of their patient.

You can position MHP on a different position on the grid based on other arguments. In this case, especially the vertical dimension of the grid leaves room for discussion. We would like to emphasize that when we would ask multiple people to position the same application on the grid we are very likely to end up with as many different positions as we asked people. We would like to argue that this is perfectly fine, because the main purpose of the grid is to help you think about certain characteristics of you application and about the implications of the positioning on the grid. In the next section we will talk about some of these implications of the different positions on the grid.

The examples discussed did not have a form of caregiver involvement that in our argumentation would be positioned in the upper halve of the grid. An example that would be positioned in the upper left corner of the grid could be an application that only is used in a face-to-face consult with a caregiver in which the 'patient' and caregiver together work through a fixed number of modules in a fixed order. An example that would be positioned in the upper right corner of the grid could be an application in which you could consult your caregiver whenever you have question about your health problem.

III. Implications

Positioning on the grid has several implications for the terms 'usage' and 'user' and for the (re)design and evaluation of eHealth applications. In this section, we will discuss some of these implications. We will start with the implications on the terms 'user' and 'usage' after which we will discuss implication for (re)design and evaluation.

A. User

Defining when someone is a user is quite clear when you are dealing with applications that are on the left-hand side of the grid. A person that uses the application is a user, and one who does not is not a user. With applications that leave the usage up to the user the way people use the application can vary widely, which leads to a discussion about the term 'user' in this context. We will discuss some questions around the term user, after which we will give our view on the answers.

An important question is: when does a person become a user of the application? There are several possible answers to this. We could argue that a person who uses the application is a user, but is there a minimum amount of usage before that person becomes a user, or is 10 seconds enough? And what about someone who does not use the application for a long period of time? Is that person not a user during this period? And could we define certain activities in the application that a person must have done before that person is marked as a user?

For applications that focus on monitoring health or increasing health awareness (mostly positioned on the right-hand side of the grid), we can argue that by only becoming aware of such an application a person could potentially be triggered to become more involved in his own health. This means that, in order to have an effect on a person, it does not automatically require that person to use that application. Is this person then a user? We might argue that this person is not a user of the application because he/she did not interact with it. However, the application could still have an effect. In this case the person is not a user in the most common sense of the word, but due to the effect that the application had it balances on the edge of the definition of 'user'.

When we consulted the people who used MHP (Figure 3), it became clear that they had their own view on being a user. There were quite a lot of people who had used the application only a couple of times and therefore declined to join several studies (interviews, questionnaires, and usability testing) because they did not see themselves as 'users'. In their minds their definition of a user involves a certain number of reoccurring visits to the application, entering some monitoring data into the system, or participating in the program of a coach. Because they did not meet their own standards of the term user, they thought they could not participate in the study. This example shows that using the system does not equal being a user, at least not for the people who used it. People might have expectations about the intended usage of an application, it is relevant to communicate the intended use to avoid misunderstanding about the usage.

For evaluation purposes the definition of what we would call a user can focus on several aspects:

1) The percentage of registered users who see themselves as user, could be a measurement for evaluation of an application. The number of registered users who see themselves as user tells you about their involvement with an application and this in turn can show which role the application holds in their lives and whether the application helps them.
2) You do not always know beforehand who will be 'user'. With evaluations of an application it is important to define which group of people can be defined as user, this group does not always include the groups you thought of before the application was used. For one type of evaluation, questionnaire, or interview, another groups of users might be suitable.
3) An important user that is often forgotten is the caregiver. The caregiver can have his own section in the application where he can see caregiver-specific functionality, he can have its own version of the application, or he can have the same functionality as a patient user. It is important to realize that the caregiver is a user as well, a user with different needs than a 'patient'. Additionally, both users (the caregiver and the 'patient') affect each other and how they use the application, which means that both types of users should be included in the (re)design and all evaluations.

B. *Usage*

The term 'usage' (in the context of an application) can mean a lot of things, such as: How often people return to a website, paths that users follow on a website, how often certain elements on a site are used, etc. This can all be measured by logging user actions with timestamps on a website. How we can use these measurement and what they tell us about the system/application differs between the extremes on the horizontal use-structure dimension of the grid. Measuring the use of a system or application is useful and insightful for both ends of the spectrum, but evaluating this use differs and the implications/interpretations are different.

For 'railroaded' application, like 'Living to the full' (see Figure 2), usage measurements can tell you much about the applications. 'Railroaded' application, positioned at the left-hand side of the grid, often are similar or represent a therapy. The user has to follow the structure within the application, do certain actions in a certain order, and use it for a certain amount for it to be successful. Therefore, we can define 'normal', or 'ideal' use. We can compare the measured use with the way the application should be used (this can be whether someone completed the application, or the use within an application).

Achieving the goal of the application is not completely dependent on the use (the amount and which parts) for applications which leave the user free (right-hand side of the grid). The duration of usage of these applications is often longer and different situations can be seen than with the use of an 'railroaded' application. For applications positioned at the right-hand side of the grid there is no definition of 'normal' use, in quantity or in order. This is a contrast with applications positioned at the left-hand side of the grid were the 'correct' following of the structure is essential.

Because there is no prescribed use for applications at the right-hand side of the grid, we cannot measure to which extend the measured usage deviates from the optimal use. For example, the measured use of an application positioned at the right-hand side of the grid (like 'MHP' in Figure 3) could show users that were dormant for maybe months or years, after which they suddenly used it again. This is unlikely to happen in an application that is 'railroaded'. Because we cannot define 'normal use', adherence, in which we compare the occurring usage with the optimum usage, cannot be measured for applications on the right-hand side of the grid. As mentioned before, applications positioned on the left-hand side often are more treatment like, in contrast with application on the right-hand side that are more supportive and longer term. This leads to a different type of 'lessons' and therefore a different kind of use. The same difference also makes for different characteristics of the user (e.g. in type of motivation), which also leads to different use.

The occurring usage and the use-structure of an application go together. When your application is 'railroaded' and positioned at the left-hand side of the grid users have all a similar usage pattern, while with an application that leaves the use up to the user the occurring usage patterns can vary greatly.

Even though measurements like adherence are not really suitable for applications that are positioned at the right-hand side of the grid, usage measurements can still be very valuable. These measurements can tell you much about the interaction with the application, which parts are used most, which parts are often used subsequent of each other, after which part do they stop, etc. Knowing more about the interaction with the application is valuable for improving applications, but can also be valuable for finding mechanisms behind application success. Combining usage measurements with use context (what triggered the session) can be used to find a better fit of the content to the context, or improve interaction with the application. By improving the system, and better tuning it to the needs of the users (based on context en measured usage) we can increase the effect of applications.

Finally, when we are looking at the usage of an application we should not forget to observe the usage of the application by the caregiver. Caregivers play an important role in the usage of an application by their 'patients' because their usage can be driven by input of said caregiver. When a caregiver does not work with the application as intended or adequate this will influence the usage of the 'patient' user as well. When the application is mend to be used with a form of caregiver involvement and the caregiver is less involved than the 'patient' expects, the 'patient' user will experience less added value of the application.

C. Implications on (re)Design

With an existing application, the grid can be used during the redesign process. Determining the current position of the application on the grid based on the characteristics of the application can help you to reflect on your current application by facilitating the thought process about your applications and its characteristic. A first step in the redesign process is to reflect whether it is feasible to reach the objectives of the application from its current position; is it possible to accomplish the goal of the application with this position or is the position of the application on the grid not suitable for the goal of the application. The second step is to determine if the current position is the best, or if there are better alternative positions. When the current position and the desired position are known, the next step is to identify their differences. Knowing these differences, it is then possible to determine if the application should be changed and can give a indication about how the application should be changed.

Of course you can also use the grid when designing an application from scratch. The grid can help facilitate the thought process of design choices and their effects to make a better conscious decisions.

D. Implications on Evaluation

Implications on the use-structure dimension of an application are about evaluation procedures. 'Railroaded' applications can be evaluated by measuring the usage and comparing it to the optimum usage in contrast to applications on the right-hand side of the grid were usage can vary widely. Because usage can vary so widely it is also harder to link measured effects to a specific element of the application. Applications on the right-hand side of the grid are less feasible to evaluate with an Randomized Controlled Trial (RCT), because they are used for a much longer time, which makes it difficult to keep the circumstances constant. Secondly, for applications on the right-hand side of the grid it can be harder to find changes in measurements like quality of life, because they target larger groups and the changes they accomplish are often small. This does not imply that these changes are unimportant. It would be more suitable to evaluate applications that are positioned on the right-hand side of the grid on processes rather than on effects. While applications that are positioned on the left-hand side of the grid are easier to evaluate on effects, because they have a fixed setting and use-time.

IV. Discussion and Conclusion

There are different ways to classify eHealth that provide an overview of eHealth, such as device driven, based on the medium the technology uses (web-based, mobile apps, etc.), context-of-care driven (eCare, eTherapy, eAppointment, ePrevention, etc.), or actor driven (based on the interaction between the actors of such a system). The grid we propose is not meant to replace these classifications, because they provide an overview that our grid does not provide. However, our grid serves as an extension of these. The different classifications mentioned above serve a different need, while they did not serve our need for a simple way to have some guiding when (re)designing and evaluating eHealth applications. We were looking for a better way to help make a conscious choice in order to find a better fit (in (re)design and evaluation). Positioning of an eHealth application on the grid helps to become more aware of implications this has (as discussed in Section III).

Based on the position of an application on the grid we discussed that the term 'user' can include a different group of people. Those who 'use' an application do not always perceive themselves as a 'user', because they have expectations about the intended usage of an intervention. It is relevant to communicate the intended use to avoid misunderstanding about the usage. The percentage of users who see themselves as a user might be an additional measurement for evaluating an application, because it includes values about involvement with the application. An important group of 'users' that is often forgotten are the caretakers. They often also use the application, and their use or their communication about the application influences the use of the application by their patients.

Adherence is an important measurement for application positioned on the left-hand side of the application. For these applications, we can define 'normal' or intended use. Because 'normal' or intended use is often a lot harder to define for applications positioned on the right-hand side of the grid (there often is no prescribed use) and usage patterns can vary widely, the measurement of adherence is not suitable. However, usage measurements can be valuable for process evaluations and improvement of the application.

The grid can help with the (re)design process by gaining more insight facilitated by the thought process needed for placement of the application on the grid. It is important to think through whether the intended (or current) position of the application is suitable for the goal of the application, or that another position might be better.

Positioning of an application on the grid (left versus right, and with or without caregiver involvement) influences which sorts of evaluations are suitable. Evaluations can be focused on process or effects and positioning on the left-hand of the grid are more suitable for effects evaluations than positions at the right-hand side of the grid.

The grid is a tool to gain insight, facilitate thought processes, and start discussions, and is not meant to be a formal and rigid model. This tool helps making conscious choices in (re)design and evaluation of applications.

REFERENCES

[1] E. Sabat, "Adherence to long-term therapies," World Health Organization, Tech. Rep., 2003.

[2] A. Graham, N. Cobb, L. Raymond, S. Sill, and J. Young, "Effectiveness of an internet-based worksite smoking cessation intervention at 12 months," *Journal of Occupational & Environmental Medicine*, vol. 49, no. 8, pp. 821 – 828, 2007.

[3] J. Saul, B. Schillo, S. Everred, M. Luxenberg, A. Kavanaugh, N. Cobb, and A. L, "Impact of a statewide internet-based tobacco cessation intervention," *Journal of Medical Internet Research*, vol. 9, no. 3, 2007.

[4] N. K. Cobb, A. L. Graham, B. C. Bock, G. Papandonatos, and D. B. Abrams, "Initial evaluation of a real-world internet smoking cessation system," *Nicotine & Tobacco Research: Official Journal of the Society for Research on Nicotine and Tobacco*, vol. 7, no. 2, pp. 207 – 216, 2005.

[5] D. S. Kelders, "Understanding adherence to web-based interventions," Ph.D. dissertation, Universiteit Twente, Enschede, 2012, [accessed: 2016-03-08]. [Online]. Available: http://doc.utwente.nl/81967/

[6] M. L. Bolier, M. M. Haverman, P. J. Kramer, D. G. J. Westerhof, P. H. Riper, P. J. A. Walburg, B. Boon, and P. E. Bohlmeijer, "An internet-based intervention to promote mental fitness for mildly depressed adults: randomized controlled trial," *Journal of medical internet research*, vol. 15, no. 9, pp. 1 – 18, 2013, [accessed: 2016-03-08]. [Online]. Available: http://doc.utwente.nl/83748/

[7] L. Donkin, I. B. Hickie, H. CHristensen, S. L. Naismith, B. Neal, N. L. Cockayne, and N. Clozier, "Rethinking the dose-response relationship between usage and outcome in an online intervention for depression: Randomized controlled trial," *Journal of Medical Internet Research*, vol. 15, no. 10, p. e231, 2013.

[8] D. J. van Gemert-Pijnen, D. S. Kelders, and P. E. Bohlmeijer, "Understanding the usage of content in a mental health intervention for depression: An analysis of log data," *Journal of medical internet research*, vol. 16, no. 1, p. e27, 2014, open access, [accessed: 2016-03-08]. [Online]. Available: http://doc.utwente.nl/88658/

[9] G. Andersson and P. Cuijpers, "Pros and cons of online cognitive–behavioural therapy," *The British Journal of Psychiatry*, vol. 193, no. 4, pp. 270–271, 2008.

[10] ——, "Internet-based and other computerized psychological treatments for adult depression: A meta-analysis," *Cognitive Behaviour Therapy*, vol. 38, no. 4, pp. 196–205, 2009, [accessed: 2016-03-08]. [Online]. Available: http://dx.doi.org/10.1080/16506070903318960

[11] Z. Hilvert-Bruce, P. J. Rossouw, N. Wong, M. Sunderland, and G. Andrews, "Adherence as a determinant of effectiveness of internet cognitive behavioural therapy for anxiety and depressive disorders," *Behaviour Research and Therapy*, vol. 50, no. 78, pp. 463 – 468, 2012, [accessed: 2016-03-08]. [Online]. Available: http://www.sciencedirect.com/science/article/pii/S0005796712000708

[12] P. Musiat and N. Tarrier, "Collateral outcomes in e-mental health: a systematic review of the evidence for added benefits of computerized cognitive behavior therapy interventions for mental health," *Psychological Medicine*, vol. 44, pp. 3137–3150, 11 2014.

[13] V. Spek, P. Cuijpers, I. Nyklícek, H. Riper, J. Keyzer, and V. Pop, "Internet-based cognitive behaviour therapy for symptoms of depression and anxiety: a meta-analysis," *Psychological Medicine*, vol. 37, pp. 319–328, 3 2007.

[14] J. D. Piette, A.-M. Rosland, D. Marinec, N. S.and Striplin, S. J. Bernstein, and M. J. Silveira, "Engagement with automated patient monitoring and self-management support calls: Experience with a thousand chronically-ill patients," *Medical Care*, vol. 51, no. 3, pp. 216 – 223, 2013, [accessed: 2016-03-08]. [Online]. Available: http://doi.org/10.1097/MLR.0b013e318277ebf8

[15] E. T. Bohlmeijer and H. Monique, *Voluit Leven. Mindfulness of de kunst van het ervaren, nu als praktisch hulpboek.* Amsterdam: Boom, 2008.

[16] J. Kabat-Zinn, *Full catastrophe living: using the wisdom of your body and mind to face stress, pain, and illness.* New York: Delacorte Press, 1990.

[17] ——, *Wherever you go, there you are: Mindfulness meditation in everyday life.* New York: Hyperion, 1994.

[18] S. C. Hayes, K. D. Strosahl, and K. G. Wilson, *Acceptance and commitment therapy: An experiential approach to behavior change.* The Guilford Press, 1999, [accessed: 2016-03-08]. [Online]. Available: http://books.google.com/books?hl=en&lr=&id=CTgSzAdxc8cC&pgis=1

[19] M. F. Sieverink, D. S. M. Kelders, D. L. M. Braakman-Jansen, and D. J. E. van Gemert-Pijnen, "The added value of log file analyses of the use of a personal health record for patients with type 2 diabetes mellitus: Preliminary results," *Journal of diabetes science and technology*, 2014, published online before print, [accessed: 2016-03-08]. [Online]. Available: http://doc.utwente.nl/89290/

[20] M. F. Sieverink, D. L. M. Braakman-Jansen, Y. Roelofsen, S. H. Hendriks, R. Sanderman, H. J. Bilo, and D. J. E. van Gemert-Pijnen, "The diffusion of a personal health record for patients with type 2 diabetes mellitus in primary care," *International journal on advances in life sciences*, vol. 6, no. 3&4, pp. 177 – 183, 2014, [accessed: 2016-03-08]. [Online]. Available: http://doc.utwente.nl/93390/

Smartphone-Based Collaborative System for Wounds Tracking

Francis Faux

ISIS, Université de Toulouse, INU Champollion, France
F81 100 Castres, France
Email: francis.faux@univ-jfc.fr

Rémi Bastide
and Nathalie Souf
and Rita Zgheib

IRIT, Université de Toulouse, INU Champollion, France
F31 062 Toulouse, France
Email: firstname.lastname@irit.fr

***Abstract*—Tracking the evolution of wounds, ulcers or bedscars is one of the important tasks of the care personnel in medical institutions. We present an innovative, smartphone-based software application aimed at supporting this task. By using a standardized protocol complemented by image processing techniques, the application allows for the reliable acquisition of wound photographs, the assessment of wounds evolution, as well as for the annotation of photographs with relevant metadata. We detail the user-centered, scenario-based design process of this application that has been undertaken in the Connected Health Lab (CHL), our state-of-the-art usability lab dedicated to e-Health applications.**

Keywords–Wounds tracking; smartphone; collaborative system.

I. Introduction

Tracking the evolution of wounds is an important medical issue, and concerns health care professionals as well as patients or their family. This paper presents a mobile application attempting to go beyond the standard smartphone use, and enabling medical experts to remotely view, analyze, and track the wounds evolution. Chronic wounds are frequent. Among them, foot or leg ulcers affect million of persons suffering with type 2 diabetes. Bedsores (also called pressure sores or pressure ulcers) [1], are one of the dangerous diseases that an elder can face. They are a localized injury resulting from prolonged pressure on the skin and plague persons who have a reduced ability to move and change positions, and stay in bed or a wheelchair most of the time. In oncology, chronic wounds concern any wound even small, prone to chronicize because of the cancerous disease and its treatment. Cancerous wounds have the peculiarity to evolve according to treatments efficiency instead of following a normal healing process.

Thanks to the popularity of mobile devices in clinics and hospitals the need for a mobile collaborative wound tracking application has arisen as the image remote analysis becomes an attractive option. A smartphone application linked to a normalized shooting protocol allows a chronological and iconographic tracking of chronic wounds with a simple, practical, and connected tool. For medical teams, the visual examination of the healing process evolution has to be made easier and more reliable, using successive pictures taken in a standardized manner. For caregivers, this is to improve inter-professional cooperation by securing the skills transfer. In the field of telecare, remote wound tracking improves the patients' quality of life as patients only visit the wound clinic if necessary. Health care costs are thus reduced thanks to a limited number of both wound check appointments and travels to the clinic requiring special transportation. Many general telehealth requirements are concerned, such as quality of data transmission, privacy and security, interoperability with legacy healtchcare systems, and clear patient identification. The process of wound assessment is based on visual examination of picture by physicians, as well as on wound descriptions which provide qualitative objectives parameters (color phase peri-lesional state) and quantitative ones (wound surface and depth). However, tracking a wounds healing process across consecutive travel remains a difficult task for both clinicians and patients. Our project aims to support the following information cycle (Figure 1):

- data registering using the smartphone,
- data analysis in order to share information,
- decision, advice and feedback.

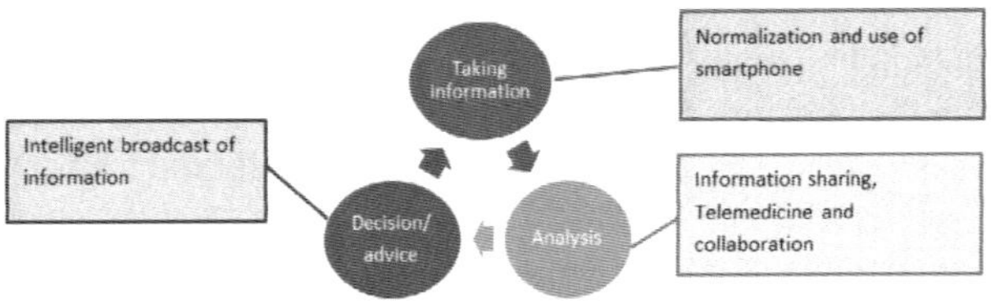

Figure 1. Information cycle

In this paper, wounds tracking is considered in a global and interdisciplinary approach, in relation with the patients referring doctor, in order to reduce the impact of the wound on the patients physical and psychological well-being. The presented application also has an impact on the patients environment. Thanks to an active communication with health carers, the family is taking a more active part and consequently feels reassured.

The remainder of this paper is organized as follows: Section 2 presents the related works, Section 3 outlines the key project objective. Section 4 presents the wound tracking application and Section 5 the standardized context of wound description. Future works and conclusion are presented in Section 6.

II. Related works

Smartphones are finding their way into Healthcare systems and one of the important fields of research today is wounds assessment and tracking applications. In [2] [3], authors prove smartphones ability to ensure accurate data. This implies a more robust report of the safety of care and a better prevention and monitoring of pressure ulcer. M-health can improve efciency and patient satisfaction. Based on high resolution cameras integrated in smartphones, several research projects are focusing on image acquisition and analysis of chronic wounds. Topics raised in these projects are addressed in accordance with two purposes: designing a wounds processing application and defining a collaborative system fostering patient-doctor interaction. Processing wounds applications consists mainly on the analysis of wound photos in order to detect and recognize wounds size and color that are essential factors for the evaluation of wounds. Several recent papers, such as [4] aiming to automate chronic wound processing and classification, propose algorithms for size and color detection of chronic wound images taken via smartphones. Their solution is based on *Mask Image* components within capturing photos and *Camera calibration* components within the segmentation phase in order to determine the relative and absolute size of the wound. These algorithms can be useful during the wound stage recognition. Other approaches for image analysis have been proposed like size determination from camera focus and zoom [5] or detecting wound size via several different types of sensors (sensor fusion approach) for increased accuracy. In [6] the wound image analysis system for diabetic foot ulcers covers size and color topics by performing image segmentation and noise removal. The authors also present a region-detection method for the prediction of wound boundaries, and they assess the healing status based on a color evaluation method.

Several research works such as [7] [8] focus on the collaborative facilities which may improve treatment of wounds in telemedicine. The proposed solutions aim at enhancing patient-doctor interaction offering dynamic features that may reduce the need for hospitalization and further facilitate a real-time advice for patients. In [7], the authors present an integrated approach of publicly available chronic wounds databases through a telemedicine platform. It offers the possibility to take high resolution images of wounds via smartphones. They can then be stored in a database with the associated metadata in order to support their discovery, access and usability. With a unique id, the patient can get prescribed medication from doctors, which provides a smooth interaction between patient and doctors. Patient history is easily accessible remotely by several doctors, which enhances the diagnosis, treatment and routine check ups. In [8] the main focus of the authors' is the real-time interaction via web technologies, enabling doctors to virtually collaborate with their patients and colleagues via applications working on mobile devices providing a collaborative annotation of Digital imaging and communications in medicine (DICOM) images. This research work addresses the issue of converting DICOM images to a compact version that can run on mobile devices before annotating and documenting images. Several smartphone applications such as UrgoExpert [9], INFINYS [10], Telap [11], Comedi-e [12] have been developed in order to document chronic wounds and wound care and follow-up the wound evolution. The main features of these applications are the following:

- Allowing nurses to define the wound's class (ulcer, bedsore) and its localization in the human body (Figure 2, right), as well as to visualize the wounds evolution (exudat level, healing status (wet necrosis, budding)) and the history of the treatment.

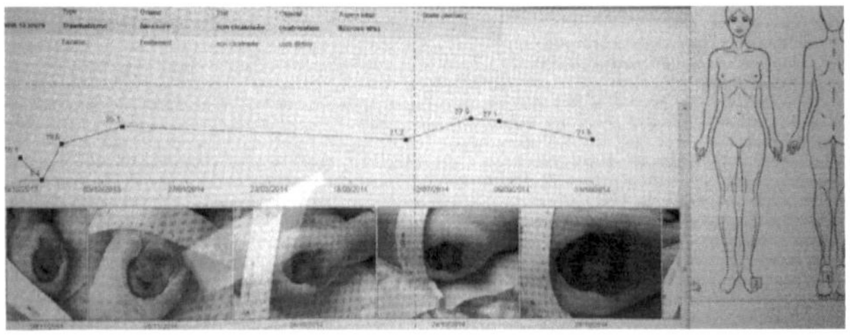

Figure 2. Comedie-e smartphone application: heel bedsore

- Photographs are captured via the smartphone camera, classified (Figure 2, bottom) and recorded in the smartphone storage card.
- Contains measurement tools. The application generates wound histories in graphical and text-based formats (Figure 2, top).
- Enable accessing a medical prescription module or consulting a protocol of treatment.
- Transfer the photographs ans associated data via a secure network to the patient health record.

III. PROJECT OBJECTIVE

We propose a new software application for wound tracking. The present paper focuses on the participatory design of the application, with a practical goal of usability and efficiency. We have taken a user-centered design process, taking advantage of the facilities offered by the Connected Health Lab (CHL), a state-of-the-art usability lab dedicated the e-Health applications, located in the premises of the authors' school of Engineering. In the framework of this user-centered design process, wounds pictures have been simulated using texture modeling, in order to keep them replicable. The mobile application is prototyped and assessed in the controlled environment provided by our usability lab, that faithfully replicates the working environments where the application will be deployed and used (hospital patient's rooms, or conventional home rooms for patients under home care). Then we describe the important points on which we have to focus to allow patients and their caregivers to take a more active role in daily wound care using a collaborative annotation functionality.

IV. WOUND TRACKING

In real life conditions, wound pictures are captured each time the nurse treats the wound and changes the dressing.

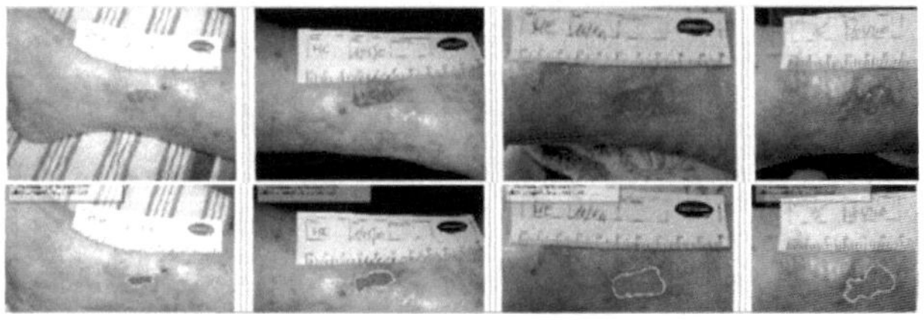

a) b) c) d)

Figure 3. Telap smartphone application, top: consecutive pictures , bottom: wound contour measurement

To achieve pictures, basic recommendations have been written and proposed to the health care professionals who are taking wound pictures. The main ideas for these recommendations are to improve the quality of the pictures and consequently the quality of treatment:

- Put a uniform background to better detect the wound area.
- No zoom to keep good resolution.
- No move to avoid fuzzy picture.
- Pacing a ruler reference in order to be able to recalculate the size of the wound.
- Placing reference color table in order to recalculate the colorimetry of the picture [13].

However, it is clear that difficulties remain to normalize shooting to enable a simpler and more reliable assessment of the wounds evolution. Pictures are often captured in an empirical manner without maintaining a constant angle of shooting (Figure 2, 3).

The difficulties mentioned by the healthcare professionals we are working with are:

- It is difficult to take different pictures with the same condition as the wound is changing and in on 3D surfaces (for example a leg). As the patient moves it is difficult to retrieve a similar shoot through the different patient's visits (Figure 3). The wound evolves and the lighting conditions are varying (Figure 3 a and b).
- It is very difficult to find reference points on which to align a picture as of course wound is evolving but moreover the wound environment is not stable. For example in case of ulcer wound, bad blood irrigation induces leg swelling and even the contour lines of the leg is not stable.

a) b) c) d)

Figure 4. Overlay: a) mask (picture 5), b) c) d) mask overlaid

In order to preserve an almost unchanging angle of shooting, we choose to overlaid a transparent mask image on the current capture in our smartphone application. The same method is used by T. K. Poon in [4]. The advantage of this approach is that no peripheral devices to the smartphone camera are necessary. We choose not to add temporary or permanent tattoo of dots or lines around the wound or skin marker reference points, which can serve as a pattern to ensure alignment of successive mask. The transparent mask image which is only the previous shot of the wound (Figure 4 a) is overlaid on the current capture. When using the application, first the nurse positions the smartphone in order to approximatively recover the previous shooting angle (Figure 4 b) then moves the smartphone (Figure 4 c) to align as well as possible (Figure 4 d) the mask on the current wound. The mask area page is visualized right after a new image acquisition. The outcome of the application is a set of consecutive wound pictures captured with same size and orientation.

A. Experimental simulation

Experiments have been realized in the environment of simulation of the Connected Health Lab (CHL). The use case of a heel bedsore evolution was chosen as simulation process in the context of a this Living Lab which simulate the patient route. The scenario of the study considers a seriously hurt person because of a car accident (or an accident at work). The person suffers of a polytrauma and is in a coma.

Weeks: 1, 2 3, 4 5, 6 7

Figure 5. Use case: convalescence in the Living Lab

The figure 5 presents the patient's convalescence of a duration of 7 weeks and which includes four main phases:

1) After managing the emergency (operation, stabilization), the patient is hospitalized in intensive care for two weeks, Heel pressure ulcer appears after a week of hospitalization (week 1). Then the bedsore deteriorates because of the patient's immobility (week 2).
2) As soon as the patient leaves the coma and wakes up, he is transferred to an orthopedic service. However the bedsore continues to deteriorate because of the lack of mobility (weeks 3 and 4).
3) The patient goes home with nursing home monitoring (week 5). His bedsore begins to heal in week 6.
4) The patient finding increasingly mobility, in particular with the setting in the chair, the bedsore continues to heal (week 7).

In order to realise a reproductive study, i.e., replicable model of an ulcer pressure evolution, we decided to use a serie of built picture of wounds, to serve as a benchmark for our scenario and for the evaluation of the proposed prototype. We synthesize the wound texture using Efros & Leung's texture synthetis algorithm [14]. This algorithm is chosen because of its simplicity (Figure 6). Three reference images serve as model for different stages of bedsore evolution. A rectangular

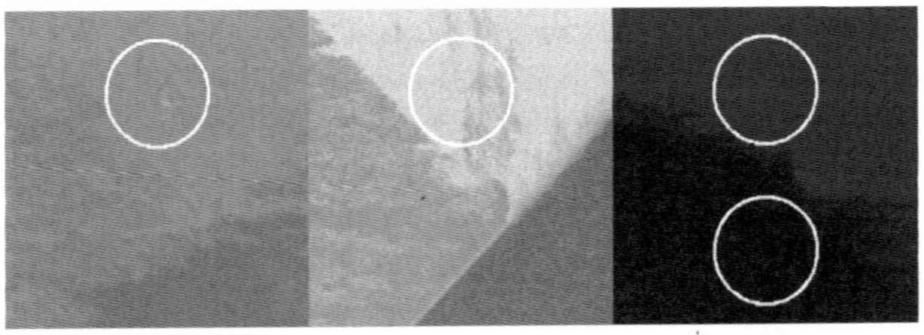

Stage 1 Stage 2 Top: Stage 3
Bottom: Stage 3+

Figure 6. Wound texture synthesize and disk modelling the heel bedsore

window (in green on Figure 7) manually selected in each image serves as texture source for the algorithm. After various tests, a 3×3 window is used as algorithm parameter because of the light granularity of the wound texture.

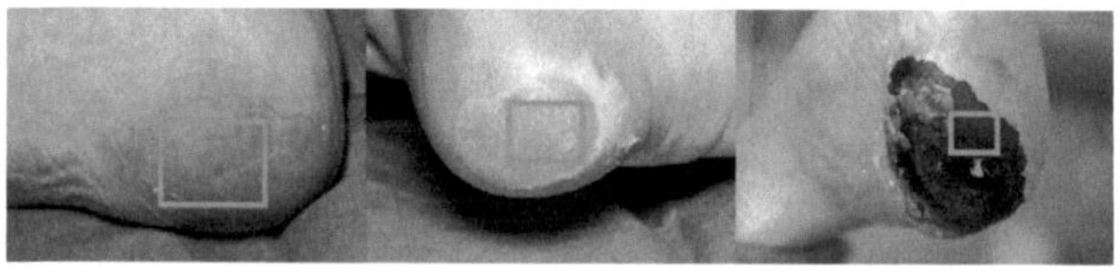

Stage 1 Stage 2 Stage 3

Figure 7. Stages of heel bedsore used for the texture synthesize

A simple disk of 16 mm of radius (area 8 mm^2) models the heel pressure ulcer. For stages 1 and 2 we select the zone which looks like the real texture. The stage 3 is segmented in two levels (called 3 and 3+) in function of level of necrosis in the wound. The bedsore evolution is summarized in the table I.

TABLE I. WOUND EVALUATION

CHL ZONE	INT. CARE		ORT. ROOM		HOME BED.		ARMCH.
WEEK	1	2	3	4	5	6	7
BEDSORE STADE	1	1	2	3	3+	3	2

B. Analysis and results

A subjective analysis of the results (Figure 8) shows that the application is relevant in medical context. Consecutive pictures are captured with almost same scale and orientation along the seven weeks of study and for the four rooms of measurement.

The quantitative results indicate a variation of about 40% of the area and 25% of the perimeter (Figure 9). The reason of these variations comes from the difficulty to align the mask over the current wound picture especially in the intensive care room or the hospital room when the patient's mobility is reduced. At home bedroom the shooting is easier when the patient can move its leg more easily.

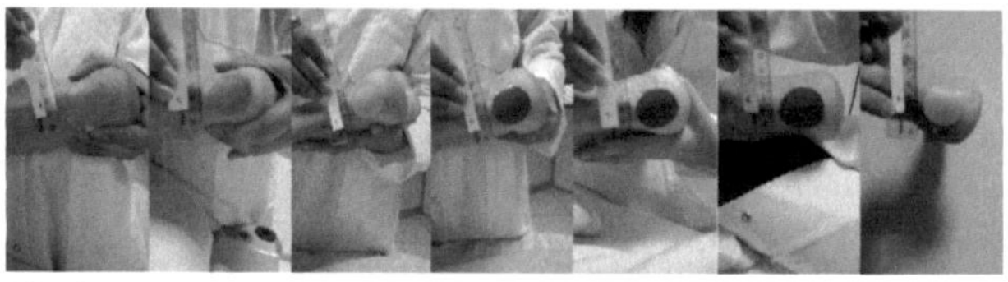

Week: 1 2 3 4 5 6 7

Figure 8. Result of tracking with the mask overlay

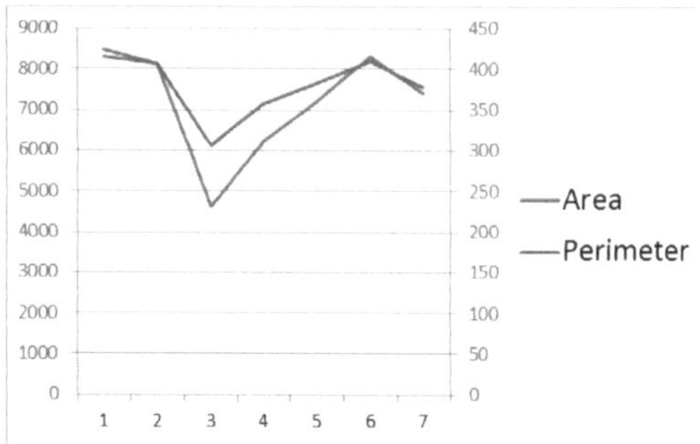

Figure 9. Area and perimeter measurements

As conclusion from these experiments in the Living Lab, the application usability is clearly insufficient in such a context. An evolution of the application is necessary. The shooting must be easier and above all automatic. As future work, we will study the possibility to introduce the correlation level between the mask overlay and the current image in order to facilitate the alignment between these two pictures. Moreover, this study will be continue for various kinds of wounds (cancerous wounds, ulcers, others bedsores) with varying areas ans complex forms.

V. Standardized Context of Wounds Description

A. Introduction

In order to improve the informative status of the wound pictures, it is important to add descriptions to the picture itself. In this project, one of our objectives is to determine recommendations to provide efficient annotation tools able to improve efficient communication between the different actors involved in the wound care (health care professionals as well as patient or family), in the context of smartphone use. In this paper we will present some of the lines of thinking we have spotted. In order to communicate, health professionals complete the information of the medical record but also use other media as phone, mail, annotations and so on. Annotation tasks are often used in the medical area as for example post-it used for the asynchronous communication of health care professionals. As mentioned in [15] "*practitioners use annotations to act: either to enrich the annotated document or as transitory support of knowledge used to create new knowledge (recorded or not in a document). Annotations are useful not only to keep pertinent knowledge in the health record, but also to help practitioners to collaborate when the documents show their limits*"(p. 4).

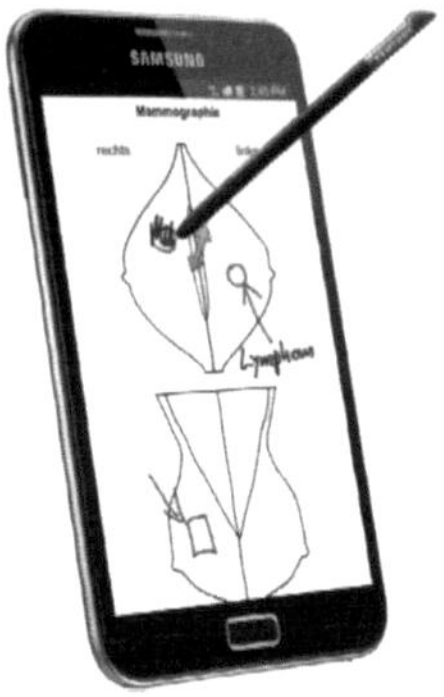

Figure 10. Wounds annotations of Pange et al.

Annotation systems are proposed in many telemedicine systems using smartphones, as shown in Figure 10 for mammography (annotation from [16]). As friendly annotation interfaces are difficult to provide through smartphone devices, in order to get usable and efficient tools, we are convinced that it is extremely important to better understand how and why the annotation process should be developed.

B. Analyse

Three main issues for annotation process were detected and are presented in this paper:

Why are the annotations required. In a wound annotation system, the objectives of the proposed annotations could be very different. On the one hand, some annotations are used to inform on general items such as Patient ID or Patient Name. These data are not directly involved with wound characteristics. Such annotations must be re-entered and duplicated on the smartphone application when the interoperability between smartphone and health care record are difficult to perform. On the other hand, some of the data bring specific awareness on an aspect of the wound : a highlighted array of interest gives the subjective point of view of a health care professional. Some annotations could then be automatically performed while improving the connection between system, adding connected objects into the wound care process (such as automatic detection of the ID of the patient) while some annotations are specifically involved on the care expertise. Being aware of these specificities will help to better understand the annotation tasks and to focus on the needed interfaces.

When and how annotations are completed. Annotations should be fed when the picture is taken or could be completed during an asynchronous task of wound image description. Items used for annotation can be from different natures: standardized description, visual annotation such as colored arrows, specific descriptions obtained after specific treatments such as wound surface calculation or colorimetry indices. By better understanding the annotation process, we hope to be able to propose efficient systems able to adapt to the different contexts of use.

What is concerned with the annotation task. Annotation tasks could bring information on three different levels of knowledge.

- The first level concerns the picture itself and the shooting environment. It could be important to remember the parameters linked to the shooting so that the quality on the picture could be memorized. Quality of context description could help to qualify a further picture comparison. Metadata of picture description used in DICOM standard are some of the information to be kept, but the context of shooting referring to the recommendation can also be used.
- The second level is used in order to memorized the context of the individual who is concerned by the picture: for instance name of the patient, part of body concerned.
- Last level deals with specific wound characteristics. The goal is to establish a standardized context of wounds description in order to ensure interoperability of information systems and to improve exchanges between healthcare systems actors.

By better identifying the need of information on each level, we will be able to better determine the needed information.

The project will then address the issues related to picture description, annotation of wound pictures, and modelling of factual wounds tracking (be able to reinterpret picture in order to build meta-data useful for caregivers). We will provide a model of annotation for wound. Using the different point of view that we have mentioned on wound care annotation, we want to propose an ontological framework.

Analysis of existing annotation systems and previous work such as the one performed by the wound ontology [17], or by the openEHR Clinical Knowledge Manager working group [18], as mentionned by Gallaguer [19] will be considered and extended by confronting them with the expertise of health professionals involved in this project. Constraints due to smartphone use, inducing mobility, small screen, distribution of information will be also detailed.

VI. Conclusion and Perspective

We have presented a mobile, smartphone-based e-health application that supports caregivers in their important activity of monitoring the evolution of wounds. The application combines innovative mobile interaction techniques with image processing and analysis in order to allow for a reliable and usable clinical use. The design of this application has followed a user-centered process, where both the standardized photograph-taking procedure and the image processing steps have been incrementally designed and validated in a controlled environment, the CHL living lab dedicated to e-Helath applications. The application is currently a validated prototype. Our future steps will be to perform a clinical validation and assessment of the application, where it will be used by care professionals in a real-life setting, with real patients. We also plan to improve the image-processing components, and to take into account the interoperability of the application with legacy health information systems (e.g., electronic health records).

Acknowledgment

Bruno Roussel, research engineer, for his involvement in the test procedure in the Connected Health Lab and Laurent Grgoire, International coordinator, for his proofreading and editing of the English language.

References

[1] R. Zgheib, R. Bastide, and E. Conchon, "A semantic web-of-things architecture for monitoring the risk of bedsores," in Proceedings of the International Conference on Computational Science and Computational Intelligence (CSCI), Las Vegas, USA. IEEE, december 2015.

[2] K. Wac, "Smartphone as a personal, pervasive health informatics services platform: literature review," Yearbook of Medical Informatics, no. 2, 2012, pp. 83–93.

[3] E. M. Quinn et al., "Clinical Unity and Community Empowerment: The Use of Smartphone Technology to Empower Community Management of Chronic Venous Ulcers through the Support of a Tertiary Unit," PLoS one, vol. 8, no. 11, 2013, p. e78786.

[4] T. W. K. Poon and M. R. Friesen, "Algorithms for Size and Color Detection of Smartphone Images of Chronic Wounds for Healthcare Applications," IEEE Access, vol. 3, 2015, pp. 1799–1808.

[5] P. J. F. White, W. P. Blake, and M. R. Friesen, "Algorithms for Smartphone and Tablet Image Analysis for Healthcare Applications," IEEE Access, vol. 2, 2014, pp. 831–840.

[6] L. Wang et al., "Smartphone-Based Wound Assessment System for Patients With Diabetes," IEEE Transactions on Biomedical Engineering, vol. 62, no. 2, 2015, pp. 477–488.

[7] C. Chakraborty, B. Gupta, and S. K. Ghosh, "Mobile metadata assisted community database of chronic wound images," Wound Medicine, vol. 6, 2014, pp. 34–42.

[8] M. F. Pasha et al., "An android-based mobile medical image viewer and collaborative annotation: Development issues and challenges," JDCTA: International Journal of Digital Content Technology and its Applications, vol. 6, no. 1, 2012, pp. 208–217.

[9] "UrgoExpert," URL: http://www.urgoexpert.fr[accessed: 2016-20-03].

[10] "INFINYS," URL: http:// www.sffpc.org[accessed: 2016-20-03].

[11] "telap," URL: http://www.telap.org [accessed: 2016-20-03].

[12] "Comedi-e," URL: http://www.esante-picardie.com [accessed: 2016-20-03].

[13] S. Van Poucke et al., "Automatic colorimetric calibration of human wounds," BMC medical imaging, vol. 10, no. 1, 2010, p. 7.

[14] A. A. Efros et al., "Texture synthesis by non-parametric sampling," in Proceedings of the Seventh IEEE International Conference on Computer Vision, vol. 2. IEEE, 1999, pp. 1033–1038.

[15] N. Bricon-Souf et al., "Informal notes to support the asynchronous collaborative activities," International Journal of Medical Informatics, vol. 76, 2007, pp. S342–S348.

[16] A. Prange and D. Sonntag, "Smartphone sketches for instant knowledge acquisition in breast imaging," Proceedings of the Sketch: Pen and Touch Recognition Workshop in conjunction with IUI14, 2014.

[17] "Introduction to the wound ontology consortium," URL: http://www.slideshare.net/SvenVanPoucke/woundontology1 [accessed: 2016-20-03].

[18] "Open EHR, Clinical knowledge Manager, Inspection of open Wound," URL: http://openehr.org/ckm/ [accessed: 2016-20-03].

[19] B. A. Gallagher, "Wound bed assessment using calibrated images and representation in openehr," Master's thesis, University of Dublin, Ireland, 2012.

A Short-Term Assessment of Cardiac Output by Using Instantaneous Pulse Rate Variability

Pei-Chen Lin
Institute of Biomedical Engineering
National Chiao Tung University
Hsinchu, Taiwan R.O.C.
e-mail: linpc1006@gmail.com

Chia-Chi Chang
Department of Electronics Engineering and Institute of Electronics, Institute of Biomedical Engineering, Biomedical Electronics Translational Research Center
National Chiao Tung University
Hsinchu, Taiwan R.O.C.
e-mail: ccchang.cs@gmail.com

Hung-Yi Hsu
Department of Neurology
Chung Shan Medical University
Section of Neurology, Department of Internal Medicine, Tungs' Taichung Metro Harbor Hospital, No.699, Sec. 8, Taiwan Blvd., Wuqi Dist., Taichung City 435
Taichung, Taiwan R.O.C.
e-mail: hungyihsu@gmail.com

Tzu-Chien Hsiao*
Department of Computer Science, Institute of Biomedical Engineering, Biomedical Electronics Translational Research Center
National Chiao Tung University
Hsinchu, Taiwan R.O.C.
e-mail: labview@cs.nctu.edu.tw

Abstract—**A hemodynamically unstable patient has the risk of going into general shock. As fluid therapy is the primary treatment for shock, the fluid responsiveness (FR) of the patient should be evaluated before volume expansion. However, conventional methods predict FR by analyzing the variation of blood pressure signal in time domain which is a nonstationary problem and makes it difficult to provide a stable index for FR. Instantaneous pulse rate variability (iPRV) is a cardiovascular assessment in frequency domain. Furthermore, iPRV uses ensemble empirical mode decomposition (EEMD), which could solve the nonstationary problem and overcome the frequency limitation in power spectrum of heart rate variability (HRV). iPRV provides a new indication in very high frequency (VHF) range (0.4-0.8Hz) of spectrum for peripheral responses. The aim of this study was to verify the ability of iPRV to indicate VHF for cardiac output assessment. Twenty-six healthy participants participated in this study and the acquired signal was recorded in supine baseline, during head-up tilt (HUT), and passive leg raising (PLR), which induces variation of venous return and helps the quantitative assessment of cardiac output individually. The result showed that the normalized power of VHF in HRV was small and there was no corresponding variation in different postures. In contrast, the normalized power of VHF in iPRV presented relative trend with different venous return changes. Overall, iPRV provides a novel and short-term cardiac output assessment in frequency domain and it has potential to evaluate FR.**

Keywords- fluid responsiveness (FR); instantaneous pulse rate variability (iPRV); ensemble empirical mode decomposition (EEMD); head-up tilt (HUT); passive leg raising (PLR).

I. INTRODUCTION

To study blood flow and circulation in humans, hemodynamics is an important part of Physiology. Hemodynamic monitoring plays a key role in hospitals for hemodynamically unstable patients. When they recover after surgery or after a suffered injury, they usually need intensive care and continuous observation of hemodynamic parameters is needed. Furthermore, a patient with unstable hemodynamics is at risk of general shock.

Fluid therapy is the primary treatment for shock. Moreover, the stroke volume of a patient increases by 12% after fluid loading which means a patient responds to fluid administration. However, only half the patients respond to fluid loading in intensive care unit (ICU). Volume expansion for non-responder patients may only exert adverse effects without having any hemodynamic benefit [1][2]. Thus, the fluid responsiveness (FR) of a patient should be evaluated before volume expansion. After fluid loading, the expectation is that it will increase cardiac output significantly. Nevertheless, stroke volume is intrinsically controlled by cardiac preload, so conventional methods evaluate FR by assessing cardiac preload or output. Several ways to evaluate FR are based on static indices and dynamic indices [3][4]. Previous studies demonstrated that static index of cardiac preload can predict FR by pressure and volume markers, but it is affected by respiratory or systolic function [5]. For this reason, dynamic indices have been developed to assess cardiac output by calculating the short-time variation of arterial pulse pressure waveform in mechanical ventilation patient. But dynamic indices are also affected by spontaneous respiration [6], and both indices are based on time domain which present the nonstationary problem in signal and make it difficult to provide a stable index for FR.

In frequency domain, one of the important cardiovascular assessment is heart rate variability (HRV). However, HRV studies are restricted by the feasibility and the reproducibility with an inconvenient measurement [7]. In addition, the maximum frequency of power spectrum is restricted at 0.5 Hz by tachogram [8]. Pulse rate variability (PRV) was proposed as a substitute measurement of HRV. PRV uses pulse wave, which collected from photoplethysmography (PPG), to replace ECG recording in HRV and has been examined as a surrogate of HRV during nonstationary conditions in a previous study [9]. Besides, the arterial pulse wave is regulated by complex physiological controls which make PRV provide much more peripheral information than HRV [9]. However, the frequency band indication in power spectrum of PRV is still a limitation that needs to be overcome. Instantaneous pulse rate variability (iPRV) is a novel method to extend frequency limitation [10]. It adopted the frequency range extension method and IF projecting technique help for PRV spectral analysis. Furthermore, iPRV used ensemble empirical mode decomposition (EEMD), which could solve the nonstationary problem. Also, iPRV provides a new indication, named very high frequency (VHF) band (0.4-0.8 Hz), in the power spectrum for more information. The literature has proposed that VHF of HRV is a novel index of left ventricular function evaluation [11]. VHF has the ability to indicate cardiac function, venous return, and FR. However, the variation and interpretation of VHF in iPRV have to be further explored and examined. The aim of this study was to verify the ability of iPRV to indicate VHF for cardiac output assessment. It thought of an experiment of change in venous return such as passive leg raising (PLR) and head-up tilt (HUT). PLR as a volume challenge induces a translocation of venous blood, and HUT induces blood pooling in the legs which can decrease venous return and cardiac output. These two positions can offer variation of venous return to verify cardiac output assessment as iPRV. The remainder of the paper is structured as follows. The next section presents the experiment design, data collection, introduction of iPRV analysis and comparison methods. Section III illustrates the similarity between iPRV and RRI and also explains the results of power spectrum in iPRV and HRV during different postures. Section IV provides the discussion about mechanism of different frequency band indication in iPRV. Moreover, it presents iPRV is a potential short-term FR evaluation. Conclusion and future work are given in the last section.

II. MATERIAL AND METHOD

A. Experiment design and subjects

The procedure of this study contained five steps. First, participants were resting in supine position with 10-minute recording as a baseline. Second, participants were tilting up passively (HUT) on the automatic tilting table and kept in tilt-up position for 10 minutes. Then, participants were back to the supine position with 5 minutes for recovering to baseline. Finally, participants were raising leg passively (PLR) for 10 minutes. After PLR, participants were resting with 5 minutes for recovering to steady state.

The signal acquisition was as follows: (i) electrical activity of ECG signal was recorded by BEST-C-04056 (BioSenseTek Corp., Taiwan); (ii) peripheral information of PPG signal was recorded by Nonin 8500 (Nonin Medical Inc., Plymouth, MN). The signal acquisition with sampling frequency is 200Hz.

Thirty healthy participants (male: 15; age: 24±1.8 years old) participated in this study. Twenty-six participants had no history of cardiovascular disease and no uncontrollable distortion in acquired data. Four exclusive participants contained distortion in certain data, such as motion artifacts from PPG signal and ECG abnormality. All measurements were performed in a quiet temperature-controlled room and the experiment was approved by the institutional review board (IRB) of Tungs' Taichung Metro Harbor Hospital.

B. Instantaneous pulse rate variability

The process of iPRV includes two parts (Figure 1). The first part is decomposition, the pulse wave component was searched from PPG by sifting process in EEMD. Sifting process is an iteratively detrending operation which is used to compute finite set of components, named intrinsic mode functions (IMFs), from source nonstationary data. Moreover, before the sifting process, EEMD provides noise-assisted method into original data for eliminating multiple characteristic problem in IMFs. After mixtures of added noise and source data, the detrending operation contains several steps. First, local extrema of data $x(t)$ are identified by peak-valley detection. The upper envelope $U(t)$ and lower envelope $L(t)$ are generated by cubic spline interpolation according to the local maxima and local minima. The trend in current timescale is computed by calculating the mean of $U(t)$ and $L(t)$, as $M(t)$.

$$M(t) = \frac{U(t)+L(t)}{2} \tag{1}$$

The new timescale $H(t)$ is the representation after detrending operation by data $x(t)$ subtracting the trend.

$$H_k(t) = H_{k-1}(t) - M_k(t), k \geq 1 \tag{2}$$

where $H_0(t) = x(t)$.. After k times detrending operation, if the trend of $H_k(t)$ satisfies the criterion as the steady constant trend, then the components $H_k(t)$ were extracted from $x(t)$ as IMF. After n sifting process, $x(t)$ was decomposed into n IMFs, $IMF_1(t){\sim}IMF_n(t)$, and one residue $r(t)$.

$$x(t) = \sum_{i=1}^{n} IMF_i(t) + r(t) \tag{3}$$

Since IMFs were decomposed from different mixtures, the ensemble IMFs are computed by averaging each corresponding IMF.

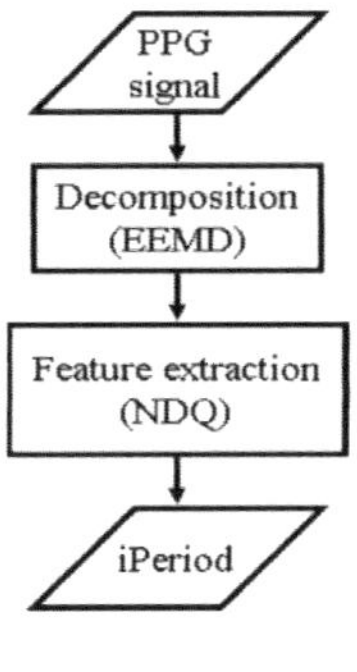

Figure 1. The flow illustration of the process of iPRV.

The second part is feature extraction. The continuous-time heartbeat rhythm can be extracted by EEMD. To resent the frequency of oscillation in IMF at a specific time instant, iPRV proposed normalized direct quadrature (NDQ) to calculate IF of relative IMF. The algorithm of NDQ is shown in Figure 2. First, the amplitude modulation of the main component IMF_{main} was eliminated by iteratively normalization for conquering Bedrosian' s theorem [12]. Then, the empirical frequency modulation signal $F(t)$ of IMF_{main} is assumed to be cosine function, and its quadrature $sin\emptyset(t)$ can be computed directly.

$$sin\emptyset(t) = \sqrt{1 - F^2(t)} \tag{4}$$

The instantaneous phase $\emptyset(t)$ is calculated by taking arctangent of $F(t)$ and its quadrature, then the IF is obtained from divided derivative of $\emptyset(t)$ by 2π.

$$\emptyset(t) = tan^{-1}(\frac{\sqrt{1-F^2(t)}}{F(t)}) \tag{5}$$

The instantaneous period (iPeriod) was estimated by the inversion of IF of IMF_{main} in order to indicate time series of heartbeat rhythm as RR intervals (RRI) in HRV.

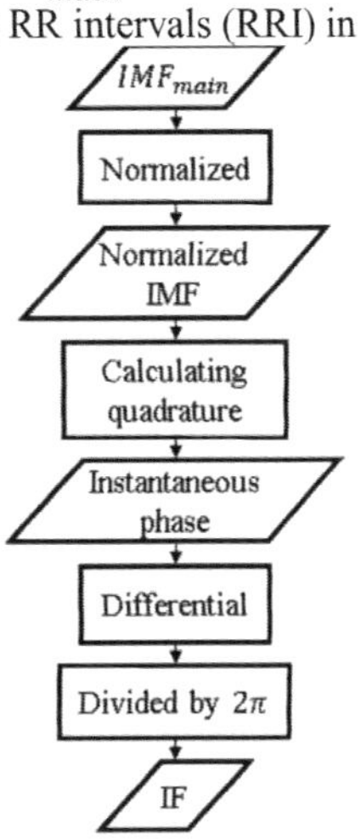

Figure 2. The flow illustration of the algorithm of NDQ.

C. Comparison method

iPeriod was calculated by feature extraction, and the analysis of iPeriod included frequency domain analysis and time domain analysis. Frequency domain analysis adopted power spectrum to compare power variation in different frequency band with RRI. The procedure of frequency domain analysis includes two steps. Firstly, fast Fourier transform (FFT) was performed as the spectrum analysis in each frequency band of iPeriod and RRI. The partition of power spectrum in HRV and iPRV is shown in Figure 3. The frequency band of HRV is divided into low frequency (LF) band (0.04-0.15 Hz), high frequency (HF) band (0.15-0.4 Hz) and very high frequency (VHF) band (0.4-0.5 Hz). Frequency band of iPRV is divided into LF (0.04-0.15 Hz) and HF (0.15-0.4 Hz), and VHF (0.4-0.8 Hz). Then, the power of each frequency band is calculated by integration for comparison of variation during different positions.

The purpose of time domain analysis is to examine the reliability of iPeriod. Before time domain analysis, there is a preprocessing for filtering high frequency band of iPeriod by EEMD in order to be compared with RRI. The procedure of preprocessing is shown in Figure 4. At first, iPeriod was filtered by EEMD into IMFs with high frequency and residue. Then, residue was extracted as low pass of iPeriod ($iPeriod_{LP}$, $f(t)$).Time domain analysis adopted cross correlation coefficient to ensure similarity between iPRV and HRV. Cross correlation coefficient also offered time lag information to discuss changes in different position. RRI series $g(t)$ is an interpolation of RRI by using cubic spline. The calculation of cross correlation coefficient shows in (6).

$$r(d) = \frac{\sum_{t=0}^{N}[(f(t)-\bar{f}(t))*(g(t)-\bar{g}(t))]}{\sqrt{\sum_{t=0}^{N}(f(t)-\bar{f}(t))^2}*\sqrt{\sum_{t=0}^{N}(g(t)-\bar{g}(t))^2}} \tag{6}$$

where d is the time lag. $\bar{f}(t)$ and $\bar{g}(t)$ are the mean of the corresponding series.

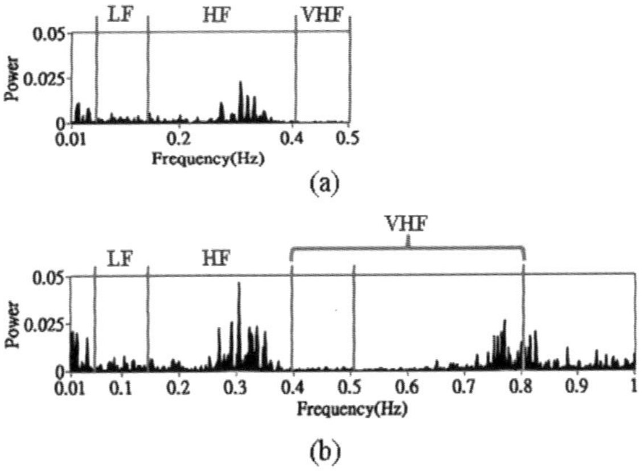

Figure 3. (a) is power spectrum in HRV with LF, HF and VHF, and (b) is power spectrum in iPRV with LF, HF, NHF and FHF.

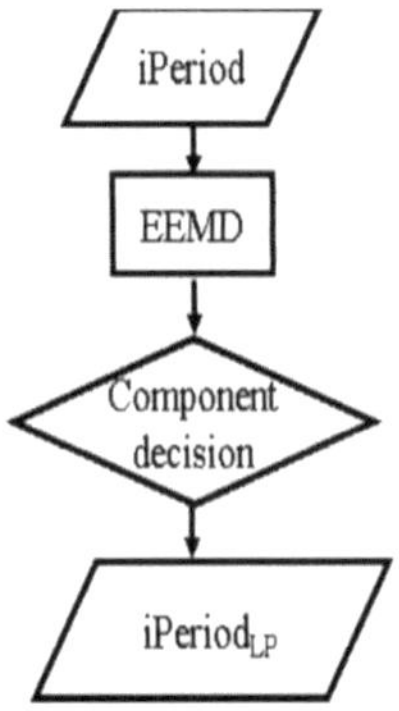

Figure 4. The procedure of preprocessing in time domain analysis.

D. *Normalized power in frequency band*

For summarizing spectral analysis with whole database during different condition, this study adopted two normalized power calculations as follows: (i) for comparing with HRV, the following normalization formula was used:

$$Power_{total}\ (\mathrm{TP}) = Power_{LF} + Power_{HF}$$

$$\rightarrow \frac{\mathrm{TP}}{\mathrm{TP}} = \frac{Power_{LF}}{\mathrm{TP}} + \frac{Power_{HF}}{\mathrm{TP}} = \mathrm{nLF} + \mathrm{nHF} = 1 \qquad (7)$$

Where nLF is normalized power of LF and nHF is normalized power of HF; (ii) for exploring power of VHF information, the following normalization formula was used:

$$Power_{total}\ (\mathrm{TP}) = Power_{LF} + Power_{HF} + Power_{VHF}$$

$$\rightarrow \frac{\mathrm{TP}}{\mathrm{TP}} = \frac{Power_{LF}}{\mathrm{TP}} + \frac{Power_{HF}}{\mathrm{TP}} + \frac{Power_{VHF}}{\mathrm{TP}}$$

$$= \mathrm{nLF} + \mathrm{nHF} + \mathrm{nVHF} = 1 \qquad (8)$$

Where nLF is normalized power of LF, nHF is normalized power of HF and nVHF is normalized power of VHF.

III. RESULT

A. *Comparison result in frequency domain*

Table I shows the comparison result with HRV. The normalized power of HRV and iPRV had the same trend between supine and other postures. The sympathetic activities increased (nLF increasing) when the venous return decreased (HUT). The parasympathetic activities increased (nHF increasing) when the venous return increased (PLR). Different venous return activated the autonomic nervous system (ANS), normalized power of iPRV responded relatively as HRV.

TABLE I. COMPARISON VARIATION WITH HRV IN LF AND HF DURING DIFFERENT POSTURES.

		nLF (0.04-0.15 Hz)	**nHF (0.15-0.4 Hz)**
HRV	Supine	0.51±0.19	0.49±0.19
	HUT	0.54±0.20	0.46±0.20
	PLR	0.49±0.17	0.51±0.17
iPRV	Supine	0.41±0.17	0.59±0.17
	HUT	0.44±0.18	0.56±0.18
	PLR	0.40±0.16	0.60±0.16

The form is (mean ± standard deviation). Color in red means increasing of normalized power. Color in green means decreasing of normalized power.

Table II shows the spectral analysis with venous return changes. nVHF of iPRV decreased when the venous return decreased (HUT), and nVHF of iPRV increased when the venous return increased (PLR). nVHF of iPRV is effective to indicate venous return changes during different postures. nVHF of HRV was not only small but also can't indicate venous return decreasing. However, nHF of iPRV shows different trend with HRV in Table II.

TABLE II. COMPARISON VARIATION WITH HRV IN LF, HF AND VHF DURING DIFFERENT POSTURES.

		nLF (0.04-0.15 Hz)	**nHF (0.15-0.4 Hz)**	**nVHF (0.4-0.8 Hz)**
HRV	Supine	0.49±0.19	0.46±0.18	0.05±0.05
	HUT	0.52±0.19	0.44±0.18	0.05±0.03
	PLR	0.46±0.18	0.47±0.16	0.08±0.10
iPRV	Supine	0.28±0.17	0.37±0.13	0.35±0.17
	HUT	0.31±0.17	0.38±0.13	0.31±0.11
	PLR	0.25±0.14	0.36±0.12	0.39±0.15

The form is (mean ± standard deviation). Color in red means increasing of normalized power. Color in green means decreasing of normalized power.

B. *Comparison result in time domain*

The results in the time domain analysis with the whole database are summarized in Table III. Before preprocessing for filtering high frequency band of iPeriod, all participants' *r* value between iPeriod and RRI series showed low correlation. In contrast, after filtering high frequency band of iPeriod, *r* value between $\mathrm{iPeriod}_{LP}$ and RRI series there was middle positive correlation (0.667±0.109 in baseline; 0.672±0.096 in HUT; 0.675±0.105 in PLR). Moreover, *r* value of $\mathrm{iPeriod}_{LP}$ and RRI series at all postures was significant difference with value of iPeriod and RRI series ($p<0.01$ at all postures). Time lag at all postures was not significant difference ($p = 0.14$ in baseline; $p = 0.23$ in HUT;

$p = 0.33$ in PLR) which means increasing of r value was only influenced by high frequency band.

IV. DISCUSSION

nVHF in iPRV showed the ability of evaluating cardiac output from PPG signal. This study adopted NDQ, which eliminated amplitude modulation, as an IF estimation to calculate periodic changes of pulse waveform for presenting cardiac output information. However, NDQ as a suitable IF estimation for pulse waveform has to be further explored. Previous studies proposed that component for NDQ needs to be a single-term expansion which absorb some of the properties [13]. The composition of PPG signal includes pulse waveform and reflected waveform, and pulse waveform is a sinusoid-like function which satisfy condition of single-term expansion. Furthermore, the pulse waveform extracted by EEMD with frequency band around 1 Hz. Overall, pulse waveform is suitable component to NDQ for calculating IF.

TABLE III. THE RESULT OF CROSS CORRELATION COEFFICIENT IN THE TIME DOMAIN ANALYSIS

	Supine	**HUT**	**PLR**
RRI series vs. iPeriod			
r	0.240±0.096	0.255±0.102	0.275±0.096
Time lag	2.09±1.39	2.07±1.76	2.27±1.76
RRI series vs. iPeriod_{LP}			
r	0.667±0.109*	0.672±0.096*	0.675±0.105*
Time lag	1.93±1.46	2.37±1.70	2.42±1.61

The form is (mean ± standard deviation). * defines as $p<0.01$ relative to r between RRI and iPeriod

TABLE IV. COMPARISON VARIATION WITH HRV IN LF, HF, NHF AND FHF DURING DIFFERENT POSTURES.

		nLF (0.04-0.15 Hz)	**nHF (0.15-0.4 Hz)**	**nNHF (0.4-0.5 Hz)**	**nFHF (0.5-0.8 Hz)**
HRV	Supine	0.49±0.19	0.46±0.18	0.05±0.05	
	HUT	0.52±0.19	0.44±0.18	0.05±0.03	
	PLR	0.46±0.18	0.47±0.16	0.08±0.10	
iPRV	Supine	0.37±0.17	0.53±0.15	0.10±0.09	
	HUT	0.41±0.18	0.52±0.17	0.08±0.04	
	PLR	0.36±0.15	0.54±0.15	0.10±0.06	
iPRV	Supine	0.28±0.17	0.37±0.13	0.06±0.04	0.29±0.15
	HUT	0.31±0.17	0.38±0.13	0.05±0.02	0.25±0.11
	PLR	0.25±0.14	0.36±0.12	0.06±0.03	0.32±0.14

The form is (mean ± standard deviation). Color in red means increasing of normalized power. Color in green means decreasing of normalized power.

After obtained iPeriod by using NDQ, FFT transformed iPeriod into power spectrum for comparing with HRV in frequency domain. Nevertheless, the normalized power of iPRV showed different trend with HRV in HF. Previous study claimed that fluctuations in the PPG signal in a normal person, which synchronous with ventilation, with a frequency in the 0.2-0.45 Hz [14]. Therefore, HF in iPRV is a complex frequency band with respiration and parasympathetic activities. In addition, VHF contains much more information than HRV, and it also affects the nHF in iPRV to present parasympathetic activities. To find out more detail, VHF was separated into near high frequency (NHF) band (0.4-0.5 Hz) and far high frequency (FHF) band (0.5-0.8 Hz). Table IV revealed that iPRV and HRV had the same trend in each frequency band when calculation in normalized power without FHF. However, there is different result if we calculate normalized power with FHF. The nHF expressed different trend with HRV, as seen in the result in Table II. As a result, cardiac output information could be narrowed down into FHF. Moreover, the power of FHF had higher proportion in total power which affected normalized power calculation.

This study showed that the iPRV spectrum can be assessed by simple PPG measurement. Moreover, it has potential to be a short-term evaluation. Conventionally, HRV needs long-term (greater than 5 min) data to evaluate autonomic nervous

TABLE V. THE NVHF OF IPRV IN SUPINE AND PLR WITH DIFFERENT DATA LENGTH.

	Data length	**Supine**	**PLR**
nVHF	10 minutes	0.35±0.17	0.39±0.15
	130 seconds	0.34±0.17	0.41±0.19

The form is (mean ± standard deviation).

activities and left ventricular information in VHF. Furthermore, conventional evaluations also adopt 5 min data to assess cardiac output. Then, iPRV has the potential to be a short term evaluation by using IF estimation method. On the other hand, iPeriod has the ability to indicate enough information with short data length in PPG signal. This study used fragment data of PPG signal in supine and PLR. Table V showed that PPG signal with 130 seconds was no significant difference ($p = 0.28$ in supine; $p = 0.19$ in PLR) with whole data length, and it also had relative trend between supine and PLR. This result demonstrated that iPRV has potential as a short-term cardiac output evaluation.

V. CONCLUSION

nVHF in iPRV had relative variation to venous return change in different postures, and it showed that iPRV is a potential evaluation for assessing cardiac output. Furthermore, result in time domain comparison illustrated that iPeriod contained high frequency information, and iPeriod_{LP} had positive correlation with RRI series. In conclusion, iPRV is not only able to present ANS activity but also reliable to evaluate cardiac output. iPRV has potential to predict FR as short-term assessment and to be a reliable system for ICU in order to avoid delaying definitive therapy or produce additional damage patients. In the future, iPRV in FR assessment has to be further explored and examined with clinical patient.

ACKNOWLEDGMENT

This work was supported by the Taiwan Ministry of Science and Technology under grant numbers MOST-103-2218-E-009-016, MOST-103-2221-E-009-139, MOST 104-2220-E-009-003, and MOST 105-2914-I-009-011-A1, and in part by 'Aim for the Top University Plan' of the National Chiao Tung University and Ministry of Education, Taiwan, R.O.C. This work was also supported in part by the UST-UCSD International Center of Excellence in Advanced Bioengineering sponsored by the Taiwan National Science Council I-RiCE Program under grant number NSC-101-2911-I-009-101.

REFERENCES

[1] J. H. Boyd, J. Forbes, T. A. Nakada, K. R. Walley, and J. A. Russell, "Fluid resuscitation in septic shock: A positive fluid balance and elevated central venous pressure are associated with increased mortality*," Critical Care Medicine, vol. 39, 2011, pp. 259–265.

[2] F. Michard, et al., "Relation between respiratory changes in arterial pulse pressure and fluid responsiveness in septic patients with acute circulatory failure," American Journal of Respiratory and Critical Care Medicine, vol. 162, 2000, pp. 134–138.

[3] M. M. Krausz, A. Perel, D. Eimerl, and S. Cotev, "Cardiopulmonary effects of volume loading in patients in septic shock," Annals of Surgery, vol. 185, 1977, pp. 429–434.

[4] D. A. Reuter, et al., "Stroke volume variations for assessment of cardiac responsiveness to volume loading in mechanically ventilated patients after cardiac surgery," Intensive Care Medicine, vol. 28, 2002, pp. 392–398.

[5] P. E. Marik, M. Baram, and B. Vahid, "Does central venous pressure predict fluid responsiveness?: a systematic review of the literature and the tale of seven mares," CHEST Journal, vol. 134, 2008, pp. 172–178.

[6] S. Soubrier, et al., "Can dynamic indicators help the prediction of fluid responsiveness in spontaneously breathing critically ill patients?," Intensive Care Medicine, vol. 33, 2007, pp. 1117–1124.

[7] M. V. Højgaard, N. H. Holstein-Rathlou, E. Agner, and J. K. Kanters, "Reproducibility of heart rate variability, blood pressure variability and baroreceptor sensitivity during rest and head-up tilt," Blood Pressure Monitoring, vol. 10, 2005, pp. 19–24.

[8] G. G. Berntson, et al., "Heart rate variability: origins, methods, and interpretive caveats," Psychophysiology, vol. 34, 1997, pp. 623–648.

[9] E. Gil, et al., "Photoplethysmography pulse rate variability as a surrogate measurement of heart rate variability during non-stationary conditions," Physiological Measurement, vol. 31, 2010, pp. 1271–1290.

[10] C. C. Chang, T. C. Hsiao, and H. Y. Hsu, "Frequency range extension of spectral analysis of pulse rate variability based on Hilbert–Huang transform," Medical & Biological Engineering & Computing, vol. 52, 2014, pp. 343–351.

[11] S. Babaeizadeh, et al., "A novel heart rate variability index for evaluation of left ventricular function using five-minute electrocardiogram," Computers in Cardiology, vol. 34, 2007, pp. 473–476.

[12] N. E. Huang, et al., "On instantaneous frequency," Advances in Adaptive Data Analysis, vol. 1, 2009, pp. 177–229.

[13] H. T. Wu, "Instantaneous frequency and wave shape functions (I)," Applied and Computational Harmonic Analysis, vol. 35, 2013, pp. 181–199.

[14] D. J. Meredith, et al., "Photoplethysmographic derivation of respiratory rate: a review of relevant physiology," Journal of Medical Engineering &Technology, vol. 36, 2012, pp. 1–7.

Telemonitoring Protocol for Prevention and Comorbidity Screening, in Paediatric Patients with Cystic Fibrosis and/or Diabetes, by HVR index

Irene Tagliente [1,2], Fabrizio Murgia[1].
[1]Scientific Direction, Research Area of Clinical-Healthcare and Management Innovations,
Bambino Gesù Children's Hospital, Rome – Italy
[2] Department of Human Sciences, Society and Health-University of Cassino and Southern Lazio.
email: Irene.tagliente@opbg.net;
fabrizio.murgia@opbg.net

Riccardo Schiaffini, Antonella Lorubio
[3] Endocrinology Department,
Bambino Gesù Children's Hospital, Rome – Italy
Email: Riccardo.schiaffini@opbg.net;
antonella.lorubio@opbg.net

Fabio Majo, Sergi Bella
Specialized Medicine Department, Complex Operating Unit of Cystic Fibrosis
Bambino Gesù Children's Hospital, Rome – Italy
Email: Fabio.majo@opbg.net; Sergio.bella@opbg.net

Marcello Chinali
Unit of Pediatric Cardiology and Arhythmology,
Department of Pediatric Cardiology and Cardiac Surgery.
Bambino Gesù Children's Hospital, Rome – Italy
Email: Marcello.chinali@opbg.net

***Abstract* - This pilot project is one of the first applications of Telemedicine solution (Telemonitoring) for comorbidity and translational study. A total of 20 patients with Diabetes and/or Cystic Fibrosis were enrolled to a telemedical intervention and assigned to different Groups. The patients, who enrolled voluntarily, were as follows: 5 with consecutive Type1 Diabetes Mellitus and assigned to Group 1; 5 with Cystic Fibrosis already under Telemonitorig protocol and assigned to Group 2; 5 with consecutive Type 1 Diabetes Mellitus and Cystic Fibrosis already followed at Bambino Gesù Children's Hospital, Unit of Endocrinology and Diabetes and by Telemedicine group of Cystic Fibrosis Unit and assigned to Group 3; and 5 voluntary without disease assigned to Group 4 (Control group). The Aim of this study was to analyze possible variation of Heart Rate Variability, depending on the glycaemia value or Forced Expiratory Volume in 1 second in adolescent and adult patients with diabetes and/or Cystic Fibrosis by telemedicine protocol and analyze the correlation between compliance and patients technology background. In the first four months of Telemonitoring, we received 855 glycaemia transmissions and 378 spirometry test transmissions. We show a good compliance trend, especially in the patients with technology background. For the patients or patient family members without technology background, we offered telephonic assistance to verify the home procedures, and to store and download the data. Preliminary analysis of data showed no overall significant differences in Heart Rate Variability parameters among the three groups. More months of observation are needed to show possible correlation between Forced Expiratory Volume in 1 second, glycaemia value and Health Rate Variability. Various Telemonitorig solutions could be important tools for new international comorbidity research, but an easy methodology to share the data from Home to Hospital has to be taken into consideration when planning a Telemedicine protocol assistance.**

Keywords- Heart rate variability; Diabetes; Cystic Fibrosis; eHealth; Telemonitoring; Prevention; Equipment.

I. INTRODUCTION

Heart Rate (HR) can appear static and regular at rest, during exercise or recovery after exercise. However, HR is constantly adjusted due to factors such as breathing, blood pressure control, thermoregulation and the renin-angiotensin system, leading to a more dynamic response that can be quantified using Heart Rate Variability (HRV).

HRV is defined as the deviation in time between successive normal heart beats and the total HR variation in a certain interval can be measured by non-invasive methods [2].

HRV can serve as measure of autonomic activity of sino-atrial node [1]. The clinical relevance of heart rate variability was first appreciated in 1965 when Hon and Lee [3] noted that fetal distress was preceded by alterations in interbeat intervals before any appreciable change occurred in the heart rate itself [4].

The clinical importance of HRV became apparent in the late 1980s when it was confirmed that HRV was a strong and independent predictor of mortality following an acute myocardial infarction [4].

A high degree of HRV is found in compensated hearts with good function, whereas HR variability can be decreased with severe coronary artery disease, congestive heart failure, aging and diabetic neuropathy [5].

Diabetes is a hormonal disorder that affects around three hundred million people worldwide [6]. Several therapies have been proposed and a good compliance to treatment offers patients a good quality of life. Diabetes is a disease characterized by a group of metabolic disorders caused by defects of insulin's secretion and / or activity.

The first condition is represented by a pancreas inability to produce insulin and the latter is characterized by incorrect use of normal secreted insulin (insulin resistance). Literature data concerning bronchial reactivity in diabetic patients is controversial [7]. Changes in the respiratory system are little known, tardive complications of diabetes. The mechanisms which lead to dysfunctions of the respiratory system in patients with diabetes might be: microangiopathy of pulmonary vessels, changes of alveoli's structure, vegetative neuropathy of the respiratory system, changes of bronchi's reactivity and dysfunctions of the mobility of the thorax [8].

The prevalence of complications such as micro and macro angiopathy involving heart, kidney, eyes, Central Nervous System (CNS) are also increasing, causing severe economic and social burden [9].

Uz-Zaman, Salim et al. [9] show significant changes of Forced Expiratory Volume in 1 second (FEV1), and related pulmonary function indicators, in Type-2 diabetes patients and it has been correlated with poor glycemic control. The above pattern of changes are possibly due to hyperglycemia induced non enzymatic glycosylation of tissue proteins and chronic diabetic microangiopathy causing basement membrane thickening (capillaries and endothelium) leading to reduction in strength and elasticity of connective tissues and reduced pulmonary blood volume.

Another chronic disease with impact on pulmonary function is Cystic Fibrosis (CF). CF is characterized by progressive lung destruction, caused by obstruction of the airways due to dehydrated thickened secretions, which results in endobronchial infection, and an exaggerated inflammatory response leading to development of bronchiectasis and progressive obstructive airways disease [10].

In these patients, spirometry shows a reduction in Forced Expiratory Volume in the first second (FEV1), and in Current Volume (FVC), (around 2 % of the expected yearly value) [11]. In previous study [12], we have shown how the Telemedicine can improve the management of CF disease and prevent complications.

In the Pediatric Hospital Bambino Gesu', Telemedicine is also offered in the follow up of diabetes patients as additional service. In previous trial research [13][14], we have also shown the impact of long-term use of eHealth systems in adolescents with Type 1 diabetes. We demonstrated a favorable impact of monthly tele-assistance (as phone call) on treatment compliance.

We have shown how patients receiving frequent feedback provided by the medical/multidisciplinary team, on telemonitoring procedure, were more compliant in self-management of diabetes. In particular, Telemonitorig protocol can help the medical team to promptly give feedback on behavioral errors, and insulin therapy adjustments. The aim of this study was to find the correlation between HVR and FEV1 and glycemia variations of patients with diabetes and/or CF and control group in a pilot study of 4 months.

The main aim of this research was to understand if HRV could be a cardiovascular prevention indicator also in patients with diabetes and/or CF followed with additional Telemonitoring assistance. The rest of this paper is organized as follows. Section II describes our methods, including the enrolling criteria. In Section III, we present our results. Section IV presents our conclusions and ideas for future works.

II. METHODS

We have performed an observational study with 20 volunteer patients already followed at Bambino Gesù Children's Hospital. The patients enrolled had different diseases: Diabetes and/or CF. They were assigned to a telemedical intervention and divided in 3 Groups, plus one Control group with volunteers, without disease. The voluntary patients enrolled were as follows: 5 with consecutive T1DM (SAP-treated) and assigned to Group 1; 5 with CF already under Telemonitorig protocol and assigned to Group 2; 5 with consecutive T1DM and CF already followed at Bambino Gesù Children's Hospital, Unit of Endocrinology and Diabetes and by Telemedicine group of Cystic Fibrosis Unit and assigned to Group 3; and 5 assigned to Group 4 (Control group).

The Telemonitorig intervention guaranteed tele-assistance and tele-interaction between the medical team, engineers and the patients/families.

All the enrolled patients were monitored and followed for a study-period. Patients with a Tanner Stage <IV (pre-pubertal) were excluded from this study. The Tanner System describes the sequence of changes in secondary sexual characteristics and is the staging system utilized most frequently in children and adolescents.

Moreover, in order to exclude a potential effect of duration of disease on diabetes compliance and management capacities, patients with diabetes duration <1 year were excluded from the randomization. Furthermore, the mean HbA1c level in the year before randomization was evaluated for each study group.

A. Standard protocol

During the whole study, all patients had a regular standardized protocol training of education about correct diabetes and CF control provided by a multidisciplinary team (diabetologist, specialized doctor, nurse, dietician, and psychologist). All the patients and their families were given instruction in carbohydrate counting, spirometry and RR test (RR interval variations present during resting conditions represent beat-by-beat variations in cardiac autonomic inputs) procedures and were recommended to follow a balanced nutritional program with a calorie intake regularly distributed between carbohydrate (55%), protein (15%) and lipids (30%).

Moreover, all of the enrolled subjects followed a similar and regular aerobic physical activity program for a total commitment of three hours per week with or without oxygen, depending on the clinical conditions. All patients were equipped with a Glycaemia kit consisting of an Android Tablet [Noesis- Infosolution Spa].

In addition, all patients were equipped with a watch Polar V800 and heart rate sensor able to record RR value for the HRV analysis, for one week a month for 4 months.

Patients in the telemedicine group and control group were asked to do a glycaemia control 5 times/day and an RR measure for 5 minutes/day for one week a month for 4 months and spirometry.

The Telemonitorig window period for this pilot study was fixed for only four months because of the median stability of HVR of each patient, in the short period.

Also, patients with CF were asked to perform a spirometry test every day, as usual, using the CF Telemonoring protocol. All the other enrolled patients not having CF were asked to perform a spirometry test during a hospital visit with Spirobank-MIR, during the equipment meeting, and not less than once per month. All spirometer tests were analysed with WinSpiroPro Software (MIR developer).

A personal online website profile on Polar, Noesis.Infosolution and dedicated internal Network EAD1 (Food Diabetes Education), was edited in order to receive regular feedback provided by the medical team during the virtual sessions at one-month intervals, by the engineer with one week intervals to have information about the quality of the transmissions or about equipment information. They were also educated and periodically (at 4 week intervals) retrained to use App updates.

B. Data Analysis

All data received by mobile devices was stored on the Noesis web server and exported from the local server.

The spirometry file could be exported directly on .xls, otherwise the data related on fitness activity could be exported on .csv and it was converted into .xls format afterwards. The same was done for the first App release and so for glycaemia data.

C. Data Transfer

The patients were able to share the data in different ways, depending on the device. RR value was stored by USB connection on the PolarFlor software on each patient profile.

The data transfer, from spirometry to hospital, was done by an integrated SIM for internet connection.

The glycaemia values were transferred by USB connection from glucometer to Android tablet and shared onto the Noesis server thanks to SIM for Internet connection. The RR value was acquired by Polar V800, downloaded by Polarflow software, exported in .csv file and analysed with HRV-Kubioshrv Software.

D. Data Storage

All patients accepted the analysis of the data in line with the international guideline for the privacy of the data.

The data was stored on Polar web-site (PolarFlow); for the spirometry data was stored on MIR server and at Bambino Gesù Children's Hospital. The glycaemia data was stored on Noesis server and shared with the hospital as .csv file.

III. RESULT

Patients with Diabetes and/or CF were enrolled, assigned to a telemedical intervention and divided into 4 groups.

In the first two months of Telemonitoring, we received 322 transmissions. We recived a total of 855 glycaemia transmissions in the last month of Telemonitoring.

In the first two months of monitoring, the most compliant with the protocol were 2 patients with only diabetes and 3 patients with diabetes and CF. The glycemia transmissions from most complaint patients are shown in Figure 1 and Figure 2.

For the spirometry test, we received data from CF patients only by Spirotel II-MIR. During the 1th period of trail, 3 CF patients were hospitalized, so they did not send the data and 1 patient asked to be enrolled in a different time period. Nevertheless, we received 378 transmissions with spirometry test.

-All RR data, downloaded on PolarFlow (Figure 4), was analysed with HRVkubioshrv Software and exported as .pdf file, as shown in Figure 3. All values were acquired with Polar V800 (Figure 5).

For this project we chose an observational window of 4 months because 3/4 weeks of HRV screening was valued enough for this kind of patients, to show some cardiac dysfunction or index of cardiovascular risk associated on glycaemia value.

Preliminary analysis of data showed no overall significant differences in HRV parameters among the three groups. In detail, no overall significant reduction in HRV parameters could be observed in DM1 and CF patients as compared to normal subjects, suggesting the lack of significant abnormalities in our selected group of patients. It is important to highlight that a significant reduction in (Root Mean Square of Successive Differences) RMSSD was found in CF patients when reduction in FEV1 was reported, often preceding acute respiratory complications. This finding is in accordance to the notion that acute increase in the patient inflammatory status reduces HR variability, through a mean increase in heart rate. Nonetheless, in our study population, this finding might be of interest as it shows that acute reduction in HRV in these high-risk patients might represent, similarly to what is already known for acute reduction in FEV1, a preclinical sign of incoming complications, thus suggesting more aggressive monitoring and need for prompt clinical evaluation.

We did not observe any learning difference depending on the age of the patients, but we showed some learning difficulty, depending of the technology background.

IV. CONCLUSIONS AND FUTURE WORK

This pilot project is one of the first applications of Telemedicine solution (Telemonitoring) for comorbidity and translational study.

In our study, we started a new Telemedicine protocol for cardiologic prevention screening in-patient with CF and or Diabetes to show the possible correlation between FEV1, glycaemia value and HRV index.

We show a good compliance rate, especially for patients with technology background. For the patients or patients family members without technology background, we offered telephonic assistance to verify the home procedures, and to store and download the data.

A long period of time of several months of Telemonitoring is needed to find more co-morbidity correlation between decrease in FEV1 and RMSSD.

After 4 months of Telemonitoring, it was possible to detect some important reduction of RMSSD in association with FEV1 reduction, before unpredictable hospitalisation.

More months of observation are needed to show possible correlation between FEV1, glycaemia value and HRV.

The pilot study will be extended for patients with CF and diabetes, so with higher risk of cardiovascular complications.

More pilot studies are needed using the Telemonitoring program for comorbidity analysis. Care must be taken when purchasing devices in order to choose the most accurate, intuitive and with an accepted communication protocol.

V REFERENCES.

[1] GRAD C. "Heart rate variability and heart rate recovery as prognostic factors". Clujul Medical. 2015;88(3):304-309. doi:10.15386/cjmed-498.

[2] American Society of Pacing and Electrophysiology. "Heart rate variability. Standards of measurement, physiological interpretation, and clinical use". Task Force of the European Society of Cardiology and the North Eur Heart J. 1996 Mar; 17(3):354-81.

[3] Hon EH, Lee ST. "Electronic evaluations of the fetal heart rate patterns preceding fetal death, further observations". Am J Obstet Gynec 1965; 87: 814–26.

[4] Task Force of The European Society of Cardiology and The North American Society of Pacing and Electrophysiology . "European Society of Cardiology. Guidelines. Heart rate variability. Standards of measurement, physiological interpretation, and clinical use." European Heart Journal (1996) 17, 354–381

[5] Kleiger R.E., Miller J.P.,. Bigger Jr. J.T, Moss A.J. Decreased heart rate variability and its association with increased mortality after acute myocardial infarction. The American Journal of Cardiology, Volume 59, Issue 4, 1 February 1987, Pages 256-262

[6] Danaei G, Finucane MM, Lu Y, Singh GM, Cowan MJ, Paciorek CJ et al. "National, regional, and global trends in fasting plasma glucose and diabetes prevalence since 1980: systematic analysis of health examination surveys and epidemiological studies with 370 country-years and 2.7 million participants". Lancet, 2011, 378(9785):31–40.

[7] Pieron. M., Scheen A. J., Corhay J.L., Radermecker M.F.,Lefebyre P.J. "Bronchial reactivity in diabetic patients". Revue des Maladies Respiratoires, (1997). 14, 379–385.

[8] Pieniawska A., Horodnicka-Józwa A., Petriczko E., Walczak M. "Evaluation of respiratory function tests in children and adolescents with Type 1 diabetes". Pediatric Endocrinology, Diabetes, and Metabolism, (2012). 18, 15–20.

[9] Uz-Zaman, Salim et al. "Assessment of Lung Function by Spirometry and Diffusion Study and Effect of Glycemic Control on Pulmonary Function in Type 2 Diabetes Mellitus Patients of the Eastern India." Journal of Clinical and Diagnostic Research : JCDR 8.11 (2014): BC01–BC04. PMC. Web. 3 Dec. 2015.

[10] Mogayzel P.J Jr., Naureckas E.T.,. Robinson K.A, Mueller G., Hadjiliadis D., et al. "Cystic Fibrosis Pulmonary Guidelines", American Journal of Respiratory and Critical Care Medicine, Vol. 187, No. 7 (2013), pp. 680-689.doi: 10.1164/rccm.201207-1160OE

[11] Meystre S. "The current state of telemonitoring: a comment on the literature". Telemed J E Health 2005;11:63-9

[12] Murgia F., Corona B., Bianciardi F., Romano P., Tagliente I., Bella S. "The application of telemedicine in the follow-up of lung transplantation in a patient with cystic fibrosis". Clin Ter. 2014;165(5):e382-3. doi: 10.7417/T.2014.1769.

[13] Tarvainen MP1, Niskanen JP, Lipponen JA, Ranta-Aho PO, Karjalainen PA. Kubios HRV--heart rate variability analysis software. Comput Methods Programs Biomed. 2014;113(1):210-20. doi: 10.1016/j.cmpb.2013.07.024. Epub 2013 Aug 6.

[14] Tagliente I, Ullmann N, Ritrovato M, Trujillo F.J.L and Riccardo Schiaffini, "Benefit of Telemedicine for patients with diabetes mellitus" The Seventh International Conference on eHealth, Telemedicine, and Social Medicine, eTELEMED 2015. IARIA, 2015. ISBN: 978-1-61208-384-1.

[15] Schiaffini R, Tagliente I, Carducci C, Ullmann N, Ciampalini P, Lorubbio A, Cappa M. "Impact of long-term use of eHealth systems in adolescents with Type 1 diabetes treated with sensor-augmented pump therapy". J Telemed Telecare. 2015 Aug 18. pii: 1357633X15598425. [Epub ahead of print]

[16]

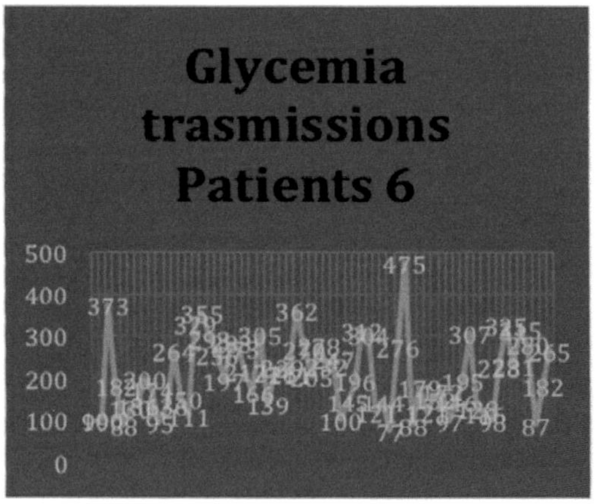

Figure 1.Glycemia's transmissions from most compliant patients.

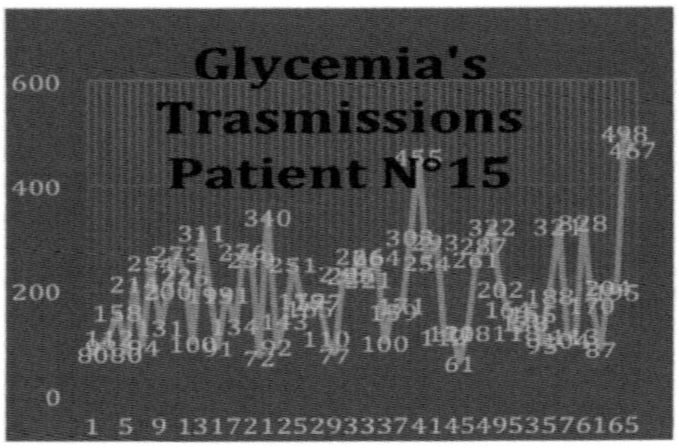

Figure 2. Glycemia's transmissions from compliant patients

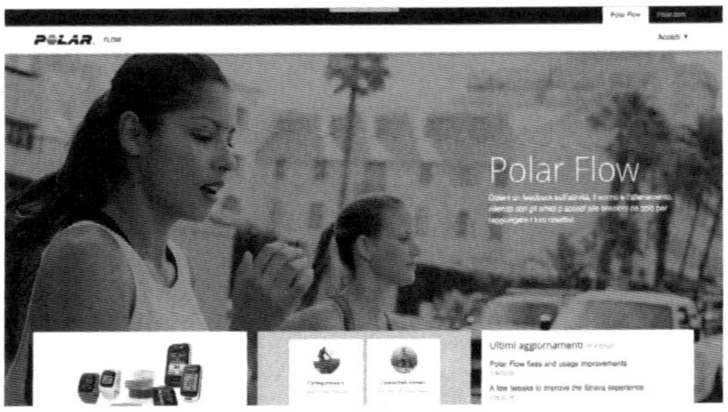

Figure 4. PolarFlow Web-site

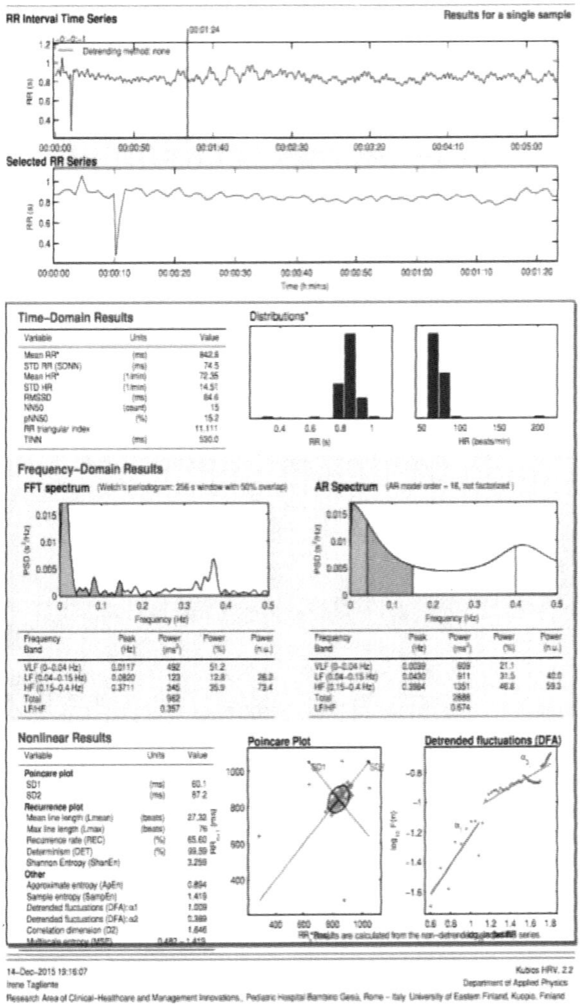

Figure 3. RR Analysis

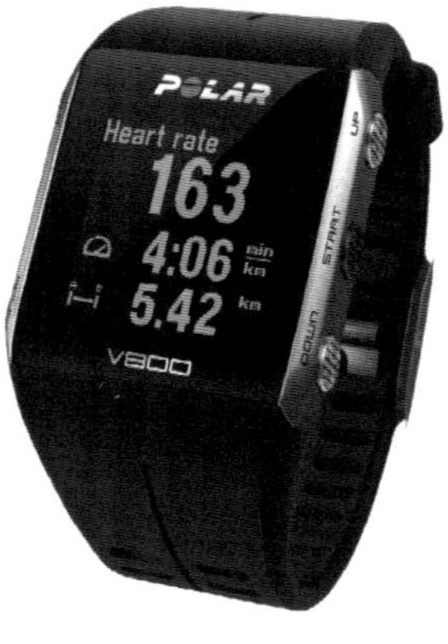

Figure 5. PolarV800

Combining Personal Health Records and Relevant External Data Sources: A Way for Achieving New Outcomes for Personal Healthcare

Juha Puustjärvi
University of Helsinki
Helsinki, Finland
email: juha.puustjarvi@cs.helsinki.fi

Leena Puustjärvi
The Pharmacy of Kaivopuisto
Helsinki, Finland
email: leena.puustjarvi@kolumbus.fi

Abstract—**Personal Health Records (PHRs) have the potential to dramatically contribute to healthcare as they enable patient to become more involved and engaged in their care. However, PHRs are rather limited in that they assume all its content to be restricted on health-oriented personal data. Yet there are a lot of related data that are stored in other systems, and which use together with PHRs' data would produce outcomes that could not be achieved by functioning independently. Using these data sources together with PHRs' data we can achieve new outcomes. How it can be carried out by using modern Semantic Web technologies, such as Resource Description Framework (RDF), Web Ontology Language (OWL) and SPARQL, and our designated ontologies is the topic of this paper. We also introduce the notion of SPARQL-affinity domain, which allows the sharing of PHRs and other relevant data in a controlled way over the Internet.**

Keywords - Personal Health Records; Semantic Web; Open Data Sources; SPARQL

I. Introduction

Issues with combining heterogeneous data sources, under a single query interface have existed from the early 1980s, when computer scientists began designing systems for interoperability of heterogeneous databases. Nowadays, at the advent of the Semantic Web the issues with combining data sources is still equally relevant as the Semantic Web paradigm involves a broad set of modern technologies such as Resource Description Framework (RDF) [1], Web Ontology Language (OWL) [2] and SPARQL [3] that tackle these issues.

A relevant issue is how these technologies can be used in combining relevant external data sources with Personal Health Records (PHRs) [4]. A PHR is a record of a consumer that includes data gathered from different sources such as from health care providers, pharmacies, insures, the consumer, and third parties [5]. It includes information about medications, allergies, vaccinations, illnesses, laboratory and other test results, and surgeries and other procedures [6]. It is accessible to the patient and to those authorized by the patient.

A problem of current PHRs is that they assume all its content to be restricted on health-oriented personal data. However, there are a lot of related data, which use together with PHR data would produce outcomes that could not be achieved by functioning independently.

Examples of such PHR-related personal data sources include gyms, smart homes and personal note books. Gyms store data that is gathered by sensor and training equipment. Smart homes store a lot of data related to heating, air conditioning, and personal well fare such as weight measurements. Personal note books may include a variety of useful information concerning working hours, meals and location data.

Using these data sources together with PHRs' data we can achieve new outcomes: For example, a person may be interested to know his or her blood pressures when his or her weight had maximal and minimum values. Also, a person might be interested to know his or her cholesterol values grouped by his or her daily training hours.

There are also a lot of public data sources, which use together with personal data would produce outcomes that would not be achieved by using only personal data. For example, personal data may indicate the vaccinations of a person, while public data source can augment this information by more informal descriptions of the vaccinations. Further, as public data sources are increasingly linked among themselves according to the notion of the Linked Data [7], the data sources that can be reached from PHRs is increasing all the time. For example, the open data sources dealing with medicines or clinical guidelines [8] are particularly useful when used with PHR-data.

Especially, clinical guidelines will have a key role in self-care. They are documents with the aim of guiding decisions and criteria regarding diagnosis, management, and treatment in specific areas of healthcare [9]. They are based on an examination of current evidence within the paradigm of *evidence-based medicine*, which is one of the most important developments in the clinical use of information over the last decades [10]. Thus, the ability to reach clinical guidelines from PHRs enable patient to become more involved and engaged in their self-care.

However, a problem in combining external data sources with PHRs is data heterogeneity: there is a variety data models on which these data sources can be based on. For example, a data source may be a relational database or an XML-file. Further, the schemas of PHRs' XML-files may be ad hoc (e.g., the CCR standard of the American Society for Testing and Materials (ASTM) [11]) or based on the Reference Information Model (RIM) (e.g., the CCD standard of the HL7 [12]).

Our solution for this heterogeneity problem is the use of smart data. *Smart data* refers to data that is application-independent, and part of a larger information ecosystem. Furthermore RDF [1] is the key for representing smart data.

RDF is a directed, labeled graph data format for representing information in the Web. It is not a data format, but a data model with a choice of syntaxes for storing data files [13]. In RDF, we can express facts with tree-part statements known as triples. The subject identifies the thing being described, predicate is a property name, and object is property value. So, each triple is like a little sentence that states a fact [14].

However, RDF in itself does not bring smartness. It depends on the expression power of the used vocabulary. By a vocabulary we refer to a set of ontologies, which formally specifies the used terms and their semantics. The key point here is that shared ontologies provide the ability of two or more systems to exchange information and to use the information that has been exchanged [15].

We will present ontologies that enable PHR system to interoperate with other relevant data sources. We illustrate these ontologies in a graphical way as well as in OWL. Data sources (RDF-files) are queried by SPARQL and these queries are processed by SPARQL processors [16].

The rest of the paper is organized as follows: First, in Section II, we introduce the notion of the SPARQL-affinity domain, which is the key concept in our designed system. The ontologies of the SPARQL-affinity domain are introduced in Section III. Then, in Section IV we present our used method for designing ontologies from XML-schemas. Especially we present how we have developed the Personal Health Ontology from the XML-schema of the Continuity of Care (CCD) documents. We present the Personal Health Ontology in a graphical way as well as in OWL. How we can query PHR-data and other external RDF-formatted data in one SPARQL query is considered in Section V. An example of presenting an ontology instance in RDF is given in Section VI. Finally, Section VII concludes the paper.

II. SPARQL-Affinity Domain

SPARQL is an RDF query language to retrieve and manipulate data stored in RDF format. The name SPARQL is a recursive acronym for SPARQL Protocol and RDF Query Language, which is described by a set of specifications from the W3C [3]. SPARQL Protocol refers to the rules for how a client program and a SPARQL processor exchange SPARQL queries and results. There is a variety of SPARQL processors available for running queries against data both locally and remotely [16].

In our architecture, RDF formatted data stores that agree to work together for data sharing are called a *SPARQL-affinity domains* (Figure 1). Its data stores agree on a common set of policies such as how the data stores are accessed by web services, how users are identified, and how the access is controlled. However, the used policy is data store specific. For example, in the case of personal data, (such as with PHR-data, welfare data, and personal note book) a strict access policy is followed while in the case of public data sources (such as with public medicine data) no access control is needed. The access policy of smart home data can be defined case-by-case.

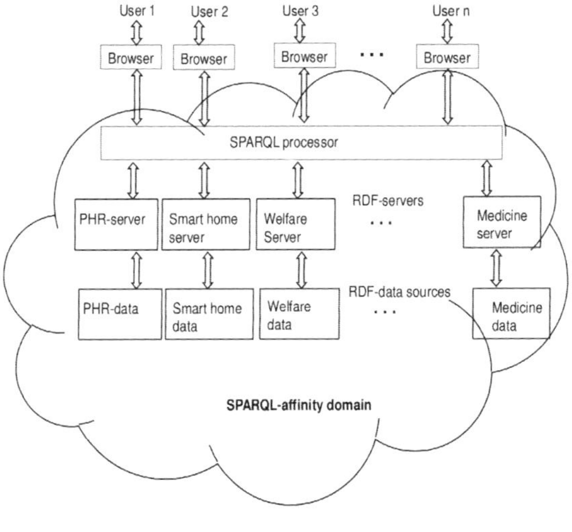

Figure 1. The architecture of the SPARQL affinity domain.

Each server provides a SPARQL endpoint [3]. A *SPARQL endpoint* is a web service that accepts SPARQL queries, runs the queries, and then returns the results. The way how multiple endpoints can be remotely processed in a SPARQL query is more detailed considered in Section V.

III. Ontologies in SPARQL-Affinity Domain

Each data source in the SPARQL-affinity domain is comprised of RDF-triples. These data sources are based on an ontology, i.e., these ontologies provide a vocabulary for these triples.

In computer science, an ontology is a general vocabulary of a certain domain, and it can be defined as "an explicit specification of a conceptualization" 17]. Essentially the used ontology must be shared and consensual terminology as it is used for information sharing and exchange.

Essentially ontology tries to capture the meaning of a particular subject domain that corresponds to what a human being knows about that domain [18]. It tries to characterize that meaning in terms of concepts and their relationships. It is typically represented as classes, properties, attributes and values. Depending on the generality level of conceptualization, different types of ontologies are needed [19]. Each type of ontology has a specific role in information sharing and exchange.

As an example consider a simplified Welfare Ontology, which is graphically presented in Figure 2. It comprises a vocabulary that a person can use in describing his or her personal welfare information. Hence, we do not assume that a person uses all the terms of the vocabulary (ontology). For example, datatype properties Father and Mother are included in the vocabulary, but the person does not have to give values for these properties. Neither the person needs class Swimming, if swimming is not included in his or her hobbies.

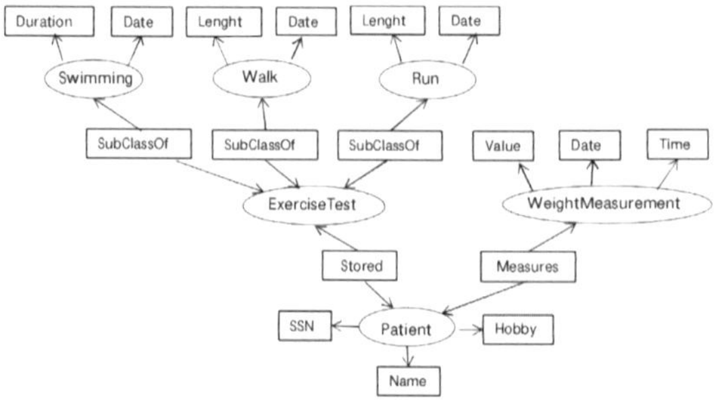

Figure 2. A graphical presentation of a portion of the Welfare Ontology.

As an example of an ontology of a public data source consider the Medicine Ontology in Figure 3. This ontology can be used in storing information about the medicines and their manufacturers. For example, it provides links to data sources (ProductInfoUrl), i.e., to the web pages, that gives detailed information about medicines.

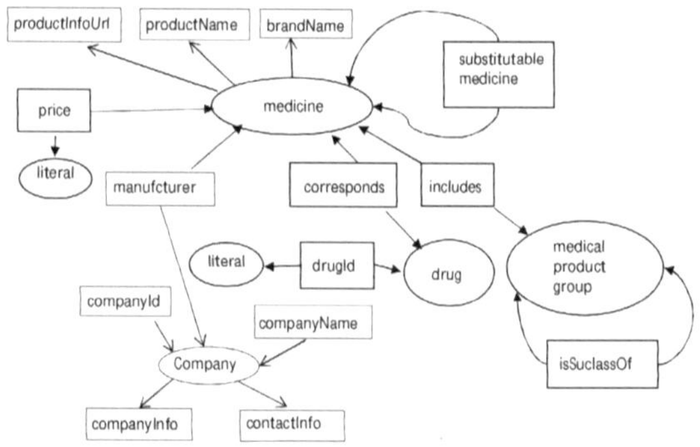

Figure 3. A graphical presentation of a portion of the Medicine Ontology.

IV. TRANSFORMING THE CCD-SCHEMA INTO PHR-ONTOLOGY

If an original data source of a SARQL-affinity domain is not in RDF, then we have to developed an appropriate ontology, and then transform the original data in the form, which is consistent with the developed ontology. For example, the data of most PHRs are based on the Continuity of Care Record (CCR) -standard [11] or CCD-standard [12], and therefore, we have developed an appropriate ontology, called the PHR-Ontology for these standards (XML-schemas).

Both CCR and CCD standards represent two different XML schemas designed to store patient clinical summaries. Both schemas are identical in their scope in the sense that they contain the same data elements such as demographics, medications laboratory results [20]. However, the structures the two XML schemas are quite different. Anyway the use of XML assures that the data contained in CCR or CCD documents can be expressed in multiple media formats that are friendly to both consumers and providers.

The CCD specification is a constraint on the HL7 CDA standard. The CCD standard has been endorsed by HIMMS (Healthcare Information and Management Systems Society Though) [21] and HITSP (Healthcare Information Technology Standards Panel) [22] as the recommend standard for exchange of electronic exchange of components of health information.

Although the original purpose of the CCD documents was to deliver clinical summaries between healthcare organizations, nowadays it increasingly used for other types of messages: it is increasingly considered as set of templates because all its parts are optional, and it is practical to mix and match the sections that are needed [23].

In transforming the XML schema of the CCD file to OWL-ontology we have used the following rules:

1. The complex elements of the XML-schema are transformed into OWL classes.
2. The simple elements of the XML-schema are transformed into OWL data properties such that the complex element is the domain of the da-ta properties.
3. The attribute of the XML-schema are transformed into OWL data properties.
4. The relationships between complex elements must be named and transformed to OWL object properties.

To illustrate this transformation consider the following example of a CCD document.

```
<SimplifiedCCDfile>
  <DocumentID>DOC_123</DocumentID>
  <Patient>
    <PatientID>AB-12345></PatientID>
    <PatientName>Tim Jones></PatientName>
  </Patient>
  <Medications>
    <Medication>
      <MedicationID>Medication.567</MedicationID>
      <DateTime>
        <ExactDateTime>2012-03-01TO12:00</ExactDateTime>
      </DateTime>
      <Source>
        <Actor>
          <ActorID>Pharmacy of Kaivopuisto</ActorID>
          <ActorRole>Pharmacy</ActorRole>
        </Actor>
      </Source>
      <Description>
        <Text>One tablet three times a day</Text>
      </Description>
      <Product>
        <ProductName>Voltaren</ProductName>
        <BrandName>Diclofenac</BrandName>
      </Product>
      <Strenght>
        <Value>50</Value>
        <Unit>milligram</Unit>
      </Strenght>
      <Quantity>
        <Value>30</Value>
        <Unit>Tabs</Unit>
      </Quantity>
    </Medication>
  </Medications>
</SimplifiedCCDfile>
```

Figure 4. A simplified example of a CCD document.

In order to illustrate the transformation rules, let us consider the graphical OWL-ontology in Figure 5, which is derived from the elements presented in the document presented in Figure 4. In the figure, ellipses represent classes, and rectangles represent data type properties and object properties. Data type properties relate objects to datatype values while object properties relate objects to other objects.

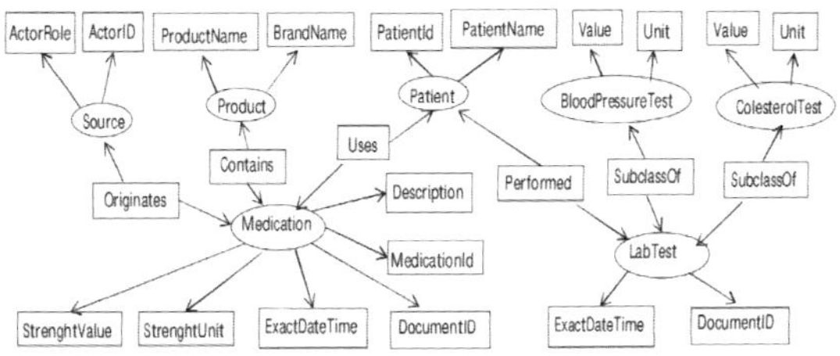

Figure 5. A simple graphical PHR-Ontology.

The graphical ontology of Figure 5 is presented in OWL in Figure 6. Due to the space limits, we have omitted the specifications of the data properties such as PatientName, ProductName and BrandName.

```
<rdf:RDF
        xmlns:rdf=http://www.w3.org/1999/02/22-rdf-syntax-nsl#
        xmlns:rdfs=http://www.w3.org/2000/01/rdf-schema#
        xmlns:owl=http://www.w3.org/2002/07/owl#>
        <owl:Ontology rdf:about="ProfileCCDontology"/>
        <owl:Class rdf:ID="Patient/">
        <owl:Class rdf:ID="Medication/">
        <owl:Class rdf:ID="Source/">
        <owl:Class rdf:ID="Product/">
        <owl:Class rdf:ID="LabTest/">
        <owl:ObjectProperty rdf:ID="Uses">
                <rdfs:domain rdf:resource="#Patient"/>
                <rdfs:range rdf:resource="#aMedication"/>
        </owl:ObjectProperty>
        <owl:ObjectProperty rdf:ID="Contains">
                <rdfs:domain rdf:resource="#Medication"/>
                <rdfs:range rdf:resource="#Product"/>
        </owl:ObjectProperty>
        <owl:ObjectProperty rdf:ID="Originates">
                <rdfs:domain rdf:resource="#Medication"/>
                <rdfs:range rdf:resource="#Source"/>
        </owl:ObjectProperty>
</rdf:RDF>
```

Figure 6. A simple PHR -Ontology in OWL.

Note that Figure 5 and Figure 6 represent only the portion of the PHR-Ontology that correspond the Medications section of the CDD. The whole PHR-Ontology is comprised of the integration of the ontologies derived from all sections of the CDD. In such integration we do not have to take care of semantic heterogeneity (i.e., one term is used in different meanings, or two terms are used in same meaning) as the all the elements in CDD documents are based on the HL7 RIM [24].

Note that querying PHR-data together with Medication data instead of querying only PHR-data we can achieve more expressive queries on patient's medication. For example, we can query the links of the web pages that provide information about the medications that are included in patient's medication. The ways multiple data sources can be queried in a SPARQL-affinity domain is the topic of the next section.

V. Querying Data Sources by SPARQL

A typical SPARQL query specifies the pieces of information that meets the stated conditions. The conditions are described with triple patterns, which are similar to RDF triples but may include variables to add flexibility in how they match against the data.

SPARQL provides two ways for querying remotely: using FROM keyword or using SERVICE keyword [3]. In the former way the FROM keyword names a dataset to query that may be local or remote file. In the latter way, instead of pointing at an RDF file somewhere, a SPARQL endpoint is pointed. A SPARQL endpoint is a web service that accepts SPARQL queries, runs the queries, and then returns the result.

Federated Queries in SPARQL allow searching multiple datasets with one query [16]. For each dataset it is created a subquery which access datasets by using SERVICE keywords. That is, federated SPARQL queries make use of subqueries and SERVICE keywords. To illustrate this consider the federated SPARQL query presented in Figure 7, which accesses two data sets The query is based on the PHR-Ontology presented in Figure 5, and on the Medicine Ontology presents in Figure 3. Prefix *med* in the query refers to the PHR-Ontology while prefix *phr* refers to the Medicine Ontology.

```
PREFIX owl: <http://www.w3.org/2002/07/owl#>
PREFIX rdf: <http://www.w3.org/1999/02/22-rdf-syntax-ns#>
PREFIX rdfs: <http://www.w3.org/2000/01/rdf-schema#>
PREFIX phr: http://www.cs.helsinki.fi/PHROntology#>
PREFIX med: <http://www.cs.helsinki.fi/medicineOntology#>

SELECT ?drugId ?price
WHERE
{
 SERVICE <http://phrRegistry/sparql>
 { SELECT ?medicineId
   WHERE
   {
    Nancy Smith phr: uses ?medicineId
   }
 }
 SERVICE <http://medicineRegistry/sparql>
 { SELECT ?drugId ?price
   WHERE
   {
     ?medicineId med: corresponds med: drugId
     ?med: drugId substitutable_drug ?drug
     ?drug med: price ?price
   }
 }
}
```

Figure 7. A simple federated SPARQL query.

The use case behind the query is the following: Nancy Smith is interested to know whether her medicines are substitutable with cheaper ones. Processing this query requires first to retrieve Nancy's used medicines from her PHR, and then querying the Medicine data set (a public RDF-formatted data source based on the Medicine Ontology).

As illustrated in Figure 7, the first subquery returns the medication identifications (medicineIds) of Nancy's medication, which in turn is the input parameter for the second subquery. This subquery first finds the active substance of Nancy's medicines, and then checks which other medicines include the same active substance (i.e., are substitutable). Finally, the main query outputs the medication identifications and the prices of these medicines.

This kind of cross-referencing feature is very useful in the SPARQL affinity domain as there is a variety of needs to cross-reference data from multiple data sources.

VI. Representing Smart Data by RDF

In order that RDF data can be represented and transmitted it needs a concrete syntax, which is given in XML, i.e., RDF statements are usually coded in XML. Hence, RDF inherits the benefits associated with XML. However, other syntactic representations (e.g., Turtle [25]) are also possible, meaning that XML-based syntax is not a necessary component of the RDF model.

One RDF description may contain one or more RDF statements about an object. For example, in Figure 8, the description concerning Mary Taylor's weight measurement (identified by "weightmeasurement100820151028") contains five RDF statements: the first states that its type in the Welfare Ontology is WeightMeasurement, and the second states that it measures Mary Taylor.

```
<rdf:RDF
    xmlns : rdf="http://www.w3.org/1999/02/22-rdf-syntax-ns#"
     xmlns : po=http://www.helsinki.fi/Welfare_Ontology#>
   <rdf:Description rdf:about="weightmeasurement100820151028">
         <rdf:type rdf:resource="&po;WeightMeasurement"/>
          <po : Measures>Mary Taylor</po : Measures>
          <po : Date >10:08:2015</po:Uses>
           <po : Time >10:28</po:Time>
            <po : Value >68.7</po:Value>
      </rdf : Description>
  </rdf:RDF>
```

Figure 8. An instance of the Welfare Ontology in RDF.

VII. Conclusion

Internet has changed the way people work, bank and shop, but a similar change in health care has been small-scale. However, the use of Internet-based e-health tools is rapidly increasing. These tools cover many fields including electronic health records, personal health records, telemedicine, evidence based medicine, information therapy and disease management.

Still a problem is that the each e-health tool has its own interfaces and data sources. By integrating the e-health tools we can achieve two gains: simplify user interaction and provide new more advanced services. In particular there are a lot of related data that are stored in other systems, and which use together with PHRs' data would produce outcomes that could not be achieved by functioning independently. Using these data sources together with PHRs' data we can achieve new outcomes. Further, the Semantic Web paradigm involves a broad set of modern technologies such as RDF, OWL and SPARQL that can tackle the issues with combining heterogeneous data sources.

In this paper, we have restricted on considering the SPARQL affinity domain as a technical infrastructure. However, to succeed the SPARQL affinity domain should not be considered just as a technical infrastructure but rather as ecosystems having many interconnected parts. Other key components of the whole ecosystem include governance regulations, financing and stakeholders. In our future work, we will restrict ourselves on analyzing the dependencies of these components.

References

[1] "RDF – Resource Description Language". Available at: http://www.w3.org/RDF/ Retrieved: 2, 2016.

[2] "OWL – WEB OntologyLanguage", Available at: http://www.w3.org/TR/owl-features/ Retrieved: 2, 2016.

[3] "SPARQL Query Language for RDF", Available at: http://www.w3.org/TR/rdf-sparql-query/ Retrieved: 2, 2016.

[4] C. Angst, R. Agarwal, J. Downing, "An empirical examination of the importance of defining the PHR for research and for practice," Proceedings of the 41st Annual Hawaii International Conference on System Sciences. 2008.

[5] J. Puustjärvi and L. Puustjärvi, "Capturing Semantics for Personal Health Records" In Poc. Of the 6th International Conference on Information Technology and Applications (ICITA 2009), 2009, pp. 26-31.

[6] J. Puustjärvi, and L. Puustjärvi. "The Role of Care Pathways in Personal Health Records," In the proc. of the International Conference on eHealth, Telemedicine, and Social Medicine (eTELEMED 2015). 2015.

[7] DuCharme. Learning SPARQL. O'Reilly Media.

[8] N. Mulyar, M. van der Aalst, M. Peleg, "A pattern-based analysis of clinical computer interpretable guideline modeling languages," Journal of American Medical Information Association, 2007, vol. 14, no. 6, pp.781–787.

[9] M. Peleg, S. Tu, P. Ciccarese, J. Fox, "Comparing computer-interpretable guideline models," A case-study approach. Journal of American Medical Information Association, 2003, pp.10–58.

[10] T. Leaon, K. Kaiser, S. Miksch, "Free and open source enabling technologies for patient-centric,guideline-based clinical decision support," A survey. Methods of Information in Medicine, 2007, pp. 74–86

[11] "Continuity of Care Record (CCR) Standard", Available at: http://www.ccrstandard.com/ Retrieved: 2, 2016.

[12] CCD, "What Is the HL7 Continuity of Care Document?", Available at: http://www.neotool.com/blog/2007/02/15/what-is-hl7-continuity-of-care-document/ Retrieved: 2, 2016.

[13] J., Davies, F. Fensel, D. Harmelen. "Towards the semantic web: ontology driven knowledge management," West Sussex: John Wiley & Sons.2002.

[14] G. Antoniou, and F. Harmelen, "A semantic web primer," The MIT Press. 2004.

[15] J. Puustjärvi, and L. Puustjärvi, "Developing a Data Infrastructure for Patient-Centered Telemedicine",

International Journal on Advances in Life Sciences (IJALS). Vol. 6, No.3&4, 2014, pp. 229-239.

[16] "SPARQL Query Language for RDF', Available at: http://www.w3.org/TR/rdf-sparql-query/ Retrieved: 2, 2016.

[17] T.R. Gruber, "Toward principles for the design of ontologies used for knowledge sharing." Padua workshop on Formal Ontology, March 1993.

[18] M. Daconta, L. Obrst, K. Smith, "The semantic web. :A Guide to the Future of XML, Web Services, and Knowledge Management," John Wiley & Sons. 2003

[19] J. Puustjärvi, and L. Puustjärvi. "Ontology-Based Competence Management in Pharmacy." International Journal of Adaptive, Resilient and Autonomic Systems (IJARAS). 4(2), 2013, pp 94-112.

[20] T. Benson, "Principles of Health Interoperability HL7 and SNOMED," Springer. 2010.

[21] HIMMS (Healthcare Information and Management Systems Society, http://himms.org/

[22] Healthcare Information Technology Standards Panel, http://hitsp.org/ Retrieved: 2, 2016.

[23] K. Boone, " The CDA Book,". Springer. 2001.

[24] R. Dolin, L. Alschuler, C. Beerb, P. Biron, S. Boyer, E. Essin, T. Kimber,. "The HL7 Clinical Document Architecture", J. Am Med Inform Assoc, 2001:8(6), pp. 552-569.

[25] "Turtle - Terse RDF Triple Language," Available at: http://www.w3.org/TeamSubmission/turtle/. Retrieved: 2, 2016.

Big Data for Personalized and Persuasive Coaching via Self-monitoring Technology

Lisette van Gemert-Pijnen, Floor Sieverink, Liseth Siemons, Annemarie Braakman-Jansen
j.vangemert-pijnen@utwente.nl, f.sieverink@utwente.nl, l.siemons@utwente.nl, l.m.a.braakman-jansen@utwente.nl
Center for eHealth and Wellbeing Research: dept., Psychology, Health and Technology
University of Twente
Enschede, The Netherlands

Abstract—**Big Data are the key towards personalized healthcare. However, this confronts us with new challenges. What are the hopes, challenges and dangers for using Big Data to develop personalized and persuasive coaching systems? The aim of this paper is to prompt a discussion on Big Data to personalize healthcare, using views and ideas from experts, and to develop an agenda for future research.**

Keywords-big data; self-monitoring; personalized healthcare; coaching.

I. Introduction

Many different definitions are available for "Big Data". Where Mayer-Schönberger and Cukier [1] focus on the new insights and economic value that can be obtained from Big Data in contrast to traditional smaller settings, Wang & Krishnan [2] refer to Big Data as complex and large data sets that can no longer be processed using the traditional processing tools and methods. Yet another definition comes from Laney [3], who defines Big Data according to 3 assets (often referred to as the 3V-model) that require new, cost-effective forms of information processing to promote insight and decision making, including: 1) high-volume (i.e., the quantity of data), 2) high-velocity (i.e., the speed of data generation and processing), and 3) high-variety (i.e., the amount of different data types).

Marr [4] expanded this 3V model to the 5V model by adding 2 additional Vs: veracity (i.e., the accuracy or trustworthiness of the data) and maybe the most important asset: value (i.e., the ability to turn the data into value).

Though this is just a grasp out of all the definitions available, there is one thing they have in common: The use of Big Data for analysis and decision making requires a change of thought from knowing "why" to knowing "what". We focused on small, exact datasets and causal connections (i.e., knowing "why") in the past; now we focus on gathering or linking a large amount of (noisy) data with which we can demonstrate the presence of (unexpected) correlational connections (i.e., knowing "what") [1]. As a result, we will obtain (and apply) new insights that we did not have before. Insights that can not only be lifesaving, but that can also open the door towards more personalized medicine [5-7] to tailor medical decisions, medications and/or products to the individual's personal profile instead of to what is best for a group of patients. For example, the use of genetic biomarkers in pharmacogenetics can be used to determine the best medical treatment for a patient [5] or the analysis of data from thousands of patients that have been treated in the past can be used to determine what treatment best fits the individual patient that is under treatment now, e.g., in terms of expected treatment effects and the risk for severe side-effects given the patient's personal characteristics like age, gender, genetic features, etc. This shift towards more personalized healthcare is reflected in the change of focus from a disease-centered approach towards a patient-centered approach, empowering patients to take an active role in the decisions about their own health [7]. As a result, an increasing number of technologies (e.g., Personal Health Records) are being launched by (insurance) companies to support chronically ill people in the development of self-management skills. Furthermore, the past decades have shown a rapid growth in the amount of (personal) data that is digitally collected by individuals via wearable technologies for self-monitoring and that may or may not be stored on online platforms for remote control [1],[5-8], and shared via other online sources like social media. Social media have become socially accepted and used by a growing group of people [9]. They use it, for example, to share data collected by activity, mood, nutrition and sleep trackers on a variety of online platforms (such as Facebook, Twitter, blogs or forums). This data provides new opportunities for personalizing and improving healthcare [10-12]. Also, the data and messages shared via these tools provide insight in vast amounts of valuable information for scientific purposes. Nagar and colleagues [12], for example, used the data from Twitter to predict flu trends and De Choudhury, Counts, and Horvitz [13] used social media as a measurement tool for the identification of depression. The information gleaned from social media has the potential to complement traditional survey techniques in its ability to provide more fine-grained measurement over time while radically expanding population sample sizes [13]. The challenge is how to predict mobility (behaviors; infections) in a trustworthy and reliable way?

Furthermore, the combination of clinical data with personal data on, for instance, physical activities, eating and sleeping patterns, might be used to tailor the treatment and coaching purposes to the needs of patients even better and are therefore seen as the key towards a future with optimal medical help [5]. However, it also confronts us with new technological and societal challenges that require more sophisticated data collection and management tools and data-analysis techniques. Concerns have been raised about security, purpose limitation, liability, safety, profiling, and data ownership, to mention just a few [1], [5], [6] but perhaps the most well-known concern bears upon our privacy [5], [6]. For a great deal, these privacy concerns are associated with potential misuse of data by, for instance, insurance companies [5], [8]. If these privacy concerns are not dealt with appropriately, the public's trust in technological applications might diminish severely [8].

In sum, we face new technological and societal challenges using Big Data:

- Amount of data is growing explosively. We have to create DATA wisdom, to enhance the value of data for personalized healthcare at individual, community and society level. This requires a new model for enhancing the value of data (in particular the social value of data),
- Transparency in decision making using large amounts of data from various sources and different qualities. How to understand automated decision rules (algorithms) to support people in making adequate choices?
- Trust, will become a key issue in data driven healthcare. The ultimate question is not privacy and security per se but how to create faith and trust in data management and data maintenance? Privacy issues become particularly relevant when the linkage of anonymous datasets leads to re-identification. Encryption of the data might prevent identification of individuals, but transparency is not always possible (e.g., when analyzing query logs with search terms). In the end it is all about creating trust to overcome uncertainty or anxiety for a digital world.
- Mobility is growing explosively, and health related issues threaten mobility (global health; infections/resistance; food, water, climate change etc. How to ensure a healthy life and to promote wellbeing?
- Our behavior models might not be up to the task to provide just-in-time, interactive and persuasive feedback, and to develop intuitive and adaptive technologies to improve self-care and to transform healthcare.
- Our business models should enable production of open source solutions in healthcare and technology

To better understand how Big Data impacts our life and healthcare this vision paper aims to generate new ideas and thoughts, to set a research agenda to address and solve the aforementioned problems. There are three sections. Section I describes the background of Big Data in literature. Section II describes the procedure of the meetings with experts (Focus group; individual meetings). Section III presents topics about Big Data that are based on the meetings with experts and literature, and Section IV describes the implications for research using Big Data in healthcare.

II. METHODS: FOCUS GROUP MEETING

To gain a more in depth picture about the pros and cons of using big data in personalized healthcare, a focus group was organized with a multidisciplinary panel consisting of 6 experts in Big Data research and quantified self-monitoring from different scientific disciplines: psychology, philosophy, computer science, business administration, law, and data science. Individual face to face meetings were conducted to validate the focus group results. Participants from the University of Twente were selected based on their societal impact, expertise and experiences with conducting Big Data research.

Many potential issues regarding the use of Big Data have already been mentioned in the literature, newspapers, social media, or debates and panel discussion websites. However, many of these media sources do not specifically address the healthcare setting and only focus on a limited set of issues at a time (e.g., the privacy and security issues). With the focus group meeting, we therefore aimed to gain more insight into the scientific and societal issues that play a role in using and managing Big Data to support the growing needs for personalized (and cost-effective) healthcare.

We used literature and information from (social)media to prepare the focus group discussions. During the discussion experts were asked to write down many issues as possible that might become relevant using Big Data for healthcare. Flap overs were used to express the issues and experts had to categorize these issues into concepts that cover the issues. These concepts are presented in this vision paper. The focus group was audio taped and transcripts were made by authors of this paper, and discussed with experts individually.

The intent of the focus group was to understand and go in-depth into issues that are related to the use of Big Data in healthcare and to provide insights into how these issues are approached by different scientific disciplines.

III. TOPICS BIG DATA: RESULTS FOCUS GROUP

A. *Empowerment*

What does it mean when you monitor your activities, food intake, or stress 24 hours a day using technologies, like smart wearables? What drives us to use these 24 hour monitoring devices and what do we need to understand the data generated by these systems? Do we understand the algorithms that are used to capture our behaviors and moods in pictures and graphs? Who owns the data and how to control the maintenance of that data? How to avoid a filtered scope on our lives ignoring others that are out of our affinity groups? The concept of empowerment captures topics as autonomy, freedom, having control.

Big data evokes a discussion about freedom and autonomy. Autonomy concerns our critical view on how to use technology, while freedom is more about our way of living and thinking. It might, therefore, be more important to focus on freedom instead of autonomy: understanding how you are being influenced and taking a stance against that instead of trying to keep everything away. The focus group made a distinction between positive freedom and negative freedom; two common concepts within the field of philosophy. Positive freedom is the freedom to do something yourself (e.g., to decide for yourself that you want to share your data), whilst negative freedom is the freedom to keep things away, protecting yourself (e.g., when you do not give permission to companies to link your data with other sources). Not losing control, being able to use, share and understand your data will be one of the topics to discuss freedom, self-efficacy using self-monitoring technologies.

Empowerment forces us to think about having control, who has the power through the use of Big Data? There might be just a small elite that understands the algorithms and with the increasing complexity, this elite will become even smaller in the future. This can create a division between people who can access and understand the algorithms and people who do not.

Empowering by personalization is one of the aims of the participatory society. Big data can be a leverage to realize this by creating a personal profile, providing the right information, at right moments to enable just in time coaching. Though it can be useful to put people in a profile, the danger of profiling is that you can never leave the assigned group again; once assigned to a group means always assigned to that group. Profiling might be suffocating to people because it creates uncertainty about what people know about you, what data are being collected, and for what purposes. Also, it is often unclear how to determine the norm to which people are compared when assigning them to a group (i.e., standardization, losing freedom). Furthermore, being assigned to a profile might lead to discrimination and certain prejudices/biases. Questions that arise are: How can profiling be used in a sensible/sound way? And who is responsible when mistakes are being made based on a certain profile?

B. Trust

Trust will become a key concept in a data driven society. This concept captures more than privacy and security issues. Trust refers to topics as how to create faith in data management and data maintenance, and how to make sense of these data for humans.

Privacy issues become particularly relevant when the linkage of anonymous datasets leads to re-identification. Encryption of the data might prevent identification of individuals, but transparency is not always possible (e.g., when analyzing query logs with search terms). In the end it is all about creating trust to overcome uncertainty or anxiety for a digital world.

People often give consent to institutions to use their data for certain purposes in return for the (free) use of the product or service. However, data can be (re)used for other purposes as well or can be sold to other interested parties, even though that is not always allowed. This leads to great concerns: e.g., healthcare insurance companies who use treatment data for other purposes on a more personal level (for instance, for determining a personalized health insurance premium based on your personal data about your health and lifestyle). It is not that people do not want to share data, they already do this using Facebook or Google services, but they want to understand what happens with the data, in particularly when it concerns the health domain.

Self-monitoring technologies, with no doctors or nurses involved in the caring process, are provided more and more by institutions. Smart algorithms can be applied to personalize data in such a way so you can manage your health and wellbeing yourself. However, these algorithms decide what information you get to see, based on information about you as a user (e.g., search history, Facebook friends, location). This will influence trust in the healthcare system, using data from your device compared to personal advices given by your doctor or nurse.

C. Data Wisdom

There is a rapid growth of self-monitoring technologies, but little is known about the reliability and validity of these systems. The lack of evidence for causality can lead to unreliability as well. Furthermore, how can you tell what you are actually measuring? How can we validate the correlations that we find? Does it really say what we think it says or are it just assumptions? How to match qualitative or experience data with quantitative data generated by your device?

Data wisdom is the concept that captures scientific and societal topics.

Scientific refers to how to create data wisdom, in several ways. Those who generate data are not the ones that have the knowledge to analyze, those who analyze lack domain insight (technologies, behaviors). Different kinds of expertise will be needed in the future to deal with Big Data. For instance, expertise to analyze Big Data, expertise to develop and understand the working of algorithms, or expertise in data interpretation and visualizations. The use of data to personalize healthcare demands for new knowledge to support critical and creative thinking to understand data driven decisions and to watch the impact on science, health and society. We all know the disaster with google flu trends, but we have to learn from these failures to set the agenda for future research in using several sources of data (geospatial data, medical data, technology device data) to develop predictive models about health and wellbeing. We have to search for new models, methods to deal with huge datasets, search for patterns rather than testing hypotheses based on small data. Results are not causal-driven but correlational-driven. This requires a change in thought. The golden rule for Randomized Clinical Trials will no longer be the ultimate format for health sciences. New methods are needed to get a grip on "big", how many data (critical mass) is needed and how rich and mature should data be to make meaningful decisions? How to add qualitative experiences and expertise to Big data? Numbers do not tell the whole story, and a clinical eye is important to interpret data in the context of individual health and wellbeing.

Societal refers to the implications for healthcare, addressing topics as ethics, values for a meaningful life. How to avoid a division between people who can access and understand the data and analytics that rule the decisions about treatments and lifestyle advices? Knowledge and skills are needed to empower people, and people should participate in debates about the values of data for self-regulations on the level of individuals, communities and society. Transparency and Trust are the key-topics in that debate. Digging into Data starts with a scientific and societal debate on the vales of data for a smart and healthy society.

IV. Discussion and Conclusion

A. *Discussion: Food for Thoughts*

What wisdom is needed to keep control, to enable self-organization in a Data Driven Society?

We want to share some food for thoughts, based on the focus group results and current debates on digging into data. We think radically new concepts, models (ICT and behavior based) and products are needed to promote self-organization, and persuasive coaching to enable intuitive and empathic designs to support decisions without nudging people into predefined formats for improving health and wellbeing.

Redefining environment in research. Radically new concepts on technology support: community driven, non-obtrusive, intelligent and intuitive ICT environments.

Redefining evidence in research, introducing value based health concepts and ecological measurements. Redefining the collection and analysis of Big Data combining social sciences with geo-health informatics and engineering to enable accessing, integrating and analysing spatiotemporal data, patient data and behavioural data.

B. *Conclusion: Future work*

This vision paper is aimed to discuss Big Data topics for personalized healthcare. The topics will be further investigated in research to 1) develop new methods, models to better measure, aggregate, and make sense of previously hard-to-obtain or non-existent behavioral, psychosocial, and biometric data, taken into account the topics from this introduction paper, 2) to develop an agenda for Big Data research to transform and improve healthcare. For example topics below will be explored in current and future research projects:

- Health analytics; advanced methods (machine learning) and models to analyze Big Data.
- Predictive modelling; to set up smart models to predict behaviors, to prevent diseases and to personalize healthcare.
- Visualization of data; how to present data meaningful to support decision making?
- Integration of (mobile) tech with data-platforms to enable automated services, to tailor feedback.
- Disruptive models (new actors, role-players in data driven systems).

Acknowledgment

There are no parties or institutes involved in this publication that require acknowledgements.

References

[1] V. Mayer-Schonberger and K. Cukier: Big Data: A Revolution That Will Transform How We Live, Work and Think. New York, NY: Houghton Mifflin Harcourt; 2013.

[2] W.Wang and E. Krishnan: Big Data and Clinicians: A Review on the State of the Science. JMIR Med Inform. 2014;2(1):e1. doi:10.2196/medinform.2913

[3] D. Laney: 3D Data Management: Controlling Data Volume, Velocity, and Variety. Stamford, CT: META Group Inc.; 2001.

[4] B. Marr: Big Data: Using SMART Big Data, Analytics and Metrics To Make Better Decisions and Improve Performance. John Wiley & Sons; 2015.

[5] S. Klous and N. Wielaard: We are big data. The future of the information society [Wij zijn big data. De toekomst van de informatiesamenleving]. Amsterdam: Business Contact; 2014.

[6] T.B. Murdoch and A.S Detsky: The inevitable application of big data to health care. JAMA. 2013;309(13):1351-2.

[7] N.V. Chawla and D.A. Davis: Bringing big data to personalized healthcare: a patient-centered framework. J Gen Intern Med. 2013;28(3):S660-5.

[8] P. Kamakshi: Survey on big data and related privacy issues. International Journal of Research in Engineering and Technology. 2014;3(12):68-70.

[9] A.M. Kaplan and M. Haenlein: Users of the world, unite! The challenges and opportunities of Social Media. Business Horizons. 2010;53(1):59-68.

[10] R. Batool, W.A. Khan, M. Hussain, J. Maqbool, M. Afzal and S. Lee, editors: Towards personalized health profiling in social network. Information Science and Service Science and Data Mining (ISSDM), 2012 6th International Conference on New Trends in; 2012: IEEE, 760-765.

[11] F. Greaves, D. Ramirez-Cano, C. Millett, A. Darzi and L. Donaldson: Harnessing the cloud of patient experience: using social media to detect poor quality healthcare. BMJ quality & safety. 2013;22(3):251-5.

[12] R. Nagar, Q. Yuan, C.C. Freifeld, M. Santillana, A. Nojima, R. Chunara and J.S. Brownstein: A Case Study of the New York City 2012-2013 Influenza Season With Daily Geocoded Twitter Data From Temporal and Spatiotemporal Persppectives. J Med Internet Res 2014;16 (10):e236,. doi:10.2196/jmir.3416

[13] M. De Choudury, S. Counts, E. Horvitz, editors: Social media as a measurement tool of depression in populations. Proceedings of the 5th Annual ACM Web Science Conference; 2013:47-56

Wearable Recognition System for Emotional States Using Physiological Devices

Ali Mehmood Khan
TZi-AGKI, Universität Bremen
Bremen, Germany
e-mail: akhan@tzi.de

Michael Lawo
TZi-AGKI, Universität Bremen
Bremen, Germany
e-mail: mlawo@tzi.de

Abstract— **Recognizing emotional states is becoming a major part of a user's context for wearable computing applications. The system should be able to acquire a user's emotional states by using physiological sensors. We want to develop a personal emotional states recognition system that is practical, reliable, and can be used for health-care related applications. We propose to use the eHealth platform [1] which is a ready-made, light weight, small and easy to use device for recognizing a few emotional states like 'Sad', 'Dislike', 'Joy', 'Stress', 'Normal', 'No-Idea', 'Positive' and 'Negative' using decision tree (J48) classifier. In this paper, we present an approach to build a system that exhibits this property and provides evidence based on data for 8 different emotional states collected from 24 different subjects. Our results indicate that the system has an accuracy rate of approximately 98%. In our work, we used four physiological sensors i.e. 'Blood Volume Pulse' (BVP), 'Electromyogram' (EMG), 'Galvanic Skin Response' (GSR), and 'Skin Temperature' in order to recognize emotional states (i.e. stress, joy/happy, sad, normal/neutral, dislike, no-idea, positive and negative).**

Keywords- Emotional states; Physiological devices; International Affective Picture System; Machine learning classifier; User studies.

I. Introduction

It is hard to express your own emotions; no one can accurately measure the degree of his/her emotional state. According to Darwin, "....the young and the old of widely different races, both with man and animals, express the same state of mind by the same movement" [16]. According to Paul Ekman, there are seven basic emotions which are fear, surprise, sad, dislike, disgrace, disgust and joy [14]. The concept behind emotional states (also known as affective computing) was first introduced by Rosalind Picard in 1995 [2]. Since then the affective computing group have produced novel and innovative projects in that domain [3]. Emotional states recognition has received attention in recent years and is able to support the health care industry. Emotions and physical health have a strong link in influencing the immune system too [15]. Due to untreated, chronic stress; occurrence of an emotional disorder is more than 50% [6]. According to Richmond Hypnosis Center, due to stress; 110 million people die every year. That means, every 2 seconds, 7 people die [4]. According to American Psychological Association, in 2011 about 53 percent of Americans claimed stress as a reason behind personal health problems [5]. According to WebMD, intense and long term anger causes mental health problems including anxiety, depression, self-harm, high blood pressure, coronary heart disease, colds and flu, stroke, gastro-intestinal problems, and cancer [6]. The Occupational Safety and Health Administration (OSHA) reported that stress is a threat for the workplace. Stress costs American industry more than $300 billion annually [6]. According to Dr. Alexander G. Justicz, in the 21st century, stress is a huge problem for men [9]. Stress affects our health negatively, causing headaches, stomach problems, sleep problems, and migraines. Stress can cause many mouth problems, the painful TMJ (temporomandibular joint) syndrome, and tooth loss [7]. "Stress has an immediate effect on your body. In the short term, that's not necessarily a bad thing, but chronic stress puts your health at risk" [8]. Long term and intense anger can be caused of mental health problems including depression, anxiety and self-harm. It can also be caused of "high blood pressure", "cold and flu", "coronary heart disease", "stroke", "cancer" and "gastro-intestinal problems"[13]. "If you have a destructive reaction to anger, you are more likely to have heart attacks" [12] whereas "an upward-spiral dynamic continually reinforces the tie between positive emotions and physical health"[17].

Modern day lifestyle has led to various physical and mental diseases such as diabetes, depression and heart diseases as well. Although the negative effects of stress are known to people, they choose (deliberately or otherwise) to ignore it. They need to be forcefully notified, that they must shrug off negative emotions; either by sending them calls or some video clips/text messages/games [10]. Emotions are the feelings which influence the human organs. According to number of studies, negative thinking or depression can adversely affect your health [19]. Probably automatic and personal applications can be very helpful if it can monitor one's emotional states and persuade people to come out of negative emotional states. According to William Atkinson; "*The best way to overcome undesirable or negative thoughts and feelings is to cultivate the positive ones*" [18]. Emotional recognition technology can tackle this problem as it is able to monitor an individual's emotional states. This kind of system can also send an alarming call to a person when he is in a negative emotional state for long time or notify the caregivers or family members. The system can also log an individual's emotional states for later analysis. In some cases, especially in heart diseases, emotional states are also required along with the physical activities and physiological information for doctors in order to examine their patient's conditions when he is away from the doctor's clinic [11].

Emotional computing is a field of human computer interaction where a system has the ability to recognize emotions and react accordingly. We want to develop a system for recognizing emotional states using physiological sensors which should be able to identify a few emotional states like sad, dislike, joy, stress, normal, no-idea, positive and negative. In our research we want to prove that it is possible to recognize the aforementioned emotional states by using physiological sensors.

II. Related Work

Recognizing emotional states by using automated systems have increased in recent years. Researchers developed systems for recognizing emotional states using speech [23, 24, and 25], facial expressions [26, 27, and 28] and physiological devices [20, 21, 22, 29, and 30]. In this research, we want to recognize different emotional states using body worn physiological devices (EMG, BVP, GSR and temperature). Researchers used physiological devices in order to recognize for different emotional states like sad [20, 21, 22, 30], joy/happy [20, 21, 22, 30, 31], normal/neutral [21, 30, 31], negative [29] etc. However, the aforementioned researches have used different physiological devices in their work. For example; some researchers recognized emotional states using EEG (Electroencephalogram), GSR and pulse sensor and they recognized joy, anger, sad, fear and relax. Audio and visual clips were used as a stimulus for eliciting the emotions [20]. Some researchers recognized emotional states using ECG (Electrocardiography) and they recognized 'Happiness', 'Sad', 'Fear', 'Surprise', 'Disgust' and 'Neutral'. Audio and visual clips were used as a stimulus for eliciting the emotions [21]. Some researchers recognized emotional states using ECG, EMG, skin conductance, respiration sensor and they recognized Joy, anger, Sadness and Pleasure. Music songs were used as a stimulus for eliciting the emotions [22]. In another case, researchers gathered the data from the "blood volume pulse", "electromyogram", "respiration" and the "skin conductance sensor". They conducted 20 experiments in 20 consecutive days, testing around 25 minutes per day on each individual. They figured out neutral, anger, hate, grief, love, romantic, joy and reverence emotion states from the data. They got 81% classification accuracy among the eight states [31]. Different techniques can be used as a stimulus for eliciting the emotions i.e. pictures, video clips, audio clips, games etc. In our work, we used International Affective Picture System (IAPS) for stimulation. IAPS is widely used in experiments studying emotion and attention. The International Affective Picture System (IAPS) provides normative emotional stimuli for emotion and attention under experimental investigations. The IAPS (pronounced eye-aps) is being produced and distributed by the Center for Emotion and Attention (CSEA) at the University of Florida [32]. In our previous work, we took two physiological sensors (i.e. BVP and GSR) for the analysis, IAPS were used as a stimulus and our system was able to recognize few emotional states with good accuracy [44]. In this paper, we used four physiological sensors in order recognize few emotional states. The above mentioned researchers used different parts of the body but in our research we used only left arm for the sensor placement.

III. Hypothesis

The physiological data measured by wearable devices (EMG, blood volume pulse, temperature and skin conductance sensor) indicate a person's emotional state ('Sad', 'Dislike', 'Joy', 'Stress', 'Normal', 'No-Idea', 'Positive' and 'Negative') using a machine learning classifier.

IV. Experimental Methodology

We developed the following systems for the user study.

A. *eHealth platform and application*

We used eHealth platform [1] in order to recognize emotional states (Figure 1) and connected Raspberry Pi [41] to eHealth platform as shown in Figure 2.

Figure 1. eHealth platform

Figure 2. Raspberri pi with eHealth platform

The eHealth sensor comes with few sensors like 2D Accelerometer sensor, Blood pressure sensor (Breathing), Pulse and oxygen in blood sensor, body temperature sensor, airflow sensor, Electrocardiogram sensor (ECG), Electromyography sensor (EMG) and Galvanic skin response sensor. We used Galvanic skin response sensor, body temperature sensor, Electromyography sensor (EMG) and we used another blood volume pulse sensor [40] as shown in Figure 3.

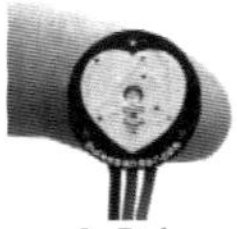

Figure 3. Pulse sensor

We connected 'GSR, 'EMG', 'BVP' and 'body temperature sensor' to the board. We wrote a piece of code which reads the values from the aforementioned sensors and writes it to a network port.

B. *Application for reading sensors from eHealth platform*

We wrote an application in Java which reads the sensed data from a network port and stores it to a text file with a timestamp in the following structure for post analysis.

Time_stamp|emg, bvp, gsr, temperature

C. *IAPS and its application (Application for Stimulus)*

We got access to IAPS [32] images and these images are already used by several researchers for emotional computing [33,34,35,36,37,38,39]. We implemented an application in C#.net that shows participants' IAPS images in a sequence in order to change participants' emotional states and also states the starting and ending time for each IAPS image during experiments. After showing participants five different images from each group, our application used to ask participants about their current emotional state by using the Likert scale (Figure 4 (b)) approach. We chose 100 IAPS images from different categories and presented them in the order shown in Figure 4(a).

Sad (5 images) Questionnaire Dislike (5 images) Questionnaire Joy (5 images Questionnaire Stress (5 images) Questionnaire

Dislike (5 images) Questionnaire Joy (5 images Questionnaire Stress (5 images) Questionnaire Sad (5 images) Questionnaire

Joy (5 images Questionnaire Stress (5 images) Questionnaire Sad (5 images) Questionnaire Dislike (5 images) Questionnaire

Stress (5 images) Questionnaire Sad (5 images) Questionnaire Dislike (5 images) Questionnaire Joy (5 images Questionnaire

Stress (5 images) Questionnaire Joy (5 images Questionnaire Dislike (5 images) Questionnaire Sad (5 images) Questionnaire

Figure 4(a). Chosen IAPS images

The images were shown as a slide show with a timer of 5 seconds for each image. For the questionnaire we used radio buttons and participants had to choose one emotional state. It also stores the participants' personal information i.e. age, gender, height and weight.

Figure 4(b). Questionnaire form

D. *Experiment setup*

Experiments were conducted in a calm room with normal temperature; there was no noise or distraction. To make sure the readings from GSR were accurate we asked the participants to dry their hands with a dryer before beginning with the experiment. Since GSR measures sweat glands as well, moist hands would result in an erroneous result. To ensure full concentration from the participants, the light in the room was kept very low and we also asked them to turn off their mobile phones during experiments. Participants were asked to wear sensors on their left arms, palms and fingers (Figure 5). They were also required to perform the experiments twice; the first experiment was useful in getting the participants to familiarize themselves with the setup, while the second attempt was actually used for analyzing their data.

Figure 5. Participant is wearing sensors

We recruited 26 participants (21 males, 5 females) for our experiment setup; two of them could not complete the experiments so we ended up with 24 participants (19 males, 5 females). The range of participants' age was from 20 to 44 (mean 26.17, SD 5.14) and ranged in BMI (body mass index) from 18.7 to 26.6 (mean 21.44, SD 2.17).Participants were required to do it twice in different days.

E. First experiment

As described earlier, the intention behind the first experiment was only familiarization with the setup. This was done to accommodate all first time participants, as they were somewhat nervous due to physiological devices and long cables and this could adversely influence our data. For this reason, the results from the first experiment were never used for analysis.

F. Second experiment

In the second experiment, all participants already knew about the setup and they were not hesitating with the sensors, they performed the task with confidence and their data was stored for later analysis. We used same settings for both experiments but IAPS images were different. We showed participants different images (IAPS) for changing their emotions to sad, dislike, joy and stress. After showing a set of images; our application used to show them the questionnaire forms for their emotional states. Physiological data was logged to a laptop with a time stamp and on the other hand image application was also logging the participants' feedback to the same laptop with timestamp. After that we merged both files to generate a single file for post analysis.

V. RESULTS AND ANALYSIS

Our experimental setup was able to change participants' emotional states; only four of the participants chose all of the given emotional states. This was due to the fact that it was hard for the participants to distinguish between sad, dislike and stress. Also being asked to distinguish between joy and normal during experiments was not a straightforward task. That also explains why some emotional states were ignored by participants. "*As everyone knows, emotions seem to be interrelated in various but systematic ways: Excitement and depression seem to be opposites; excitement and surprise seem to be more similar to one another; and excitement and joy seem to be highly similar, often indistinguishable*" [42]. Therefore, we generated another dataset from our experimental data; we categorized emotional states into two collections:

- ➢ Positive {Joy, Normal}
- ➢ Negative {Sad, Dislike. Stress}; 'No-Idea' is excluded

Now, we have the following types of datasets:

- Type1: It contains {Normal, Sad, Dislike, Joy, Stress and No-Idea}
- Type2: It contains {Positive and Negative}

Due to the fact that it was a huge dataset, it was not possible for WEKA [43] application to process the data of all 24 participants together. Therefore, we divided our datasets into six groups, each group consisting the data of four participants (as shown in Table 2); we grouped the four participants who chose all emotional states together and put them in Group-1, others were assigned to remaining groups in alphabetic order.

TABLE I. GROUPS

Groups	Age	Gender	Chosen Emotional states
Group 1	25, 24, 25, 26	3 Males, 1 Female	Normal(4), Sad(4), Dislike(4), Joy(4), Stress(4) and No-Idea(4)
Group 2	24,25,25, 38	4 Males	Normal(0), Sad(3), Dislike(4), Joy(4), Stress(4) and No-Idea(2)
Group 3	24,24,25, 44	3 Males, 1 Female	Normal(3), Sad(3), Dislike(4), Joy(4), Stress(4) and No-Idea(1)
Group 4	20,25,25, 33	2 Males, 2 Females	Normal(2), Sad(4), Dislike(4), Joy(4), Stress(2) and No-Idea(1)
Group 5	22,24,24, 25	3 Males, 1 Female	Normal(3), Sad(3), Dislike(4), Joy(4), Stress(3) and No-Idea(2)
Group 6	24,25,25, 27	4 Males	Normal(2), Sad(4), Dislike(4), Joy(4), Stress(3) and No-Idea(0)

We received values from sensors i.e. EMG, BVP, GSR and Temperature where the sample rate was around 650Hz. We analyzed both types (i.e. Type 1 and Type 2) in the following three different ways:

A. *Individuals*

We applied J48 classifier [43] on the dataset of each participant.

B. *Group-wise*

We divided the participants in 6 groups (as shown in Table 2) and applied J48 classifier on the dataset of each group.

C. *Portioned data*

As mentioned earlier due to the limitations of processing huge datasets in WEKA toolkit, we chose small portions of data randomly pertaining to each emotion from each participant in Figure 6(a) and Figure 6(b) below.

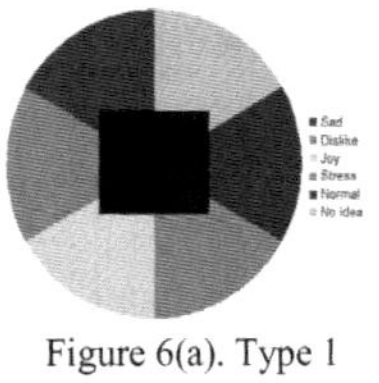

Figure 6(a). Type 1

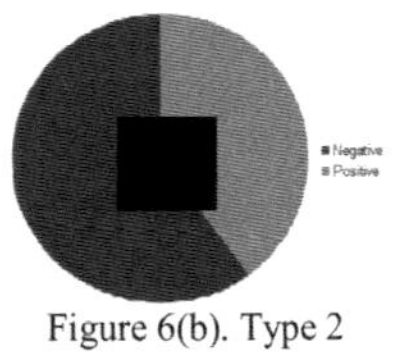

Figure 6(b). Type 2

D. *Analysis structure*

We got two types of data i.e. "Two-Class" and "Six-class"; each type was analyzed on "Individual", "Group" and "Portioned" basis. We applied J48 classifier with 10-fold cross validation.

E. Two-Class

1) Individuals: The outcome from the J48 classifier represents the average data of 24 participants where it correctly classified the instances with the accuracy of 99.4%; Min: 97.72%; Max: 99.67% and SD: 0.45.

a	b	<-- classified as
2169370	24518	a = Positive
18038	4896951	b = Negative

We took the confusion matrices from all participants and summed them all. Our results show the summation of all confusion matrices and accuracy of each emotional state where 'Positive' and 'Negative' emotional states were predicted with the accuracy of 98.88% and 99.63% by J48 classifier respectively.

2) Group wise: We took an average of correctly classified instances from all groups in order to figure out the variation amongst them. Our result shows that there is not a high variation among the groups and the average result was 99.3%; Min: 99.06%; Max: 99.45%; SD: 0.14.

a	b	<-- classified as
2181321	29401	a = Positive
19447 4	804414	b = Negative

We took confusion matrices from each group and summed them up. Our results show the summation of all confusion matrices from the groups and accuracies of emotional states where positive and Negative emotional states were predicted with the accuracy of 98.67% and 95.6% by J48 classifier respectively.

3) Portioned data: Our results show that J48 was able to correctly classify the instances with the accuracy of 99.33% and it was also able to predict positive and Negative emotional states with the accuracy of 98.56% and 99.67% respectively.

a	b	<-- classified as
409236	5982	a = Positive
3095	934177	b = Negative

We compared the accuracy between the categories i.e. 'Individual', 'Group' and 'Portioned' as shown in Figure 7 which shows that there is not much difference in results among them.

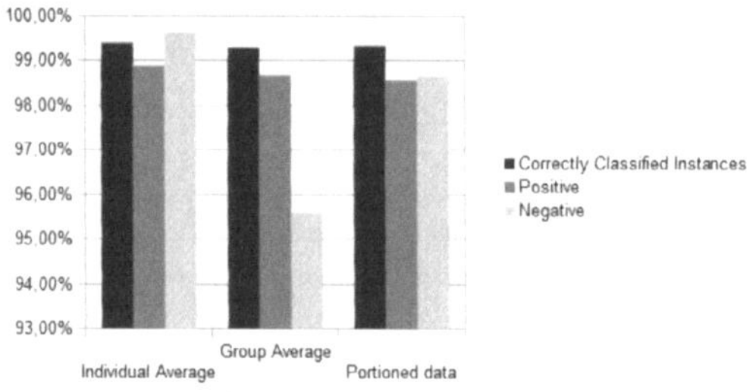

Figure 7. Comparison

F. Six-Class

1) Individuals: The outcome from the J48 classifier represents the average data of 24 participants where it correctly classified the instances with the accuracy of 99.13%; Min: 98.39%; Max: 99.52% and SD: 0.25.

a	b	c	d	e	f	<-- classified as
1298734	7019	3629	2091	962	571	a = Sad
6760	2152047	5540	3829	2211	931	b = Dislike
4074	5775	1455566	3288	1008	524	c = Joy
2139	3890	2896	1329053	1092	467	d = Stress
915	2267	974	1120	734834	377	e = Normal
885	1214	602	502	450	341361	f = NoIdea

We took confusion matrices from all participants and summed them up. Our results show the summation of all confusion matrices and accuracy of each emotional state where 'Sad', 'Dislike', 'Joy', 'Stress', 'Normal' and 'No-Idea' emotional states were predicted with the accuracy of 98.99%, 99.11%, 99%, 99.22%, 99.24% and 98.94% by J48 classifier respectively.

2) Group wise: We took an average of correctly classified instances from all groups in order to figure out the variation amongst them. Our result shows that there is not a high variation among the groups and the average result was 98.67%; Min: 98.29%; Max: 99.04%; SD: 0.26.

a	b	c	d	e	f	<-- classified as
1293196	9506	4521	3550	1500	733	a = Sad
8791	2144771	7418	5817	3248	1273	b = Dislike
5020	8403	1449781	4449	1733	849	c = Joy
3484	6295	4432	1323013	1747	566	d = Stress
1619	3706	1749	1924	730989	500	e = Normal
1039	1708	897	684	609	340077	f = NoIdea

We took confusion matrices from each group and summed them up. Our results show the summation of all confusion matrices from the groups and accuracies of emotional states where 'Sad', 'Dislike', 'Joy', 'Stress', 'Normal' and 'No-Idea' emotional states were predicted with the accuracy of 98.49%, 98.78%, 98.61%, 98.76%, 98.72% and 98.57% by J48 classifier respectively.

3) Portioned data: Our results show that J48 was able to correctly classify the instances with the accuracy of 98.47% and it was also able to predict 'Sad', 'Dislike', 'Joy', 'Stress', 'Normal' and 'No-Idea' emotional states with the accuracy of 98.24%, 98.75%, 98.41%, 98.34%, 98.62% and 97.99% respectively.

a	b	c	d	e	f	<-- classified as
244303	2123	1016	800	278	168	a = Sad
1571	428263	1594	1481	507	288	b = Dislike
983	1977	275773	1020	303	162	c = Joy
913	1762	1142	250641	252	170	d = Stress
328	706	413	307	133139	107	e = Normal
210	509	241	199	130	62872	f = NoIdea

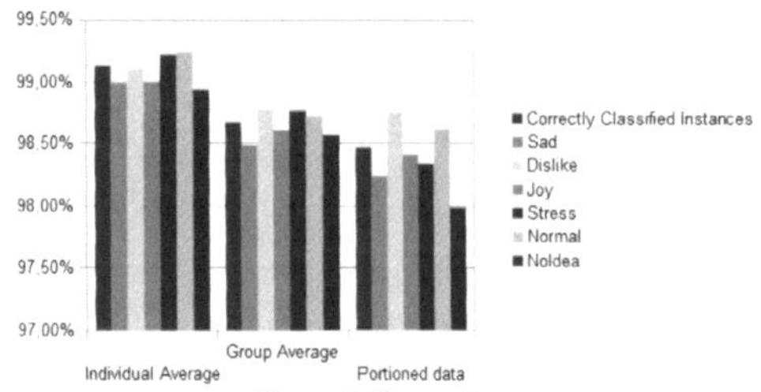

Figure 8. Comparison

We also compared the accuracy between the categories i.e. 'Individual', 'Group' and 'Portioned' as shown in Figure 8 which shows that there is not much difference in results among them.

VI. Conclusions and Future Work

We used the following approaches for analyzing the data

1. We took data of each participant and applied J48 classifier and then took an average of 'Individual' data.
2. We took integrated data from six participants, applied J48 classifier and then took an average of 'Group' data.
3. We took a small portion of data randomly from each participant and applied J48 classifier on the data

We categorized data into the following collections:

- Six emotional states i.e. 'Sad', 'Dislike', 'Joy', 'Stress', 'Normal' and 'No-Idea'.
- Two emotional states i.e. 'Positive' and 'Negative'.

Our system was able to recognize the aforementioned emotional states by using physiological devices and J48 (decision tree) classifier with high accuracy. Results have shown that few physiological devices are enough for recognizing required emotional states ('Sad', 'Dislike', 'Joy', 'Stress', 'Normal', 'No-Idea', 'Positive' and 'Negative'). This prototype is only a "proof of concept" and our results show that our approach can identify the above mentioned emotional states independent of BMI (body mass index) and age group. The physiological sensor has to be fixed properly on the participants' skin in order to predict their emotional states successfully. We will conduct more user studies where we will use physiological data and facial expressions for recognizing these emotional states.

References

[1] Cooking Hacks: e-Health Sensor Platform V2.0 for Arduino and Raspberry Pi [Biometric / Medical Applications]. http://www.cooking-hacks.com/documentation/tutorials/ehealth-biometric-sensor-platform-arduino-raspberry-pi-medical#step4_9. [Last Accessed 24th March, 2016].

[2] Professor Rosalind W. Picard. http://web.media.mit.edu/~picard/index.php. Last Accessed 24th March, 2016].

[3] Affective Computing: Publications. http://affect.media.mit.edu/publications.php .Last Accessed 24th March, 2016].

[4] Richmond Hypnosis Center. http://richmondhypnosiscenter.com/2013/04/12/sample-post-two/. Last Accessed 24th March, 2016].

[5] American Psychological Association: The Impact of Stress. http://www.apa.org/news/press/releases/stress/2011/impact.aspx . Last Accessed 24th March, 2016].

[6] WebMD: Stress Management Health Center. http://www.webmd.com/balance/stress-management/effects-of-stress-on-your-body.Last Accessed 24th March, 2016].

[7] Krifka, S., Spagnuolo, G., Schmalz, G., Schweikl, H. A review of adaptive mechanisms in cell responses towards oxidative stress caused by dental resin monomers. Biomaterials. 2013;34:4555–4563.

[8] Healthline : The Effects of stress on the Body. http://www.healthline.com/health/stress/effects-on-body. Last Accessed 24th March, 2016].

[9] Miamiherald: Chronic stress is linked to the six leading causes of death. http://www.miamiherald.com/living/article1961770.html. Last Accessed 24th March, 2016].

[10] Online Stress Reliever Games. *http://stress.lovetoknow.com/Online_Stress_Reliever_Games*. Last Accessed 24th March, 2016].

[11] Ali Mahmood Khan. (2011). Personal state and emotion monitoring by wearable computing and machine learning. BCS-HCI 2012, Newcastle, UK

[12] WebMD: Stress Management Health Center. http://www.webmd.com/balance/stress-management/features/how-anger-hurts-your-heart. Last Accessed 24th March, 2016].

[13] BetterHealth: Anger - how it affects people. http://www.betterhealth.vic.gov.au/bhcv2/bhcarticles.nsf/pages/Anger_how_it_affects_people. Last Accessed 24th March, 2016].

[14] Ekman, Paul (1999), "Basic Emotions", in Dalgleish, T; Power, M, Handbook of Cognition and Emotion (PDF), Sussex, UK: John Wiley & Sons.

[15] Health and Wellness: Are Happy People Healthier? New Reasons to Stay Positive. http://www.oprah.com/health/How-Your-Emotions-Affect-Your-Health-and-Immune-System.Last Accessed 24th March, 2016].

[16] Darwin, C. (1872). The expression of the emotions in man and animals. John Murray, London

[17] Kok, B.E., Coffey, K.A., Cohn, M.A., Catalino, L.I., Vacharkulksemsuk, T., Algoe, S., Brantley, M. & Fredrickson, B. L. (2013). How positive emotions build physical health: Perceived positive social connections account for the upward spiral between positive emotions and vagal tone. Psychological Science, 24(7), 1123-1132.

[18] Atkinson, William Walker (1908). Thought Vibration or the Law of Attraction in the Thought World.

[19] Rush, A. J., Beck, A. T., Kovacs, M. & Hollon, S. D. Comparative efficacy of cognitive therapy and pharmacotherapy in the treatment of depressed outpatients. Cognit. Ther. Res. 1, 17-38.

[20] Remarks on Emotion Recognition from Bio-Potential Signals: research paper (2004)

[21] Emotion Recognition from Electrocardiogram Signals using Hilbert Huang Transform (2012)

[22] Emotion Pattern Recognition Using Physiological Signals (2014)

[23] Walter Sendlmeier Felix Burkhardt, Verification of Acoustical Correlates of Emotional Speech using Formant-Synthesis, Technical University of Berlin, Germany.

[24] RECOGNIZING EMOTION IN SPEECH; Frank Dellaert, Thomas Polzin and Alex Waibel

[25] RECOGNIZING EMOTION IN SPEECH USING NEURAL NETWORKS; Keshi Dai, Harriet J. Fell and Joel MacAuslan

[26] Recognizing Emotion From Facial Expressions: Psychological and Neurological Mechanisms Ralph Adolphs, University of Iowa College of Medicine.

[27] Analysis of emotion recognition using facial expressions, speech and multimodal information. Proceedings of the 6th International Conference on Multimodal Interfaces, ICMI 2004, State College, PA, USA, October 13-15, 2004

[28] Recognizing Emotions from Facial Expressions Using Neural Network; Isidoros Perikos, Epaminondas Ziakopoulos, Ioannis Hatzilygeroudis.

[29] Emotions States Recognition Based on Physiological Parameters by Employing of Fuzzy-Adaptive Resonance Theory; Mahdis Monajati, Seyed Hamidreza Abbasi, Fereidoon Shabaninia, Sina Shamekhi.

[30] Classification of emotional states from electrocardiogram signals: a non-linear approach based on hurst; Jerritta Selvaraj, Murugappan Murugappan, Khairunizam Wan and Sazali Yaacob

[31] Jennifer Healey and Rosalind W. Picard (2002), Eight-emotion Sentics Data, MIT Affective Computing Group, http://affect.media.mit.edu."

[32] THE CENTER FOR THE STUDY OF EMOTION AND ATTENTION: http://csea.phhp.ufl.edu/Media.html#topmedia. Last Accessed 24th March, 2016].

[33] Cuthbert BN, Schupp HT, Bradley MM, Birbaumer N, Lang PJ.; Brain potentials in affective picture processing: covariation with autonomic arousal and affective report. Biol Psychol. 2000 Mar; 52(2): 95-111.

[34] Keil A, Bradley MM, Hauk O, Rockstroh B, Elbert T, Lang PJ.; Large-scale neural correlates of affective picture processing.;Psychophysiology. 2002 Sep; 39(5):641-9.

[35] Lang PJ, Bradley MM, Fitzsimmons JR, Cuthbert BN, Scott JD, Moulder B, Nangia V.; Emotional arousal and activation of the visual cortex: an fMRI analysis. Psychophysiology. 1998 Mar; 35(2):199-210.

[36] Bradley MM1, Sabatinelli D, Lang PJ, Fitzsimmons JR, King W, Desai P.; Activation of the visual cortex in motivated attention.; Behav Neurosci. 2003 Apr; 117(2):369-80.

[37] Sabatinelli D, Bradley MM, Fitzsimmons JR, Lang PJ.; Parallel amygdala and inferotemporal activation reflect emotional intensity and fear relevance.; Neuroimage. 2005 Feb 15;24 (4):1265-70. Epub 2005 Jan 7.

[38] Sabatinelli D, Lang PJ, Keil A, Bradley MM; Emotional perception: correlation of functional MRI and event-related potentials; Cereb Cortex. 2007 May; 17(5):1085-91. Epub 2006 Jun 12.

[39] Bradley MM, Codispoti M, Lang PJ.; A multi-process account of startle modulation during affective perception; Psychophysiology. 2006 Sep; 43(5):486-97.

[40] Maker Shed: Pulse Sensor AMPED for Arduino. http://www.makershed.com/products/pulse-sensor-amped-for-arduino .Last Accessed 24th March, 2016].

[41] Raspberry Pi. https://www.raspberrypi.org/ .Last Accessed 24th March, 2016].

[42] Russel and Bullock, Multidimensional scaling of emotional facial expressions, Journal of Personality and Social Psychology (1985), pp. 1290-1298

[43] Mark Hall, Eibe Frank, Geoffrey Holmes, Bernhard Pfahringer, Peter Reutemann, Ian H. Witten (2009); The WEKA Data Mining Software: An Update; SIGKDD Explorations, Volume 11, Issue 1

[44] Ali Mehmood Khan, Michael Lawo 2016, "Recognizing Emotion from Blood Volume Pulse and Skin Conductance Sensor Using Machine Learning Algorithms", MEDICON 2016, Paphos-Cyprus

Attribute-Based Authenticated Access for Secure Sharing of Healthcare Records in Collaborative Environments

Mohamed Abomhara and Huihui Yang

Department of Information and Communication Technology
University of Agder, Grimsatd, Norway
Email: {mohamed.abomhara, huihui.yang }@uia.no

***Abstract*—This study highlights authorization matters in cooperative engagements with complex scenarios in the collaborative healthcare domain. Focus is mainly on collaborative activities that are best accomplished by organized groups of healthcare practitioners within or among healthcare organizations with the objective of accomplishing a specific task (a case of patient treatment). In this study, an authorization schema is proposed that is suitable for collaborative healthcare systems to address the issue of information sharing and information security. The proposed scheme is based on attribute-based authentication (ABA), which is a way to authenticate users by attributes or their properties. The main goal is to provide an authorization model that strikes a balance between collaboration, flexible access to patient information and safeguarding sensitive patient information.**

Keywords–Electronic health records; Authorization; Authentication; Data security and privacy; Attribute-based authentication; Collaboration.

I. Introduction and Motivation

Electronic health records (EHRs) are widely adopted by healthcare providers and patients to efficiently and effectively create, manage and access patient healthcare information [1], [2], [3]. Integrated use of EHRs seem promising in enhancing healthcare services due to a number of attractive features, such as improving the quality and delivery of health services by giving healthcare providers access to information they require to provide rapid patient care [1]. In addition, reducing the cost of care by facilitating easy collaborative support from multiple parties to fulfill the information requirements of daily clinical care [4], [5].

Typically, rapid patient care requires the collaborative support of different parties including primary care physicians, specialists, medical laboratory technicians, radiology technicians and many other medical practitioners [6], [7]. Moreover, collaboration among healthcare organizations is required for patients being transferred from one healthcare provider to another for specialized treatment [8]. However, security control over information flow is a key aspect of such collaboration where sensitive information is shared among a group of people within or across organizations. In this study, focus is mainly on authorization issues when EHRs are shared among healthcare providers in collaborative environments with the objective of accomplishing a specific task.

A. Problem Definition

EHRs system is considered in this study. Multiple owners (referring to patients who have full control of their EHRs) and healthcare providers, such as physicians, nurses, family, and relatives, among others, who require access to these EHRs to perform a task. Healthcare providers can only access and perform actions (e.g., read and/or write) to patients' EHRs, for which they are responsible. For example, doctors have access to their own patients' data, but not the data of another doctor's patients, while nurses or personal assistants have access to the information of the patients for whom they are responsible [9]. On the one hand, healthcare services need the collaborative support of multiple healthcare professionals and administrators in order to deliver rapid patient care. Therefore, multiple healthcare providers (e.g., doctors and nurses) may require access to patient information to perform tasks. On the other hand, EHRs contain sensitive information about patients, including demographic information, medical history, laboratory tests and radiologic images that call for appropriate authorization mechanisms in place to ensure that information is accessible only to those authorized to have access [10].

The main concern with EHRs sharing during collaborative support is having an authorization mechanism with flexibility to allow access to a wide variety of authorized healthcare providers while preventing unauthorized access. Since healthcare services necessitate collaborative support from multiple parties and healthcare teamwork occurs within a dynamic group, dynamic authorization is required to allow team members to access classified EHRs.

B. Study Objective

The main objectives of this work is to design an attribute-based group authorization model that is suitable for collaborative healthcare systems to address the concern with information sharing and information access. The proposed model ensures that access rights are dynamically adapted to the actual needs of healthcare providers. Healthcare providers can access the resources associated with a work task, but only while the work task is active. Once the task is completed, access rights should be invalidated.

C. Structure of the Study

The remaining parts of this study are organized as follows. In Section II, a brief description of the EHRs system, the usage scenario and security requirements are provided. Proposed scheme and security analysis are presented in Section III. Finally, conclusions and aspects for future work are given in Section IV.

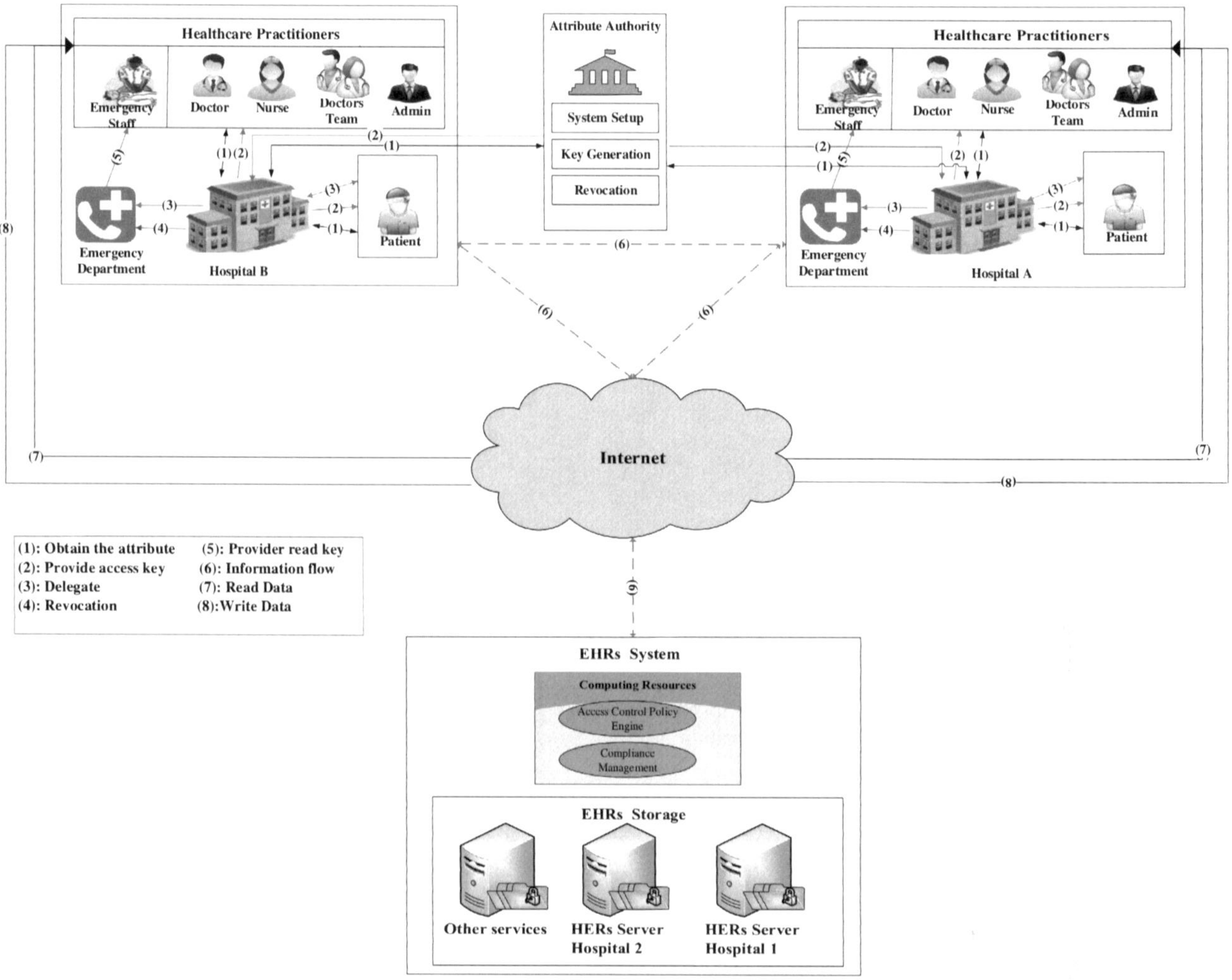

Figure 1. Reference system architecture overview

II. EHR SYSTEM

In this section, relevant work underlying the current study is discussed. First, the system architecture is briefly introduced, followed by the usage scenario and an overview of security requirements.

A. Systems architecture

In Figure. 1, the architecture of the reference system is illustrated. The reference system includes the following main domains:

1) **EHRs**: The medical records are collected, stored and provisioned by the electronic health records system to achieve the features of low cost operation, collaborative support and ubiquitous services. The EHRs can reside in a centralized or distributed systems depending on the deployment needs [11]. Authorized healthcare providers, including hospitals and healthcare practitioners can access EHRs through different services such as web portals and health apps [12]. The access to medical files is controlled via the requirements of attributes. For each medical file, the access policy is represented by a combination of attributes. When a user (patients and healthcare providers) requires to access (read, write, etc) the file, it should show an evidence that it satisfies the required attributes. Only if the evidence is valid that the user's access requirement can be granted. This process will be implemented by an attribute-based authentication (ABA) scheme presented in Section III-A.
2) **Trusted authority**: A fully trusted authority such as the Ministry of Health is responsible for key generation, distribution and management of users' keys. The main responsibilities of the trusted authority include the following:
   a) Generate the main system public and private keys.
   b) Generate user keys for each user.
   c) Generate public and attribute keys for each attribute in the system.
   d) Generate attribute keys for attributes possessed by each user.

As for implementation, it is possible to have different authorities perform these responsibilities separately, such that the compromise of one authority will not lead to the compromise of the whole system. More specifically, healthcare delivery organizations (e.g., hospitals) perform as a registration center with a certain qualification certified by the trusted authority. Healthcare delivery organizations are responsible for checking their healthcare practitioners' professional expertise and send their attributes to the trusted authority to issue the corresponding attribute-based credentials.

3) **Healthcare providers**: Healthcare providers from various domains, such as doctors, nurses, radiology technicians and pharmacists, to name a few, require access to patients' records to perform a task. Once a new healthcare practitioner join a system, the healthcare delivery organization must send healthcare practitioner' attributes to the trusted authority to obtain attributes based credentials. Healthcare practitioners apply their authentication credentials obtained from the trusted authority to access classified EHRs through authorization mechanisms in the EHR aggregator. In case of group collaboration, multiple EHRs have to be shared with various healthcare providers and practitioners. A group manager is responsible for registering healthcare practitioners to form a group. The hospital's (registration center) responsibility is to verify the authenticity of each healthcare practitioners in the group based on the professional expertise and required access, and send it to the trusted authority to issue the corresponding group credentials for the group.

B. *Usage scenario of work-based authentication*

In this Section, a typical use case scenario adopted from [4] is presented. As shown in Figure. 2 , a patient named Alice is recently diagnosed with gastric cancer. Surgical removal of the stomach (gastrectomy) is the only curative treatment. For many patients, chemotherapy and radiation therapy are given after surgery to improve the chances of curing. Alice entered a cancer-treatment center at her chosen hospital (e.g., hospital A in Figure.1). Alice has a general practitioner (Dean) who she regularly visits. Upon entering the hospital, Alice also sees an attending doctor (Bob) from the hospital. Alice's health condition has caused some complications, so her attending doctor would like to seek expert opinions and consultation regarding Alice's treatment from different hospitals (e.g., hospital B in Figure.1), including Alice's specific general practitioner who is fully informed about Alice's medical history. Note that the invited practitioners are specialized in different areas, where some are specialists and others are general practitioners. In such group consultation, every participant needs to obtain the medical records they request based on the health insurance portability and accountability act (HIPAA) [13] minimal disclosure principle.

Furthermore, the consultation results, such as diagnosis and treatment suggestions, should be signed and certified by this group of specialists and practitioners. The medical certificate with their signatures is sent to Alice.

In this case, the act of managing the collaborative work

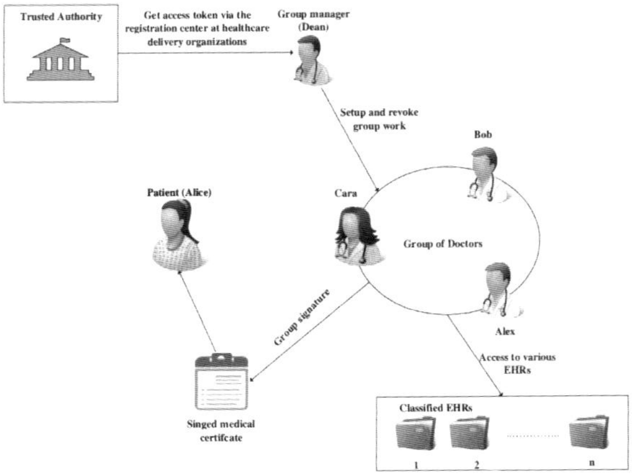

Figure 2. An EHRs usage scenario

must be clearly defined. By default, only the main practitioner (Dean) should be aware of the patient's personal information. The three other medical practitioners with supporting roles are given information based on their contributing roles (need-to-know principle) [14]. For instance, if the supporting party is included solely for consultation purposes concerning the disease, only information essential for diagnosis is provided. It is not necessary to allow perusal of personal information related to the patient. In this way, improper access to the patient's sensitive information can be prevented.

Hospital personnel roles are often simplistically split into medical practitioners, nurses and administrators [15], [16]. However, in [17] (paper by one of us), we further categorized personnel roles into a total of nine roles per group, which are classified into main, action, strategic and management roles, as shown in Figure. 3.

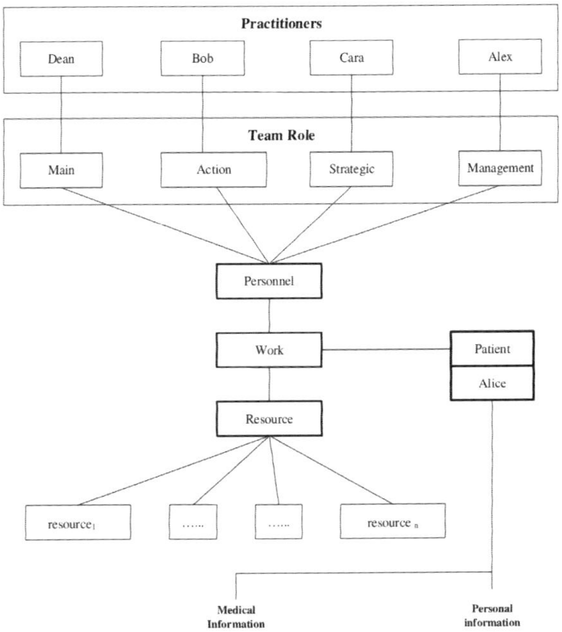

Figure 3. EHRs usage scenario

The workflow of every healthcare practitioner is as follows:

1) The general practitioner (Dean) serves as the group manager. He is responsible for initiating the work (treatment of Alice's case) and choosing the practitioners (group of doctors) who may be required to attend Alice's consultation and treatment. Moreover, the general practitioner must revoke the group upon completion of the patient's diagnosis consultation.
2) Bob helps Dean with the operational part of the case. Operation refers to a series of responsibilities that entail interaction with the patient. Bob needs to see Alice on a face-to-face basis to perform various tasks that are related to her recovery. In this respect, there is a need for Bob to know personal and medical information about Alice to perform his duty effectively.
3) Cara has more of a strategy role. She is responsible for helping Dean solve the medical case. There is no need for Cara to meet Alice personally on a day-to-day basis. In fact, Cara is only required to analyze the medical situation and suggest a possible solution. Cara's strategic role within the team implies a rather clear indication of the access that she needs. Since Cara is predominantly preoccupied with diagnosing the disease, there is no urgent need for her to know the patient's personal information. As such, she is only given access to the patient's medical information as per her strategic team role.
4) With the increasing number of physicians working on Alice's case, their interaction can become more complex. For instance, if there exists a competition between conflicting diagnoses given by Bob and Cara, which would gain priority? This is where Alex comes in. He contributes to the team by coordinating the interaction of the other members by taking on the team management role. To work effectively, Alex does not really need to know the patient's personal information. However, he must be aware of the patient's medical information to enable coordination.

In addition, Alice may have some historical health information (e.g., mental illness or sexual issues, etc.), to which the group (or some of the group) of specialists and practitioners do not have to have access. We assume that each resource (EHR files) in the system are divided into type, mainly shareable and non-shareable during the collaborative work. The collaborative resources required for work are enumerated in table form as proposed by Abomhara in [17]. Each shared resource is tied to the set of collaborative roles or team roles that can access it. In effect, the selected roles will determine the extent of collaborative access.

C. Security requirements

Some of the requirements of a well-designed attribute-based authentication system were presented by Yang [18], [19]. According to our usage scenario, the system should fulfill the following requirements:

- **Confidentiality**: Unauthorized users who do not possess enough attribute satisfying the authorization policy should be prevented from reading EHR documents.
- **Minimum attributes leakage**: To be authenticated, healthcare provider only need to provide required attributes rather than the whole package of attributes it possesses.
- **Signature**: The final medical report of Alice's treatment should be signed by appropriate practitioners using digital signatures.
  Alice should be able to verify the authenticity of the consultation results through the practitioner's digital signature. Note that the practitioner's digital signature can be opened (reveal the practitioner's identity) depending on the requirements. In some cases, practitioners do not want to reveal their identities when participating in group treatment.
- **Unforgeability**: An adversary who does not belong to the group should not be able to impersonate a group member and forge a valid signature to get authenticated.
- **Coalition resistance**: Group members should not be able to pile up their attributes to forge a signature to help a member to get authenticated.

III. PROPOSED SCHEME

In this section, the system setup and security analysis are presented.

A. System setup

System setup, including key generation, distribution and revocation are explained in this subsection. As mentioned before (Section II-A), the trusted authority is responsible for users' key and attribute key generation. For each user in the system, the trusted authority will generate a unique user key that represents the user's identity information and will be used to trace users' identities if necessary. The proposed scheme is based on bilinear mapping [20], [21].

Definition 1: **[Bilinear Mapping]** [22] Let G_1, G_2 and G_3 be cyclic groups of prime order p, with $g_1 \in G_1$ and $g_2 \in G_2$ as the generators. e is an efficient bilinear map if the following two properties hold.

1) Bi-linearity: equation $e(g_1^a, g_2^b) = e(g_1, g_2)^{ab}$ holds for any $a, b \in \mathbb{Z}_p^*$.
2) Non-degenerate: $e(g_1, g_2) \neq 1_{G_3}$, where 1_{G_3} is the unit of G_3.

Firstly, the proposed ABA scheme needs to set up the system, which is considered as a preparation for the phase of signature generation, verification and opening. During system setup, the system main parameters, such as main public and private keys set will be generated by the trusted authority. Based on the main private and public keys set, the trust authority will generate system attribute keys and users' keys. More importantly, the trusted authority will authorize Dean the power to generate attribute keys for group members. This is how Dean gains the control over the group.

Assume k_0 is the system security parameter. G_1, G_2 are two multiplicative groups of prime order p with $g_1 \in G_1$ and $g_2 \in G_2$ as their generators. Let $e : G_1 \times G_1 \rightarrow G_2$ be a bilinear mapping. Select $h \in G_1$, $\xi_1, \xi_2 \in \mathbb{Z}_p^*$, where $\mathbb{Z}_p^* = \{a \in \mathbb{Z}_p | gcd(a, p) = 1\}$ is a multiplicative group modulo a big prime number p. Set $u, v \in G_1$ such that $u^{\xi_1} = v^{\xi_2} = h$. Select $x_0, \beta_0 \in \mathbb{Z}_p^*$ as the top secret and compute $w_0 = g_1^{x_0}$, $f_0 = g_1^{1/\beta_0}$ and $h_0 = g_1^{\beta_0}$. The public key set of the trusted authority is denoted by $MPK =< G_1, G_2, g_1, g_2, h, u, v, f_0, h_0, w_0 >$ and the private key set is $MSK =< x_0, \beta_0, \xi_1, \xi_2 >$, where the pair $< \xi_1, \xi_2 >$ is handed to the opener as its tracing key tk.

Then the system setup proceeds as follows.

1) **Dean authorization**: Dean described in our usage scenario can be considered as an attribute domain authority in the scheme proposed in [23]. To authorize Dean, first, the trusted authority selects a secret $x_d \in \mathbb{Z}_p^*$ and computes $A_d = g_1^{(x_0+x_d)/\beta_0}$ and $w_d = g^{x_d}$. The pair $DSK =< A_d, x_d >$ is the Dean's private key and A_d should be registered in the opener's database for identity tracing. $DPK =< w_d >$ as the Dean's public key.
2) **User key generation**: All users in the system should register themselves and obtain their users' key from the trusted authority. Assume there are N users in the EHRs usage case. To generate the secret key of user U_i ($1 \leq i \leq N$), the trusted authority randomly selects $x_i \in \mathbb{Z}_p^*$ and computes $A_i = g_1^{(x_0+x_i)/\beta_0}$. $bsk_i =< A_i, x_i >$ is U_i's secret key base and A_i should be handed to the opener.
3) **Attribute key generation**: Assume the attribute set owned by all members in the EHRs usage case is denoted by $\Psi = \{att_1, \cdots, att_{N_a}\}$ ($N_a = |\Psi|$). To generate a pair of private and public attribute key for an attribute $att_j \in \Psi$ ($1 \leq j \leq N_a$), the trusted randomly selects $t_j \in \mathbb{Z}_p^*$ as its private attribute key and computes $apk_j = g_1{}^{t_j}$ as its public attribute key.
4) **Attribute key authorization**: The trusted authority authorize attribute keys to Dean. For attribute att_j, the trusted authority selects $r_j \in \mathbb{Z}_p^*$ and computes $T_{d,j} = g_1^{(x_0+x_d)/\beta_0} H(att_j)^{t_j+r_j}$ and $apk_{dj} = g_1^{r_j}$ as Dean's private and public attribute keys for attribute att_j respectively.
5) **User attribute key generation**: To be active in the EHRs usage case described above, each member should gain their attribute keys from Dean. Assume the attribute set possessed by user U_i is denoted by $\Psi_i = \{att_{i1}, \cdots, att_{iN_i}\}$. Assume attribute att_{ik} ($1 \leq k \leq N_i$) corresponds to $att_j \in \Psi$. For simplicity, we will use att_j to represent att_{ik} instead. To generate a private attribute key of att_j ($1 \leq k \leq N_i$) for U_i, Dean interacts with U_i and computes $T_{i,k} = f_0^{x_i} T_{d,j} = g_1^{(x_0+x_d+x_i)/\beta_0} H(att_j)^{t_j+r_j}$ as U_i's private attribute key for attribute att_j.

All these attribute keys are only active during the period of a specific workload. When this workload is finished, all attribute keys of users in this group should be revoked. This requirement can be realized by combining these attribute keys with a timing token. Thus, these attribute keys are only valid during this fixed time period.

B. Signature generation, verification and opening

After the system setup, all entities in the group of the EHRs usage case have obtained their users' keys and attribute keys for authentication. As described before, each medical file is bound with access policies represented by a combination of attributes. More specially, this combination of attributes is represented by an attribute tree [18]. An attribute tree is a tree structure that represents the logical relations among required attributes, based on which a user generates a signature as a proof of possessing the required attributes.

The user can only be authenticated when the signature is valid. However, it is also possible that the user's access request is reject even though the signature is valid because of other factors, such as system time, locations and so on.

Assume that U_i is a user to the authenticated, V is the verifier and f is the file that U_i wants to access. The verifier here can be the access system or another entity that is responsible for users' authentication. It depends on the specific enforcement of the system. The authentication phase proceeds as follows:

1) (U_i) **access request sending**: U_i sends a request to the verifier V wants to access file f.
2) (V) **attribute requirement embedding**: In this step, the verifier embeds a secret key K_s and the attribute requirements in an attribute tree and sends related parameters to U_i. The details are as follows:
Once V receives the access request, it retrieves the access policy related to the requested access and file f. Next, V will generate an attribute tree Γ with root value $\alpha_r \in \mathbb{Z}_p^*$ for root r to represent the access requirement as described in [18]. The same as in [23], we use $q_{Node}()$ to denote the polynomial bound to an interior node $Node$. For a leaf node y whose parent is interior node $Node$, $q_y(0)$ is computed by $q_{Node}(0)$. Thereafter, the verifier computes
$$K_s = (e(f_0, w_0)e(g_1, w_d))^{\alpha_r} = e(g_1, g_1)^{(x_0+x_d)\alpha_r/\beta_0}.$$
Let $L(\Gamma)$ be the leaf node set of the attribute tree Γ. V computes $\forall y \in L(\Gamma), C_y = g_1^{q_y(0)}$ and $C'_y = H(y)^{q_y(0)}$ and sends $\{\Gamma, g_1^{\alpha_r}, \forall y \in Leaf(\Gamma) : C_y, C'_y\}$ to U_i.
3) (U_i) **signature generation**: In this step, U_i recovers the embedded secret key K_s as K_v first if it owns all the required attributes. Next it generates a signature as a proof that it possesses the required attributes and to provide traceability, which means that an opener can trace the identity information of U_i given this signature.
The details are as follows. Assume U_i possesses all the required attributes represented by attribute tree Γ and att_{ik} owned by U_i is the attribute related to leaf node y in attribute tree Γ. After U_i receives the message from V, it computes
$$DecryptNode(T_{i,k}, C_y, C'_y, y) = \frac{e(T_{i,k}, C_y)}{e(apk_j apk_{dj}, C'_y)} = e(g_1, g_1)^{(x_0+x_d+u_k)q_y(0)/\beta_0}.$$
If x is an interior node, $DecryptNode(T_{k,j}, C_y, C'_y, y)$ proceeds as follows: for all x' children z, $DecryptNode(T_{k,j}, C_y, C'_y, y)$ is called and the output is stored as F_z. Assume S_x is the subset of all x's children z and $ind(x)$ is the index of node x. We

define

$$\Delta_{S_x, ind(z)} = \prod_{l \in \{S_x - ind(x)\}} \frac{l}{ind(z) - l}.$$

Then we have

$$\begin{aligned} F_x &= \prod_{z \in S_x} F_z^{q_z(0)\Delta_{S_x, ind(z)}} \\ &= \prod_{z \in S_x} (e(g_1, g_1)^{(x_0+x_d+x_i)q_z(0)/\beta_0})^{\Delta_{S_x, ind(z)}} \\ &= \prod_{z \in S_x} (e(g_1, g_1)^{(x_0+x_d+x_i)q_{par(z)}(ind(z))/\beta_0})^{\Delta_{S_x, ind(z)}} \\ &= e(g_1, g_1)^{(x_0+x_d+x_i)q_x(0)/\beta_0}. \end{aligned}$$

U_i calls $DecryptNode(T_{i,k}, C_y, C'_y, y)$ for the root and gets the result

$$F_r = e(g_1, g_1)^{(x_0+x_d+x_i)\alpha_r/\beta_0}.$$

Next U_i computes

$$K_s = F_r / e(g_1^{x_i}, g_1^{\alpha_r}) = e(g_1, g_1)^{(x_0+x_d)\alpha_r/\beta_0} = K_v.$$

Until here, U_i has successfully recovered the embedded secret key K_s as K_v. In the following, U_i generate a signature to provide traceability.

The signer randomly selects ζ, α, β, r_ζ, r_α, r_β, r_x, r_{δ_1}, $r_{\delta_2} \in \mathbb{Z}_p^*$ and calculates

$$\begin{aligned} &C_1 = u^\zeta, C_2 = v^\beta, C_3 = A_i h^{\zeta+\beta}, \\ &\delta_1 = x_i\zeta, \delta_2 = x_i\beta, \\ &R_1 = u^{r_\zeta}, R_2 = v^{r_\beta}, R_4 = C_1^{r_x} u^{-r_{\delta_1}}, R_5 = C_2^{r_x} v^{-r_{\delta_2}}, \\ &R_3 = e(C_3, g_1)^{r_x} e(h, w_d)^{-r_\zeta - r_\beta} e(h, g_1)^{-r_{\delta_1} - r_{\delta_2}}, \\ &c = H_{K_s}(M, C_1, C_2, C_3, R_1, R_2, R_3, R_4, R_5) \in \mathbb{Z}_p^* \\ &s_\zeta = r_\zeta + c\zeta, s_\beta = r_\beta + c\beta, s_\alpha = r_\alpha + c\alpha, \\ &s_x = r_x + cx_i, s_{\delta_1} = r_{\delta_1} + c\delta_1, s_{\delta_2} = r_{\delta_2} + c\delta_2. \end{aligned}$$

Finally, the signer sends the signature $\sigma =< M, C_1, C_2, C_3, c, s_\zeta, s_\beta, s_\alpha, s_{\delta_1}, s_{\delta_2} >$ to the verifier.

4) (V) **signature verification**: V computes

$$\begin{aligned} R'_1 &= u^{s_\zeta} C_1^{-c}, R'_2 = v^{s_\beta} C_2^{-c}, R'_4 = u^{-s_{\delta_1}} C_1^{s_x}, R'_5 = v^{-s_{\delta_2}} C_2^{s_x}, \\ R'_3 &= e(C_3, g_1)^{s_x} e(h, w_d)^{-s_\zeta - s_\beta} e(h, g_1)^{-s_{\delta_1} - s_{\delta_2}} \left(\frac{e(C_3, w_d)}{e(g_1, g_1)}\right)^c \end{aligned}$$

and $c' = H_{K_v}(M, C_1, C_2, C_3, R'_1, R'_2, R'_3, R'_4, R'_5)$. If c' equals to c that V has received from U_i, V believes that U_i owns the required attributes and the authentication succeeds.

5) **(The opener) signature opening**: The opener computes $A_i = C_3/(C_1^{\varepsilon_1} C_2^{\varepsilon_2})$, where A_i was registered in the opener's database as U_i's identity information during system setup.

C. Group operations

As described in Section II-B, Bob needs to read patients' personal and medical information, but Cara only needs to have access to patients' medical records. To achieve this goal, we first express these access policies based on attributes. When group members want to access the documents, they generate a signature based on the required attributes defined in the access policies. If their signature is valid, we believe that they satisfy the access policies and will be granted with the required access.

In addition, Dean needs to revoke this temporary group and the privileges granted to group members during this workload. There are two possible solutions. The first solution is to combine all keys generated for this temporary workload with a time token, but it requires a precise estimation about the time period how long this task will last. If the time period is too short, all keys will be revoked before the task is finished and the system has to be set up again. To the contrary, if the time period is too long, group members will still be able to access to patients' documents after the task is completed, which may cause security and privacy problems. The second solution is too add the temporary attribute public keys in a revocation list. Before signature verification, the verifier firsts check whether the related attribute public keys are valid. If not, the verifier will abort the signature verification, and group members will not gain additional access privileges when the temporary task finishes.

D. Security requirement analysis

Since our proposed model is a specific application of the general attribute-based HABA scheme from [23], it follows the same correctness and security requirements of the attribute-based HABA scheme. Hence we can draw the conclusion that the ABGA scheme proposed in this paper is correct. Meanwhile, it also provides anonymity and traceability. From the description from [23], we know that once an ABA scheme is fully anonymous and traceable, it also provides the security requirement of unlinkability, unforgeability and coalition resistance. These three security requirements are provided by the ABA scheme proposed in Section III.

- **Unlinkability**: Only the group manager (Dean) has the capability of revoking and discover the signers identity. This ensures that an adversary can not trace any member of the team because he or she unable to establish linkage between identities and attribute sets.
- **Unforgeability**: To get an access to EHRs, any member of the team must gain an attribute key from the team manger. Without this key, he or she will not be able to access EHRs. Therefore, an adversary cannot impersonate a group member unless he or she was assigned by the group manager.
- **Coalition resistance**: As mentioned earlier, users in the group cannot pile up their attributes to generate a signature. Therefore, they cannot conclude and cheat the system to get authenticated if a single user does not possess the required attributes.

IV. CONCLUSIONS AND FURTHER WORK

In this work, an authorization scheme was proposed for collaborative healthcare system to address the problem of information sharing and information security. The proposed scheme provides an efficient solution to security challenges related to authorization. The security analysis has showed that our proposed scheme is unforgeable, coalition resistant, and traceable.

In the future, the plan is to develop and prototype the functionality to be implemented as well as evaluate the validity of the scheme based on its efficiency and practicality. Efficiency is the scheme's performance in terms of resource consumption, e.g., time and computational capability. Practicality denotes the possible difficulties in managing the model during actual implementation. The motivation behind studying the issue of efficiency and practicality is to simplify decentralized administrative tasks, and enhance the practicability of authorization in dynamic collaboration environments. It is very important to design system to not only ensure shared information confidentiality but also to avoid administration and management complexity.

EU countries are seeking new ways to modernize and transform their healthcare systems using information and communications technology in order to provide EU citizens (patients) with safe and high quality treatment in any European Union country [24], [25] (EU directive 2011/24/EU framework on cross-border health care collaboration in the EU [26], [27], [28]). The proposed scheme will be further investigated towards cross-border healthcare collaboration. The plan is to evaluate the validity of the scheme to provide solutions to improve healthcare quality, provide access to a high-quality healthcare system to all EU citizens around Europe, and support close cooperation between healthcare professionals and care providers from different organization.

ACKNOWLEDGMENT

The authors would like to thank Geir M. Køien for the support in investigating and writing this work.

REFERENCES

[1] N. Menachemi and T. H. Collum, "Benefits and drawbacks of electronic health record systems," Risk management and healthcare policy, vol. 4, 2011, p. 47.

[2] M.-H. Kuo, "Opportunities and challenges of cloud computing to improve health care services," Journal of medical Internet research, vol. 13, no. 3, 2011, p. e67.

[3] R. Wu, G.-J. Ahn, and H. Hu, "Secure sharing of electronic health records in clouds," in Collaborative Computing: Networking, Applications and Worksharing (CollaborateCom), 2012 8th International Conference on. IEEE, 2012, pp. 711–718.

[4] R. Zhang and L. Liu, "Security models and requirements for healthcare application clouds," in Cloud Computing (CLOUD), 2010 IEEE 3rd International Conference on. IEEE, 2010, pp. 268–275.

[5] S. Silow-Carroll, J. N. Edwards, and D. Rodin, "Using electronic health records to improve quality and efficiency: the experiences of leading hospitals," Issue Brief (Commonw Fund), vol. 17, 2012, pp. 1–40.

[6] B. Fabian, T. Ermakova, and P. Junghanns, "Collaborative and secure sharing of healthcare data in multi-clouds," Information Systems, vol. 48, 2015, pp. 132–150.

[7] M. Li, S. Yu, Y. Zheng, K. Ren, and W. Lou, "Scalable and secure sharing of personal health records in cloud computing using attribute-based encryption," Parallel and Distributed Systems, IEEE Transactions on, vol. 24, no. 1, 2013, pp. 131–143.

[8] O. Moonian, S. Cheerkoot-Jalim, S. D. Nagowah, K. K. Khedo, R. Doomun, and Z. Cadersaib, "Hcrbac–an access control system for collaborative context-aware healthcare services in mauritius," Journal of Health Informatics in Developing Countries, vol. 2, no. 2, 2008.

[9] M. Abomhara, M. Gerdes, and G. M. Køien, "A stride-based threat model for telehealth systems," Norsk informasjonssikkerhetskonferanse (NISK), vol. 8, no. 1, 2015, pp. 82–96.

[10] K. T. Win, "A review of security of electronic health records," Health Information Management, vol. 34, no. 1, 2005, pp. 13–18.

[11] D. Patra, S. Ray, J. Mukhopadhyay, B. Majumdar, and A. Majumdar, "Achieving e-health care in a distributed ehr system," in e-Health Networking, Applications and Services, 2009. Healthcom 2009. 11th International Conference on. IEEE, 2009, pp. 101–107.

[12] S. de Lusignan, F. Mold, A. Sheikh, A. Majeed, J. C. Wyatt, T. Quinn, M. Cavill, T. A. Gronlund, C. Franco, U. Chauhan et al., "Patients online access to their electronic health records and linked online services: a systematic interpretative review," BMJ open, vol. 4, no. 9, 2014, p. e006021.

[13] S. J. Dwyer III, A. C. Weaver, and K. K. Hughes, "Health insurance portability and accountability act," Security Issues in the Digital Medical Enterprise, vol. 72, no. 2, 2004, pp. 9–18.

[14] R. S. Sandhu and P. Samarati, "Access control: principle and practice," Communications Magazine, IEEE, vol. 32, no. 9, 1994, pp. 40–48.

[15] M. A. Valentine and A. C. Edmondson, "Team scaffolds: How mesolevel structures enable role-based coordination in temporary groups," Organization Science, vol. 26, no. 2, 2015, pp. 405–422.

[16] N. Meslec and P. L. Curşeu, "Are balanced groups better? belbin roles in collaborative learning groups," Learning and Individual Differences, vol. 39, 2015, pp. 81–88.

[17] M. Abomhara and G. M. Køien, "Towards an access control model for collaborative healthcare systems," in proceedings of the 9th International Joint Conference on Biomedical Engineering Systems and Technologies (BIOSTEC 2016), vol. 5, 2016, pp. 213–222.

[18] H. Yang and V. A. Oleshchuk, "A dynamic attribute-based authentication scheme," in Codes, Cryptology, and Information Security. Springer, 2015, pp. 106–118.

[19] ——, "An efficient traceable attribute-based authentication scheme with one-time attribute trees," in Secure IT Systems. Springer, 2015, pp. 123–135.

[20] T. Okamoto, "Cryptography based on bilinear maps," in Applied Algebra, Algebraic Algorithms and Error-Correcting Codes, ser. Lecture Notes in Computer Science. Springer Berlin Heidelberg, 2006, vol. 3857, pp. 35–50.

[21] R. Sahu and S. Padhye, "Efficient ID-based signature scheme from bilinear map," in Advances in Parallel Distributed Computing, ser. Communications in Computer and Information Science. Springer Berlin Heidelberg, 2011, vol. 203, pp. 301–306.

[22] Y. Qin, D. Feng, and X. Zhen, "An anonymous property-based attestation protocol from bilinear maps," in 2009 International Conference on Computational Science and Engineering (CSE '09), vol. 2, Aug 2009, pp. 732–738.

[23] H. Yang and V. A. Oleshchuk, "Traceable hierarchical attribute-based authentication for the cloud," in Communications and Network Security (CNS), 2015 IEEE Conference on. IEEE, 2015, pp. 685–689.

[24] D. Byrne, Enabling Good Health for All : A Reflection Process for a New EU Health Strategy. Commission of the European Communities, 2004.

[25] M. Wismar, W. Palm, J. Figueras, K. Ernst, E. Van Ginneken et al., "Cross-border health care in the european union: mapping and analysing practices and policies." Cross-border health care in the European Union: mapping and analysing practices and policies, 2011.

[26] E. Commission, "Expert panel on effective ways of investing in health: Cross-border cooperation," 2015. [Online]. Available: http://ec.europa.eu/health/expert_panel/opinions/docs/009_crossborder_cooperation_en.pdf

[27] ——, "Overview of the national laws on electronic health records in the eu member states and their interaction with the provision of cross-border ehealth services," EU Health Programme (2008-2013), 2013. [Online]. Available: http://ec.europa.eu/health/ehealth/docs/laws_report_recommendations_en.pdf

[28] I. Passarani, "Patient access to electronic health records," Report of the eHealth Stakeholder Group, 2013. [Online]. Available: http://ec.europa.eu/health/expert_panel/opinions/docs/009_crossborder_cooperation_en.pdf

Fuzzy One-Decision Making Model with Fuzzified Outcomes in the Treatment of Necrotizing Fasciitis

Elisabeth Rakus-Andersson
Department of Mathematics and Natural Sciences
Blekinge Institute of Technology
37179 Karlskrona, Sweden
e-mail: Elisabeth.Andersson@bth.se

Janusz Frey
Department of Surgery and Urology
Blekinge County Hospital
37185 Karlskrona, Sweden
e-mail: Janusz.Frey@ltblekinge.se

Abstract—**By proposing a new approach to fuzzy decision making, we try to support the medical decision, concerning recommendations for the treatment with hyperbaric oxygen (HBO). This treatment can be used for patients, suffering from necrotizing fasciitis. Due to the disease rarity, it sometimes is difficult for a physician to determine, if a single patient needs the treatment with HBO. We thus identify the decision with a linguistic variable, equipped with treatment recommendation levels. The choice of the appropriate level is based on values of clinical symptoms, found in the patient. To extract the optimal recommendation level for the treatment with HBO, we involve fuzzy set techniques in the decision model. In the paper, we mainly concentrate on designs of fuzzy sets, standing for clinical symptoms and recommendation levels. The levels act as the outcomes, dependent on the cumulative input of the patient's clinical markers. Since the focus is laid on a parametric structure of the outcomes, then we can categorize the model as robust approach to algorithmic modeling of outcomes, being part of eHealth data records.**

Keywords-fuzzy one-decision making; fuzzy sets; families of membership functions; s-functions; necrotizing fasciitis; treatment with hyperbaric oxygen.

I. INTRODUCTION

Necrotizing fasciitis (NF) is a rare, but deadly soft tissue infection. The disease is known from Hippocratic times, but has been newly rediscovered in modern times as an "infection with flesh eating bacteria" by Jones in 1871 [1]. More specifically, the illness was described in 1952 by Wilson [2], who also renamed these types of infections as necrotizing fasciitis. The NF group contains various types of infections, usually treated with antibiotics and surgery [3]. In some cases, the treatment with hyperbaric oxygen (HBO) is the adjunct of treatments, mentioned above [4]. Blekinge County City Hospital in Karlskrona, Sweden, has the possibility of providing HBO. Therefore, we serve the treatment to NF patients, who live in the south-eastern part of Sweden.

From the clinical point of view, we want to know, if the patient has a good prognosis of recovery without recommendations for the specialized treatment with HBO or he/she needs the HBO supplement.

To make this prognosis, a physician has to rely on his experience. Nevertheless, the number of patients is not so large, which makes difficult to solve routinely the problems of HBO dosing.

Therefore, we initialize the mathematical model of fuzzy decision making, which considers only one decision (indications of treating the patient with HBO).

From our design, we have excluded a utility matrix, which constitutes the main part in most of fuzzy decision making models [5]-[9]. The entries of the matrix are stated as numerical or verbal utilities, assigned to pairs (decision, state). When using the utility matrix, shown in Section II, the researchers developed different decision methods, like, e.g., unequal objectives or minimization of regret in order to extract an optimal decision [10][11].

The theoretical designs of fuzzy decision making [5]-[11] were benefited in practical applications like, e.g., medical decisions [12], making decisions in nutrition [13] or making decisions in stock market [14].

The authors applied fuzzy decision making with the utility matrix to select the most efficacious treatment having an effect on a collection of clinical symptoms. In our proposals, the pair (decision, state) was interpreted as (treatment, symptom) [15]-[17]. The results of determining the optimal decision-treatment had a general nature, and were not adapted to the health state of a single patient.

The decision, concerning the treatment with HBO, is differentiated in recommendation levels. These create a scale of hints, telling us, if the health condition of the patient agrees with the decision of giving HBO to him/her or not. For one decision "treatment with HBO", we arrange a verbal recommendation range of stages in two families of terms, namely, "stages of non-indication" versus "stages of indication". The conversion of these terms in two families of fuzzy sets with parametric membership functions is planned as a substantial contribution in the model proposed. The procedure of establishing two common formulas of parametric membership functions of fuzzy sets, representing "stages of non-indication" and "stages of indication", should prevent us from determining the boundary values of fuzzy sets in an intuitive manner. Another task to fulfill will concern the introduction of fuzzy sets, assigned to symptoms. By cumulating the symptoms intensities, we wish to find the patient's clinical characteristics. To accept the most convincing recommendation for HBO dosing, the cumulated characteristics of the patient will be tested in all decision levels.

We recall the classical fuzzy decision making model with the utility matrix in Section II. Our proposition of the one-decision model is sketched in Section III. Section IV contains the descriptions of constructions of clinical entry data. The structure of fuzzified outcomes will be engineered in Section V. The case study, referring to the treatment with HBO, will be tested in Section VI. We will formulate some concluding remarks in Section VII.

II. THE MODEL OF FUZZY DECISION MAKING WITH THE UTILITY MATRIX

Let us recall the definition of a fuzzy set.

If X is a collection of objects denoted generically by x, then the fuzzy set A in X is a set of ordered pairs $A = \{(x, \mu_A(x)): x \in X\}$, where $\mu_A(x) \in [0,1]$ [18].

Each element x gets a membership degree $\mu_A(x)$, which expresses the strength of the relationship between x and A. Membership degrees, equal to 1, inform about the total relation between the element and the set. The function $\mu_A: X \to [0,1]$ is called "*the membership function*" of A.

In classical fuzzy decision making, we introduce the notions of a space of states (e.g., symptoms) $X = \{x_1,...,x_n\}$ and a decision space (e.g., treatments) $D = \{d_1,...,d_d\}$. The utility matrix U, given by

$$U = \begin{array}{c} \\ d_1 \\ \vdots \\ d_d \end{array} \begin{array}{c} \begin{matrix} x_1 & \cdots & x_n \end{matrix} \\ \begin{bmatrix} u_{11} & \cdots & u_{1n} \\ \vdots & \ddots & \vdots \\ u_{d1} & \cdots & u_{dn} \end{bmatrix} \end{array}, \tag{1}$$

has the entries u_{bj}, $b = 1,\ldots,d$, $j = 1,\ldots,n$ [5]-[9]. Each u_{bj} is the fuzzy utility of applying decision d_b to state x_j. In most of applications, u_{bj} are evaluated intuitively as values belonging to interval [0, 1], e.g., utility of $(d_1,x_1) = 0.7$. Some users prefer determining the utilities as fuzzy sets, e.g., utility of (d_1,x_1) = "large".

The aggregated utility U_{d_b} of d_b was estimated as $U_{d_b} = \sum_{j=1}^{n} u_{bj}$ in the early trials of adapting fuzzy decision making to practical solutions.

The operation $\max(U_{d_1},...,U_{d_d})$ allowed selecting the optimal d_b, satisfying the maximum criterion. Later on, utilities U_{d_b} have been calculated with a more complicated precision.

III. THE OUTLINE OF FUZZY ONE-DECISION MAKING

Before discussing our conception of fuzzy decision making, let us add other useful definitions.

The support of a fuzzy set A, supp(A), is not a fuzzy set (it is a crisp set) of all $x \in X$, such that $\mu_A(x) > 0$ [18].

The α-cut of a fuzzy set A, A_α, is a non-fuzzy set of all $x \in X$ such that $\mu_A(x) \geq \alpha$ [18].

The Euclidean distance $d(P_1, P_2)$ between points $P_1 = (x_1,\mu(x_1))$ and $P_2 = (x_2,\mu(x_2))$, in the two dimensional system with x- and $\mu(x)$-axes, is estimated as $d(P_1, P_2) = \sqrt{(x_2 - x_1)^2 + (\mu(x_2) - \mu(x_1))^2}$.

In the current trial of fuzzy decision making, we suppose that set $D = \{d\}$ consists of only one decision, i.e., the treatment with HBO. Decision d, made for patient P_i, $i = 1,\ldots,p$, is constructed as a linguistic variable, whose verbal values are term-sets L_l, $l = 1,\ldots,m$. These terms are edited as recommendation levels of the treatment with HBO. The levels graduate indications of the treatment, scaled from the most contraindicated to the most advised by a physician.

We still keep the set of symptoms $X = \{x_1,...,x_n\}$.

For patient P_i, $i = 1,\ldots,p$, we need to sample the characteristics s_i, informing about presence or absence of symptoms X_j, $j = 1,\ldots,n$, in the patient. Symptoms X_j are typical of necrotizing fasciitis. We test the behavior of s_i on all levels L_l to select this L_l, for which the patient characteristics matches best.

Let us suppose that each clinical marker X_j, $j = 1,\ldots,n$, is replaced by a fuzzy set, also named X_j. If a marker value $x_{i,j}$ for symptom X_j is found in P_i, then the membership degree $\mu_{X_j}(x_{i,j})$ will be assigned to $x_{i,j}$. The way of designing membership functions $\mu_{X_j}: X_j \to [0,1]$ will be evolved in Section IV.

The importance weights w_j of symptoms X_j are added to the formula of s_i to emphasize X_j's harmful influence on the disease course. Due to the professional experience, the physician suggests the placement of X_j in the sequence $X_1 \succ \ldots \succ X_n$, where "$\succ$" means "$X_j$ emerges more dangerous impact on the patient health state than X_h, $j, h = 1,\ldots,n$. We state $w_1 > \ldots > w_n$ and want $\sum_{j=1}^{n} w_j = 1$.

The collected patient characteristics s_i (the numerical knowledge about the symptoms), made via all $x_{i,j}$ and w_j, will be derived for patient P_i as

$$s_i = \sum_{j=1}^{n} \mu_{X_j}(x_{i,j}) \cdot w_j, \; i = 1, \ldots, p. \tag{2}$$

We note that the minimal value of s_i is 0 since, for all minimal $\mu_{X_j}(x_{i,j}) = 0$, we obtain $s_i = \sum_{j=1}^{n} 0 \cdot w_j = 0$, $i = 1, \ldots, p$.

The maximal value of s_i will reach 1 if, for all maximal $\mu_{X_j}(x_{i,j}) = 1$, $s_i = \sum_{j=1}^{n} 1 \cdot w_j = 1 \cdot \sum_{j=1}^{n} w_j = 1 \cdot 1 = 1$, $i = 1, \ldots, p$.

Hence, $s_i \in [0, 1]$, $i = 1,\ldots,p$.

The term-sets L_l, $l = 1,\ldots,m$, are designed as a collection of fuzzy sets, assisting recommendation levels of decision d. Sets L_l have their supports allocated in a common non-fuzzy reference set $L = [0, 1]$ in compliance with the domain of s_i ($s_i \in [0, 1]$).

We prove the action of s_i, found in P_i, in each L_l by computing $\mu_{L_l}(s_i)$, $l = 1,\dots,m$. We adopt the optimal decision level L_l of d as level L^*, satisfying the condition $\mu_{L^*}(s_i) = \max_{1\le l\le m}(\mu_{L_l}(s_i))$.

Equation 2 shows the decision process for patient P_i as a procedure

$$\begin{bmatrix} \mu_{X_1}(x_{i,1})\cdot w_1 \\ \vdots \\ \mu_{X_n}(x_{i,n})\cdot w_n \end{bmatrix} \to s_i = \sum_{j=1}^{n} \mu_{X_j}(x_{i,j})\cdot w_j \to \begin{pmatrix} \mu_{L_1}(s_i) \\ \vdots \\ \mu_{L_m}(s_i) \end{pmatrix} \quad (3)$$

$$\to d = L^* \text{for which } \mu_{L^*}(s_i) = \max_{1\le l\le m} \mu_{L_l}(s_i).$$

We emphasize that the decision level, selected for P_i, is patient-tailored.

In Section IV, we construct the entries of the model.

IV. The Construction of Entry Data

Symptoms X_j are recognized as quantitative and qualitative features. We assign fuzzy sets X_j, $j = 1,\dots,n$, to both types. As the rising order of symptom values (real values or codes) is associated with the growing states of the disease threat then, as a consequence, the membership functions of X_j will be constructed as ascending functions.

For the measurable symptoms X_j, taking values $x_{i,j}$ in interval $[\alpha, \gamma]$ continuously, we have prepared the membership function $\mu_{X_j}(x_{i,j})$ as a parametric s-function $s(x_{i,j}, \alpha, \beta, \gamma)$, yielded by [18]

$$\mu_{X_j}(x_{i,j}) = s(x_{i,j}, \alpha, \beta, \gamma) =$$

$$\begin{cases} 0 & \text{for} \quad x_{i,j} \le \alpha, \\ 2\left(\frac{x_{i,j}-\alpha}{\gamma-\alpha}\right)^2 & \text{for} \quad \alpha < x_{i,j} \le \beta, \\ 1-2\left(\frac{x_{i,j}-\gamma}{\gamma-\alpha}\right)^2 & \text{for} \quad \beta < x_{i,j} \le \gamma, \\ 1 & \text{for} \quad x_{i,j} > \gamma, \end{cases} \quad (4)$$

where $\beta = \frac{\alpha+\gamma}{2}$, $j = 1,\dots,n$, $i = 1,\dots,p$, $x_{i,j} \in \text{supp}(X_j)$.

Example 1

Symptom "*age*"= X_2 is a fuzzy set, constrained by the membership function $s(x_{i,2}, 18, 59, 100)$. For, e.g., $x_{i,2} = 76$, we estimate $\mu_{X_2}(76) = 1 - 2\left(\frac{76-100}{100-18}\right)^2 = 0.828$ in accordance with the condition 59<76<100.

We adopt the own procedure [19] to calculate the membership degrees for compound qualitative symptoms X_j, characterized by a list of codes $C_{X_j} = \{0,\dots,k,\dots,z\}$, where $k = 0,\dots,z$, are non-negative integers. Let us assume that z is an even integer. The codes k mark alternative answers to a question, investigating the intensity of symptom X_j in P_i. We suppose that answer 0 denies the presence of X_j, whereas value z confirms X_j's critical stage. Code value $\frac{0+z}{2}$ indicates the uncertain symptom status as "*medium intensity*", "*difficult to say*", and the like.

Let us first set up a function $g(k)$, which starts with $g(0) = -1$ and terminates with $g(z) = 1$. In general,

$$g(k) = g(0) + k \cdot \frac{g(z)-g(0)}{z} = -1 + k \cdot \frac{2}{z} \quad (5)$$

for $k = 0,\dots,z$.

Interval $[-1, 1]$, containing discrete values $g(k)$, constitutes a support of fuzzy set X_j, assisting the compound qualitative symptom. In order to estimate membership degrees of $g(k)$, where $k = 0,\dots,z$, we use, as the membership function of X_j, the s function

$$\mu_{X_j}(g(k)) = s(g(k), -1, 0, 1) =$$

$$\begin{cases} 2\left(\frac{g(k)+1}{2}\right)^2 & \text{for} \quad -1 \le g(k) \le 0, \\ 1-2\left(\frac{g(k)-1}{2}\right)^2 & \text{for} \quad 0 \le g(k) \le 1. \end{cases} \quad (6)$$

After examining in detail the properties of (6), we note that: the lack of the symptom $g(0) = -1$ is characterized by membership 0, and the critical condition of the symptom $g(z) = 1$ is tied to membership 1. The value $\frac{g(0)+g(z)}{2} = \frac{-1+1}{2} = 0$, assigned to an uncertain appearance of X_j, is furnished with membership 0.5. These features of (6) logically agree with medical expectations for symptoms coded.

Example 2

The levels of symptom "*medical state*" = S_1 are coded as: "*comfortable*" = 0, "*satisfactory*" = 1, "*stable*" = 2, "*critical but stable*" = 3, and "*critical*" = 4. In accordance with (5), for $k = 0,\dots,4$, $g(0) = -1 + 0\cdot\frac{2}{4} = -1$, $g(1) = -0.5$ $g(2) = 0$, $g(3) = 0.5$, and $g(4) = 1$. The membership degrees, found for $g(k)$, $k = 0,\dots,4$, are, by (6), numbers: $\mu_{X_1}(g(0)) = 0$, $\mu_{X_1}(g(1)) = 0.125$, $\mu_{X_1}(g(2)) = 0.5$, $\mu_{X_1}(g(3)) = 0.875$, and $\mu_{X_1}(g(4)) = 1$.

In the last part of Section IV, let us solve the problem of assigning the importance weights w_j to symptoms X_j. By "importance" we mean the strength of X_j's adverse and harmful power in the running process of the illness diagnosed. We bring into light another own mathematical algorithm, allowing the estimation of weights [19].

Generally, if we consider n symptoms X_j to find importance weights for them, we will wish to arrange them in the sequence $X_1 \succ ... \succ X_n$ in accordance with the expert's opinion. We want the sum of all weights w_j, assisting X_j, $j = 1,...,n$, to be 1. Therefore,

$$n \cdot r + (n-1) \cdot r + ... + 2 \cdot r + 1 \cdot r = 1 \qquad (7)$$

where r is a quotient dependent on n.

Further,

$$w_j = (n - j + 1) \cdot r \qquad (8)$$

for $j = 1,...,n$.

Example 3

The decisive symptoms for the recognition of necrotizing fasciitis are listed in the importance order, decided by the physician, as "*medical state*" = $X_1 \succ$"*age*" = $X_2 \succ$"*risk factors*" = $X_3 \succ$"*crp*" = $X_4 \succ$*wbc* = $X_5 \succ$"*temperature*" = X_6. The abbreviation "*crp*" stands for C-reactive proteins and "*wbc*" – for white blood cells. In conformity with (7), equation $6r+5r+4r+3r+2r+r = 1$ provides $r = 0.0476$. After employing (8), we receive, in turn for $j = 1,\ldots,6$, the weights: $w_1 = (6-1+1)\cdot 0.0476 = 0.2856$, $w_2 = 0.238$, $w_3 = 0.1904$, $w_4 = 0.1428$, $w_5 = 0.0952$, and $w_6 = 0.0476$.

V. The Structure of Fuzzified Outcomes

As (3) recommends, we should now generate a sample of output recommendation fuzzy levels L_l of decision d, $l = 1,\ldots,m$. The supports of L_l cover parts of [0, 1], as proved in Section III ($s_i \in [0, 1]$). To calculate the membership degrees of signal s_i in L_l, $i = 1,\ldots,p$, we need to derive a formula of the membership function of each L_l. The largest value $\mu_{L_l}(s_i)$, $l = 1,..,m$, points out the optimal recommendation level of decision d, advised for P_i.

Theoretically, m can be either an even or an odd positive arbitrary integer. An own procedure [20], expanded in this paper, helps us to derive membership functions of L_l. These are dependent only on two parameters, namely, a number m of term-sets in a list of decision d and a width E of the common reference set L, containing all supports of L_l.

In the medical problem discussed, m is supposed to be the even number, as the differentiation of non-indication and indication levels of the HBO treatment is bipartite. Our intention is to derive two common formulas of membership functions of sets L_l, separated in two families. Then, we do not need to predetermine the boundary values of supports of fuzzy sets in an intuitive or a random way. By the way, it is important to emphasize that the procedure can be easily computerized.

Due to the definition of the α-cut set of a fuzzy set (introduced in Section III), we denote by $L_{l,\alpha}$ a set of $s_i \in L = [0, 1]$, for which $\mu_{L_l}(s_i) \geq \alpha, l = 1, \ldots, m$.

As a pattern of membership function of L_l, the s-function $s(s_i, \alpha_{L_l}, \beta_{L_l}, \gamma_{L_l}) = \mu_{L_l}(s_i)$ is arranged in accord with (4).

If we wish to narrow domains of functions $s(s_i, \alpha_{L_l}, \beta_{L_l}, \gamma_{L_l}) = \mu_{L_l}(s_i)$ and, consequently, to narrow supports of fuzzy sets L_l, then we will modify $s(s_i, \alpha_{L_l}, \beta_{L_l}, \gamma_{L_l})$ as $s(s_i, \alpha_{L_l} \cdot \delta, \beta_{L_l} \cdot \delta, \gamma_{L_l} \cdot \delta)$.

Function $s(s_i, \alpha_{L_l} \cdot \delta, \beta_{L_l} \cdot \delta, \gamma_{L_l} \cdot \delta)$ is expanded by

$$s(s_i, \alpha_{L_l} \cdot \delta, \beta_{L_l} \cdot \delta, \gamma_{L_l} \cdot \delta) = \begin{cases} 0 & \text{for } s_i \leq \alpha_{L_l} \cdot \delta, \\ 2\left(\frac{s_i - \alpha_{L_l} \cdot \delta}{(\gamma_{L_l} - \alpha_{L_l})\delta}\right)^2 & \text{for } \alpha_{L_l} \cdot \delta \leq s_i \leq \beta_{L_l} \cdot \delta, \\ 1 - 2\left(\frac{s_i - \gamma_{L_l} \cdot \delta}{(\gamma_{L_l} - \alpha_{L_l})\delta}\right)^2 & \text{for } \beta_{L_l} \cdot \delta \leq s_i \leq \gamma_{L_l} \cdot \delta, \\ 1 & \text{for } s_i \geq \gamma_{L_l} \cdot \delta. \end{cases} \qquad (9)$$

Values $0 < \delta < 1$ have an effect of narrowing domains in (9). Value $\delta = 1$ allows returning to (4).

Theorem 1

Let us suppose that term-sets L_l, $l = 1,\ldots,m$, have supports included in the common non-fuzzy reference set L, where $\min(L) = 0$. Patient characteristics s_i belongs to L and the width of L is E.

If m is even, then we divide all fuzzy sets L_l in two families. A family of "*left*" sets $L_1,\ldots, L_{\frac{m}{2}}$ contains L_t sets, where $t = 1,\ldots, \frac{m}{2}$. A family of "*right*" sets $L_{\frac{m+2}{2}},\ldots,L_m$ is composed of $L_{\frac{m+2}{2}+t-1}$ sets for $t = 1,\ldots \frac{m}{2}$.

We assume that sets $L_{1,0.5}$, $L_{t,0.5}-L_{t-1,0.5}$, $t = 2,\ldots, \frac{m}{2}$, $L_{\frac{m+2}{2}+t-1,0.5} - L_{\frac{m+2}{2}+t,0.5}$, $t = 1,\ldots, \frac{m-2}{2}$, and $L_{m,0.5}$, established by α-cuts of $L_1,\ldots, L_{\frac{m}{2}}$ and $L_{\frac{m+2}{2}},\ldots,L_m$ for $\alpha = 0.5$, have the same width. Suppose further that the membership functions of the last "*left*" set $L_{\frac{m}{2}}$ and the first "*right*" set $L_{\frac{m+2}{2}}$ have the intersection point on membership level 0.5. Hence, the common formulas for membership functions of L_l in their families are given by

$$\mu_{L_t}(s_i)=$$
$$\begin{cases} 1 & \text{for } s_i \le \frac{(m-2)E}{2(m-1)}\delta(t), \\ 1-2\left(\frac{s_i-\frac{(m-2)E}{2(m-1)}\delta(t)}{\frac{E}{(m-1)}\delta(t)}\right)^2 & \text{for } \frac{(m-2)E}{2(m-1)}\delta(t)\le s_i \le \frac{E}{2}\delta(t), \\ 2\left(\frac{s_i-\frac{mE}{2(m-1)}\delta(t)}{\frac{E}{(m-1)}\delta(t)}\right)^2 & \text{for } \frac{E}{2}\delta(t)\le s_i\le \frac{mE}{2(m-1)}\delta(t), \\ 0 & \text{for } s_i \ge \frac{mE}{2(m-1)}\delta(t), \end{cases} \tag{10}$$

where $\delta(t)=\frac{2}{m}\cdot t$, $t=1,\ldots,\frac{m}{2}$, for the "*left*" family and, for the "*right*" family,

$$\mu_{L_{\frac{m+2}{2}+t-1}}(s_i)=$$
$$\begin{cases} 0 \text{ for } s_i \le E-\frac{mE}{2(m-1)}\varepsilon(t), \\ 2\left(\frac{s_i-\left(E-\frac{mE}{2(m-1)}\varepsilon(t)\right)}{\frac{E}{(m-1)}\varepsilon(t)}\right)^2 \\ \text{for } E-\frac{mE}{2(m-1)}\varepsilon(t)\le s_i\le E-\frac{E}{2}\varepsilon(t), \\ 1-2\left(\frac{s_i-\left(E-\frac{(m-2)E}{2(m-1)}\varepsilon(t)\right)}{\frac{E}{m-1}\varepsilon(t)}\right)^2 \\ \text{for } E-\frac{E}{2}\varepsilon(t)\le s_i\le E-\frac{(m-2)E}{2(m-1)}\varepsilon(t), \\ 1 \text{ for } s_i\ge E-\frac{(m-2)E}{2(m-1)}\varepsilon(t), \end{cases} \tag{11}$$

if $\varepsilon(t)=1-\frac{2}{m}\cdot(t-1)$, $t=1,\ldots,\frac{m}{2}$. The membership functions are derived on the basis of (4) and (9).

Proof:

We start with the assumption: $L_{1,0.5}$, $L_{t,0.5}–L_{t-1,0.5}$, $t=2,\ldots,\frac{m}{2}$, $L_{\frac{m+2}{2}+t-1,0.5}-L_{\frac{m+2}{2}+t,0.5}$, $t=1,\ldots,\frac{m-2}{2}$, and $L_{m,0.5}$ have the same width. It results in making the partition of reference set L in $m–1$ subintervals with the same width equal to $\frac{E}{m-1}$.

We estimate *Euclidean distance* between points $(\alpha_{L_{\frac{m}{2}}},1)$ and $(\frac{E}{2},1)$ as $\frac{E}{2(m-1)}$ (half a width of the middle subinterval lying along set L).

We compute

$\alpha_{L_{\frac{m}{2}}}=\frac{E}{2}-\frac{E}{2(m-1)}=\frac{(m-2)E}{2(m-1)}$ and $\gamma_{L_{\frac{m}{2}}}=\frac{E}{2}+\frac{E}{2(m-1)}=\frac{mE}{2(m-1)}$ to assure that the membership function of $L_{\frac{m}{2}}$ intersects the membership function of $L_{\frac{m+2}{2}}$ in point $(\frac{E}{2},0.5)$.

If $\beta_{L_{\frac{m}{2}}}=\frac{E}{2}$, then $\mu_{L_{\frac{m}{2}}}(s_i)=1-s(s_i,\frac{(m-2)E}{2(m-1)},\frac{E}{2},\frac{mE}{2(m-1)})$.

We employ (4) to get the membership function of $L_{\frac{m}{2}}$ as a formula

$$\mu_{L_{\frac{m}{2}}}(s_i)=$$
$$\begin{cases} 1 & \text{for } s_i\le\frac{(m-2)E}{2(m-1)}, \\ 1-2\left(\frac{s_i-\frac{(m-2)E}{2(m-1)}}{\frac{E}{(m-1)}}\right)^2 & \text{for } \frac{(m-2)E}{2(m-1)}\le s_i\le\frac{E}{2}, \\ 2\left(\frac{s_i-\frac{mE}{2(m-1)}}{\frac{E}{(m-1)}}\right)^2 & \text{for } \frac{E}{2}\le s_i\le\frac{mE}{2(m-1)}, \\ 0 & \text{for } s_i\ge\frac{mE}{2(m-1)}\delta(t). \end{cases} \tag{12}$$

The "*left*" family of fuzzy sets from (10) will be generated, when we add function $\delta(t)=\frac{2}{m}\cdot t$, $t=1,\ldots,\frac{m}{2}$, to (12) in accordance with (9).

The modifier $\delta(t)$, $0<\delta(t)\le 1$, is inserted in (12) to cause narrowing effects of supports of L_t, $t=1,\ldots,\frac{m}{2}$. Function $\delta(t)$ reveals the properties: $\delta(\frac{m}{2})=1$ (no impact on the support of the last "*left*" set $L_{\frac{m}{2}}$) and $\delta(1)=\frac{1}{m/2}=\frac{2}{m}$ (the largest scale value 1 is divided by the number of left sets). If we suppose that $\delta(t)=a\cdot t$, then the solution of equation $a\cdot\frac{m}{2}=1$ will provide $a=\frac{2}{m}$. Hence, $\delta(t)=\frac{2}{m}\cdot t$.

We now construct the membership function of the first right fuzzy set $L_{\frac{m+2}{2}}$ as a reverse membership function of $L_{\frac{m}{2}}$. We find

$$\mu_{L_{\frac{m+2}{2}}}(s_i)=$$
$$\begin{cases} 0 & \text{for } s_i\le E-\frac{mE}{2(m-1)}, \\ 2\left(\frac{s_i-\left(E-\frac{mE}{2(m-1)}\right)}{\frac{E}{m-1}}\right)^2 & \text{for } E-\frac{mE}{2(m-1)}\le s_i\le E-\frac{E}{2}, \\ 1-2\left(\frac{s_i-\left(E-\frac{(m-2)E}{2(m-1)}\right)}{\frac{E}{m-1}}\right)^2 & \text{for } E-\frac{E}{2}\le s_i\le E-\frac{(m-2)E}{2(m-1)}, \\ 1 & \text{for } s_i\ge E-\frac{(m-2)E}{2(m-1)}. \end{cases} \tag{13}$$

The membership functions of sets $L_{\frac{m+2}{2}},\ldots,L_m$ are initialized after inserting a new modifier $\varepsilon(t)=1-\frac{2}{m}\cdot(t-1)$, $t=1,\ldots,\frac{m}{2}$, $0<\varepsilon(t)\le 1$, in (13). The insertion matches the model provided by (9) and proves (11). For $t=1$, we get $\varepsilon(1)=1$, while $t = \frac{m}{2}$ follows $\varepsilon(\frac{m}{2}) = \frac{1}{m/2} = \frac{2}{m}$. We derive $\varepsilon(t) = 1-a\cdot(t-1)$ to ensure the equality $\varepsilon(1) = 1$.

Equation $1 - a\cdot(\frac{m}{2}-1) = \frac{2}{m}$ has solution $a = \frac{2}{m}$.

Example 4

The term list of decision d = "*recommendation for treating with HBO for patient P_i*" is stated as $d = \{L_1$ = *strong non-indication for treating with HBO*", L_2 = *moderate non-indication for treating with HBO*", L_3 ="*moderate indication for treating with HBO*", L_4 = "*strong indication for treating with HBO*". L_1 and L_2 belong to the "*left*" family of fuzzy sets, whereas L_3 and L_4 build the "*right*" family of fuzzy sets. Sets L_l have the supports included in interval [0, 1], due to the statement $s_i \in [0, 1]$. For $m = 4$ and $E = 1$, we get

$$\mu_{L_1}(s_i) = \begin{cases} 1 & \text{for } 0 \le s_i \le 0.166, \\ 1-2\left(\frac{s_i-0.166}{0.166}\right)^2 & \text{for } 0.166 \le s_i \le 0.25, \\ 2\left(\frac{s_i-0.333}{0.166}\right)^2 & \text{for } 0.25 \le s_i \le 0.333, \\ 0 & \text{for } s_i \ge 0.333, \end{cases} \quad (14)$$

and

$$\mu_{L_2}(s_i) = \begin{cases} 1 & \text{for } 0 \le s_i \le 0.333, \\ 1-2\left(\frac{s_i-0.333}{0.333}\right)^2 & \text{for } 0.333 \le s_i \le 0.5, \\ 2\left(\frac{s_i-0.666}{0.333}\right)^2 & \text{for } 0.5 \le s_i \le 0.666, \\ 0 & \text{for } s_i \ge 0.666, \end{cases} \quad (15)$$

when setting $t = 1$ ($\delta(1) = 0.5$) and $t = 2$ ($\delta(2) = 1$) in (10), respectively.

The action of placing $t = 1$ ($\varepsilon(1) = 1$) and $t = 2$ ($\varepsilon(2) = 0.5$), in (11), yields

$$\mu_{L_3}(s_i) = \begin{cases} 0 & \text{for } 0 \le s_i \le 0.333, \\ 2\left(\frac{s_i-0.333}{0.333}\right)^2 & \text{for } 0.333 \le s_i \le 0.5, \\ 1-2\left(\frac{s_i-0.666}{0.333}\right)^2 & \text{for } 0.5 \le s_i \le 0.666, \\ 1 & \text{for } s_i \ge 0.666, \end{cases} \quad (16)$$

and

$$\mu_{L_4}(s_i) = \begin{cases} 0 & \text{for } 0 \le s_i \le 0.667, \\ 2\left(\frac{s_i-0.667}{0.167}\right)^2 & \text{for } 0.667 \le s_i \le 0.75, \\ 1-2\left(\frac{s_i-0.833}{0.167}\right)^2 & \text{for } 0.75 \le s_i \le 0.833, \\ 1 & \text{for } s_i \ge 0.833. \end{cases} \quad (17)$$

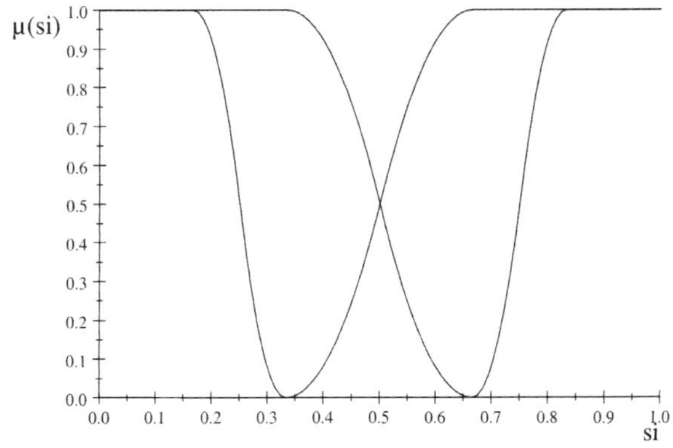

Figure 1. Fuzzy sets L_1-L_4.

Fuzzy sets L_1-L_4 are sketched in Figure 1.

Section VI is devoted to tracking the theoretical proposal by a solution of the medical query, formulated as the recommendation of the treatment with HBO. The decision is made for a single patient.

VI. The Recommendation for Treating with HBO

It has already been mentioned in Section I that the mathematical apparatus, built in Sections III-V, will be applied to select either a non-indication level or an indication level of decision d.

The data, including the values of crucial clinical markers, have been sampled for 13 patients (12 men and 1 woman) treated in the Blekinge County City Hospital in Karlskrona, Sweden, between 2006 and 2010.

The clinical symptoms, essential in NF, have been introduced in Example 3. For quantitative symptoms we adapt (4) as follows:

$\mu_{X_2=\text{"}age\text{"}}(x_{i,2}) = s(x_{i,2}, 18, 59, 100)$,

$\mu_{X_4=\text{"}crp\text{"}}(x_{i,4}) = s(x_{i,4}, 0, 250, 500)$,

$\mu_{X_5=\text{"}wbc\text{"}}(x_{i,5}) = s(x_{i,5}, 0, 15, 30)$,

and

$\mu_{X_6=\text{"}temp.\text{"}}(x_{i,6}) = s(x_{i,6}, 36, 38.5, 41)$.

In Example 2, we have already determined the membership degrees for the coded symptom X_1 = "*medical state*" as:

$\mu_{X_1}(g(0)) = 0$, $\mu_{X_1}(g(1)) = 0.125$, $\mu_{X_1}(g(2)) = 0.5$,

$\mu_{X_1}(g(3)) = 0.875$, and $\mu_{X_1}(g(4)) = 1$.

We repeat the algorithm for symptom X_3 = "*risk factors*", coded between 0 and 6, to find $g(0) = -1$, $g(1) = -0.666$, $g(2) = -0.333$, $g(3) = 0$, $g(4) = 0.333$, $g(5) = 0.666$ and $g(6) = 1$. When applying $\mu_{X_3=\text{"}risk\ factors\text{"}}(g(k)) = s(g(k), -1, 0, 1)$, $k = 0,\ldots,6$, we list:

$\mu_{X_3}(g(0)) = 0$, $\mu_{X_3}(g(1)) = 0.056$, $\mu_{X_3}(g(2)) = 0.221$,

TABLE I. PATIENT SYMPTOM VALUES AND MEMBERSHIP DEGREES IN FUZZY SETS DESIGNED FOR SYMPTOMS X_j, $j = 1,\ldots,6$

P_i	X_1	X_2	X_3	X_4	X_5	X_6
P_1	0.13/1	0.06/32	0/0	0.83/352	0.52/15.3	0/36.2
P_2	0/36.2	0.83/76	0.5/3	0.56/267	0.49/14.9	0.39/38.2
P_3	0.88/3	0.30/50	0.06/1	0.43/232	0.22/10	0.14/37.3
P_4	0.5/2	0.66/66	0.22/2	0.7/305	0.99/28.2	0.29/37.9
P_5	0.88/3	0.75/71	0/0	0.29/189	0.99/27.8	0.14/37.3
P_6	0.5/2	0.45/57	0/0	0.64/281	0.53/15.5	0.42/38.3
P_7	1/4	0.29/49	0.06/1	0.85/363	0.76/19.5	0.03/36.6
P_8	0.88/3	0.89/81	0.5/3	0.91/394	0.36/12.7	0.20/37.6
P_9	1/4	0.48/58	1/6	0.94/413	0.68/18	0.32/38
P_{10}	0.88/3	0.45/57	0.06/1	0.48/246	0.02/3.1	0/35.8
P_{11}	0.5/2	0.52/60	0.22/2	0.06/85	0.62/16.9	0.29/36.5
P_{12}	0.88/3	0.73/70	0.78/4	0.92/403	0.99/28.5	0.32/38
P_{13}	1/4	0.88/80	0.22/2	0.05/76	0.73/18.9	0.98/40.5

$\mu_{X_3}(g(3)) = 0.5$, $\mu_{X_3}(g(4)) = 0.779$, $\mu_{X_3}(g(5)) = 0.944$, and $\mu_{X_3}(g(6)) = 1$.

TABLE I contains the clinical data and assigned to them membership degrees, computed in compliance with the membership functions of X_j. The membership degree of $x_{i,j}$ in X_j appears before the dash, and the $x_{i,j}$ clinical value is placed after the dash, $j = 1,\ldots,6$.

As emerged in (2), the concatenation of membership degrees $\mu_{X_j}(x_{i,j})$ with weights w_j, evaluated in Example 3, $j = 1,\ldots,6$, will constitute a basis for the calculation of the cumulated clinical characteristics s_i for patient P_i.

Example 5

Patient P_1 is represented by
$s_1 = 0.125 \cdot 0.286 + 0.058 \cdot 0.238 + 0 \cdot 0.19 + 0.824 \cdot 0.1428 + 0.52 \cdot 0.095 + 0.003 \cdot 0.047 = 0.217$.

In order to select one of four decision levels by means of membership degrees in L_l, $l = 1,\ldots,4$, we return to (14)-(17).

We choose the decision characterized by the largest membership degree out of $\mu_{L_l}(s_i)$.

Example 6

TABLE II collects s_i, their membership degrees in L_l, $l = 1,\ldots,4$, and the physician's assertion already made. The abbreviations mean: PD HBO = the physician's decision, concerning treating the patient with HBO, N = none treating with HBO, and Y = treating with HBO.

For example, for $s_i = 0.217$ (characteristics of P_1), we get:
$\mu_{L_1}(0.217) = 1 - 2\left(\frac{0.217-0.166}{0.166}\right)^2 = 0.28$ (0.166 <0.217 <0.25), $\mu_{L_2}(0.217) = 1$ (0.217 < 0.333), $\mu_{L_3}(0.217) = 0$ (0.217 < 0.333), and $\mu_{L_4}(0.217) = 0$ (0.217 < 0.667).

The largest value of the membership degree indicates level L_2.

TABLE II. THE COMPARISON OF FUZZY DECISIONS (UNDERLINED) TO DECISIONS MADE BY THE PHYSICIAN

P_i	s_i	$\mu_{L_1}(s_i)$	$\mu_{L_2}(s_i)$	$\mu_{L_3}(s_i)$	$\mu_{L_4}(s_i)$	PD HBO
P_1	0.217	0.81	1	0	0	N
P_2	0.58	0	0.13	0.87	0	Y
P_3	0.42	0	0.86	0.14	0	N
P_4	0.558	0	0.25	0.75	0	Y
P_5	0.57	0	0.17	0.83	0	Y
P_6	0.41	0	0.89	0.11	0	N
P_7	0.56	0	0.21	0.79	0	Y
P_8	0.73	0	0	1	0.29	Y
P_9	0.80	0	0	1	0.93	Y
P_{10}	0.44	0	0.8	0.2	0	N
P_{11}	0.39	0	0.94	0.06	0	N
P_{12}	0.81	0	0	1	0.97	Y
P_{13}	0.66	0	0	1	0	Y

In the future research, we plan to test the model with an odd number of decisions levels, where the middle level "*wait and see*" will be assigned to values about 0.5.

VII. CONCLUSION AND FUTURE WORK

By suggesting modifications in the classical fuzzy decision making, we have used our model to advise the treatment with hyperbaric oxygen. This treatment can improve the health state in patients, suffering from necrotizing fasciitis.

Instead of designing a utility matrix filled with distinct utilities of pairs (decision, state), we have introduced only one decision, designated by the list of term-sets. These express recommendation levels of the treatment as non-indications and indications. The decision levels are involved in the algorithm in its final phase. This differs the model, proposed in the current paper, from most of fuzzy decision making models, in which decisions are already active in the first stage of designing the utility matrix. It is also worth emphasizing that our decisions are made for individuals, and they have not general characters, as it often happens in other patterns of fuzzy decision making.

The input data and output recommendation levels are fuzzified by designs of own suggestions of membership functions. The membership functions of the outcomes (recommendation levels) are sampled in two common formulas. The formulas depend only on a number of recommendation terms and the width of a reference set, linking all supports of recommendations. The functions are derived in the way, which allows entering an arbitrary number of recommendation levels. This extends the decision scale of linguistic expressions without making changes in formulas.

The own procedures of estimating the importance weights of symptoms and approximating membership degrees of qualitative symptoms have also been added as contributions in imprecise mathematics.

Necrotizing fasciitis is a quite rare entity, and there is no widespread consensus regarding neither treatment nor grading. There were done several attempts of using laboratory results to facilitate grading of the severity of the disease, but as far as we know, they are not used widely. The idea of combining analysis of numerical parameters, such as body temperature, white blood cell count, age etc. with the qualitative estimations, such as, e.g., medical state, is very promising because it will reflect the real decision making progress. The model, tested above, is based on retrospective analysis of data of patients treated with hyperbaric oxygen (HBO) at the surgery department in Karlskrona, Sweden.

We realize that the proposition of making decisions in the case of the HBO dosing has weaknesses, mostly, when the group, used to check the model, has not been very numerous. In spite of this, it seems that we have been successful in selecting essential clinical and biochemical parameters for the correctness of the mathematical model. The decisions have "*softer*" character than two-valued decisions "*yes-no*". This is a result of imprecision, introduced by the overlapping effect of fuzzy sets.

In the further research, we will redefine the ordering of importance weights of symptoms more carefully to refine the results. We also plan to test the model with an odd number of decisions levels, where the middle level "*wait and see*" will be assigned to values about 0.5.

Since an emphasis is laid on the design of recommendation levels, appearing as the output of the mathematical algorithm, then we can classify the model as robust approach to algorithmic modeling of outcomes.

REFERENCES

[1] W. F. Quirk Jr. and G. Stembach, "Joseph Jones: Infection with Flesh Eating Bacteria," Journal of Emergency Medicine, vol. 14, Issue 6, Nov.-Dec. 1996, pp. 747-753, Doi: http://dx.doi.org/10.1016/S0736-4679(96)00197-7 [retrieved: March, 2016].

[2] B. Wilson, "Necrotizing Fasciitis," American Surgeon, vol. 18, Issue 4, Apr. 1952, pp. 416-431, PMID: 14915014.

[3] S. Hasham, P. Matteucci, P. R. Stanley, and N. B. Hant, "Necrotizing Fasciitis," BMJ, vol. 330, Apr. 2005, pp. 830-833, Doi: 10.1136/bmj.330.7495.830.

[4] D. Mathieu, R. Favory, J. Cesari, and F. Wattel, Necrotizing Soft Tissue Infections. Handbook on Hyperbaric Medicine, Netherlands: Springer, pp. 263-298, 2006.

[5] R. E. Bellman and L. A. Zadeh, "Decision Making in a Fuzzy Environment," Management Sci., vol. 17, Issue 4, Dec. 1970, pp. 141-164.

[6] F. Herrera and E. Herrera-Viedma, "Linguistic Decision Analysis: Steps for Solving Decision Problems under Linguistic Information," Fuzzy Sets and Systems, vol. 115, Dec. 2000, pp. 67-82, doi:10.1016/S0165-0114(99)00024.

[7] R. Jain, "Decision Making in the Presence of Fuzzy Variables," IEEE Trans. Syst. Man and Cybern., vol. 6, Oct. 1976, pp. 698-703.

[8] S. M. Chen and J. M. Tan, "Handling Multicriteria Fuzzy Decision Making Problems Based on Vague Set Theory," Fuzzy Sets and Systems, vol. 67, Issue 2, Oct. 1994, pp. 163-172.

[9] D. Vanier, S. Tesfamariam, R. Sadiq, and Z. Lounis, "Decision Models to Prioritize Maintenance and Renewal Alternatives," Joint International Conference on Computing and Decision Making in Civil and Building Engineering, Montreal, Canada, June 2006, pp. 2594-2603, NRCC-45571.

[10] R. R. Yager, "Fuzzy Decision Making Including Unequal Objectives," Fuzzy Sets and Systems, vol. 1, April 1978, pp. 87-95.

[11] R. R. Yager, "Decision Making Using Minimization of Regret," International Journal of Approximate Reasoning, vol. 36, Issue 2, June 2004, pp. 109-128, Doi: 10.1016/j.ijar.2003.10.003.

[12] V. Levashenko and E. Zaitseva, "Fuzzy Decision Trees in Medical Decision Making Support System," Federated Conference on Computer Science and Information Systems (FedCSIS 2012), IEEE, Sept. 2012, pp. 212-219, ISBN: 978-1-4673-0708-6.

[13] B. Wirsam, A. Hahn, E. Uthus, and C. Leitzmann, "Fuzzy Sets and Fuzzy Decision Making in Nutrition," Eur. J. Clin. Nutr., vol. 51, Issue 5, May 1997, pp. 286-296.

[14] M. Gunasekaran, K. S. Ramaswami, and K. P. Rajesh, "Fuzzy Decision Making System for Stock Market," International Conference on Sensors, Security, Software and Intelligent Systems, Jan. 2009, http://dx.doi.org/10.2139/ssrn.2335241 [retrieved: March, 2016].

[15] E. Rakus-Andersson, "Decision-making Techniques in Ranking of Medicine Effectiveness," in Advanced Computational Intelligence Paradigms in Healthcare 3, M. Sordo and S. Vaidya, Eds., Berlin Heidelberg: Springer, pp. 51-73, 2008.

[16] E. Rakus-Andersson and L. C. Jain, "Computational Intelligence in Medical Decision Making," in Recent Advances in Decision Making, E. Rakus-Andersson, R. R. Yager, and N. Ichalkaranje, Eds., Berlin Heidelberg: Springer, pp. 145-160, 2009.

[17] E. Rakus-Andersson and J. Frey, The Choquet Integral Applied to Ranking Therapies in Radiation Cystitis," Proceedings of the International Conference on Intelligent Systems (IS'2014), Berlin Heidelberg; Springer, Sept. 2014, pp. 443-452, ISBN: 978-3-319-11309-8.

[18] H. J. Zimmermann, Fuzzy Set Theory and Its Applications. 3rd edn, Boston: Kluwer Academic Publishers, 1996.

[19] E. Rakus-Andersson, J. Frey, and D. Rutkowska, "The Fuzzified Quasi-Perceptron in Decision Making about Treatment of Necrotizing Fasciitis," Proceedings of the 14th International Conference on Artificial Intelligence and Soft Computing (ICAISC 2015), Part II, LNAI 9120, Berlin Heidelberg: Springer, June 2015, pp. 130-141, ISBN: 978-3-319-19368-7.

[20] E. Rakus-Andersson, "Complex Control Models with Parametric Families of Fuzzy Constrains in Evaluation of Resort Management System," Journal of Advanced Computational Intelligence and Intelligent Informatics, vol. 18, Issue 3, fujipress, March 2014, pp. 271-279, https://www.fujipress.jp/jaciii/jc/jacii001800030271/ [retrieved: March, 2016].

The Value of Clinical Information Models and Terminology for Sharing Clinical Information

Rune Pedersen
Norwegian Centre for eHealth Research
University Hospital of North Norway
Telemedicine- and eHealth
University of Tromsø
Rune.pedersen@unn.no

Conceição Granja, Luis Marco Ruiz
Norwegian Centre for eHealth Research
University Hospital of North Norway
Tromsø, Norway
[conceicao.granja, luis.marco.ruiz]@telemed.no

Abstract—**This paper reports from the national strategy for OpenEHR adoption in Northern Norway Regional Health Authority encouraged by the unfolding of a national repository for OpenEHR archetypes and a national initiative to integrate clinical terminologies. The paper contributes to a qualitative longitudinal interpretive study with an effort to increase the possibility to obtain semantic interoperability (towards integrated care) and discusses SNOMED-CT and other relevant clinical terminology and Clinical Information Models (CIMs) such as OpenEHR archetypes. Terminology and archetypes are used to structure the EPR two-folded, and we discuss a general use of information models to increase interoperability extensively. A two-folded use of terminology where terminology is integrated in archetypes, or where terminology is used to structure the EPR system while using the hierarchical model of the terminology is discussed. Secondly, we discuss for what purpose OpenEHR is the choice of CIM to succeed in Norwegian healthcare.**

Keywords-eHealth medical records; electronic health records; web technology; e-health; interoperability; semantics; integrated care; OpenEHR; terminology; classification systems

I. INTRODUCTION

The increased focus on process-oriented systems across different health care organizations presupposes standardization in the form of shared terminologies and information models to enable semantic interoperability. Terminology standards have significant importance in modern medicine, and have been used to structure clinical information for different purposes [1]. Such standards also have the potential of supporting nursing terminologies such as International Classification of Nursing Practice (ICNP) or Nursing Intervention Classification (NIC) and The North American Nursing Diagnosis Association (NANDA) [2][3]. These standards have been developed, and used, to ensure consistency of meaning across time and place. On one level, nursing classifications enable day-to-day planning for local users (Primary use) where clinical terminology is used to structure information (standardized care-plans) using diagnosis and interventions from NIC and NANDA, for example. In practice, this will generate an automatic and reliable use of terminology for information that is sent and received between systems or health care deliverers. Examples of terminologies are the International Classification of Diseases (ICD), Systematized Nomenclature of Medicine Clinical Terms (SNOMED-CT), and International Classification of Nursing Practice (ICNP) [4][5][6].

Clinical Information Models (CIMs) [7] are models specified in some clinical information standard to express the clinical data entities processed by Health Information Systems (HIS). CIMs are used to appropriately maintain the consistency of clinical information structures inside a HIS, and to enable semantic interoperability across different systems and organizations. This makes CIMs a basic component for the appropriate management of patient data [8]. Moreover, with recent advances in data reuse strategies, CIMs are also playing an important role in defining the clinical information structures needed for secondary use of data [9][10].

OpenEHR is one of the specifications available to define CIMs. It relies on a meta-model based on reference model and constraint mechanisms to define formal specifications of CIMs called archetypes. Archetypes are information elements of clinical concepts where observations, options, instructions, and actions form the iterative process of treatment and care [11]. When using archetypes with a terminology binding, it becomes possible to make Electronic Patient Record (EPR) systems content structured in a multilevel modelling approach enabling semantic interoperability and the reuse of data. Using OpenEHR archetypes with a terminology binding it is possible to structure Electronic Patient Record (EPR) information models in a standard way. This enables semantic interoperability among HIS compliant with such standard [12].

For more than a decade, national initiatives towards shared and integrated care have been a focus area for the Norwegian health authorities [13][14]. This especially emphasizes the need to organize the clinical content of the EPR system in a more structured manner, to ensure interoperability across heterogeneous practices. Such need for contents organization is due to an increasing demand for rapid feedback on results, and an urge to compare organizational and / or clinical data. Structured data will allow clinicians to categorize variables in order to build meaningful reports, to extract data for quality registers and reuse EHR data in clinical research.

The definition of archetypes requires governance organizations that coordinate their definition on a broad scale. In Norway, the national initiative that deals with the contents organization has gradually gained a foothold. The OpenEHR architecture has been used to build a national repository, a.k.a. Clinical Knowledge Manager (CKM), of common Clinical Information Models. CIMs are defined collaboratively in the CKM as OpenEHR archetypes to provide vendors a library of common formal models to build their clinical information systems on. CKM archives information about how new archetypes are translated, modelled, and shared, and is planned to contain between 1000 and 2000 archetypes. The target is to build an open source repository of clinical content, based on the OpenEHR clinical information model. A precondition for success is that clinicians agree on the content of each archetype in the CKM consensus processes. Clinicians from the four Regional Health Authorities are active contributors in the process of developing archetypes. This process has been coordinated by the national editorial group, and the National Administration Office of Archetypes (NRUA).

This paper describes the national work accomplished to support the new two-levelled modelling of EPR systems with focus on the development of clinical value. We describe the work performed in: a) CIMs definition as archetypes by multidisciplinary teams of information architects and clinicians; b) the evaluation of the adequacy of adopting SNOMED-CT as reference terminology to annotate CIM. Based on this we present the following research questions: How interconnected is the choice of OpenEHR as clinical information model when the purpose is to have a library of clinical data to share between systems and healthcare levels? What is the adequacy of adopting clinical terminology to annotate any CIM?

The rest of the paper is structured as follows. In section two, we introduce the qualitative method used in the study, in section three we describe the value of information models, clinical terminology, the specifics of OpenEHR, and the national work towards the use of archetypes to structure the content of EPR systems. In section four the models used, and actions considered are discussed, and section five concludes based on the findings in the previous section.

II. Materials And Methods

The research presented herein has mainly been developed in the North Norwegian Regional Health Authority in coordination with NRUA, and the Norwegian Directorate of Health. Interpretive and ethnographically oriented qualitative methods have been applied, grounded in participation and contribution in the work accomplished [15]. Analysis of longitudinal research is a continuous and iterative process with an ever-changing intensity. The fieldwork has focused on the regional/national work accomplished, and, secondly, the forthcoming process where numerous archetypes will be tested as structured elements in the new process oriented EPR system. During the last seven years several meetings, courses, and workshops with focus on archetypes and terminology have been covered by observations, document analysis, and interviews. Conversations, discussions, reflections, and debates from these meetings are the foundation of this work. The observations and description of on-going work has been followed by the interviews with members of the regional and national initiatives. This includes six interviews on the archetype governance, 10 interviews and 180 hours of observations on the use of clinical terminology, conversations with end users of the CKM while guiding them to become users, and participants in national discussions on the consensus of archetypes. The interviews includes as said 10 clinicians, doctors and nurses that are active users of the CKM. The process of educating them to become CKM users has given valuable knowledge on how to develop the learning and recruitment strategies.

TABLE I. AN OVERVIEW OF THE DATA COLLECTION

Data source	
Interviews with contributors to the work with archetypes, and the development of new EPR.	18 open ended interviews
Participatory observation	180 hours
Participation in meetings, workshops, and informal discussions.	300 hours
Document studies: Documents from the CKM, concerning archetypes in general.	

III. Results

During the last three years, the use of OpenEHR archetypes has grown with focus on a national anchorage in Norwegian healthcare. The initiative has developed through national ICT, and an EPR vendor that holds more than 80 % of the secondary healthcare EPR systems. From the outset, a national collaborating group is working, in coordination with the aforementioned vendor, to build a national repository of archetypes. Simultaneously, such work contributes to the development of a structured EPR system that is based on OpenEHR technologies. Still, OpenEHR is only one of several comprehensive information models for standardizing and sustaining clinical content for health care.

At the same time, the Norwegian Directorate of Health has put focus on clinical terminology, and has engaged clinicians nationwide to explore the integration of SNOMED-CT in the existing ICT portfolio.

A. *The value of Clinical Information Models*

The future of Norwegian healthcare depends upon communication between ICT systems of different vendors, and between the primary and specialist health systems. The national CKM will contain the reference archetypes and guidelines to help implementers in the adoption of OpenEHR and terminologies. This makes the Norwegian CKM unambiguously unique based on the grade of consensus. In order to define CIMs that are general enough to be applied in different organizations and systems, it is necessary to define them as a collaborative effort among domain experts. The environments to carry out this work are the so called CKMs. The definition of CIMs typically encompasses two main tasks. The first is the specification of the information structure in a clinical information standard such as OpenEHR. The second is the binding of the meaningful sections of the CIM to a terminology to attach unambiguous standard descriptions to them. In the last decade, the work of initiatives to model CIMs is leading to the definition of an extensive catalogue of models publically available, upon which clinical systems implementations can be based. Nowadays, there is a considerable diversity in the standards and approaches available to define CIMs. Although most editorial teams follow similar steps, there is no published unified methodology or guideline for their definition [7]. The scope of modelling initiatives varies significantly from the local to the international level. For example, the international CKM and the Clinical Information Model Initiative (CIMI) define CIMs at an international level; the Norwegian CKM defines them at a national level; and the Intermountain Clinical Element Models (CEMs), were defined at intra-organizational level. The work in parallel of different initiatives has led to semantically equivalent models expressed in different information standards, a.k.a. iso-semantic models.

B. The use of terminology to standardize local practice

Since 2005, one of the largest hospitals in Norway, Akershus University Hospital, has used an EPR that includes a module for nursing. Along the lines of standardization, the nursing care plan, including nursing classification systems were viewed as a mean for making nursing work more effective and offering quality assurance. The classification systems are ICT-based standards integrated with the care plan. The diagnoses are represented by the international classification system NANDA, consisting of 206 nursing diagnoses [6]. The interventions are represented by the NIC system, consisting of 486 interventions. Care plans are increasingly made to replace the use of free text in the documentation, foremost to establish a common, formalized language based on the best practices. Free text documentation is whatever information the nurses share about the patient in the EPR in addition to, or without, writing formalized care plans. However, the implementation of the EPR led to a systematic use of standardized care plans. The care plan has been organized in such a manner that each diagnosis, dimension, and action is firmly attached to the plan with a start and a stop date. When standardizing these plans, the nurse can easily choose several actions from a predefined list for the applicable diagnosis. By doing this, the nurse saves time, while the standardized sentences work as a quality indicator. The purpose of using terminology as a primarily means to standardized EPR systems is challenging, still terminology has been used to structure an unstructured EPR system with success.

C. Clinical terminology, the national strategy

Terminology offers a common vocabulary for national health authorities, local researchers, and quality registers. All in all, this means that in addition to being a storing device for free text data the EPRs are capable of encoding commonly occurring data using fixed lists of multiple choices for certain purposes. Thus, data becomes more comparable and computable than free text would be. Some key examples may be found through the global World Health Organization (WHO)-based ICD. There are also terminological standards for more specific domains, such as the ICF (International Classification of Functioning, Disability and Health) for rehabilitation. In the case of ICD-9 and -10, which has been used in Norwegian healthcare systems from their origin, the primary target could have been achieved: If the clinicians had used the diagnosis codes from ICD-10 to categorize the subscription of patients in the EPR, it would clearly be of primary use (Clinical IT Manager). On the contrary, it is difficult for clinicians to be explicit and specific early in the trajectory of the patient, since diagnosis change throughout the patient pathway. Diagnosis change and the IT systems in use need to track and categorize these changes logically to support the activity coding of clinical work. The coding of activities reflects the focus of the clinical pathway, not the diagnosis of the patient. In this sense, the archetypes become valuable: as a quality assurance of the completeness of the clinical terminology, to direct the clinical content of the EPR systems and other integrated systems, and to identify relevant information and give the clinicians access to this information. On the contrary, to the local level the patient pathways fixed to clinical ICD diagnosis probably need to be determined on a national level, and based on national directives. A member of the regional archetype group stated *"The archetype is not annotated but this is a subset of the SNOMED concepts available for severities. As a maximum data set, the archetype should not restrict the "standard" set of terms agreed in terminologies. However, before doing so, I think that the implications in term of SNOMED licenses should be considered very carefully."* For instance, clinical pathways for cancer diagnosis are today organized from national cancer groups that have resulted in national guidelines that easily could be followed and connected to already existing "Pakkeforløp" standardized packages to monitor that cancer patients receive the right treatment at the applicable time. Large scale Infrastructure projects, with increasingly more focus on integrated care, put pressure on the Norwegian Directorate of Health to focus on clinical terminologies and archetypes. Recently, there has been a growing activity in the section of e-health towards increased focus on terminologies such as SNOMED-CT, and ICF, and how terminology and archetypes fit together. A selected number of personnel has, through the last 6 months, gathered resource personnel from all over the country. These are clinicians and health informatics with special interest in the use of clinical terminology. The work started in November 2015 with the purpose to map SNOMED-CT towards the most commonly used EPR functions. At the same time ICNP will be piloted in the primary healthcare services, this has been organized by the Norwegian Nursing Association that has translated the terminology, and it is acknowledged by the Norwegian Directorate of Health. SNOMED-CT and ICNP are both discussed in the new national project. Other Scandinavian countries such as Sweden and Denmark have earlier allocated significant resources both to translate and get SNOMED-CT operational for clinical practice.

The national project has focused on SNOMED-CT; should Norway become an organized user of SNOMED? How is the coverage of SNOMED-CT for the content of the clinical pathway? How is the integration of SNOMED-CT solved technically? The last question includes the use of archetypes, but also the possibility to use SNOMED directly to structure EPR content using the hierarchical model?

D. The national governance of archetypes

NRUA was established in 2013 by the National ICT with the goal of producing high quality archetypes. The NRUA has assigned six full/part time associates with an increasing number of collaborators in the Regional Health Authorities. There are between two and three members from each of the four regions. As an example, there is an increasing number of members from the North Norwegian Health Authority, one physician with special interest in health informatics, one nurse with a PhD in information Systems, one ICT-advisor, two PhD students in part time possitions and one project manager from the regional ICT development program where the new process oriented EPR is developed. NRUA also cooperates closely with global connections such as the founder of archetypes, the international governed repository, and vendors that cooperate with the Norwegian vendor. The vendors are important contributors with a mutual interdependency. In all, the governance work is important in local, national, and global environments. The overall goal with NRUA is to coordinate the development and use of archetypes on a national level, both handling translations of international archetypes as well as handling local initiatives. It is called "Do-ocracy" where doers make the decisions, but where the reviews are initiated by the Editorial Group which also covers the recruitment of the reviewers to the national Clinical Knowledge Manager. The further approval is done by the Editorial Group if the requirements are met. The requirements are factors such as the right number of clinical specialists for the right archetype (national level) where all four regions are included. One of the leaders of the international CKM stated, "*the collaboration between the international and the Norwegian CKM is unique and all activities with archetypes in Norway is followed by the international society and vice versa.*" She continued by saying, "*neither the CKM nor the consensus process is perfect and adjustments will be necessary along the way. Changes can be related to open-Source and Web based CKM/process where everything is stored open and is constantly evolving.*"

Since the beginning in January 2014, NRUA first focused on the translation of already existing archetypes and observation-archetypes like blood pressure, body weight, nutritional risk, height, and temperature. During this period of time, national consensus has for instance been reached for the archetype Blood pressure, Screening of Nutritional Risk, and Body weight. In 2015 and the beginning of 2016 more complex archetypes like Evaluation and Cluster archetypes has been defined. Clinicians have been invited to participate through the national CKM after coordination between the regional groups and the secretariat at NRUA. Archetypes are used as standards for the clinical content of the EPR and it was important for clinicians to have an essential role in defining and designing them. One clinician said: "*It is crucial to include clinicians in this work; they have the clinical knowledge and know what is important to focus on, for the archetypes to be useful standards for clinical work.*" The same clinician commented, "*If others than clinicians design the archetypes, it will be troublesome to get clinicians to accept and use them.* Other archetypes were also considered, all based on regional programs or initiatives such as a specific nursing registration scheme in the West Norway Regional Health authority, archetypes for national clinical registers, archetypes ordered by clinical work-groups with focus on the development of the new EPR system, and a number of archetypes ordered by cooperating vendors on a global level. In total, 39 archetypes have been approved in national consensus processes, and more than 100 archetypes are in process. The first archetype that reached national consensus was the Observation archetype for blood pressure; The clinical value of this archetype consists of all possible clinical values for data (systolic), state (score), events (24 hour blood pressure), protocol (Type of equipment). The archetypes in process are of different classes, observations, Actions, Compositions, Evaluations, and Clusters.

E. Terminology binding of archetypes

A key aspect of archetypes is that they can be annotated with terminological codes. Archetypes can be tagged with SNOMED-CT codes adding a standard term to each of the sections and nodes of the archetype. This includes, the archetype name to recognize it over organizational or even national borders. In turn, the information becomes interoperable for multiple purposes, and over several boundaries. This makes it essential that standardized terminologies for different domains can be integrated either in the archetype or in the EPR system. The use of a terminology like SNOMED-CT, which is widely exploited, increases the semantic interoperability on several levels, both for primary and for secondary use. The tag/code of the SNOMED-CT is enclosed in different systems and formats such as Medical Technical equipment for use in Medical Chart systems, and clinical specialist systems all bound to integrate with the EPR system of choice. This makes the terminology a mediator both on the national and global level. As an example; the Electronic Chart systems used in Norway/Europe are mostly based on structured data elements with a CIM, and the system needs to integrate structured elements from integrated systems to visualize the patient pathway, and for process and decision support. The integration with co-existing EPR systems is especially important. Medication, laboratory results, and the care plan have to be integrated in the same view to visualize the pathway. The Chart systems have a CIM for structured data elements that differs from the OpenEHR/archetype CIM intended to be used in Norway, and mapping between them demands unknown resources. In practice SNOMED-CT or other terminology can connect variables from the two systems to avoid problems with the mediation between the two different reference models if both CIM had included terminology. An internationally viable terminology like SNOMED-CT will make it possible to communicate information globally as long as it is implemented. When a patient contracts an illness when traveling abroad, healthcare personnel could get access to vital information by using the tags of SNOMED-CT to vital parameters in for instance a core health record. For secondary purposes, SNOMED-CT codes increases the scope and interoperability for clinical research, where a clinical trial could attract research communities outside the consisting parameters.

IV. Discussion

Terminology standards are used on a daily basis in health care work. Still, we know little about the processes of how these terminologies come into being, and how they are co-constructed with daily work, (for further information on: how clinical terminologies are used to categorize/structure nursing diagnosis and interventions in a standardized nursing plan in Norwegian healthcare refer to [16]; the use of medical diagnosis through the use of ICD-9 to categorize medical diagnosis refer to [4]). On the contrary, structured data for secondary purposes has gained more attention, both for the use of archetypes and terminology (See [3]). Structured EPR data will make it possible for clinicians to categorize variables to build meaningful reports, to extract data for quality registers, and for clinical research. Structured data elements will also make it possible to organize information that supports process and decision support inside an integrated EPR portfolio, and the use of OpenEHR will support clinicians with a more open, adaptive, and collaborative system, which enables modelling of clinical content. Structured data, OpenEHR based, and tagged with Clinical terminology codes is opening new possibilities to obtain integrated care. Information becomes standardized and understandable between heterogeneous local practices, such as different wards in the same hospital, nationally through the repository of archetypes, and globally through primary or secondary use of terminology and archetypes. Further, the use of clinical terminology increases the semantic interoperability to integrate different CIMs. The portfolio of different EPR systems in Norwegian health care for instance needs to communicate structured information. This regards both information between integrated systems at the hospital (between the EPR and the electronic chart system), and between EPR systems in the specialist and primarily health care services.

A. Primary use of terminology

The integration of clinical terminology for use in EPR systems to support clinical practice has proven difficult to accomplish. With the use of archetypes, and a national governance of clinical variables through a common repository for structured data elements, there are future advantages of both semantic and interoperable character. Earlier research elaborates on how the categorical use of clinical terminology to structure nursing diagnosis and interventions in standardized nursing plans has been a success for increased quality and efficiency. However, the use of clinical terminology to categorized clinical documentation for enabling process- and decision support in the EPR portfolio is limited. In this sense, clinical terminology is used to support the primary health care process giving semantic meaning to the content of the clinical processes. The use of standardized nursing plans at a large scale in a Norwegian hospital showed clear advantages both for quality and efficiency. Furthermore, when information is tagged with the purpose to categorize such as with ICD-10 and medical diagnosis, the same information becomes available for secondary purposes. On the next level, any of these tagged nodes of information could be recirculated. Archetype based elements such as blood pressure, pulse, temperature, and laboratory data can also be used for primary purposes. Local, regional, or national coordinated process - and decision support will in the future be based on templates where these variables are put together, where mathematical matrixes calculate the risk for a given condition/disease. The national repository of archetypes is structured and standardized so that terminology could become superfluous. For instance, an archetype for use in a national clinical register is harvested from different EPR systems and the nodes used, such as diastolic and systolic for blood pressure, are prefixed in a template that is produced for the register. In this sense, the archetype could be used to solve a given interoperability challenge. Under conditions where the terminology was lacking, the structure, the semantics, and the demarcation of the archetype would cover the national requirements, but all global advantages would be absent.

B. Secondary use of terminology

At the same time as the primary information becomes interoperable, both as single archetype/terminology or intervened, the information becomes semantically interoperable for use in secondary settings. As an illustration, all information that is tagged with the nursing classification ICNP, both diagnosis and interventions, becomes sharable for secondary use. All the nursing diagnosis and interventions would be an object for clinical research on a national or global level which is a relatively unexploited research arena. For patient safety when traveling abroad, specialists need to view the core health record written in another language. For this setting, the terminology for earlier diagnosis and interventions could be compared and used for treatment and care. Another secondary use purpose is the integration between the functioning EPR system and various specialist systems and medical chart systems. The structured medical chart system has for instance a CIM that differ a lot from OpenEHR. To integrate values from the two systems clinical terminology as SNOMED-CT can be used as a mapping device. When all the systems are based on structured variables it becomes important that data elements with different reference models can be shared and recognized between them. For instance, the care plan is intended to be an interdisciplinary tool for categorizing documentation in the EPR. For this to become a success it would be important that structured information from other applications is used in the care plan even if the master system for the information is for instance the medical chart.

Equal for both cases is that information used for any given purpose on a local or national level can be reused with another purpose both locally, nationally, and globally. For the archetypes that are approved in the national governance processes, increasingly in number, and in the end a repository that includes a number of archetypes that support clinical work there will be a possibility for a terminology binding. An increasing number of archetypes, more than 1000 will in the end be accessible in an open repository, and each archetype that is translated or modelled will be compared or reviewed with the purpose of being added to the global repository. For instance, the process of getting consensus on the observation archetype blood pressure started with a translation of the global standard, and ended with a new version that also is planned for the global or international repository.

C. *The clinical information model*

The diversity in standards, scope and methodology complicates the decision about adopting one standard or another for the definition of CIMs since it will influence the systems that can be deployed in the health network. Now, the preferred standards to define clinical models according to the literature are OpenEHR/ISO13606 archetypes, followed by HL7 templates. Therefore, in the near future it is expected to find an ecosystem where implementations based on different standards coexist. This may add a burden for those implementers that need to adapt from one standard to another. However, it is important to notice that the most valuable resource of a CIM is not the technical specification, but the conceptual model that it contains. The reason is that a CIM defines a way of combining clinical concepts together to build more complex conceptual structures beyond providing a format to express clinical information. For example, the archetype OpenEHR-EHR-CLUSTER.symptom_sign.v1 aggregates several granular concepts such as Body site, Episodicity, Impact etc. to build the more complex entity Symptom/Sign. This aggregation of concepts is more evident when the CIM is annotated with an international terminology. Reaching a consensus about the conceptual model of the CIM is the task that consumes most of the efforts of editorial teams since they need to coordinate professionals from different domains. Nevertheless, if the modelling work is appropriately performed, the conceptual model will be equivalent in most iso-semantic models.

As a consequence, once CIMs are defined in a particular standard, the conceptual model is clear and can be transformed to other representations/standards. In fact, that is the approach of openCIMI initiative which pursues the definition of CIMs that can be expressed in several formats such as CEMs, HL7 CDA or OWL by defining transformation functions among them. These transformations, although complex, are technical tasks that can be accomplished with much less effort than the definition of stable conceptual models. Transformations among standards vary in complexity, and the easiest case is the transformation from OpenEHR to ISO13606 that can be fully automated. In more difficult scenarios, the EU project SemanticHealthNet has provided insights to define an ontology based on the CIM conceptual model that allows the access to equivalent information hosted in repositories expressed with disparate information standards [16].

V. CONCLUSION

Currently several standards and terminologies are available for the specification and annotation of CIMs respectively. openEHR, HL7 CDA and ISO 13606 are examples of standards to define CIMs which in most cases are annotated with standard terminologies to enable their interoperability across systems.

With the parallel national initiatives running at this time in different countries, it is starting to become visible how the organization and size of countries influences their standardization efforts. On one hand, large countries with very heterogeneous health networks are aiming for the adoption of standards that allow sharing EHR information documents extracts. That is the case of Spain with ISO 13606 [17] or the US with HL7 CDA [18][19].

On the other hand, Norway is heading to the adoption of a nationwide EHR information architecture with openEHR that defines not only some relevant CIMs but the whole EHR information structure. Three are the factors that have influenced this direction of work. The first is the homogeneity in the market since only one vendor represents 80% of the market share in hospital. The second is the close collaboration between vendors and health authorities; this allows coordinating the definition of the whole information model of new systems. The third, and most determinant, is the body of knowledge already available in the international CKM that has fed the national CIMs definition pipeline with existing archetypes. This has accelerated their validation at a national level avoiding their definition from scratch.

At the moment, the Norwegian eHealth strategy has established a multidisciplinary community of vendors, governmental agencies and health organizations collaborating in order to define a nation-wide EPR information architecture. The knowledge management framework of OpenEHR supports to manage the national CKM. The OpenEHR governance model and the collaboration between the international and Norwegian CKM teams are proving to be effective to manage the definition of CIMs for the national eHealth strategy. On the technical side, the rich reference model provided by OpenEHR acts as a powerful modelling tool for the definition of CIMs. On the organizational side, the collaborative environment provided by the CKM is allowing to ensure the validity of the CIMs generated. As a result the National eHealth Department is providing the health informatics community a body of standard clinical models which allows implementers and researchers to define standard interoperable implementations on them.

The semantic interoperability gained from the use of both terminology and OpenEHR archetypes separately is a highly valuable asset. For instance, earlier studies in Norway have showed that clinical terminology has the potential to structure information of unstructured EPR systems. The ongoing national work also suggests that the combined use of archetypes and terminology further increases the semantic interoperability for connecting EPR systems on different layers of healthcare. Using for instance SNOMED "non-hierarchical" to tag the nodes of archetypes is interesting, and could be an integration advantage for vendors. It is a fact that both subjects complement each other's capacity to reach semantic interoperable.

Another "feature" that could increase the semantic interoperability is the growing possibility to use different Clinical Information Models to extract and share information from the National repository of archetypes/ clinical variables and content. The Government and the National e-health administration has decided to use different ICT systems in the primarily and specialist healthcare for several years to come. This requires a possibility to use clinical content from the national repository using another CIM specification standard than OpenEHR to extract and use semantic interoperable information. In this sense, the clinical model defined by an archetype can be represented in another standard by defining transformation rules among OpenEHR and the other standard. This way, the archetype-based repository becomes the reference common information 'ontology' or conceptual model used by different vendors regardless the standard, classes, and model they implemented.

ACKNOWLEDGEMENT

Thanks to the people at North Norwegian Health Authority, the NRUA, and the National ICT for contributing with vital information to the project.

REFERENCES

[1] G. C. Bowker and S. L. Star, Sorting Things Out: Classification and Its Consequences: MIT press, 2000.

[2] SNOMED-CT. Ihtsdo, Snomed-Ct®, the Organization. Available: http://www.ihtsdo.org/snomed-ct/

[3] S. Garde, R. Chen, H. Leslie, T. Beale, I. McNicoll, and S. Heard, "Archetype-Based Knowledge Management for Semantic Interoperability of Electronic Health Records," The XXIInd International Congress of the European Federation for Medical Informatics (MIE 2009), IOS Press, 2009, pp. 1007-1011.

[4] G. Wade and S. T. Rosenbloom, "The Impact of Snomed Ct Revisions on a Mapped Interface Terminology: Terminology Development and Implementation Issues," Journal of biomedical informatics, vol. 42, pp. 490-493, 2009.

[5] P. C. Dykes, H.-e. Kim, D. M. Goldsmith, J. Choi, K. Esumi, and H. S. Goldberg, "The Adequacy of Icnp Version 1.0 as a Representational Model for Electronic Nursing Assessment Documentation," Journal of the American medical informatics association, vol. 16, pp. 238-246, 2009.

[6] N. R. Hardiker, D. Hoy, and A. Casey, "Standards for Nursing Terminology," Journal of the American Medical Informatics Association, vol. 7, pp. 523-528, 2000.

[7] A. Moreno-Conde, D. Moner, W. D. da Cruz, M. R. Santos, J. A. Maldonado, M. Robles, and D. Kalra, "Clinical Information Modeling Processes for Semantic Interoperability of Electronic Health Records: Systematic Review and Inductive Analysis," Journal of the American Medical Informatics Association, vol. 22, pp. 925-934, 2015.

[8] V. Stroetman, D. Kalra, P. Lewalle, A. Rector, J. Rodrigues, K. Stroetman, G. Surjan, B. Ustun, M. Virtanen, and P. Zanstra. Semantic Interoperability for Better Health and Safer Healthcare [34 Pages]. Available: http://dx.doi.org/10.2759/38514

[9] L. Marco-Ruiz, D. Moner, J. A. Maldonado, N. Kolstrup, and J. G. Bellika, "Archetype-Based Data Warehouse Environment to Enable the Reuse of Electronic Health Record Data," International journal of medical informatics, vol. 84, pp. 702-714, 2015.

[10] J. Pathak, K. R. Bailey, C. E. Beebe, S. Bethard, D. S. Carrell, P. J. Chen, D. Dligach, C. M. Endle, L. A. Hart, and P. J. Haug, "Normalization and Standardization of Electronic Health Records for High-Throughput Phenotyping: The Sharpn Consortium," Journal of the American Medical Informatics Association, vol. 20, pp. e341-e348, 2013.

[11] R. Chen, P. Georgii-Hemming, and H. Åhlfeldt, "Representing a Chemotherapy Guideline Using Openehr and Rules," The XXIInd International Congress of the European Federation for Medical Informatics (MIE 2009), IOS Press, 2009, pp. 653-657.

[12] T. Beale and S. Heard, "Openehr Architecture Overview. Openehr Foundation," London, UK, 2008.

[13] Stortingsmelding 9, "En Innbygger, En Journal," Helse- og Omsorgsdepatementet, Available: http://www.regjeringen.no/nb/dep/hod/dok/regpubl/stmeld/2012-2013/meld-st-9-20122013.html?id=708609 November 2014.

[14] Nasjonal IKT, "Tiltak 48 (2012). Klinisk Dokumentasjon for Oversikt Og Læring," ed, 2009.

[15] H. K. Klein and M. D. Myers, "A Set of Principles for Conducting and Evaluating Interpretive Field Studies in Information Systems," MIS quarterly, pp. 67-93, 1999.

[16] C. Martínez-Costa, R. Cornet, D. Karlsson, S. Schulz, and D. Kalra, "Semantic Enrichment of Clinical Models Towards Semantic Interoperability. The Heart Failure Summary Use Case," Journal of the American Medical Informatics Association, vol. 22, pp. 565-576, 2015.

[17] J. E. Huerta, C. A. Villar, M. C. Fernández, G. M. Cuenca, and I. A. Acebedo, "Nhs Electronic Health Record," Ministerio de Sanidad y Politica Social, Available: http://www.msssi.gob.es/organizacion/sns/planCalidadSNS/docs/HCDSNS_English.pdf March 2014.

[18] D. Blumenthal and M. Tavenner, "The "Meaningful Use" Regulation for Electronic Health Records," New England Journal of Medicine, vol. 363, pp. 501-504, 2010.

[19] Health Level Seven. Hl7 Standards Product Brief - Cda® Release 2. Available: http://www.hl7.org/implement/standards/product_brief.cfm?product_id=7 March 2015

Three Levels of Access Control to Personal Health Records in a Healthcare Cloud

Gabriel Sanchez Bautista
and Ning Zhang

School of Computer Science
The University of Manchester
Manchester M13 9PL, United Kingdom
Email: {sanchezg, nzhang}@cs.man.ac.uk

Abstract—**We present a novel access control framework (3LAC), which supports multiple levels of access privileges. 3LAC is aimed to tackle the privacy issues in existing access control solutions to access patients' records in cloud computing environments. In 3LAC, we propose an access control framework that extends the Ciphertext-Policy Attribute-Based Encryption (CP-ABE) with the integration of secret sharing in a way that different number of shares are needed to reconstruct a level-key. In this research work, we introduce the idea of level-keys. Level-keys are used to authenticate users when requesting the generation of private keys to decrypt patients' data. Level-keys are split into different shares and users will request the shares to different level-key authorities (LKAs). The number of shares needed to reconstruct the level-key depends on the level of access privilege of the user. As the level of access privilege increases, the number of shares needed also increases. In a healthcare cloud context, 3 levels of access privileges have been identified, L_1- Access to de-identified data-objects, L_2- Access to individual data-objects and L_3- Access to a large set of data-objects of a patient. The 3LAC framework incorporates a CP-ABE based 3-level access control model and the design of 4 protocols: 1- Upload data-object (UDO) , 2- Share acquisition (SAc), 3- Private key acquisition (PrKAc) and 4- Access to data-objects (ADO).**

Keywords–Privacy; eHealth; Attribute-based encryption; Secret sharing; Access control; Multilevel.

I. INTRODUCTION

The implementation of Personal Health Records (PHRs) over cloud computing environments assumes advantages to the access of data. Different users may access data anytime and anywhere in a flexible and scalable manner. Advantages of cloud computing are scalability, flexible access to data and on-demand use of resources [1].

However, cloud computing raises an important issue in terms of privacy. Typically, PHRs are stored on the servers of the cloud service providers and patients have no control on how their records are accessed [2]. Patients have to trust that the cloud service providers only grant access to users for legitimate access purposes. Unauthorized access of data can be used to collect information of patients for other purposes such as marketing, commercial or research without the patients' consent.

Concerns about privacy in access to PHRs have been raised by different organizations and communities around the world. For example, the U.S. Department of Health and Human Services (HHS) has issued the Health Insurance Portability and Accountability Act (HIPAA) [3], which provides a set of standards and regulations for the privacy protection of PHRs.

PHRs are electronic records of patients' health data [4]. They are different from Electronic Health Records (EHRs) in the sense that PHRs are patient-centric while EHRs refer only to health records transmitted electronically [5]. PHRs are the aggregation of data-objects generated by different healthcare providers that may collaborate and contribute to the generation of data [6]. In this context, there may be different purposes to access patients' data. It may be for regular medical treatment, for secondary use (e.g., marketing or research) or for emergency situations in which access to a large set of data-objects of a patient may be needed. A data-object is defined as the most granular piece of data in a patient's records.

The rest of this paper is organised as follows: Section II presents the privacy issues of PHRs in a healthcare cloud context. Section III specifies the design requirements. Section IV gives the notations and introduces the high-level ideas in the design of 3LAC. Section V describes the 3LAC framework. Section VI presents the conclusion and future work.

II. PRIVACY ISSUES

Privacy issues of PHRs in a healthcare cloud are described as follows.

- **Unauthorised access to patients' PHRs.** Healthcare cloud service providers are responsible for the management and storage of patients' health records in the cloud. PHRs are usually processed on servers and machines in which patients have no control. This increases the possibility of theft or misuse of their data. Access to sensitive information on patients' records represents a risk of privacy when data is accessed by unauthorized users.
- **Lack of fine-grained access control.** Patients' records are often treated as a unitary piece of data. When users are granted access, they can usually see all the information on the patient's records even if the records contain information they do not need to perform their job functions. This constitutes a privacy issue. For example, a patient may not want to disclose certain information (e.g., sexual abuse) when it is not necessary for her current treatment.
- **Linkability to patients' identity when PHRs are accessed for secondary purposes.** Data contained in a patients records can become a source of information that may be used to track the identity of the patient for malicious purposes. If identifying information is

not removed, it can represent a risk to the privacy of patients' data. We have identified that there are certain access purposes in which users do not need to know the identity of the patient to perform their job functions.

III. REQUIREMENTS

This section specifies the set of requirements that 3LAC is aimed to address.

- **(R1) Patients (i.e., data owners) should be in the position to decide who accesses their data.** Patients should be able to specify who accesses their data and for what specific purposes.
- **(R2) Support fine-grained access control.** Access to patients' data should be fine-grained. Users should access only the portion of data that they need to perform their job functions.
- **(R3) Support access to data in emergency situations.** Access in emergency situations should be granted to the most complete set of data of a patient. As this is a high-privileged access, there should be an increased level of protection.
- **(R4) Support access to data for secondary use.** Access to patients' data for secondary use should be granted. As this is not a high-privileged access, it should not require a rigorous level of protection that may add extra computational costs to the system.
- **(R5) Support removal of access privileges.** Patients should be in the position to remove access privileges when they consider access to a certain data-object is no longer needed.
- **(R6) Scalability.** The access control should be scalable in terms of key management and distribution. It should support a large number of users requesting access to the data-objects.

IV. DESIGN PRELIMINARIES

A. Notations

The notations used in the design of the 3LAC framework are given in Table I.

TABLE I. NOTATIONS.

Notation	*Meaning*
LK_i^1	User i's level-key 1
LK_i^2	User i's level-key 2
LK_i^3	User i's level-key 3
$S_i^{2_{s1}}$	Share 1 of LK_i^2
$S_i^{2_{s2}}$	Share 2 of LK_i^2
$S_i^{3_{s1}}$	Share 1 of LK_i^3
$S_i^{3_{s2}}$	Share 2 of LK_i^3
$S_i^{3_{s3}}$	Share 3 of LK_i^3
$SKGA$	Private key generation authority
$RLKA$	Root level-key authority
LKA_1	Level-key authority 1
LKA_2	Level-key authority 2
LKA_3	Level-key authority 3

As can be seen in Table 1., in 3LAC there is a private key generation authority, a root level-key authority and 3 non-root level-key authorities. Similarly, it can be seen that level-keys are classified into three different levels and based on the level, a level-key may be split into different shares.

B. High-level ideas

In this section we describe the high-level ideas in the design of 3LAC.

- **3LAC supports multiple access privilege levels.** 3LAC supports different levels of access privileges for users requesting access to data-objects. Based on the analysis of different use-case scenarios, we have identified that 3 are the levels necessary to classify the access purposes that users have when requesting access to patients' health records. At the lowest level, there should be access to PHRs for secondary use purposes. At a medium level, there should be access to PHRs for regular medical treatment. At a higher level, there should be access to PHRs in emergency situations in which access to the most complete set of data-objects of a patient may be desired [7]. Based on the analysis of these different scenarios, in this work we propose 3 levels of access privileges, which are: L_1- Access to de-identified data-objects, L_2- Access to individual data-objects and L_3- Access to a large set of data-objects of a patient. Similarly, each level supports access to data-objects with different levels of sensitivity. L_1 supports access to data-objects with low level of sensitivity. L_2 supports access to data-objects with medium level of sensitivity. L_3 supports access to data-objects with high level of sensitivity.
- **Users are divided into different user-groups {G1, G2, G3}.** Each group corresponds to a level of access privilege. Users are intended to access data-objects at the assigned and the lower privilege levels. For example, users of G1 should access data-objects at L_1. Users of G2 should access data-objects at L_2 and L_1. Users of G3 should access data-objects at L_3, L_2 and L_1.
- **Level-keys are split into shares.** The level-keys (LK_i^1, LK_i^2, LK_i^3) are used to authenticate users according to their access privilege level. LK_i^1 is generated by LKA_1. LK_i^2 is generated by $RLKA$. LK_i^2 is split into two shares ($S_i^{2_{s1}}$ and $S_i^{2_{s2}}$). $S_i^{2_{s1}}$ is distributed to LKA_1 and $S_i^{2_{s2}}$ to LKA_2. LK_i^3 is generated by $RLKA$. LK_i^3 is split into three shares ($S_i^{3_{s1}}$, $S_i^{3_{s2}}$ and $S_i^{3_{s3}}$). $S_i^{3_{s1}}$ is distributed to LKA_1, $S_i^{3_{s2}}$ to LKA_2, and $S_i^{3_{s3}}$ to LKA_3.
- **The control to the access privilege level is embedded into the level-keys.** Level-keys are reconstructed from shares obtained by different $LKAs$. The number of shares needed is based on the access privilege level. As the level of access privileges increases, the number of shares needed also increases. A level-key of L_1 (LK_i^1) does not need any share, LK_i^1 itself must be obtained. A level-key of L_2 (LK_i^2) needs two shares ($S_i^{2_{s1}}$ and $S_i^{2_{s2}}$) to be reconstructed. A level-key of L_3 (LK_i^3) needs three shares, ($S_i^{3_{s1}}$, $S_i^{3_{s2}}$ and $S_i^{3_{s3}}$) to be reconstructed. In this way, to access higher sensitivity data-objects, one has to obtain more shares, and for the acquisition of each share, there will be an authentication process. This makes the impersonation and unauthorised access to more sensitive data-objects more difficult. Figure 1.

below illustrates the number of shares needed based on the group of access privileges of users.

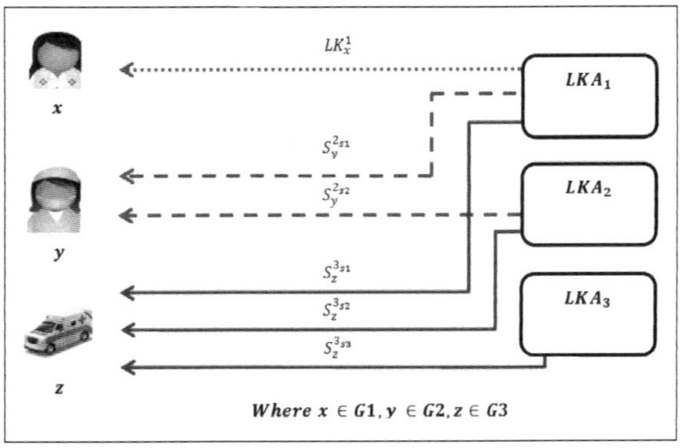

Figure 1. Different users obtaining shares from different LKAs

In the example given in Figure 1., x is a user of G1, y is a user of G2 and z is a user of G3. They obtain shares from the level-key authorities LKA_1, LKA_2 and LKA_3 in order to reconstruct their corresponding level-key that will be used for authentication when requesting the issuance of private keys. Private keys are used to decrypt patients' data-objects.

V. A Novel 3-Level Access Control Framework (3LAC)

3LAC is an access control framework that is formed of a CP-ABE-based 3-level access control model and the design of 4 protocols. 3LAC provides an access control framework that can be implemented in different applications that provide access to patients' personal health records. The front-end of 3LAC will depend on which is the application that makes use of the 3LAC framework. More details of 3LAC are given in the following subsections.

A. A CP-ABE based 3-level access control model

It is a novel access control model that supports 3 levels of access privileges (L_1, L_2, L_3). Each level supports access to data with different levels of sensitivity. L_1 supports access to low sensitive data, L_2 supports access to medium sensitive data, and L_3 supports access to high sensitive data.

3LAC also supports fine-grained access control because this is based on CP-ABE [8], an encryption scheme in which patients define access policies based on attributes to specify who has privileges to access which data-objects. Users' attributes should satisfy the access policy defined by the patient in order to decrypt the data-object. The reason it supports fine-grained access control is because users can be assigned any number of attributes. This permits a more detailed description of their identities, thus a more fine-grained access control.

Similarly, 3LAC supports revocation of access privileges because the patient can redefine the access policy at any time in order to specify the new attributes that users must have to decrypt the data-objects.

3LAC supports scalability as access to data-objects is made more efficient for different groups of users while at the same time the privacy of the data-objects is protected such that the level of protection increases with the level of sensitivity of the data-objects.

The following list presents the architectural components of 3LAC.

- $RLKA$: The root level-key authority. It is responsible to generate the level-keys (at L_1 and L_2) and split them into shares. Then, it distributes the shares to the different non-root level-key authorities.
- LKA_1: The level-key authority 1. It generates LK_i^1 and obtains the shares $S_i^{2_{s1}}$ and $S_i^{3_{s1}}$ from $RLKA$.
- LKA_2: The level-key authority 2. It obtains the shares $S_i^{2_{s2}}$ and $S_i^{3_{s2}}$ from $RLKA$.
- LKA_3: The level-key authority 3. It obtains the share $S_i^{3_{s3}}$ from $RLKA$.
- $SKGA$: The private key generation authority. It is a trusted authority responsible for generating the private keys (i.e. decryption keys) for different users based on their attributes.
- CA: The certification authority. It is a trusted authority that is responsible for signing the users' digital certificates. The public keys of the users are certified by this authority [9].
- AA: The attribute authority. It is a trusted authority that is responsible for gathering all the users' attributes. This authority passes the users' attributes to the CA. The CA takes these attributes to include them in the digital certificate of the user.
- DP: The data provider. It is the cloud service provider where the data-objects of patients are stored.

The architectural components of 3LAC are used in 5 phases as described below.

- **1: Initialisation**. In this phase, the user makes a request to the CA in order to obtain a digital certificate. The CA certifies the public key of the user and signs the digital certificate. The digital certificate also includes the attributes of the user in the extension field. The attributes are obtained from the AA. Similarly, it is during initialisation that the $RLKA$ distributes the shares to the different non-root level-key authorities (i.e., LKA_1, LKA_2, LKA_3).
- **2: Shares Acquisition**. In this phase, the user makes a request to the non-root level-key authorities in order to obtain the shares needed to reconstruct his level-key. Users of $G1$ make a request to LKA_1. Users of $G2$ make a request to LKA_1 and LKA_2. Users of $G3$ make a request to LKA_1, LKA_2 and LKA_3. Once the shares have been obtained, the user can reconstruct his level-key, which will be used for authentication during private key acquisition.
- **3: Private Key Acquisition**. In this phase, the user makes a request to the $SKGA$ in order to be issued a private key. The $SKGA$ will send a challenge to the user in order to prove that the user has been able to reconstruct his level-key. The user then responds to the challenge and encrypts the response by using his level-key. Then, the $SKGA$ decrypts the challenge response by using the level-key, which is symmetrical

and known by both the user and the $SKGA$. Once the $SKGA$ verifies the user knows his level-key, the $SKGA$ generates and encrypts the private key of the user. The private key is encrypted by using the level-key. In other words, the level-key is used to distribute the private key to the user. Then, by using his level-key, the user can decrypt his private key, which will be used to decrypt data-objects.

- **4: Patient uploading data-objects**. This is when patients (i.e., data owners) request to upload data-objects to the databases. Patients make a request to the data provider (DP). DP is the entity that stores data on its databases. The data-objects uploaded are encrypted with an access policy that specifies the attributes that users must have to decrypt them. The access policy is specified by the patient. Patients may specify an access policy for each data-object. However, patients can also encrypt a large set of data-objects under one package. In this case, one access policy can be specified for a large set of data-objects of a patient. In other words, it depends if the patient desires fine-grained or coarse-grained access for his data-objects, so he can define an access policy per data-object or per many data-objects under one access policy.
- **5: Users requesting data-objects**.
This is when users request to access data-objects of patients. Data-objects given to users are encrypted. However, it will depend if the user has a private key with the attributes necessary to satisfy the access policy embedded in the cipher-text. If this is true, the user can decrypt the data-object.

Three levels of access privileges

For proof of concept in this work, the three levels of access privileges are defined as follows.

L_1**: Access to de-identified data-objects:** L_1 supports access to data-objects for secondary use purposes (e.g., marketing or commercial). L_1 is defined as the lowest privilege level. Users of G1 are intended to access data-objects at L_1. Additionally, users of higher privilege levels are allowed to access L_1. Users of G1 need LK_i^1 to authenticate and obtain the private key. During private key generation, $SKGA$ verifies if LK_i^1 corresponds to or is higher than the user-group of the requesting user (i.e., user-group G1). Access at L_1 is granted only to de-identified data-objects. De-identified data-objects are in a separate database (DB^d) from the database that contains the original data-objects (DB). Data-objects at this level are assumed to be de-identified because users of G1 do not need to know the identities of the patients. However, in cases that users of G1 need to access the original data-objects, they may request access to data-objects at L_2. To accomplish this, they need to obtain 2 new shares to reconstruct the level-key for L_2 (LK_i^2).

L_2**: Access to individual data-objects:** L_2 supports access to data-objects for regular medical treatment. L_2 is defined as the medium privilege level. Users of G2 are intended to access data-objects at L_2. Additionally, users of a higher privilege level are allowed to access L_2. Access is granted to the specific data-objects that users need to perform their job functions. Access at L_2 is to the original data-objects database (DB). Users at L_2 need two shares ($S_i^{2_{s1}}$ and $S_i^{2_{s2}}$) to reconstruct LK_i^2. During private key generation, $SKGA$ verifies if LK_i^2 corresponds to or is higher than the user-group of the requesting user (i.e., user-group G2). If it is true, LK_i^2 can be used to distribute the private key. If users of G2 desire to access data-objects at L_3, they need to obtain 3 new shares to reconstruct the level-key for L_3 (LK_i^3).

L_3**: Access to a large set of data-objects of a patient:** L_3 supports access to data-objects in emergency situations. L_3 is defined as the highest privilege level. Users of G3 are intended to access data-objects at L_3. For a proof of concept in 3LAC, there is no higher privilege level than L_3. Access at L_3 is to the original data-objects database. Users at L_3 need 3 shares ($S_i^{3_{s1}}$, $S_i^{3_{s2}}$ and $S_i^{3_{s3}}$) to reconstruct LK_i^3. During private key generation, $SKGA$ verifies if LK_i^3 corresponds to the user-group of the requesting user (i.e., user-group G3). If it is true, LK_i^3 can be used to distribute the private key.

B. 3LAC protocol designs

In 3LAC, there are 4 different protocols: 1- UDO, 2- SAc, 3- SKAc and 4- ADO. The following list describes the protocols.

1) *Upload data-object protocol (UDO)*. It is executed when a patient requests to encrypt and upload a data-object. The patient makes a request to the data provider which is responsible to accept requests for encryption of patients' data-objects and upload them to the corresponding database.
2) *Shares Acquisition protocol (SAc)*. It is executed when a user requests to obtain a share from a LKA. The user may need to request different shares to different $LKAs$ depending on how many shares the user has to obtain in order to reconstruct his level-key. This protocol can be executed between any user and any LKA.
3) *Private key acquisition protocol (SKAc)*. It is executed when a user makes a request to obtain a private key. The request is sent to the SKGA, the authority that is responsible for generating the private keys for users so they can decrypt data-objects of patients. In order to obtain a private key, the requesting user needs to prove he has the level-key (LK) that he was able to reconstruct from the shares. The LK has the level of access control embedded within it as the number of shares needed to reconstruct it depends on the access privilege of the user which is specified on his user-group. The level-key is used to encrypt and distribute the private key generated. In other words, if the user knows his level-key then he can decrypt his private key.
4) *Access data-objects protocol (ADO)*. It is executed when a user requests access to the data-objects of a patient. The request is sent to the data provider which is responsible to authorise access to the data-objects stored in the databases.

VI. Conclusion

In this paper, we presented a novel access control framework (3LAC) which supports different levels of access privileges: L_1- Access to de-identify data-objects, L_2- Access to individual data-objects and L_3- Access to a large set of data-objects of a patient.

In 3LAC, we introduce the concept of level-keys and the idea that the level of access control for a user is embedded into his level-key. 3LAC extends CP-ABE with the integration of secret sharing [10]. With the integration of secret sharing as the level of access privilege increases, the number of shares needed also increases in order to reconstruct a level-key. With a level-key a user is authenticated when requesting the issuance of a private key. A private key is used to decrypt a patient's data-objects. In 3LAC, privacy protection is increased because for the acquisition of each share there will be an authentication process. This makes the impersonation and unauthorised access to more sensitive data-objects more difficult.

Future work includes the implementation and evaluation of 3LAC.

ACKNOWLEDGMENT

This research is supported by the National Council of Science and Technology Mexico (CONACYT).

REFERENCES

[1] M. Armbrust, A. Fox, R. Griffith, A. D. Joseph, R. Katz, A. Konwinski, G. Lee, D. Patterson, A. Rabkin, I. Stoica et al., "A view of cloud computing," Communications of the ACM, vol. 53, 2010, pp. 50–58.

[2] H. Takabi, J. B. Joshi, and G.-J. Ahn, "Security and privacy challenges in cloud computing environments," IEEE Security & Privacy, vol. 6, 2010, pp. 24–31.

[3] "U.S. Department of Health and Human Services, The Health Insurance Portability and Accountability Act," URL: http://www.hhs.gov/ocr/privacy/hipaa/understanding/special/healthit/phrs.pdf [retrieved: 03,2016].

[4] C. C. Lee, P. S. Chung, and M. S. Hwang, "A Survey on Attribute-based Encryption Schemes of Access Control in Cloud Environments," International Journal of Network Security, vol. 15, 2013, pp. 231–240.

[5] C. Wang, X. Liu, and W. Li, "Implementing a Personal Health Record Cloud Platform Using Ciphertext-Policy Attribute-Based Encryption," 4th International Conference on Intelligent Networking and Collaborative Systems (INCoS), 2012, pp. 8–14.

[6] A. Abbas and S. U. Khan, "A Review on the State-of-the-Art Privacy-Preserving Approaches in the e-Health Clouds," Biomedical and Health Informatics, IEEE Journal of, vol. 18, 2014, pp. 1431–1441.

[7] S. R. Reti, H. J. Feldman, S. E. Ross, and C. Safran, "Improving personal health records for patient-centered care," Journal of the American Medical Informatics Association, vol. 17, 2010, pp. 192–195.

[8] J. Bethencourt, A. Sahai, and B. Waters, "Ciphertext-policy attribute-based encryption," Security and Privacy, IEEE Symposium on, 2007, pp. 321–334.

[9] N. Leavitt, "Internet security under attack: The undermining of digital certificates," Computer, vol. 44, 2011, pp. 17–20.

[10] A. Shamir, "How to share a secret," Communications of the ACM, vol. 22, 1979, pp. 612–613.

Evaluation of a Context-Specific Communication System Based on Smartphones

A field study of use and nurses' expectations

Trine S Bergmo[1], Elin Johnsen[2], Monika A Johansen[1,3], Terje Solvoll[1]
[1]Norwegian Centre for E-health Research, University Hospital of North Norway, Norway
[2]Health Services Development, Innovation and Implementation, University Hospital of North Norway, Tromsø, Norway
[3]Telemedicine and E-health Research Group, Arctic University of Norway, Tromsø, Norway
Tromsø, Norway
Trine.bergmo@telemed.no
Elin.Jonsen@telemed.no
Monika.alise.Johansen@telemed.no
terje.solvoll@telemed.no

Abstract—**This project evaluates a context aware communication system designed for use in hospitals. This communication system aims to increase and improve the communication and information exchange between nurses, between nurses and physicians, and between physicians. In this paper, we focus on nurse-to-nurse communication. Nurses in the oncology department at the University Hospital of North Norway tested the system. The aim of this field study is to investigate the expectations, consequences and experiences of using the phone system in a clinical setting. The study uses a multi-method research approach including both quantitative and qualitative data. We use a before-and-after study design, with data from questionnaires, interviews and logs as well as step count measurements. In this paper, we describes the phone system and how it can be used. We also report on the nurses' expectations and presents some preliminary findings on system usage and the before-and-after step count assessment.**

Keyword- hospital communication systems; context awareness; nursing; e-health; work practices; ICT in hospitals; work effinecy

I. INTRODUCTION

Working in a hospital environment requires highly mobile personnel. Health care personnel depend on effective modes of communication and information exchange to provide high quality services [1][2][3]. In a clinical setting, gathering the right information can be complex, requiring frequent conversations and discussions [4]. However, research has shown that communication in hospitals suffers due to poor practices and inefficiency; caused by an insufficient infrastructure, such issues emerge particularly when the need for communication is urgent [4][5][6].

Most hospitals rely on a mobile communication infrastructure with devices dedicated for each role, with pagers being the most common mobile communication device. Generally speaking, mobile phones are not widely used in hospitals, with personnel using wired/wireless phones and personal digital assistants (PDA) in addition to their pagers [7]. To ameliorate communication issues, more extensive use of mobile phones could offer a solution by improving accessibility and communication [4][6][8]. Compared to pager use, important advantages can be achieved with texting and voice services. Decreasing communication delays may lead to improved patient care while reducing the risk of medical errors [6]. Studies have shown that wireless phones can overcome most of the limitations of pagers, thus facilitating communication within hospital settings [9].

Despite the advantages of mobile phones, there are also well-known downsides to the usage of such devices. For instance, the increased availability and accessibility can overload key human resources, such as senior physicians and on-call staff [2][9]. Only a few staff members carry wireless phones due to the assumption that phones are more interruptive than pagers [2][7]. Therefore, before introducing wireless phones as standard hospital equipment, it is important to evaluate usability and user satisfaction and to assess impacts on work practices.

Our study investigates how the use of one such system, the CallMeSmart (CMS) system, influences nurse-to-nurse communication and affects work practices at the Oncology Department (OD) at the University Hospital of North Norway (UNN). In this paper, we describe the technology, present the pilot study and report on preliminary results. We discuss results regarding nurses' expectations towards the phone system, patterns of use and time use, as measured by step count.

The paper now proceeds into Section II, which provides background information and includes an overview of the local context as well as the use of phone systems in clinical work. Section III outlines the materials, research setting and methodology of this study. Section IV provides an overview of our results. Section V discusses implications and limitations. Finally, conclusions and future work are detailed in Section VI.

II. BACKGROUND

CallMeSmart is a context-sensitive mobile communication system developed for hospital use. The system aims to reduce unnecessary interruptions from mobile devices in situations where disruptions should be avoided. These situations include, for instance, times when health personnel is involved in surgery (dressed in sterile clothing), doing patient examinations at outpatient clinics, or having conversations with patients/relatives in designated rooms.

Context-sensitive systems for hospitals represent a promising application domain. Hospital staff depend on a wide and reliable communication infrastructure for the exchange of different types of data, such as patient reports, lab tests and work shifts. They also depend on this infrastructure for text, voice and alarm services. Management of this information is difficult and requires avoidance of a wide variety of problems in order to properly meet the needs of hospital professionals. Context-sensitive applications for mobile communication seem to promise a valid solution that could also move some workers' activities over to computers.

The phone system focuses on context-sensitive interfaces, middleware and new interaction forms for mobile devices that support multi-modal communication in hospitals. These devices support media, such as voice services, text-messaging and paging services, in an efficient and non-interruptive manner. They also enable support for individual and role-based contact on a single device. That is, users only need to carry one device for both personal and role-based communication, which enables other users to, for example, contact someone assigned to "on-call" duty in a specific department, even if they do not know the identity of that person. At the same time, the system seeks to balance between availability and interruptions while enabling acute calls and alarms to go through.

The phone system senses the context automatically from different sensors, calendar information and work schedules. It makes changes to the health care worker's availability and the phone's profile according to the collected contextual information. At the same time, the caller is given feedback about the health care worker's availability, thereby making it possible for the caller to force through an emergency call or to forward the call to another available health care worker at the same level. The system is based on ideas from earlier studies on interruptions in combination with ideas from Solvoll (2013) [10]. A first version of the system is ready and has been tested in lab settings with physicians/nurses as test users. The tests were performed as scenarios derived from real situations. The feedback was mostly positive, and the input was used to improve and further develop the system, moving from prototype to production [10]. A more detailed description of the system, including figures that itemise the functionality, architecture, interruption reduction and compatibility with other existing systems can also be found in Solvoll (2013) [10].

Management at the Oncology Department at the University Hospital of North Norway has decided to invest in mobile communication devices to save time and improve patient care. Currently, all phone communication at the department occurs through wired/wireless landline telephones. People on call also carry pagers. The nurses have no mobile devises to help them exchange information. To ease the situation in the department, management has decided to invest in the CallMeSmart system. As part of this initiative, we proposed a study to investigate the expectations, consequences and experiences of using the phone system in a clinical setting.

The purpose of this evaluation is to assess the impact and experiences of using a context-aware communication system at a hospital department (the Oncology Department). The main research question can be put as follows: Will work practices at a hospital ward be improved by using local smart phones? We will investigate these aspects:

- System usage
- Nurses' expectations
- Changes in time use
- Changes in work practices
- Nurses' experiences

In this paper, we report results from the three first aspects outlined above. Changes in work practices and nurses' experiences are reported in a separate paper [11].

III. MATERIALS AND METHODS

In this section, we present details on the materials and methods used to investigate system usage, nurses' expectations and changes in time use. It also provides detailed information regarding how to use the phone system and the research setting.

A. *Using the phone system*

Users are provided personal accounts that they can access from any phone connected to the system. Users log on using personal login credentials. Configuration and personal data (like contact lists, messages and phone logs) are temporarily stored on the phones – only as long as the user is logged into the system. All personal data are stored on the user profile. The profile is downloaded to the phone when the user logs on and deleted from the phone when the user logs off. In all cases, the user's data are stored in the system and protected so that other users cannot access this information. Authentication is based on Lightweight Directory Access Protocol (LDAP) and is compatible with Active Directory (AD), meaning that each user can potentially use his or her ordinary username and password from the hospital information system to log in to the phone system. The users can make and receive calls in a one-to-one configuration, or in a one-to-many configuration for conference calls. A user cannot receive or start a second call without hanging up the first one.

The system silently offers 'delivered' and 'read' acknowledgement for each message. Messages are stored on the users' profiles, which means that whenever the user is logged on, his or her messages are available on the phone through the profile.

Before the nurses started to use the phone system, they were provided an introduction and training lasting approximately five minutes. The phone system's inventor visited the ward for the first two days after the first nurses started using it to provide more support if needed. The only support required involved creating accounts for new users.

B. *The setting*

The study site was the Oncology Department at the UNN. The department offers chemotherapy, radiation therapy, hormone therapy, other symptomatic treatment and palliative care. Nurses follow national guidelines for treatments with the goal of curing as many people as possible of their cancer or to prevent or delay the development of the disease.

The Oncology Department includes a ward with 25 beds and about 120 employees, including nurses, nurse assistants and medical doctors. The ward treats and cares for patients between the ages of 16 and 95 years old. Most of these patients are seriously ill and in need of heavy care. In addition, about ten patients stay at the patient hotel connected to the hospital. The nurses' work schedule is organised in three shifts: the day shift includes 10 nurses, the afternoon shift has 5 or 6 nurses, and the night shift has 3 nurses.

C. *Methods*

This evaluation study follows a multi-method research approach using both quantitative and qualitative data. We recognise that randomisation is the most robust method of avoiding systematic bias, but this method was not possible for this evaluation. To randomise the nurses into two groups would require the nurses to switch between their usual mode of communication and the new phone system. Such a switch between the intervention and usual care is unlikely to succeed [12].

We have collected data before and after the introduction of the phones. The data were collected from phone logs and we registered the nurses' step count. We utilised an anonymous survey to explore the nurses' expectations of the system. The survey was made accessible to the nurses on duty prior to implementing the communication system. The survey included a one-sided questionnaire and was placed in a cardboard box in the main corridor close to the nurses' station. The questionnaire included questions about their expected changes in work practices as a result of using the phone system. They were first asked to name three changes they anticipated in their own work. Then, they were asked to specify which of the changes they thought were most important and why. The questionnaire included questions about when they expected the phone system device to be most used (during the day or night shift) and in which situations. Furthermore, the nurses were asked if they expected the phones to be used inside or outside the ward location. All questions in the questionnaire were open ended. Two of this paper's authors categorised the answers.

To analyse if the phone system saved time for the nurses, we measured the step counts before and after introducing the phone system. Twenty-five nurses received the Fitbit step count, carrying it in their uniform pockets during their shifts. Different nurses used the step count before and after. It was not possible to measure the step count for the same nurses before and after the phone system

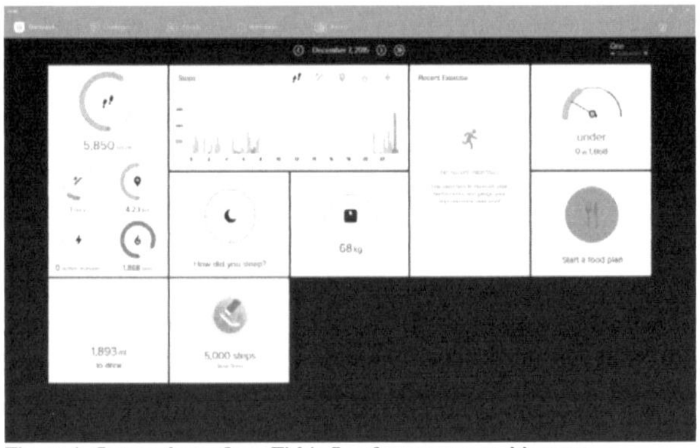

Figure 1. Screen dump from Fitbit One for a nurse working two consecutive night shifts

was introduced. The main reason for this limitation was that the shift plan for the study period was already made.

The steps were monitored one week before and one week after the introduction of the phone system (in December 2016).

Fitbit One tracks all steps taken, stairs climbed, calories burned and distances travelled (see figure 1). The results from each day were stored digitally and displayed as an activity from midnight one evening until the following midnight. This feature implies that if a nurse was working the night shift, data from two different data collection screen dumps had to be summarised to measure the steps over the whole shift. Likewise, data from one screen dump sometimes had to be divided if the nurse had two consecutive night shifts. Data from the step counts were reported in an Excel sheet. Data regarding calories burned was considered irrelevant and not reported.

We also collected logs of the system usage, including how many messages and phone calls were sent/performed on which date and at what time of day.

IV. Results

In this section, we report on the nurses' expectations and presents some preliminary findings on system usage and the before-and-after step count assessment.

A. *Expectations*

A total of 11 nurses completed the questionnaire. All respondents answered the first part of the questionnaire, except one who did not specify the change perceived to be most important. All respondents expressed positive expectations. No one voiced any concerns. Expectations revolved around two main themes: phone use would save time and reduce disruptions and interferences in daily tasks. As reasons for these expected benefits, the nurses pointed to improved information and communication flow in their work practices. They expected changes in time use as a result of spending less time walking around and searching for colleagues. Furthermore, they noted that fewer paper messages and a more "orderly" information flow would improve their work practices. They seemed to assume that fewer messages would be lost or forgotten; instead of having to use several stages to deliver messages, the process of sending questions and receiving answers/feedback would be streamlined. They expected that the system would make it easier to get in contact with colleagues and cause fewer interruptions during reports and rounds. The nurses could also "mark themselves as busy" when in a sterile setting or working with a patient, for instance.

All respondents envisaged situations where they expected to use the phone system. This response applied both when the nurse would be with 'their own patients' (patients for whom they have a special responsibility) and with other nurses' patients. They also reported that they believed the system would be useful in the following situations: during patient admittance, patient examinations, shift handovers, and while sending and receiving messages from secretaries, medical doctors and other colleagues.

Several nurses expected to use the phone system especially in the daytime, but some assumed that they would benefit most from the phone system during night shifts. Most nurses estimated that they would use the system inside the ward location, not outside.

B. Phone logs

We also collected phone log data regarding the use of both the message and phone services. Figures 2 to 5 show the total usage of the system since implementation, from December 16th to January 21st. Figure 2 shows the total usage during the day from zero (midnight) until 23:59. Figure 3 shows the total number of messages sent during the day. Nurses send messages on all shifts, with a peak during the day shift between 9 and 10 am. Figure 2 compared with Figure 3 reveals that nurses only use the message function during the night shift, omitting the call function. Figure 4 shows the total number of phone calls during the same period. We can see that there were almost no phone calls between December 19th and January 3rd. Figure 5 shows the number of messages sent per day from December 16th to January 21st. A few messages were sent on December 21st due to a false alarm, but this error was fixed the following day. The system was not used during the Christmas holidays.

C. Step count

Before the phone system was introduced, we measured the steps of 21 nurses from 80 shifts, 24 afternoon shifts and 56 day shifts. The day shift resulted in 369,589 steps over 446 hours, and the afternoon shifts in 150,600 steps over 171 hours. After the phone system was introduced, we collected steps from 13 nurses on 82 shifts, 27 afternoon shifts and 55 day shifts. The day shift resulted in 334,017 steps over 411 hours, and the afternoon shifts led to 174,445 steps over 197 hours.

We found no significant difference in step count after implementing the phone system. Using a t-test, the mean number of steps per hour before and after was 838 and 830, respectively ($P = 0.82$).

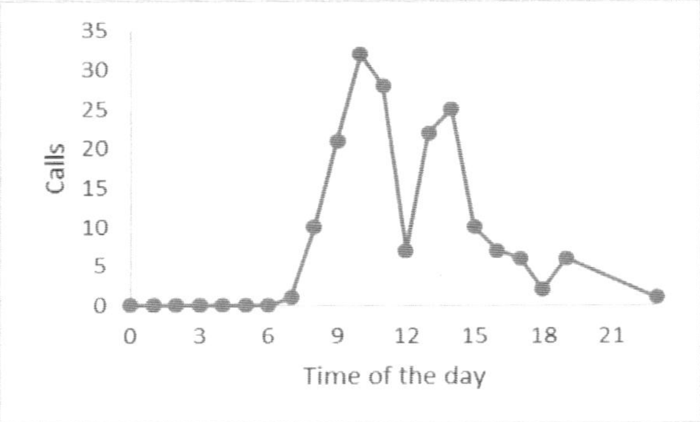

Figure 2. Total number of actual calls at certain times of the day from December 16th to January 21st

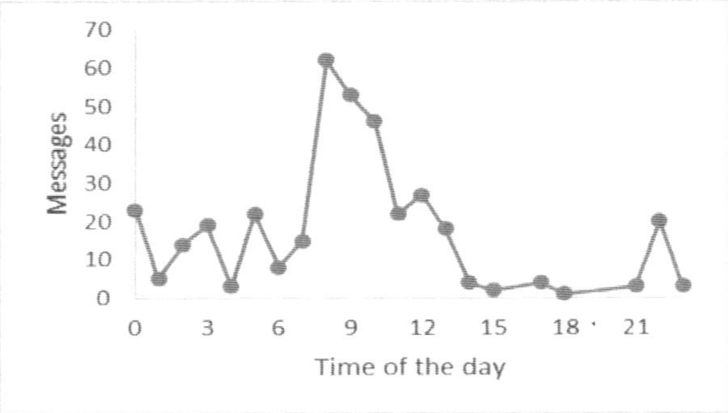

Figure 3. Total number of actual messages divided by time of the day from December 16th to January 21st

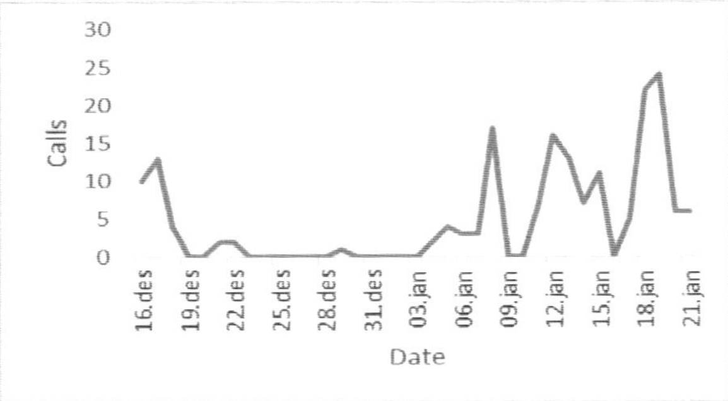

Figure 4. Total number of actual calls for each day from December 16th to January 21st

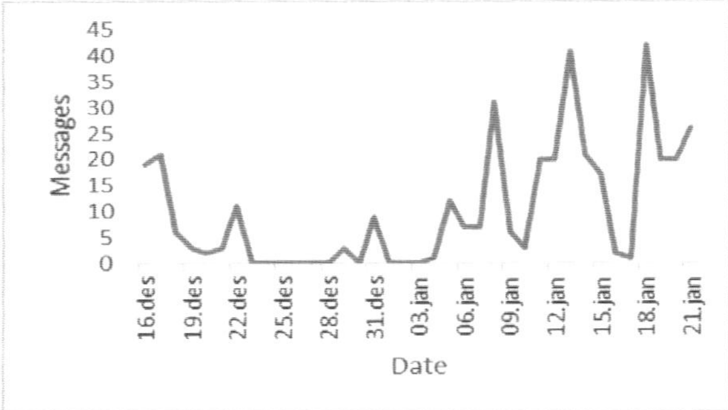

Figure 5. Total number of actual messages for each day from December 16th to January 21st

V. Discussion

In this paper, we have evaluated a hospital phone system at the Oncology Department at the main hospital in Northern Norway. We found that the nurses believed that phone use would save time and reduce disruptions and interferences during their daily work. They also envisaged fewer paper messages and a more "orderly" information exchange. The logs show that phone use increased over time and that the system was most frequently used during the day shift with a peak at 9 and 10 am. However, we did not find that nurses spent less time walking around looking for colleagues.

The results of the survey suggest that the nurses believed the phone system would make the information and communication flow more efficient while improving work practices. Furthermore, they emphasised that this improvement would not only benefit the nurses themselves by saving time and reducing disruptions, but it would also improve patient care. The nurses believed that they would spend less "unnecessary time" on getting messages and questions to the right people. Fewer interruptions might also reduce stress according to the nurses. Furthermore, they noted the stress they felt when they had "to run after" people or when they "must remember" things. They also expressed frustrations over "not finding" or "forgetting to pass on a message". It seems logical that when nurses spend less time and energy on getting and passing on information, it benefits the patients. The patients would have easier access to the nurses who would experience fewer interruptions during patient treatment and care. The nurses suggested that the phone system would make them more time efficient and thereby improve patient care.

However, we found no significant difference in step count after implementing the phone system, a result that must be seen in light of several methodological limitations. The step count was measured over a short period, one week before and one week after the system was introduced. Steps were measured in December. Perhaps more importantly, the "after" measurements were collected the last week before the Christmas holidays, which might have affected the results as normal workloads are reduced around the holidays. Furthermore, collecting the measurements right after the phones' introduction might have been too soon to capture an effect. It is reasonable to assume that the users need more time and practice to learn how to use the phone system before we can expect a change in work habits.

It should also be noted that a few messages were sent on December 21st due to an alarm going off. Caused by a bug in the system, the issue caused the nurses stop using the system for a day. The bug was fixed the same day, but due to the upcoming holidays and less activity at the department, several of the nurses were not aware that the system was up and running again. This situation might explain the system's lowered use during the holidays.

We did not record the nurses' exact workload for these two weeks; such information would include how many patients each nurse had in his or her care and how much care these patients needed. In addition, the nurses were given very little training in using the system. The phone logs show that system use has increased over time since we measured the steps, which might indicate that the nurses did not know the full functionality of the system at the time. When the nurses become more experienced in using the system, a change in work habits and practices may follow. Another possible explanation for the low use and the lack of difference in step count is that the nurses will run as much as possible independent of any technological system due to their eagerness to help as many patients as possible [13]. This finding could imply that more patients might receive better treatment and care due to the new phone system.

Another limitation is that it was not possible to measure the step count for the same nurses before and after the phone system was introduced. The main reason for this limitation was that the shift plan for the study period was already made. Some of the nurses working the first week (before data) were off work during the second week (after data). Another reason for this limitation is the short time period allotted for this pilot study.

Therefore, **future studies** are necessary. Data should be collected in different departments, and over longer periods, both before and after introducing the new system. Before the "after" data is collected, the nurses should be given time to become familiar with the phones and their functionalities. The data collected should be from periods with equal workloads, and data on step counts should be collected from the same nurses before and after. These changes would improve the validity of the results.

VI. Conclusion and Future Work

In this paper, we have reported some preliminary results from a pilot study evaluating the introduction of a phone system in one hospital department. We found that the nurses believed that phone use would save time and reduce disruptions and interferences in their daily work. The logs show that phone use increased over time and that the system was most frequently used during the day shift with a peak at 9 and 10 am. However, we did not find any time savings, as measured by reduced step count.

The next step in this evaluation is to collect interview data and measure use and step count over a longer time period. Future research questions will also investigate if the phone system influences the quality of patient care as well as the nurses' perceived safety of information exchange and productivity at the hospital ward.

Acknowledgment

We would like to thank the personnel at the Oncology Department at the University Hospital of North Norway for participating in this study. We also want to thank Stein Olav Skrøvseth for his valuable input.

REFERENCES

[1] E. Coiera, "Communication Systems in Healthcare," Clin Biochem Rev, vol. 27, pp. 89-98, May 2006.

[2] J. Scholl, P. Hasvold, E. Henriksen, and G. Ellingsen, "Managing Communication Availability and Interruptions: A Study of Mobile Communication in an Oncology Department," in *Pervasive Computing*, ed, 2007, pp. 234-250.

[3] J. Bardram and A. Doryab, "Activity Analysis: Applying Activity Theory to Analyze Complex Work in Hospitals," presented at the Proceedings of the ACM 2011 conference on Computer supported cooperative work, Hangzhou, China, 2011.

[4] E. Coiera and V. Tombs, "Communication Behaviours in a Hospital Setting: An Observational Study," BMJ, vol. 316, pp. 673-676, February 28, 1998 1998.

[5] M. A. Munoz, M. Rodriguez, J. Favela, A. I. Martinez-Garcia, and V. M. Gonzalez, "Context-Aware Mobile Communication in Hospitals," Computer, vol. 36, pp. 38-+, 2003.

[6] R. G. Soto, L. F. Chu, J. M. Goldman, I. J. Rampil, and K. J. Ruskin, "Communication in Critical Care Environments: Mobile Telephones Improve Patient Care," Anesth Analg, vol. 102, pp. 535-41, Feb 2006.

[7] K. J. Ruskin, "Communication Devices in the Operating Room," Curr Opin Anaesthesiol, vol. 19, pp. 655-9, Dec 2006.

[8] P. A. Spurck, M. L. Mohr, A. M. Seroka, and M. Stoner, "The Impact of a Wireless Telecommunication System on Time Efficiency," J Nurs Adm, vol. 25(6), pp. 21-26, Jun 1995.

[9] T. Solvoll and J. Scholl, "Strategies to Reduce Interruptions from Mobile Communication Systems in Surgical Wards," Journal of Telemedicine and Telecare, vol. 14, pp. 389-392, 2008.

[10] T. Solvoll, "From Being Interrupted by Mobile Devices to Callmesmart - a Context-Sensitive Communication System for Mobile Communication in Hospitals," University of Tromsø, Vitenskapelig artikkel 978-82-8236-100-2, 2013.

[11] E. Johnsen, T. S. Bergmo, M. A. Johansen, and T. Solvoll, "Evaluation of a Context Specific Communication System Based on Smartphones: Nurses' Uses and Experiences " in *Scandinavian Health Informatics, Shi 2016*, ed: Linköping University Electronic Press, 2016.

[12] M. Campbell, R. Fitzpatrick, A. Haines, A. L. Kinmonth, P. Sandercock, D. Spiegelhalter, and P. Tyrer, "Framework for Design and Evaluation of Complex Interventions to Improve Health," BMJ, vol. 321, pp. 694-6, Sep 16 2000.

[13] E. Jovanov, K. Frith, F. Anderson, M. Milosevic, and M. T. Shrove, "Real-Time Monitoring of Occupational Stress of Nurses," Engineering in Medicine and Biology Society, EMBC, 2011 Annual International Conference of the IEEE, 2011, pp. 3640-3643.

Designing User Interfaces for Personal Health Assessment Questionaires

A Report From a Pilot Study

Conceição Granja, Kari Dyb, Stein Roald Bolle
Norwegian Centre for eHealth Research
University Hospital of North Norway
Tromsø, Norway
{conceicao.granja, kari.dyb, stein.roald.bolle}@telemed.no

Gunnar Hartvigsen
Norwegian Centre for eHealth Research
University Hospital of North Norway
Tromsø, Norway
Department of Computer Science
UiT – The Arctic University of Norway
Tromsø, Norway
gunnar.hartvigsen@telemed.no

***Abstract*—The cancellation of elective surgeries wastes valuable hospital resources. Our site of research has identified inadequate planning due to lack of information as a main cause for cancellations. It is anticipated that the pre-operative planning process may be improved if adequate patient information is gathered at an earlier stage, before the patient is admitted at the hospital. The aim of our research is to determine how an electronic personal health assessment questionnaire interface to the patient should be designed. Participants were asked to complete the electronic personal health assessment questionnaire in the two versions of the interface. The results indicate that, contrarily to literature, in a pre-operative setting, patients prefer to be presented various questions at a time when completing personal health assessment questionnaires.**

Keywords-cancellations; surgery department; electronic communication; electronic health record; patient interfaces.

I. INTRODUCTION

The cancellation of elective surgeries wastes valuable hospital resources [1][2][3][4]. It is also reported that cancellations negatively affect the patient [5][6][7]. Accordingly, cancellations are stressful and costly, with a high level of emotional involvement before surgery [2]. However, it is reported that between 10 and 40 % of elective surgeries are cancelled [8][9][10]. Furthermore, it has been reported that 50 % of these cancellations might be avoided [2][7][11].

The eTeam-Surgery project is studying the cancelation of elective surgeries problem at the University Hospital of North Norway (UNN). At UNN, inadequate planning due to lack of information was identified as the main cause for cancellations [12]. It is anticipated that the pre-operative planning process may be improved if adequate patient information is gathered at an earlier stage, before the patient is admitted at the hospital. It has been reported previously that such patient information may be included in a personal health assessment questionnaire, and requested from the patient at an earlier stage and while the patient is still at home [12]. Such document refers to the patient medical history and includes questions on previous diseases (e.g., heart disease, high blood pressure, and diabetes), previous surgeries, general information (e.g., allergies, and smoking habits), woman specific (e.g., maternity), and use of medication, as shown in Figure 1.

Do you haver or had any of the following diseases:

	Example	**No**	**Yes**	**If yes, please specify**
1. Heart disease	Chest pain, myocardial infarction, irregular pulse, heavy breathing triggered by effort			Year?
2. High blood pressure				
3. Bleeding disorders	Blood clots, haemophilia, easy bruising / nose bleeding			
4. Neurological disease	Stroke, ischemia, brain haemorrhage, epilepsy			Year?
5. Pulmonary or respiratory diseases	Asthma, COPD (Chronic Obstructive Pulmonary Disease), snoring with apnoea			
6. Diabetes	Broadly controlled, insulin, treated by medication			
7. Thyroid disease				
8. Kidney disease				
9. Liver disease				
10. Stomach problems	Peptic ulcer, heartburn, acid reflux, oesophageal hernia			
11. Infectious diseases	Hepatitis, HIV, tuberculosis or other serious infectious diseases			
12. Psychological problems that the hospital should know				
13. Rheumatic disease	Impaired mobility or neck instability			
14. Mobility limitations				
15. Other diseases / conditions				

Figure 1. Sample of the paper version of the personal health assessment questionnaire, which provided the basis to the eTeam-Surgery electronic questionnaire.

The eTeam-Surgery project gathered the information required from the patient in a structured document, and developed a web-based tool to make it available to the patient. When designing the interface of the web-based tool the doubt arises: "How a Personal Health Assessment questionnaire should be graphically presented to the user to promote his/her participation?"

Adamson and Bachman [13] , at the Mayo Clinic, reported on a pilot study of using structured histories for patient. The authors suggested that questionnaires should be presented to the patient one question at a time [13]. Considering that both studies refer to patient reported history, it was decided to try the configuration suggested by the Adamson and Bachman study [13], despite it had been carried out in a primary care setting.

The aim of our research is to determine how an electronic personal health assessment questionnaire interface to the patient should be designed. We will explore if, in a pre-operative setting, patients prefer to be presented with one question at the time, or various questions simultaneously, when completing an electronic personal health assessment questionnaire.

This paper is divided in four sections. In the first section, the problem with surgical cancellations is introduced, and the aim of the study is described. In the second section, the data collection methodology is presented and explained. The results are disclosed and interpreted in section three. In the last section, the authors discuss the study configuration, and suggest improvements to it based on the feedback of participants.

II. Method

Two prototypes of the electronic personal health assessment questionnaire were built, each one featuring a different version of presenting the questionnaire. In Version 1, various questions were presented to the participant, grouped by subject. In Version 2, participants were presented one question at a time, following the suggestion in Adamson and Bachman's study.

Participants were asked to complete the electronic personal health assessment questionnaire in the two versions of the interface. A cross over methodology was used to ensure that the participants' choice was not influenced by the version completed first. Thus, for each participant the starting version was randomly selected.

III. Results

The fieldwork was carried out through a period of two days, in February 2014. Participants were randomly approached on the street, and a total of 11 persons agreed to participate. The sample consisted of six women and five man. Participants were given documentation where it was provided information related to the eTeam-Surgery project, and the context of their participation, and their consent to participate in the study was requested. Subsequently, the participants were given a tablet and asked to complete both versions of the electronic personal health assessment questionnaire. Upon completion the participants were asked to choose their preferred version. The results are presented in Figure 2.

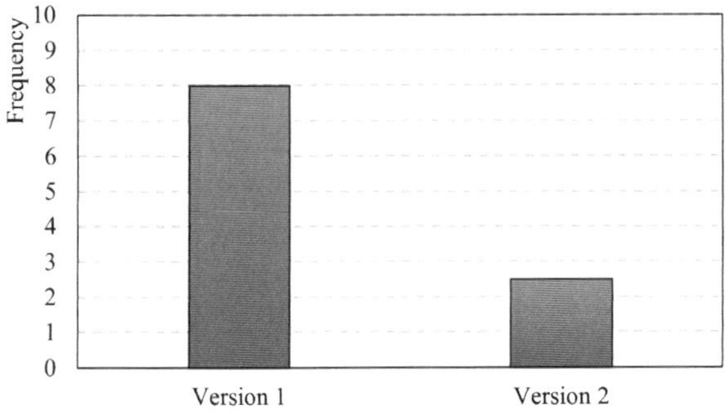

Figure 2. Frequency of the participants' preferred interface version.

IV. Conclusions and Discussion

The work presented herein was collected through the period of two days. After this period the authors' decided to terminate the study due to the low adherence, and evaluate possible measurements to improve the participation in the study.

The size of the sample achieved in this study did not meet the required statistical significance and, therefore, no conclusions may be drawn based on the results. However, the results indicate a probability that, in a pre-operative setting, patients prefer to be presented various questions at a time, that the authors which to further explore in a future study. Following, the authors explore on the reasons provided to refuse the participation in the study, as a basis to improve the configuration of future studies.

The study presents herein was terminated after two days of fieldwork by authors' decision, since the participation fell short on the expectations. The two main reasons provided by the possible participants to not take part in the study were:

A. Being uncomfortable in providing their information

It was explicit in the consent form that it was not required from the participants to provide their true medical history. This was further emphasized when approaching the participants. Nevertheless, it was still used as a reason to not take part in the study.

B. Did not want to complain on the hospital.

The authors' affiliation with the hospital lead possible participants to believe that opinions on the hospital's services quality was being collected. This could not be clarified since, once this was presented as a reason to not take part in the study participants did not want to engage in a conversation.

These behaviors lead the authors' to consider that future studies require a less constraining context. This implies that the participants should participate, e.g., from home, without the presence of the researchers, mimicking the real context of the web-based tool application. Furthermore, it is required to emphasize the main interest (i.e. the choice on the interface version), and anonymous nature of the study. This may be achieved by making it explicit in the consent form that the data on the answers was not stored, and providing non-participant related login data to the tool.

ACKNOWLEDGMENT

The authors would like to thank the regional health authority Helse-Nord for funding the research project HST 1119-13 and HST 1125-13.

REFERENCES

[1] A. Foster, "Operating Theatres: Review of National Findings," Audit Commission for Local Authorities and the National Health Service, London 2003.

[2] W. N. Schofield, G. L. Rubin, lM. Piza, Y. Y. Lai, D. Sindhusake, M. R. Fearnside, and P. L. Klineberg, "Cancellation of Operations on the Day of Intended Surgery at a Major Australian Referral Hospital," Medical Journal of Australia, vol. 182, pp. 612-615, 2005.

[3] M. J. Lacqua and J. T. Evans, "Canceled Elective Surgery: An Evaluation," American Surgeon, vol. 60, pp. 809-811, Nov 1994.

[4] B. Ivarsson, P. O. Kimblad, T. Sjoberg, and S. Larsson, "Patient Reactions to Cancelled or Postponed Heart Operations," J Nurs Manag, vol. 10, pp. 75-81, 2002.

[5] M. G. Perroca, M. d. C. Jericó, and S. D. Facundin, "Surgery Cancelling at a Teaching Hospital: Implications for Cost Management," Revista latino-americana de enfermagem, vol. 15, pp. 1018-1024, 2007.

[6] A. R. Tait, T. Voepel-Lewis, H. M. Munro, H. B. Gutstein, and P. I. Reynolds, "Cancellation of Pediatric Outpatient Surgery: Economic and Emotional Implications for Patients and Their Families," Journal of clinical anesthesia, vol. 9, pp. 213-219, 1997.

[7] T. L. Trentman, J. T. Mueller, S. L. Fassett, C. L. Dormer, and K. P. Weinmeister, "Day of Surgery Cancellations in a Tertiary Care Hospital: A One Year Review," Journal of Anesthesia & Clinical Research, vol. 1, 2010.

[8] M. Aaserud, MariTrommald, and J. Boynton, "Elektiv Kirurgi - Strykninger, Skjerming Og Effektivitet," Tidsskr Nor Lægeforen vol. 21, pp. 2516-2519, 2001.

[9] A. González-Arévalo, J. I. Gómez-Arnau, F. J. DelaCruz, J. M. Marzal, S. Ramírez, E. M. Corral, and S. García-del-Valle, "Causes for Cancellation of Elective Surgical Procedures in a Spanish General Hospital," Anaesthesia, vol. 64, pp. 487-493, 2009.

[10] S. Z. Yoon, S. I. Lees, H. W. Lee, H. J. Lim, S. M. Yoon, and S. H. Chang, The Effect of Increasing Operating Room Capacity on Day-of-Surgery Cancellation vol. 37. Edgecliff, AUSTRALIE: Anaesthesia Society of Anaesthetists, 2009.

[11] P. Sanjay, A. Dodds, E. Miller, P. J. Arumugam, and A. Woodward, "Cancelled Elective Operations: An Observational Study from a District General Hospital," J Health Organ Manag, vol. 21, pp. 54-58, 2007.

[12] C. Granja, K. Dyb, S. R. Bolle, and G. Hartvingsen, "Reduced Elective Surgery Cancellations through Patient Involvement in Pre-Operative Planning in Norway," eTELEMED 2014, The Sixth International Conference on eHealth, Telemedicine, and Social Medicine, 2014, pp. 164-69.

[13] S. C. Adamson and J. W. Bachman, "Pilot Study of Providing Online Care in a Primary Care Setting," Mayo Clinic Proceedings, Elsevier, 2010, pp. 704-710.

Core Archetypes

The Means to Build Confidence Around the Power of Structured EPR Systems?

Gro-Hilde Ulriksen
Norwegian Center for E-health Research – University hospital North Norway
Telemedicine and eHealth Research Group, Faculty of Health Sciences, Arctic University of Norway, Tromsø
e-mail: Gro-Hilde.Ulriksen@telemed.no

Rune Pedersen
University Hospital of North Norway, Section for E-health administration
Norwegian Center for E-health Research – University hospital North Norway
Telemedicine and eHealth Research Group, Faculty of Health Sciences, Arctic University of Norway, Tromsø
e-mail: Rune.Pedersen@unn.no

Abstract—**This paper reports from three different levels of working with archetypes in Norwegian healthcare. The paper contributes to a qualitative longitudinal interpretive study, connected to the development of a large-scale electronic patient record (EPR) system in the North Norwegian Health Authority. The focus of the paper is on how the notion of core archetypes could contribute to speeding up the process of developing national archetypes in Norway. Also, how core archetypes may increase clinicians understanding of structured EPR systems, based on different user experiences. We discuss when to start using archetypes in clinical practice. Also, whether constructing a prototype of a basis EPR system based on core archetypes, can contribute to a faster development of the new EPR system, and national archetypes in Norway.**

Keywords-archetypes; electronic patient records; semantic interoperability; core archetypes.

I. INTRODUCTION

An increased emphasis on cost savings, patient safety, and efficiency in healthcare, has raised the focus on seamless integration, and standardization across professional, departmental, and institutional boundaries [1]-[3]. Important efforts to achieve such goals, are improving the role of the electronic patient records (EPR), to make it a structured work tool, supporting patient pathways and decision support [4]. Semantic interoperability is a key requirement for improving EPR communication [8], to ensure that senders and recipients have the same understanding of information and standards [9]. An important element for reaching semantic interoperability is standardizing the clinical EPR content. However, standardizing clinical workpractice and routines have been difficult to accomplish [5]-[7].

In Norwegian healthcare, improving the role of the EPR has been part of national strategies and visions since the 1990's [1]-[2], to enable sharing and integrating healthcare, as well as organizing information in a more structured manner [10]. Hence, in 2012 the North Norwegian Health Authority established a large ICT (information and communication technology) project named FIKS (standardization of the regional ICT portfolio), to standardize the regional ICT portfolio. One of FIKS's most important roles was to collaborate closely with the largest EPR vendor in Norway, on developing a new openEHR based EPR system, using archetypes as core elements for standardizing the clinical content. The openEHR framework built on a two-level modelling. The intention of the first level, the technical reference model, was to increase semantic interoperability, and reuse of data [11]-[13]. The second level, contained archetypes and templates, as standards for the clinical content. The two-level model made it possible to make changes only to the clinical content of the archetypes, without having to alter the underlying open EHR information model.

The openEHR framework allowed for archetype design at different levels of healthcare organizations. In Norway however, the primarily work with archetypes was conducted at a national level, with NRUA (National Editorial group for Archetype development in Norway) as the coordinator of the work. To design optimal archetypes for the clinical content of EPR systems, it was necessary for clinicians to have a key role in both developing and approving the national archetypes. The clinical content of an archetype based EPR system had to contain numerous archetypes, to cover the clinical practice. There were 39 approved archetypes in March 2016, and even if 100, more were in the process, an estimated number of 1000-2000 was necessary, to encompass the clinical content of an EPR system. The absence of completed archetypes complicated and delayed the development of the new archetype-based EPR system. Hence, this raised a question whether to start using unapproved archetypes or not, and the consequences of using them before a national consensus was reached. In addition, it was challenging for the users included in developing the new EPR, to grasp the potential of this system based on a completely new technology. These factors might in part explain why the development process lasted much longer than expected.

We introduce the concept of core archetypes, as a contribution to solving this dilemma. NRUA has claimed it possible to use 30 core archetypes, as a foundation for about

90% of a basic EPR system [14]. The remaining 10 % represent specific archetypes related to clinical specialties, important to include for completing the EPR system. Our questions are therefore: Can a set of core archetypes provide the necessary foundation for understanding the potential of a new open-EHR based EPR system, and can focusing on these core archetypes speed up the archetype design process? Our goal was to evaluate the establishment of this core set of archetypes, through interaction with projects essential to the ongoing process. The rest of the paper is structured as follows; in section two, the method is presented, in section three, the three levels of archetype work in Norway are described. The concluding discussion in the fourth section, focuses on when to start using archetypes, and the use of core archetypes for a prototype of a basic EPR system.

II. METHOD

This paper contributes to a longitudinal interpretive study, connected to the development of a large-scale EPR system by the North Norwegian Health Authority. The methodologically positioning of the study is within a qualitative interpretive paradigm. The focus is on evolving and improving the understanding of a studied phenomenon, by looking at it from different viewpoints, within a context [15]-[16]. An advantage of using a qualitative interpretive approach is enabling complex textual descriptions of how people experience a particular matter, by providing information about the human side of a given process [16].

The fieldwork draws on the first author's role working in FIKS for two years and afterwards continuing to follow activities in the project, by participating in workshops and meetings connected to the development of the new EPR systems. The second author has contributed in the regional and national work with the new EPR and archetypes for the last seven years, participating in meetings, discussions and observations. The personal information protection commissionaire for research in the health region, and the Norwegian social science data service (NSD), approved the data collection for this study. All informants provided written consents for the interviews by e-mail. Both authors have conducted several open-ended interviews, both related to the development of the new EPR, and the national work with archetypes. The purpose of using open-ended interviews is enabling informants to tell their story, without the author's pre-perceptions getting in the way. The interviewers had prepared some questions, to make sure the interviews covered the topics they wanted to focus on. In addition, new interesting issues to include emerged in several interviews. The interviews were transcribed and analyzed separately, and as a part of a whole [Ibid]. Interesting quotes were translated into English (almost all Interviews were in Norwegian). In addition to transcribed and interpreted interviews, observations from several workshops and meetings as well as project documents from FIKS and NRUA were included in the data collection. In table I, an overview of the data collection is presented.

TABLE I. AN OVERVIEW OF THE DATA COLLECTION

Data source	
Interviews with contributors to the work with archetypes, and the development of new EPR.	30 open ended interviews
Participatory observation	250 hours
Participation in meetings, workshops, and informal discussions.	300 hours
Document studies: Documents from the CKM, concerning archetypes in general, and the problem/diagnosis archetype in particular.	

A document analysis on the consensus process of one particular archetype namely Problem/diagnosis, provided important insights for understanding the national work with archetypes.

III. WORKING WITH ARCHETYPES IN NORWAY

We have focused on three parallel processes important for the Norwegian archetype development between 2012-2016. First, the development of a new openEHR based EPR system, using archetypes as standards for the clinical content, with focus on development of the surgical pre-planning module. Second, the internal process of establishing a well-functioning NRUA organization, including a network of competent healthcare personnel to participate in the consensus process. Third, the development of local archetypes at a specialized hospital clinic, including the first attempts of using archetypes for clinical practice.

A. *Development of the new EPR system*

In 2012, the North Norwegian Health Authority completed an extensive tender, and decided to regionalize their new ICT portfolio. To carry out these changes, they established a regional project FIKS, to run from 2012-2016. FIKS was considered one of the largest ICT investments in Norwegian healthcare, with a total cost expected to exceed €100 million [17]. The main goals of the project were to establish a regional ICT portfolio, as a foundation for regionally standardized patient pathways, decision support, and integrations between clinical ICT systems.

A regionalization, including standardizing EPR work practice, was a necessary requirement to reach such goals, enabling the Health Authorities to better administrate and compare information from the hospitals in the region. The FIKS project run in close collaboration with system users from the hospitals and the vendors.

The new EPR was designed to improve the user's workdays, providing structured data including predefined content elements, and schemes for documentation, enabling an increased overview and reuse of patient data. In addition, the possibility of including patient pathways was important for improving the EPR. This was enabled by using the international openEHR architecture, standardized by CEN/ISO [18]. The openEHR architecture built on standardized information models, open source components, and highly structured clinical content, with archetypes as core building blocks. Archetypes were structured data elements of clinical concepts, where observations, evaluations, instructions, and actions, formed the ongoing process of treatment and care [19]. Archetypes were used to define how clinical data was structured, seamlessly stored, and transferred between EPR systems [20]. The intention was for archetypes to contain a maximum dataset, including evidence about knowledge objects, and relevant attributes [21]-[22]. It was possible to design both widely reusable generic archetypes, as well as specialized ones, designed for a distinct local setting [8], [14], [22-24]. The new EPR system required archetypes, as standards for the clinical content. The problem was that there were no archetypes available in Norway, when they started designing this new EPR system in 2012. The idea was for the vendor to design some generic archetypes necessary for the overall structure of the EPR. Further, the system users themselves would continue developing archetypes necessary for clinical practice, for example clinical observations such as blood pressure, body weight, clinical scorings, and schemes for procedures. One member of FIKS stated: *"At first it seemed possible for clinicians to design archetype based schemes on the fly, I don't however think this will be the case."*

The vendor started in 2012 the work with designing generic archetypes for the new EPR system. Since there were several archetypes available in the international CKM, constructed by the openEHR organization, it seemed rather straightforward to import these archetypes, and adjust them to Norwegian conditions. However, after working with archetype design for a year without achieving the desired results, it became obvious that this work was more complex than anticipated. The vendor recommended to establish a Norwegian CKM, and standardized methods for developing and maintaining Norwegian archetypes. Technology for storing and reusing archetypes, as well as defining how archetypes would relate to terminologies, were also important requirements identified.

In addition to the archetype development, a parallel process of designing functionality for the new EPR system took place. More than 100 system users, from all the hospitals in the North Norwegian Health Authority, participated in an attempt of user centered system design, where an agile method - scrum was practiced. The idea was to develop the new EPR module by module, based on user stories from the system users as a foundation for the functionality. It became difficult for the system users to grasp the potential of the forthcoming EPR system, as neither archetypes nor the EPR system functionality was finished. The needs for the new EPR, identified by end-users, mainly built on challenges connected to the existing EPR system. One clinician stated: *"When you ask clinicians today, they will outline needs related to their current work. Their starting point is the EPR they use today, that may be 20 years old. They are more likely to think small steps ahead, rather than focusing on large revolutionary changes, necessary for exploiting the potential of archetypes and openEHR."* Consequently, developing the EPR based on this approach, was time consuming, and inefficient. Since they did not have any prototype or model of the new system, it was demanding for clinicians to see the potential of it.

As an example, they have worked with the pre-surgical planning module for the new EPR from FIKS started in 2012, still close to the end of 2015, six of the 18 nationally approved archetypes necessary for this module, have reached national consensus. They were prioritized, as a collaboration between NRUA and FIKS. Without the necessary archetypes in place, it was challenging to develop and test functionality for this module, as well as knowing what to prioritize working on. To increase the pace of finishing this module, NRUA has lately been involved closer in working with the necessary archetype requirements for this module.

B. Constructing a national organization for handling archetypes

The EPR vendor, developing the new EPR system, had gained more than 70% of the Norwegian marked, over the last years [25]. Hence, National ICT, the organization responsible for coordinating ICT-related initiatives in the Norwegian specialized health care services, decided to work with Norwegian archetypes at a national level. NRUA was established in 2013, to form a national archetype repository – clinical knowledge manager (CKM). The overall goal of NRUA, was to coordinate the development, and use of archetypes on a national level, both handling the national consensus process of agreeing on archetypes, as well as supporting local initiatives for archetype design and usage. The NRUA organization consisted of five persons responsible for clinical modeling of archetypes, in addition to representatives from all the four Regional Health Authorities in Norway. NRUA established an editorial group to initiate archetype reviews, and form collaboration with clinicians. They also managed the recruitment for the national consensus process. In the consensus process clinicians used the web based CKM to review and approve archetypes, enabling asyncronically communication between participants in the process. The first years NRUA focused primarily on establishing a well-functioning organization, with a network of clinicians and other system users to work with archetypes. They provided training and support for new CKM users, established connections to the international CKM, and translated existing archetypes into Norwegian, for the clinicians to review. One of the archetype reviewers stated: *"NRUA has members with a genuine interest in archetypes and they have worked very hard to get this organization up and running."* Because of establishing the organization, the actual consensus work moved slowly the first years. The first archetype was nationally approved in 2014, five months after NRUA started their work. However, investing this much time on the organization early on, enabled NRUA to increase the pace of archetype development last year. There were 39archetypes approved in Norway in March 2016 and a 100 more was in the process. A goal is to have 200 archetypes approved by the end of 2016. In addition, NRUA had gained valuable knowledge along the way, both on how to structure archetypes, and on how to run the organization.

The extensive process of reaching archetype consensus was time-consuming for the contributors to participate in. When they first started as reviewers, it took time to get to know the CKM tool, and understand the complex clinical and technical relations of the archetypes. Each archetype had several review iterations. One clinician stated: *"A review iteration takes everything between 15min and 1,5 hour, depending on the complexity of the archetype. In addition, all archetypes go through more than one review iteration."* Recruiting participants to work with national archetypes was challenging, and the time-consuming review process led to several dropouts from the work, especially from clinicians. Since archetypes were used as standards for the clinical content of the EPR, it was important for clinicians to have an essential role in defining and designing them. One clinician said: *"It is crucial to include clinicians in this work; they have the clinical knowledge and know what is important to focus on, for the archetypes to be useful standards for clinical work."* He also commented, *"If others than clinicians design the archetypes, it will be troublesome to get clinicians to accept and use them."* It was however difficult for NRUA and the clinicians to know what archetypes to prioritize working with and how long the consensus process of each archetypes would take, to be able to plan the work with archetypes ahead. It was also problematic to ensure that the archetypes fit the clinical practice of Norwegian healthcare, since they had no way of testing them out in clinical practice.

C. Developing and trying out archetypes in clinical practice

Based on the notion of archetypes being the new standard for communication in Norwegian healthcare, several projects have been eager to start using them. Since the national design of archetypes took years to gain foothold, some local initiatives began to use archetypes for clinical practice, before they had been through the national consensus process, some even started developing local archetypes themselves. One member of NRUA stated: *"Systems that use archetypes today are not designed on nationally approved archetypes, or even international ones. They are mainly constructed by system users themselves."*

One example was a hospital clinic in the Southern and Eastern Norway Regional Health Authority that developed highly specialized archetypes for their clinical practice. This was a clinic working within a very narrow clinical field. Thus, they had a clearly defined focus area, and mainly needed specialized archetypes designed particularly for this field of expertise. When they started working with archetypes in 2014, NRUA was still in the process of establishing national consensus on their first archetype, (even if they had started the consensus process of several other archetypes as well). The clinic used approximately eight archetypes that had started the consensus process in Norway, the rest they assembled from the international CKM, and other existing repositories or clinical standards worldwide. The ones they did not find in any existing CKM, they developed themselves, in close collaboration with the EPR vendor. When the clinic started using archetypes for clinical practice, this was one of the first attempts to try out archetypes in a Norwegian clinical setting. This provided the clinic, NRUA, and the vendor, retrospective with important insight on the usability of archetypes at different levels, especially compositions, evaluations, and cluster archetypes. Testing archetypes in an actual clinical setting enabled identifying necessary requirements for improving, not only the local, but also the national archetypes, to make them suitable for both small and large-scale clinical usage. The clinic identified some challenges related to conforming these local archetypes to the national standards. First, the versioning of the local and the national archetypes clashed. Second, extensive structural differences between the local and the national archetypes, led to interoperability issues for different versions of the same archetype, with the risk of losing data when switching the local archetypes into national ones. In addition, the users experienced that creating local archetypes based on schemes from the old EPR were unpractical to use. NRUA is assisting the clinic in their further work, to enable the local archetypes to conform to the national ones.

IV. Concluding Discussion

The three examples described; A) the development of the new EPR, B) the process of establishing a national organization for developing and approving archetypes, and C) the process of starting a local initiative with hands on archetypes and solutions, indicates a need for improving and speeding up the archetype development process. In this context, there are two important issues to address. First, the question of consensus and/or the clinical value of archetypes; There is a broad agreement to only use consensus made archetypes in a production environment (use in the EPR), despite this it has been necessary to use "unapproved" archetypes in production, to speed up the processes. Secondly, and based on this, is the use of core archetypes to provide a prototype of a basic EPR system, to enable accelerating the process of archetype consensus, and the development of the new EPR system.

Based on these examples, the work with archetypes in different settings of Norwegian healthcare has raised a number of important issues. In the North Norwegian Health Authority, the strategy was not to start using archetypes before they had reached national consensus. The overall goals were to secure structured high quality archetypes, in line with the national standard, confirming that the archetypes they include in the new EPR system were compatible with other archetypes in Norway. However, since the vendor started developing archetypes for the new EPR themselves, the national consensus work related to these archetypes started too late to support the development process. This strategy contributed to delays in the development strategy. In addition, it was not possible to test the developed functionality for the modules of the new EPR, like the surgical pre-planning, without any available archetypes to structure the clinical content. They adopted this strategy, due to the unknown consequences of using archetypes that had not reached national consensus. A project leader in FIKS stated: *"Some of the consequences we dread from using unapproved archetypes are the lack of interoperability, the need for converting data, loss of historical data, all leading to increasing cost."* However, they did not want to stop developing the new EPR system completely, due to the lack of archetypes. *"There is a risk that if we are too cautious with starting to use archetypes it will make our development set to provide excessive profit related to reuse of data and clinical parameters fall way behind (project leader FIKS)."*

In the Southern and Eastern Norway Regional Health Authority, the strategy was nearly the opposite. They started to use archetypes that had been included in the consensus process. Thus, some places, as in the small clinic described, they also used archetypes not yet included in the consensus process, and even developed some new ones themselves. Starting to use unapproved archetypes in a clinical production environment, provided important insights for the vendor, the clinicians, and NRUA, on the actual usability of the archetypes that were under development nationally. NRUA has gradually undergone the hospitals very specialized structured schemes and variables. The development that this local initiative started was important for the maturity of NRUA as an organization, the archetype development, as well as for the EPR vendors. If they had waited for the regional EPR project to have all their archetypes approved, for example for the surgical pre-planning module, before testing out archetypes in clinical practice, they would still be missing the knowledge on how the archetypes actually work in a clinical setting. The understandings gained from this project, made it conceivable to improve the structure and content of the national archetypes. It also contributed to NRUA increasing their knowledge on archetypes, and capacity of assisting similar future projects. Clinical involvement, a suitable graphical interface, and integrations, are all interconnected with the development of archetypes and templates.

On the other hand, a clinical environment using archetypes that had not reached national consensus led to several interoperability challenges. The local archetypes were versioned following the same standard used by NRUA. This would consequentially lead to both a local and a national archetype with the same version number. Firstly, if the possibility to create a new version disappears there will be a loss of clinical data when converting to a consensus made archetype. Secondly, mixing up the local and the national archetypes might be a secondary problem, since the two definitely are comparable. Further, since the national archetypes were developed after the local ones were taken into use, there is a risk of a dissimilar structure of the local and national archetypes. If the deviations are too extensive, the local and the national archetypes might not be able to communicate. One of the archetype reviewers with a technological background described this potential problem: *"The local system will continue to work on its own, but if the structure of the archetypes is changed extensively to enable national consensus, they will no longer be able to communicate with the old version of the archetypes. Consequentially a system based on the local archetypes, cannot communicate with systems using national archetypes."* Accordingly, this might lead to losing historical data, or having to spend an extensive amount of time and money on converting all existing data to the new national archetype format.

To solve these complex issues, instead of trying out the archetypes in local projects with potentially complicated and expensive consequences, another approach would be to prioritize finishing the national consensus on the 30 core archetypes, and then use them to create a prototype of a basic EPR system. NRUA has defined the core archetypes as a sufficient basis for an EPR system[14], and they have composed a synthesis stating that: *"90% of the journal functions in the electronic patient record including non-specialized examinations and procedures can be represented by using 30 core archetypes"* [14]. Having a prototype based on the core archetypes for the clinicians to test, would most likely help them better grasp the potential of the new EPR, and how to continue the development process. One clinician stated: *"It is difficult for clinicians to imagine the possibilities of new EPR and not base their requirements on today's needs. Having a prototype would ensure that the archetypes cover then necessary clinical content for the modules of new EPR."* Such prototype might also provide useful information for the vendor on how to include the archetypes technically in their new system, and gain knowledge that is missing today on how to create and import templates into the EPR system based on archetypes. In addition, NRUA could profit on such a prototype, to identify which archetypes to prioritize for the national consensus process for each clinical specialty. This also includes further experience with fitting the archetypes for clinical practice. As time has gone by and NRUA's knowledge on archetype development has increased, they have already started to revise the synthesis. One member of NRUA stated: *"The synthesis was created by NRUA based on a fundamental understanding of essential elements to include in an EPR, and archetypes necessary to cover the fundamentals areas. It will probably be necessary to extend the number of core archetypes to 30-50."*

18 of the core archetypes are nationally approved and five more are in the process. Creating a basis model of an EPR system, based on core archetypes for testing the clinical usability of archetypes, could increase the pace of the national consensus work. The question is however, why these were not prioritized for the consensus processes. One of the members of NRUA said: *"We started out prioritizing the defined core archetypes. However, the work with core archetypes takes time since these are very generic and extensive concepts archetypes."* Establishing generic archetypes within an

immature organization was time consuming, since there were no experienced archetype designers, neither clinicians nor technical personnel in Norway. Hence, trying and failing was part of the process. How many clinicians to include, how to structure the archetypes, whether to translate international archetypes, or establish new ones from scratch, were some of the questions to consider. As a result, the last two years of working with archetypes in Norway on different levels of healthcare, NRUA has gained the necessary level of competence to fulfill their role as an organization that coordinate the national work with archetypes, as well as supporting local initiatives. Consequentially, future local initiatives should include NRUA at an early stage, to avoid the type of challenges that the clinic in the Southeast health region experienced from using local archetypes. Based on this, an important conclusive remark is; yes, the core archetype is a promising tool for future accomplishments of archetypes, and speeding up the development of the new EPR. Still the process described, and the following maturation of the national environment has been a necessary process. Now, NRUA and the national consensus work have reached the required maturity level, to exploit the possibilities that constructing a prototype/model by using core archetypes can provide for the future work with archetypes, and the new EPR system in Norwegian healthcare.

REFERENCES

[1] Norwegian Ministry of Health and Care Services, White Paper, Report 9, "One citizent one journal," national goverment 2012.

[2] National ICT's action 41, Use of archetype methodology for the definition, availabiliy and use of clinical information models in health-based Information Systems, Norway, 2012.

[3] National ICT's action 48, Clinical documentation for overview and learning, Norway, 2012.

[4] The office of the Auditor General, The office of the Auditor General's investigation of the ICT in hospitals and electronic interaction in health services, 2008.

[5] S. Timmermans and M. Berg, "Standardization in Action: Achieving Local Universality through Medical Protocols," Social Studies of Science, vol. 27, issue 2, pp. 273-305, 1997.

[6] R. Pedersen and G. Ellingsen, "Standardizing nursing work: works in practice but not in theory?" Studies in health technology and informatics, vol. 157, pp. 91-96, 2010.

[7] G. Ellingsen, E. Monteiro, and G. Munkvold, "Standardization of work: Co-constructive practice," The Information Society, vol 23, issue 5, pp. 309–26, 2007.

[8] C. Rinner, et al., "Creating ISO/EN 13606 archetypes based on clinical information needs." Proceedings of EFMI Special Topic Conference" e-Health Across Borders Without Boundaries; pp. 14-15, 2011.

[9] P. Gibbons, et al., "Coming to terms: Scoping interoperability for health care" Health level seven, 2007.

[10] T. Meum, G. Ellingsen, E. Monteiro, G. Wangensteen, and H. Igesund, "The interplay between global standards and local practice in nursing," International journal of medical informatics, volum 82, issue 12, pp. 364-74, 2013.

[11] R. Chen, et al., "An archetype-based testing framework," Studies in health technology and informatics, vol.136, pp. 401-406, 2008.

[12] S. Garde, P. Knaup, E. Hovenga, and S. Heard, "Towards semantic interoperability for electronic health records," Methods Inf Med, vol. 46, issue 3, pp. 332-343, 2007.

[13] T. Beale and S. Heard, "Open EHR Release 1.0.2," Architecture Overview. London, 2008.

[14] National editorial board for archetypes. NRUA: *Corearchetypes.*[Online]. Available from: http://wiki.arketyper.no/wiki/Kjernearketyper, 2015 2015.12.09.

[15] HK. Klein and MD. Myers, "A set of principles for conducting and evaluating interpretive field studies in information systems," MIS Quarterly, vol.23, issue 1, pp. 67-94, 1999.

[16] G. Walsham, "Interpretiv case studies in IS research: nature and methode," European Journal of information systems, vol. 4, pp. 74-81, 1995.

[17] B. Christensen and G. Ellingsen, editors, "Standardizing Clinical Pathways for Surgery Patients through ICT," European Workshop on Practical Aspects of Health Informatics (PAHI), Edinburgh, Scotland. http://ceur-ws.org http://ceur-ws.org. 2013.

[18] ISO: *Health informatics - Electronic health record communication, Part 1 Reference model CEN.* [Online]. Available from: http://www.iso.org/iso/catalogue_detail.htm?csnumber=40784.European comitte for standardization, 2007, 2015.12.07.

[19] R. Chen, P. Georgii-Hemming and H. Åhlfeldt, "Representing a Chemotherapy Guideline Using openEHR and Rules," Studies in Health Technology and Informatics, vol. 150, pp. 653–657, 2009.

[20] A. Moreno-Conde et al., "Clinical information modeling processes for semantic interoperability of electronic health records: systematic review and inductive analysis," Journal of the American Medical Informatics Association, pp. 925-934, 2015.

[21] MR. Santos, MP. Bax and D. Kalra, "Dealing with the archetypes development process for a regional EHR system," Appl Clin Inform, vol. 3, issue 3, pp. 258-275, 2012.

[22] M. Marcos, JA. Maldonado, B. Martínez-Salvador, D. Boscá and M. Robles, "Interoperability of clinical decision-support systems and electronic health records using archetypes: a case study in clinical trial eligibility," Journal of biomedical informatics, vol. 46, issue 4, pp. 676-689, 2013.

[23] E. Hovenga, S. Garde, T. Carr and C. Hullin, "Innovative approaches and processes for capturing expert aged care knowledge for multiple purposes," Electronic Journal of Health Informatics, vol. 2, issue 1, 2007.

[24] H. Leslie and S. Heard, Archetypes 101, Ocean Informatics 2006.

[25] DIPS-ASA, About us, Available from: http://dips.com/eng/about-us.2015, 2015.10.16.

Concept Testing Toward a Patient-Validated Information Architecture

Prototype Development of Healthtalk Norway

Tore Høgås
Centre for Quality and Development
University Hospital of North Norway
Tromsø, Norway
e-mail: tore.hogas@telemed.no

Torsten Risør
Faculty of Health Sciences
UiT The Arctic University of Norway
Tromsø, Norway
e-mail: torsten.risor@uit.no

Marianne Trondsen, Line Lundvoll Warth, Kari Dyb and Hege Andreassen
Norwegian Centre for E-health Research
University Hospital of North Norway
Tromsø, Norway
e-mail: marianne.trondsen@telemed.no, line.lundvoll.warth@telemed.no, kari.dyb@telemed.no
and hege.andreassen@telemed.no

***Abstract*—A Norwegian research group is adopting the Database of Individual Patient's Experience of Illness (DIPEx) international methodology standards for collecting qualitative research into people's health experiences and disseminating it on a web site. We are in the concept phase of developing the web site, and decided to build a topical ambiguous taxonomy together with a more clinically influenced taxonomy with top-level labels "Health and lifestyle" and "Illness" for the information architecture of the web site. In this paper, we report from usability testing of the top-level label of the topical taxonomy. We ran qualitative and quantitative A/B tests on wireframe concept sketches. The two top-level labels were a generic variant, "Topic", tested against the control variant, "Everyday life". Both qualitative and quantitative tests indicate better results for "Everyday life" as the top-level label for the topical ambiguous taxonomy of the web site. While not fully conclusive, the results provide reasonable confidence in the more descriptive label "Everyday life" at this early stage. It is preferable in that it both seems to create a more coherent set of expectations amongst the users, and more closely matches the content of the web site. The concept test is therefore deemed a useful first step in a rigorous testing program to ensure that the development process is informed by a patient-validated information architecture.**

Keywords- usability; information architecture; A/B testing; health experiences; qualitative research.

I. Introduction

The Internet is an increasingly important source of information for health purposes [1]. Evidence suggests that peer-to-peer exchange of health experiences has been one of its most transformational features [2] and is most likely to engage site users [3][4]. However, many web sites and other media sources present only a few anecdotal accounts or are skewed toward heroic or exceptional testimonials. Hence, there is a need for comprehensive research based collection and dissemination of patient accounts that includes the multiplicity of subjective everyday experiences.

To meet this need, the project "Healthtalk Norway" will pilot the Database of Individual Patient's Experience of Illness (DIPEx) methodology of health experiences research [5][6] in Norway, with the aim to "promote excellence of qualitative research into people's experiences of health and illness" [7]. DIPEx, developed by the Health Experiences Research Group (HERG) at the University of Oxford, is a qualitative research methodology based on in-depth interviews with patients and carers, for developing, producing and systematizing knowledge on peoples' health experiences. The core of the DIPEx methodology is a web site disseminating these health experiences to patients, carers, students, health professionals, health care services as well as the general public. Extracts from interviews are presented on the web site as video, audio, or text.

Through the DIPEx International network, researchers in 11 other countries have adopted the methodology, and at the time of writing, there are nine active web sites globally. A research group based at the Norwegian Centre for E-health Research (NSE) and with members from two Norwegian universities (UiT and NTNU) is currently working on a Norwegian Healthtalk web site. The first stage in this work is a feasibility study where one of the activities is to develop a prototype of the web site. Web development is an iterative process of creating and testing, often in four main phases: Concept, Prototype, Build, and Implement. To validate choices in functionality, content, structure, and design, it is important to employ a wide variety of tests throughout all four phases. In the following, we present results from usability testing in the very beginning of this process: concept development.

The starting point for the concept phase was a competitive audit of the nine active DIPEx web sites. The original United Kingdom site healthtalk.org was established by HERG in 2001 and has grown organically over the years, currently comprising more than 80 sections covering various conditions, diagnoses and health topics through approximately 250 interview excerpts. This means that its navigation is comprehensive, but has usability challenges because of scale. The other eight sites, in contrast, have very rudimentary navigational structures where individual diagnoses such as stomach cancer or epilepsy are listed as top-level labels, lacking any overarching categories. Therefore, they may face challenges as they grow. Hence, rather than duplicating existing web sites, we decided to start work on the information architecture for our site anew.

The research question in this paper is in what way quantitative and qualitative concept testing can be a useful first step toward a patient-validated information architecture. Our objective is to present a model for information architecture that will be both suitable and scalable, as well as user friendly. To this end, Section 2 discusses information architecture considerations for this project in the context of expected user categories or idioms. Section 3 describes the method of concept testing in the usability field and how we applied it in our tests. In Section 5, the results from these tests are discussed, and conclusions from these tests are drawn in Section 6.

II. INFORMATION ARCHITECTURE IDIOMS AND CONTEXTS

In the context of web development, information architecture can be described as the discipline of organizing a web site's information so that its users can find the right answers to their questions [8]. The challenge for a web site for the general public is to find organization schemes that are meaningful to a large and heterogeneous target audience, using labels that are relevant to and resonate with the user's own categories. At the same time, these schemes must be robust enough to accurately represent current content and scalable enough that we can reasonably expect them to represent future content.

In our case, there is an additional challenge: Users as patients are known to have several cognitive domains regarding health and present different parts of these domains in different contexts; collectively known as illness idioms [9]. This is often the case even for health care professionals who find that, as patients, their medical knowledge is not as helpful in resolving the relational and everyday challenges associated with being a patient (particularly with a chronic illness). Thus, while at a clinic, specific questions about symptoms and medicine may take center stage in communication, whereas questions about managing cooking, driving a car or using the bathroom may be much more central in a home context.

The content of the web site in question consists of people's health experiences, which encompass both the clinical experiential field and most areas of everyday life. Its categories must therefore reflect the multiplicity of patients' illness idioms. It is our aim to provide an alternative to official health information services and give prominence to patients' everyday experiences. We therefore decided to implement a dual taxonomy organization scheme both in site navigation and in faceted search results to cater to search-dominant users.

First, to facilitate known-item searching we will use *diagnoses/illnesses* and predefined *health and lifestyle* topics as the primary taxonomy. This approximates an *exact organization scheme*, as there is a large degree of consensus around diagnostic classification. Moreover, this taxonomy is already in use at the official Norwegian health portal helsenorge.no. The Norwegian Directorate of Health has based the diagnoses/illnesses part of the taxonomy on the Medical Subject Headings (MeSH) controlled vocabulary and used a variety of methods, including usability testing and web statistics analysis, to increase its usability for the Norwegian public and adding the health and lifestyle dimension. Additionally, since official health information is increasingly routed through this portal, many in the target audience will be familiar with this taxonomy when exposed to it on our web site.

In addition, we decided on supplementing the primary taxonomy with a *topical ambiguous organization scheme*. This has the disadvantage of adding cognitive load for the user, i.e., added mental resources required to use the site [10]. However, it was judged necessary to account for the fact that so much of the information we want to convey from patient experiences does not fall within the clinical field as defined in the primary taxonomy, but is part of a different idiom for the same illness. Examples of such topics range from how one deals with getting a diagnosis to how one's illness or health condition affects sex and intimacy. Additionally, this will support a common serendipitous mode of searching where the user has not necessarily formed a clear idea of what she is looking for.

When designing a topical organizational scheme it is crucial to develop a typology that has a strong topical relevance relationship to its content. The stronger the topical relevance, the lower the cognitive load for the user. Topical relevance can be arrayed in three facets: the functional role of information, how information contributes to the user's reasoning about a topic, and how information connects to a topic semantically [11]. When evaluating the topical relevance of a term in a taxonomy, we have to evaluate all three facets.

III. TOP-LEVEL TOPIC LABEL TESTING

Topical relevance is more important in the top level of a taxonomy than further down in the hierarchical structure, since the top-level label often also serves as a user's navigational entry point. For the exact organization scheme, we appropriated the top-level labels from helsenorge.no as "Sykdom", or "Illness" in English for the MeSH-based structure, and "Helse og livsstil" ("Health and lifestyle") for other categories such as pregnancy, nutrition or smoking.

We then needed to determine which top-level label would be the most appropriate for our topical ambiguous organization scheme. The first candidate was the generic label "Tema" ("Topic" in English). This label was judged strong in the facet of functional role, and has the advantage of flexibility through being generic. The other was "Hverdag" ("Everyday life" in English). While less flexible, it was judged stronger in both the user reasoning and semantic facets of topical relevance. However, it was not a clear-cut decision as to which candidate would ultimately have the highest topical relevance.

To find out in practice which candidate would result in the least amount of added cognitive load for users of the site we decided to run qualitative and quantitative A/B tests on wireframe concept sketches. An A/B test is a randomized experiment where users are exposed to one of two variants of a web site design. The variants are identical except for the one variation that is being tested, in this case the top-level label of the secondary taxonomy. Figure 1 is the wireframe of the generic ("Topic") variant we tested, formally the control variant, i.e., the null hypothesis for statistical evaluation. Figure 2 is the wireframe of the treatment A ("Everyday life") variant.

As this is only the first phase in a longer development process of iterative designs, we used the Notable web-based test platform to conduct remote testing. For each A/B test, we only need to recruit a single pool of test users, and the platform itself randomizes which of the variants is shown to the users.

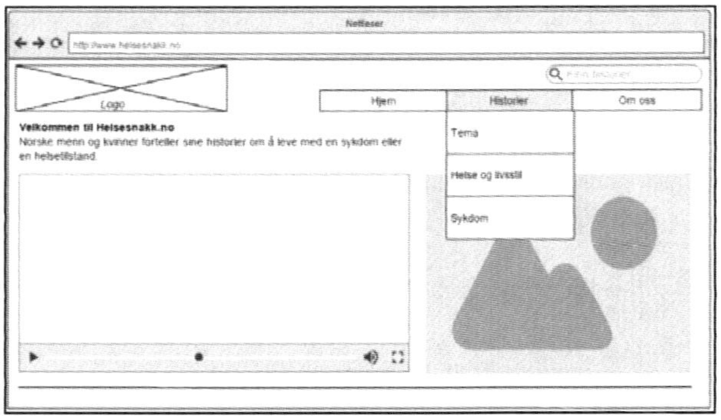

Figure 1. Wireframe of the variant with the "Tema" ("Topic") label.

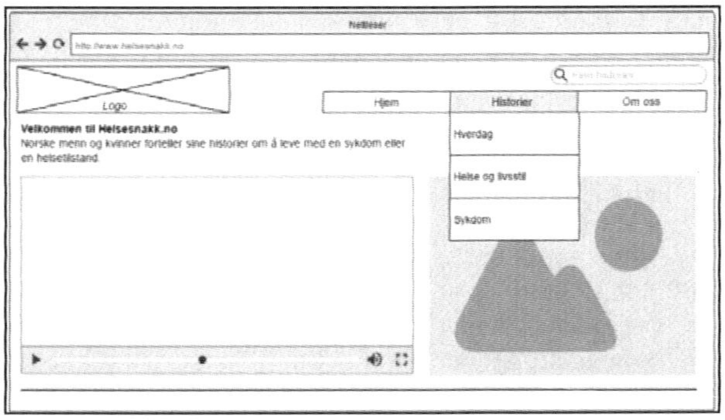

Figure 2. Wireframe of the variant with the "Hverdag" ("Everyday life") label.

A. *Qualitative A/B Testing*

For the qualitative test we recruited users via email, phone calls, and social media, limiting self-selection bias by recruiting users of both genders, of varying ages and levels of education, and excluding users working in either the health sector or technology/new media. As the web site in question has a large target audience, i.e., the general public of Norway, additional profiling criteria were not deemed necessary for either of the two tests. The recruitment method, which was largely online, as well as the online test delivery itself, ensured a certain minimum of Internet expertise in the user pool.

The minimum recruitment pool was set at five users per variant, following Jakob Nielsen's findings, which indicate that usability testing yield the best results when conducted in an iterative process with only five users in each test [12].

The test started with contextualizing the task: "You will see a sketch of a new web site with patient experiences. The site navigation will have three categories of experiences, and we would like your feedback on one of these options." The test then showed users the wireframe sketch with the label we were testing for highlighted, and the instructions: "Consider the highlighted menu option on this sketch. What do you think you would see if you clicked on that option?" We designed this open-ended question to determine topical relevance from the extent to which their responses match the intended meaning of the label.

B. *Quantitative A/B Testing*

For the quantitative test, we used social media for self-recruiting, i.e., asking a large number of people to volunteer for testing without applying any kind of selection criteria. Self-recruiting tests inherently run the risk of self-selection bias toward users who are more Internet-savvy, especially when online. However, since A/B tests are a form of multivariate research, this risk does not apply [13]. Even if responders are above-average experienced Internet users, their bias applies equally to the two variants, so we still get meaningful data.

The minimum requirement pool was set at twenty users, again following recommendations from Jakob Nielsen based on the findings that testing with twenty users gives you a confidence interval of maximum +/- 19%, while you would need as many as 76 users to reach +/- 10%. [14]

While the qualitative test was designed to determine topical relevance through users' written expression of reasoning about the label, the quantitative test focused on behavior in interacting with the label. Thus, while the contextualizing introduction was similar, the task and instruction were different. Test users were shown the wireframe sketch without any markings and the instructions: "Imagine that you have a serious illness and need a practical question answered. Look at this image. Click where you would have clicked to find information on how this illness affects driving a car."

IV. TEST RESULTS

A. *Results from Qualitative A/B Test*

The qualitative test received six responses for each variant, one more than the minimum requirement. The responses are translated and reproduced below with original punctuation preserved.

Responses to the control variant, i.e., what respondents expected to find under the "Topic" label:

1. What kind of illness
2. Heart-warming stories
3. Various diagnoses? (not obvious)
4. Gender
5. Meeting the doctor, monitoring after illness
6. Symptomless. Now what?

Responses to the treatment variant, i.e., what respondents expected to find under the "Everyday life" label:

1. How the illness has changed my everyday life?
2. How the illness affects my daily life
3. This is how I live with my illness
4. Living with the illness in my everyday life, 'trying' to live a normal life
5. This is how I feel
6. (Blank response)

B. *Results from Quantitative A/B Test*

The quantitative test received a total of 62 responses; 32 for the control variant and 30 for the treatment variant. These results are summarized in the contingency tables below. We have defined "success" as a click on the label we were testing for, "Topic" and "Everyday life" respectively. Our definition of "failure" is any other click on the image. Table 1 shows all collected results. Table 2 displays results filtered on responses where the user spent more than 10 seconds looking at the image before clicking.

TABLE I. ALL RESULTS

	"Topic"	**"Everyday life"**	**Marginal Row Totals**
Success	10	14	24
Failure	22	16	38
Marginal Column Totals	32	30	62

TABLE II. SPEED FILTERED RESULTS

	"Topic"	**"Everyday life"**	**Marginal Row Totals**
Success	3	9	12
Failure	19	20	39
Marginal Column Totals	22	29	51

V. DISCUSSION

A. *Qualitative Analysis*

Out of six responses to the "Topic" variant, three were irrelevant to the planned content of this taxonomy: "What kind of illness", "Various diagnoses? (not obvious)", and "Heart-warming stories." The two first are obvious misunderstandings of what might be under the "Topic" label, since there already is a separate top-level label for "Illness." One of them calls attention to the user's uncertainty by adding the question mark and the parenthesis stating explicitly that the label is not obvious. The third answer seems to expect the type of content other sites skew towards; heart-warming stories of heroic endurance. The sites built with the DIPEx methdlology are not about such stories; they are about the unheroic and unfiltered experiences of regular people.

The other three responses to this variant were relevant to varying degrees. Responses 5 and 6 in particular go beyond the clinical experiential field to what happens after an illness. Nevertheless, in sum these six responses indicate that the flexibility of the "Topic" label turns to plasticity, with widely varying understandings of its meaning.

Turning to the six responses to the "Everyday life" label, we see that with the exception of the one blank response, they are all relevant to the intended meaning of the label. Three of them are even paraphrasing the label: "How the illness has changed my everyday life?", "How the illness affects my daily life" and "Living with the illness in my everyday life, 'trying' to live a normal life." As many as four responses make an explicit connection between this label and the adjacent "Illness" label of the other taxonomy, which indicates that the users' reasoning is that the content behind this label is connected to, but different from, illnesses and diagnoses. Indeed, the interplay between the three top-level labels seems to be much more productive when the "Everyday life" label is used, than with the "Topic" label.

It is also valuable input that two of the respondents to this variant introduces the subject "I" in their responses: "This is how I live with my illness" and "This is how I live." In combination with the leading demonstrative pronoun "this", it indicates that they have formed a quite strong identification with the level on a personal, emotional level. The same may be said of the response that contrasts "living with the illness in my everyday life" and "trying to live a normal life." A diagnosis often represents a biographical disruption in the patient's life, and we know that restoring normality and a coherent biography in times of illness is demanding emotional work [15].

The fact that four out of twelve responses in total are irrelevant or blank may indicate that they did not fully understand the task, which is a weakness in unmoderated online tests. Nevertheless, this weakness is the same for both variants, yet there is a clear difference in responses to two variants tested. While users' free-form responses to the "Topic" label go in different directions, the "Everyday life" label elicits responses about relationships between illness and daily life, with indications of both an emotional component and changes induced in the patient's life. The label "Everyday life" thus creates a more coherent set of expectations in users along the facets of both users' reasoning about the label and its semantic value. These expectations more closely match the content that we will publish on the web site, resulting in higher topical relevance.

Seen as a whole, the results from the qualitative test also provides valuable insight into what kind of content prospective users would expect or even want from our web site. These expectations cover the entire spectrum from first meeting the doctor and getting a diagnosis, through the illness progression, and beyond.

B. *Quantitative Analysis*

The results from the quantitative test corroborate the indication that the "Everyday life" label has higher topical relevance for users tested. From the results in table 1 we can calculate a success rate of *0.31* for the control variant, while the treatment variant has a success rate of *0.47*. The difference in success rates is *0.16*, which is clearly in favor of the treatment variant; "Everyday life."

The speed filtered results in Table 2 are of interest based on the assumption that variance in cognitive load affects slow users, defined as users spending more than 10 seconds on deciding where to click, more strongly than fast users. In other words, if we are to reduce cognitive load as much as possible, responses from slow users are more important than responses from fast users, because its effect is amplified.

These results show that the control variant has a success rate for slow users of *0.14*, while the treatment variant has a success rate of *0.31*. As expected, success rates are lower for slow users of both variants, yet the difference in success rates is slightly greater for slow users at *0.17*. Thus, both contingency tables indicate higher topical relevance for the "Everyday life" label.

However, there are two caveats to this indication. First, if we use the Fisher exact test to calculate the p-values for the contingency tables from the quantitative A/B test, they are *0.297548* for the full results and *0.192494* for the speed filtered results. In both cases, the values are higher than the significance level of 0.1. or 10%. Therefore, while at least the speed filtered results are within the +/-19% confidence interval we have deemed acceptable for this test, they are technically not statistically significant.

Second, an analysis of the failure clicks on click maps generated by the Notable test platform shows that there are more responses to the control variant that cannot be categorized as navigation-dominant behavior (i.e., focusing on navigational elements) or search-dominant behavior (i.e., focusing on the search box). Figure 3 shows the click map for the control variant with the "Topic" label. In contrast, figure 4 shows a more focused click map for the treatment variant with the "Everyday life" label. While the control variant received 11 clicks that are not on a navigational element or the search box, the treatment variant received only 4.

A reasonable hypothesis for behavior that is neither navigation-dominant nor search-dominant in a task such as this is that these users did not fully understand the task. Their

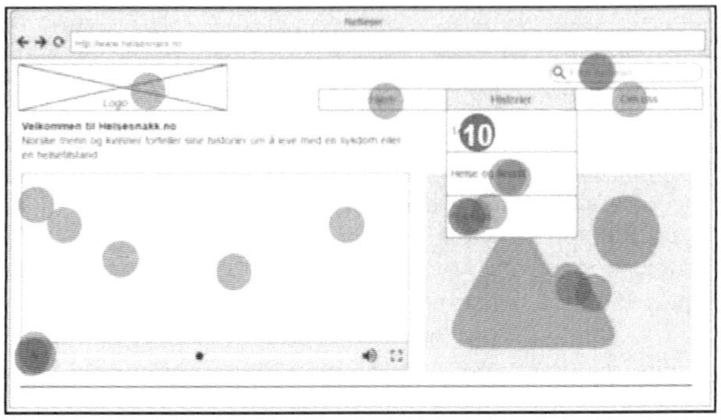

Figure 3. Click map of the "Topic" variant.

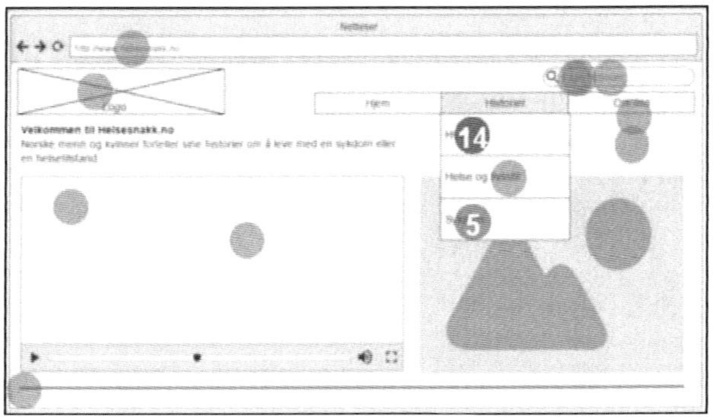

Figure 4. Click map of the "Everyday life" variant.

TABLE III. RESULTS WITH OUTLIERS EXCLUDED

	"Topic"	"Everyday life"	Marginal Row Totals
Success	10	14	24
Failure	11	12	23
Marginal Column Totals	21	26	47

behavior does not correspond with the expected patterns of users who are actually trying to complete the task as given. If this is the case, we may have to define all these responses as outliers and exclude them from the contingency tables. Table 3 shows this modified table of results, with failures defined as outliers excluded.

From these results, we can calculate a success rate for the control variant of *0.48*, while the treatment variant has a success rate of *0.53*. Although the latter still has a higher success rate, the difference is minimal at *0.06*.

VI. CONCLUSION AND FUTURE WORK

In this paper, we have described an early web-based A/B test set of information architecture concept sketches. We found an indication in both quantitative and qualitative tests that "Hverdag" ("Everyday life") is most likely a better choice than "Tema" ("Topic") as the top-level label when creating a topical ambiguous taxonomy for a patient-oriented web site. The data suggest that this label has a higher topical relevance and therefore results in lower cognitive load for users.

We note that this indication is tempered by the two caveats discussed above. The collected data from the quantitative test is not statistically significant, and excluding outliers reduces the difference between the two variants dramatically. Regardless, the qualitative test gives a much stronger indication since it gives us insight into users' reasoning about the top-level labels. In fact, that is why qualitative methods are often preferred in usability testing; because they do not only show that a given design is problematic, but provide insights into *why* there is a problem as well as *how* you can solve it [16]. Therefore, although test results are not fully conclusive, they have some value in providing reasonable confidence in the "Everyday life" top-level label at this early concept stage.

Going forward in the web development process, we will continue to employ a variety of quantitative and qualitative test methods to achieve patient validation of our information architecture and design decisions for the "Healthtalk Norway" web site prototype. While this first set of tests was delivered online for speed and convenience, we recognize the limitations of this delivery method and will in the rest of the process conduct usability tests in person as well. This experiment has shown that it is necessary to ensure that respondents understand tasks and to improve response registration.

This rigorous testing program will be of importance for further refinement of our model for online presentation of qualitative research on people's health experiences.

REFERENCES

[1] H. K. Andreassen et al, "European citizens' use of E-health services: A study of seven countries," BMC Public Health 7, 2007, p. 53.

[2] S. Ziebland and S. Wyke, "Health and illness in a connected world: How might sharing experiences on the internet affect peoples' health?" Milbank Quarterly, 90(2), 2012, pp. 219-249.

[3] E. F. France, S. Wyke, S. Ziebland, V. A. Entwistle, and K. Hunt, "How personal experiences feature in women's accounts of use of information for decisions about antenatal diagnostic testing for foetal abnormality," Social Science & Medicine, 72(5), 2011, pp. 755-762.

[4] E. Sillence, P. Briggs, P. R. Harris, and L. Fischwick, "How do patients evaluate and make use of online health information?" Social Science & Medicine, 64(9), 2007, pp. 1853-1862.

[5] C. Pope, S. Ziebland, and N. Mays, "Analysing qualitative data," BMJ 320, 2000, pp. 114-116.

[6] S. Ziebland and A. McPherson, "Making sense of qualitative data analysis: an introduction with illustrations from DIPEx (personal experiences of health and illness)," Medical Education 40(5), 2006, pp. 405–414.

[7] H. K. Andreassen, K. Dyb, M. V. Trondsen, and L. L. Warth, "Study protocol: Health talk Norway" Proceedings of the 13th Scandinavian Conference on Health Informatics, June 2015, Tromsø, Norway.

[8] L. Rosenfeld and P. Morville, Information Architecture for the World Wide Web. O'Reilly & Associates, pp. 22-31, 1998.

[9] M. B. Risør, "Illness explanations among patients with medically unexplained symptoms: different idioms for different contexts," Health, vol. 13 no. 5, Sept. 2009, pp. 505-521, doi:10.1177/1363459308336794.

[10] K. Whitenton, "Minimize Cognitive Load to Maximize Usability." Available from: http://nngroup.com. 2013.12.22. [retrieved: 3, 2016]

[11] X. Huang, "Developing a Cross-Disciplinary Typology of Topical Relationships as the Basis for a Topic-Oriented Information Architecture," 20th Annual ASIS SIG/CR Workshop: Bridging Worlds, Connecting People: Classification Transcending Boundaries, 2009, doi: 10.7152/acro.v20i1.12884.

[12] J. Nielsen, "Why You Only Need to Test with 5 Users" Available from: http://nngroup.com. 2000.03.19. [retrieved: 3, 2016]

[13] R. Dewey, Introduction to Psychology. Wadsworth Publishing, 2004. Available from: http://www.intropsych.com. [retrieved: 3, 2016]

[14] J. Nielsen, "Quantitative Studies: How Many Users to Test?" Available from: http://nngroup.com. 2006.06.26. [retrieved: 3, 2016]

[15] M. Bury, "Chronic illness as biographical disruption," Sociology of Health and Illness, vol. 4, no. 2, 1982, pp. 167-182.

[16] C. Rohrer, "When to Use Which User-Experience Research Methods" Available from http://nngroup.com. 2014.10.12. [retrieved: 3, 2016]

Investigating Factors Determining the Use of the Clinical Care Module by Nurses Through the UTAUT Model

Munyaradzi Zhou[1]
School of Computing, College of Science and Engineering
University of South Africa
Florida Campus, Johannesburg, South Africa
E-mail: zhoum@msu.ac.zw

Marlien Herselman[2]
CSIR, Meraka
Pretoria, South Africa
MHerselman@csir.co.za

Alfred Coleman[3]
School of Computing, College of Science and Engineering
University of South Africa
Florida Campus, Johannesburg, South Africa
Email: Colema@unisa.ac.za

Abstract **- Nurses can be empowered in the decision making process if provided with objective diagnostic data. This research applied the Unified Theory of Acceptance and Use of Technology model to investigate the usage of electronic health systems and analyse the factors affecting the intention to use the clinical care module which supports decision making during point of care. Despite the efforts by governments, donors and international partners in rolling out electronic health systems, these systems are 'partially' or not being used by healthcare workers during point of care. The deployed applications are incompatible with nurses' routine clinical practice as they're not customized. There is need to move from using applications which solely supports capturing patients' demographic data to systems customized for clinical care practice. Semi-structured questionnaires were deployed in three hospitals to 200 nurses who have been randomly selected in maternity section. For triangulation purposes, focus group interviews have been conducted in these hospitals. Interviewees were purposively selected from respondents who completed the questionnaires. The results showed that the key constructs for use of the clinical care module in their order of importance are facilitating conditions, performance expectancy and social influence. Healthcare institutes authorities in Zimbabwe must improve conditions that facilitate the use of the clinical module. To derive value from electronic health systems being adopted, the work processes involved need to be redefined and adequate information has to be provided to healthcare workers.**

Keywords - UTAUT; clinical care module; nurses; clinical care

I. INTRODUCTION

Informed decision making is made possible by merging nurses' decisions (based on the nurse's clinical experience and patient's narration) and the knowledge that informs nurses. Healthcare workers' clinical practice may be improved through the use of clinical decision support systems (clinical care module). The clinical care module is an application which aids Public Health Workers (PHWs) to make valuable decisions and to collaborate with all authorized healthcare professionals during patient point of care [18][33]. The clinical care module allows completion of tasks electronically such as making appointments, referrals, accessing and updation of patient records, and making laboratory test requests. Decisions which can be made using the module include writing the correct prescription for the client, deciding whether to refer the case to another provider such as transferring a client to caesar ward or to make re-referral. The clinical care module also support sending of medication alerts messages and aids nurses in carrying out evidenced based care. Nurses are key decision makers and are directly involved in patient care in a healthcare set-up and hence they have to be equipped with knowledge during clinical care. Delay in patient care might be caused by lack of relevant and accurate information for decision making.

[14] pointed out that *'Healthcare involves the use and management of an abundance of information that must be collected, managed, reviewed, processed, and mined. High-quality patient care relies on careful documentation of every patient's medical and family history, health status, current medical conditions, and treatment plans. A clinical decision based on information that has been efficiently managed and processed lends itself to quality care outcomes.'*

The Ministry of Health and Child Care (MoHCC) and the hospital health authorities have managed to deploy hardware and software requirements for the implementation of Electronic Health applications together with the necessary technical expertise [27]. The District Health Information System (DHIS) has been rolled out since 2010 and the platform allows integration of other healthcare systems such as mobile health Short Message Service (SMS) and android based platforms. The DHIS has mainly been used as a tool for capturing patient details, reporting disease surveillance statistics (reporting and analysis needs) by the Health Information System (HIS) personnel and the administrative healthcare authorities. This platform does not have the clinical care module functionality [29]. The

Electronic Health (E-Health) projects rolled out do not offer the clinical services module to aid clinicians during decision making and this has been a trend even in mHealth projects deployed in South Africa [30][35]. The MoHCC also piloted a SAP healthcare solution application at Chitungwiza Central Hospital which is mainly being used as a patient records management system (patients' demographics, and billing purposes). The SAP healthcare solution supports management of patient demographics data, billing of patients and maintenance of clinical records; provides real-time access to relevant patient and clinical information at the point of care; and improves care collaboration [22]. Irrespective clinical care delivery module SAP Healthcare solutions a support, the module is not being by nurses, they're are using the software for management of patient's information [28]. The focus of this research is to relate the UTAUT model to the use of the clinical care module in E-Health projects deployed in Zimbabwe since this might result in improved quality care. The unified theory of acceptance and use of technology (UTAUT) model was used to determine the intention to use the clinical module by nurses without compromising workers' performance and to explore the usage of E-Health systems by nurses. Technology adoption will not improve a firm's competitiveness unless the adopted technology ends up being used [13].

II. Background

The MoHCC (Zimbabwe) is supporting health information technology by deploying DHIS and SAP Healthcare solution applications in hospitals and there is need to harness the clinical services module during clinical care delivery. Adoption and use of the clinical module is important for value addition in the healthcare system. This study applied the UTAUT model to investigate the factors that affect the use of clinical care module in deployed E-Health applications and the use of the applications. The DHIS platform is mainly used by the Health Information System (HIS) department in all districts where it has been rolled out. Nurses do not have direct access to patient's information on this platform. The Health Information System (HIS) department personnel such as clerks are the ones with authorization rights to use the system for capturing and retrieving patient information. Nurses obtain patient information upon request to the HIS department personnel who are responsible for accessing the DHIS. There has been adoption of these systems but they are not being used at all (or 'very little use') by nurses in clinical practice [29].

Most researches focused on the factors affecting adoption and diffusion of E-health systems while applying the UTAUT and TAM models. In addition, several studies hinged on adoption and use of IT in healthcare and how to deploy mHealth projects, while little attention has been given to the use of the clinical care module in E-Health applications adopted and used [1][25][34]. However, this research focuses on the use of the clinical care module in applications rolled out and to explore the extent of usage of projects deployed by nurses. The UTAUT model has been applied in Health Informatics researches but little or no research has been made on the UTAUT model as a tool for ensuring the use of the clinical module of nursing [19]. The use of the clinical care module might result in improved quality care.

III. Literature Review

Improving health outcomes such as access to high-quality healthcare is important for national development and can be achieved through a more effective health force [21]. Nurses' work processes need to be explicitly defined and avoid overlapping. Inter-linking of processes and having good processes results in IT complementing the work processes [2]. The clinical module of nursing is a decision support application designed to improve nursing care [5]. These applications are used in a variety of ways by clinicians such as documentation of patient information, monitoring patients' progress and validating decisions made using paper-based clinical notes [7].

A. *The UTAUT Model*

The UTAUT model is behavioural factors centric and focuses on four constructs namely: performance expectancy, effort expectancy, social influence, and facilitating conditions to give an overview of problems related to Information Systems/Information Technology (IS/IT) adoption and diffusion. The UTAUT theory was adopted in this research since it mapped and integrated opportunities of eight dominant theories and models including the Technology Acceptance Model (TAM) [32]. Validation and the use of the UTAUT model has been done to investigate diffusion and use of E-Health [1][24][34] while this research applies the UTAUT model to investigate the usage of E-Health systems and determining the factors affecting the intention to use the clinical care module.

Performance expectancy is defined as the extent to which using a technology will provide benefits to nurses in their work processes; effort expectancy is viewed as the degree of ease-of-use of technology by nurses; social influence is the extent to which nurses perceive that local health authorities, the MoHCC, patients and other stakeholders believe they should use the clinical module; and facilitating conditions refer to nurses' perceptions of the resources (ICT infrastructure) and support needed and available to use a technology [3][31]. The UTAUT model has four moderators that influence the perception of the four constructs of the model namely gender, age, experience, and voluntariness of use which are applied in this research [16][31]. UTAUT model constructs were applied to ensure use of the clinical care module. [6] coined that '*Emerging information technology cannot deliver improved organizational effectiveness if it is not accepted and used by potential users.*'

B. District Health Information System (DHIS) and Systems, Applications and Products in data processing (SAP) Healthcare solution

The DHIS is a useful tool for the collection of aggregate health data from all levels of healthcare which is transferred to national server. The analysis of data collected and reports generation is based on national indicators. The challenge with DHIS data is the ability to follow patient cohorts, for which an Electronic Health Record (EHR) should be used. Most EHRs have a functionality which aids in decision making and this concept has to be applied to E-Health systems deployed. EHRs have an integrated view of the patient across health facilities and a Master Patient Index (MPI) for cohort analysis. The use of the clinical module seems to be lagging behind in healthcare institutions, while its use might result in best practices [15][17][30]. New technologies are sometimes adopted and then used very little or not at all [14]. The DHIS is being used by HIS department staff but do not have the clinical care module functionality.

SAP Healthcare solution is SAP's IT software for the healthcare industry. The SAP healthcare solution supports management of patient demographic data, billing of patients and maintenance of clinical records; provides real-time access to relevant patient and clinical information at the point of care; and care collaboration [22]. An integrated system consists of grouped functionalities called modules thus both DHIS and SAP Healthcare solution consists of modules [4][10]. DHIS consists of modules but do not have a module specifically tailored to provide clinical decision support solutions while the SAP Healthcare solution has this functionality but the module with this functionality has to be bought as part of the package. The clinical module supports capturing of patient demographic and clinical health information, clinical decision making and sharing of information between authorized healthcare entities and assist in direct patient care and can be linked with other systems from other sections such as the laboratory and pharmacy [26].

IV. METHODS

The Mixed method approach was employed with the use of questionnaires for numerical data collection and narrative data collection which was mainly done through the use of focus group discussions. 200 nurses in maternity section were randomly selected from three health institutions and have been given questionnaires. 176 nurses returned the questionnaires giving us a response rate of 88 percent. Three FGDs have been held with purposively selected respondents of six on average from each health institution. The interviewees mainly consisted of healthcare workers who completed the questionnaires. The selected hospitals have been selected as pilot sites for implementation of Electronic government projects in healthcare sector. A Sequential Explanatory Design was used, thus the use of questionnaires followed by FGDs for triangulation purposes [12]. The research focused on investigating the factors affecting use of the clinical care module in perinatal care services. Data was analysed using Statistical Package for the Social Science (SPSS) and the UTAUT constructs tested are performance expectancy, effort expectancy, facilitating conditions, and social influence using descriptive statistics. Thematic analysis was used to analyse data collected using Focus Group Discussion (FGDs). The results were later on merged with the identified statistical relationships from questionnaire data and then interpretation was done. The clinical care module was described to nurses as an application which aids nurses to make informed decisions during patient care.

V. RESULTS AND DISCUSSION

A. Participant profile demographics

Health institution A constitutes 36.4 percent, B 36.9 percent, and C 26.7 percent of the respondents. The median experience was more than 5 years and the median age is 38 years. The majority of the respondents (71.0%) are registered general nurses with midwifery.

B. Mobile devices and applications used

SAP Healthcare solution has the clinical care delivery module which is not being used by healthcare workers since the module was not part of the package when the solution was bought. In addition, the healthcare solution platform was not customized to suit user requirements. There is need for integrating the clinical care module with the DHIS and the SAP platforms since there is a potential to improve care processes and patient care outcomes.

Users of applications such as Google Play store, Whatsapp and Facebook have a more positive perception of the use of technology in comparison to the non-users as they already have experience in computer use. At work PHWs generally use desktop applications while at home 98% use mobile devices to access the internet hence there is need for use of mobile devices such as tablets, laptops at when carrying out clinical activities.

There is need for integrating the DHIS with the nurses' work processes to improve patient care (avoid delay in patient care) and this can be achieved through a clinical module for nursing. For example blood tests might be sent to a laboratory where it will take three or more days to receive the results which results in delaying patient care, hence a link with laboratory services is important to receive the results as soon as they have been processed. PHWs had the opinion that there is need for employing more specialist doctors to avoid unnecessary delays as they also provide informed decisions such as referring patients to appropriate wards. Resources such as personnel, drugs and instruments, etc. for respective departments are essential for good processes to be in place and then complement them with a clinical module for nursing. The customisation of the clinical module for nursing has to involve users' requirements analysis so that the healthcare workers will not worry about issues such as security, for example, through sharing of summarized clinical notes with authorized healthcare professionals thereby supporting confidentiality [9].

The MoHCC authorities and the hospital's Chief Executive Officers support the deployment and use of healthcare applications by nurses. PHWS concurred that the MoHCC and the local health authorities expect them in using the clinical care module since they have already deployed the DHIS and SAP Healthcare solution and are partnering with local mobile network operators, software development houses and international partners such as European GSM networking companies [8].

Nurses agreed that the use of the clinical module for healthcare processes is useful in their jobs and will augment the accomplishment of tasks more quickly, for example, the hassle of searching a patient file is eliminated. In general, the use of healthcare applications directly linked to patient care was applauded as it was suggested it simplifies documentation of patients' outcomes, reduces medical errors and promotes quality care outcomes [32].

C. *User requirements analysis*

There is lack of user training since maintenance of clinical records in not comprehensive (medical history is either missing or missing), which results in fear of exploring new concepts by nurses as this might result in errors such as inaccuracy of patient records. The project leader of the company which deployed SAP for Healthcare solution confirmed that users were not trained on use of the clinical care module and its customisation to suit specific wards work processes was not done. Even though other modules of the SAP Healthcare solution are being used, the clinical part might not be user-friendly to nurses. The IT staff is using the platform for capturing patient data rather than supporting nurses in using the system. The use of the clinical module for nursing must actually be done by nurses while the IT staff gives support [11][23]. Effective use of the patient demographic information module will result in easy deployment and use of the clinical care module. The clinical module becomes obsolete are there is no proper user centred requirements gathering exercise hence nurses might protest against the use of the solution as they feel they were not involved and the application is not user-friendly. User requirements gathering is essential for successful deployment of an application. Installation and training of users is essential after deployment of the application. This must be followed by a transition phase of familiarizing with the new system and then data migration will be done by the healthcare workers while the consultants and IT staff play a monitoring role to address any challenges. The consultants must gradually withdraw their support unless there is need for help. The IT staff must provide support only where necessary while software providers must address major challenges which can be beyond the scope of the IT personnel [9][20].

D. *UTAUT constructs outcomes*

The UTAUT model was used to measure the variables facilitating conditions, performance expectancy, attitude towards the use of the platform, social influence and effort expectancy. To measure internal consistency (reliability of a series of items), Cronbach's Alpha technique was used and obtained a coefficient of 0.763 which matches with the acceptance benchmark of at least 0.70 [31]. On average, the variable performance expectancy resulted in a mean of 4.227 and a standard deviation of 0.923, social influence resulted in a mean of 4.19 and a standard deviation of 0.849, effort expectancy resulted in a mean of 4.09 and a standard deviation of 0.919 and facilitating conditions resulted in a mean of 3.803 and a standard deviation of 1.351. The results from the three constructs performance expectancy, social influence and effort expectancy show that respondents had a common opinion (95% Strongly Agree) while on facilitating conditions respondents had varied opinions.

95 percent strongly agree that the use of the clinical module is easy since there will be training of users before the use of the system and 'some' of their colleagues are already using desktop applications (DHIS/SAP Healthcare solution), thus they have the necessary skills in using healthcare applications. ICT professionals will be helpful in training them on how to use health informatics products and services though the PHWs feel that some nurses must be trained as IT Experts who reside in wards where the applications run rather than being dependent on the HIS Department personnel. The nurses must be trained to acquire certificates, diplomas or degrees in the Health Informatics field to directly support their work processes. The course in IT must be implemented in nursing curriculum and even during their post basic education.

10.2 percent of the respondents were male and all concurred to the use of the clinical module during patient care while 80.28 percent of females had the same notion as men. This conforms to previous researches that men perceive usefulness of a technology more than women [1][19]. 66.5 percent of the respondents had the qualification Post Basic Diploma of which 33% of them agree and 55.7% strongly agree to use of the module by nurses. Health workers who are above the age of 43 proved not to easily adopt and use clinical decision support systems as they felt it might exert pressure on how they normally carryout their duties, thus showing the element of resistance to change. Middle aged to younger generations who have the passion to learn and have higher social status tend to use a technology early compared to the elderly as proved by other researches Nurses with an experience of at least 2 years had a positive attitude towards the use of technology [19][34].

VI. Conclusion

The UTAUT model constructs predicted a moderate to high level of user acceptance and use of the clinical care module. The principal factors that affect the use of the clinical care module are facilitating conditions, social influence and performance expectancy. Healthcare institute authorities must improve healthcare staffs' performance expectancy towards use of the clinical care module and facilitating conditions for use of the platform. The clinical module has to be used as a blue-print for decision making though a clinician must give the final recommendation based on experience. Training of both nurses and HIS personnel is essential for the use of the clinical module after conducting thorough user requirements analysis. The system must have comprehensive patient data unlike the current platforms which have basically demographic data which is incomplete and do not have the clinical notes and other patient medical details. Nurses must be trained to attain professional courses in Nursing Information System related qualifications. Time taken while documenting patient profiles and other related information must be benchmarked not to conflict with daily work processes operations. Management buy-in (MoHCC, hospital Chief Executive Officers and Medical Superintendents) might result in successful implementation and use of the applications as they craft policies and strategies for the use of the clinical module [2].

Teamwork of all parties concerned is important such as the project manager of the consultancy firm delivering the SAP healthcare solution and the project manager of the healthcare institute thus creating the spirit of ownership and total commitment. Key decision makers must be role models in the use of the technology thus encouraging their subordinates. In future other Technology Acceptance Models can be blended with the UTAUT model to ensure full utilization of Clinical Decision Support Systems. There is need to explicitly apply the UTAUT model for the use of the clinical care module in perinatal care services as a software solution while the clinical areas are assessed and benchmarked for improving work processes.

The users of the system must be able to use it with ease and the applications must not be technically user centric, otherwise the platform must provide a detailed user manual. The customisation of applications must be in terms of patients' characteristics (such as age, gender and experience) and societal values centred and to match their level of aptitude (users of the system). Healthcare personnel needs to be trained and receive refresher courses on the clinical module applications upgrades and deployment of new programs as this might help in reducing resistance to change. The necessary infrastructure for rolling out of HIS is important and also government policies and regulations concerning the use of decision support systems, needs to be considered.

E-health applications are partially used by healthcare workers. Midwives do not have direct access to the DHIS application, they only get patient information they need from the HIS department staff. The midwives need to be authorized to use the system independently in wards since the HIS department closes at 5pm while wards are open 24/7. Nurses who are using the SAP Healthcare solution are making use of it for patient demographic data. There is need for value addition on the DHIS and SAP Healthcare Solution for the use of the clinical care module of nursing. The E-Health maturity model need to be applied for use of mobile devices during clinical healthcare delivery with the aid of the clinical care module and strategies have to be explored for reduced transition period between phases up to the transaction and transformation phases. In addition, a study on how clinical decision support systems can be simulated to reduce delay in healthcare delivery (patient flow) has to be done.

Acknowledgments

I would like to acknowledge my academic supervisors for their inestimable guidance and encouragement and for consistently being a source of reassurance and support. I would also like to thank the Medical Research Council of Zimbabwe and the Ministry of Health and Child Care (MoHCC) for their approval to carryout research in health institutions. My gratitude also goes to authorities and participants in the hospitals which took part in this study for giving their time despite having busy work schedules. Special mention goes to Mr. T.G Rebanowako for assistance with language editing.

References

[1] J. T. Ami-Narh, and P. Williams, 'A Revised UTAUT Model to Investigate E-health Acceptance of Health Professionals in Africa.' Journal of Emerging Trends in Computing and Information Sciences, 3(10), 2012, pp.1383–1391.

[2] C.J. Anumba, and K. Ruikar, e-Business in Construction. 1 st ed., West Sussex, United Kingdom: John Wiley & Sons, 2008.

[3] S.A. Brown, and V. Venkatesh, 'Model of Adoption of Technology in the Household: A Baseline Model Test and Extension Incorporating Household Life Cycle.' MIS Quartely, 29:4, 2005, pp.399–426.

[4] L.M.C. Buydens, and P.J. Schoenmakers, Intelligent Software for Chemical Analysis, 1993.

[5] M.M. Cruz-Cunha, I.M. Miranda, and P. Goncalves, Handbook of Research on ICTs and Management Systems for Improving Efficiency in Healthcare and Social Care. IGI Global, 2013.

[6] F.D. Davies, and V. Venkatesh, 'Measuring user acceptance of emerging information technologies: anassessment of possible method biases.' Proceedings of the Twenty-Eighth Annual Hawaii International Conference on System Sciences, vol 4, 1995.

[7] D. Dowding, N. Mitchell, R. Randell, R. Foster, V. Lattimer,and C. Thompson, 'Nurses' use of computerised clinical decision support systems: a case site analysis.' Journal of clinical nursing, 18(8), 2009, pp.1159–1167.

[8] S. Emmitt,and K. Ruikar, Collaborative Design Management, New York, USA: Routledge, 2013.

[9] J.O. Grady, System Requirements Analysis, New York, USA: Academic Press, 2006.

[10] P. Jalote, An Integrated Approach to Software Engineering, Springer Science & Business Media, 2013.

[11] M. Khosrow-Pour, Encyclopedia of Information Science and Technology, 1st ed., Pennsylvania, USA: Idea Group Inc (IGI), 2005.

[12] L.N. Kirk, M.L. Lewis, and J. Gaugler, Neuropsychiatric Symptoms in Mild Cognitive Impairment: Development and Testing of a Conceptual Model, ProQuest, 2008.

[13] G. Lanzolla, and F.F, Suarez, 'Closing the Technology Adoption-Use Divide: The Role of Contiguous User Bandwagon.' Journal of Management, 38(3), 2009, pp.836–859.

[14] D.F. McHaney, Information Management and technology, 2009.

[15] Medical Information Science Reference, E-Health and Telemedicine: Concepts, Methodologies, Tools, and Applications: Concepts, Methodologies, Tools, and Applications, Pennsylvania, USA: IGI Global, 2015.

[16] V. Michell, D.J. Rosenorn-Lanng, S.R. Gulliver, and W. Currie, Handbook of Research on Patient Safety and Quality Care through Health Informatics, Pennsylvania: IGI Global, 2013.

[17] E. Nikoi, K. Boateng, Collaborative Communication Processes and Decision Making in Organizations, IGI Global, 2013.

[18] K. Ong, Medical Informatics: An Executive Primer, 2nd ed, Chicago, IL, USA: HIMSS, 2011.

[19] S.T.M. Peek, E.J.M Wouters, J. van Hoof, K.G. Luijkx, H.R. Boeije, and H.J.M. Vrijhoef, 'Factors influencing acceptance of technology for aging in place: A systematic review. International Journal of Medical Informatics.',83(4), 2014, pp.235–248. Available at: <http://linkinghub.elsevier.com/retrieve/pii/S1386505614000173>.

[20] D. Peters, Building a GIS: System Architecture Design Strategies for Managers, Calfornia, USA: ESRI, Inc, 2008.

[21] C. Rokx, J. Giles, E.Satriawan, P. Marzoeki, P. Harimurti, and E. Yavuz, New Insights Into the Provision of Health Services in Indonesia: A Health Workforce Study, World Bank Publications, 2010.

[22] SAP, Transform patient care with our healthcare industry software, 2014. Available at: <http://go.sap.com/solution/industry/healthcare.html> [Accessed 12 Dec. 2015].

[23] G. Shelly, G. Gunter, and R. Gunter, Teachers Discovering Computers: Integrating Technology in a Connected World, 7th ed., Boston, USA: Cengage Learning, 2012.

[24] T.A.M. Spil, and R.W. Schuring, E-Health Systems Diffusion and Use: The Innovation, the User and the Use IT Model: The Innovation, the User and the Use IT Model, Idea Group Inc (IGI), 2006.

[25] Y. Sun, N. Wang, X. Guo, and Z. Peng, 'Understanding the Acceptance of Mobile Health Services: a Comparison and Integration of Alternative Models.' Journal of Electronic Commerce Research, 14(2), 2013, pp.183–200.

[26] J. Tan, and F.C. Payton, Adaptive Health Management Information Systems: Concepts, Cases, & Practical Applications: Concepts, Cases, & Practical Applications, 3rd ed., Jones & Bartlett Publishers, 2010.

[27] TechZim, Zim hospitals introduce a new e-Health system: How does it work?, 2014, Available at: <http://www.techzim.co.zw/2014/10/zim-hospitals-introduce-new-e-health-system-work/> [Accessed 12 Dec. 2015].

[28] The Herald, Chitungwiza Hospital now computerised, 2012, Available at <http://www.herald.co.zw/chitungwiza-hospital-now-computerised/> [Accessed 12 Dec. 2015].

[29] UNDP Zimbabwe, Promoting timely delivery of health services, 2014, Available at: <http://www.zw.undp.org/content/zimbabwe/en/home/presscenter/articles/2014/06/05/supporting-zimbabwe-to-deliver-timely-health-services-at-a-click-of-a-button.html> [Accessed 16 Jan. 2016].

[30] UNDP, Newsletter of The Global Fund Programme in Zimbabwe, March 2014.

[31] V. Venkatesh, M.G. Morris, G.B. Davis, and F.D. Davis, 'User Acceptance of Information Technology: Toward a Unified View.' MIS Quartely, 27:3, 2003, pp.425–478.

[32] V. Venkatesh, J.Y.L. Thong, and X. Xu, Consumer acceptance and use of information technology: extending the unified theory of acceptance and use of technology. MIS Quartely, 36(1), 2012, pp.157–178.

[33] N. Wickramasinghe, J.N.D. Gupta, and S.K. Sharma, Creating Knowledge-based Healthcare Organizations. Idea Group Inc (IGI), 2005.

[34] M. Wills, O. El-Gayar, and D. Benett, Examining Healthcare Professionals' Acceptance of Electronic Medical Records Using UTAUT. Issues in Information Systems, IX(2), 2008, pp.396–401.

[35] R. Wootton, and L. Bonnardot, Telemedicine in Low-Resource Settings, Frontiers Media SA, 2015.

An Integrated and Collaborative eHealth System for the Mental Health Services

Claudio Dario, Claudio Saccavini, Silvia Mancin, Giulia Pellizzon
Veneto's Research Centre for eHealth Innovation
Arsenàl.IT - Treviso, Italy
{presidente, csaccavini, smancin, gpellizzon}
@consorzioarsenal.it

Gerardo Favaretto
Local Health Authority n. 9
Mental Health Department, Cà Foncello Hospital
Treviso, Italy
gfavaretto@ulss.tv.it

Abstract—**As reported by the World Health Organization, depression is a disorder that can be reliably diagnosed and treated in primary care, with preferred treatment consisting of basic psychosocial support combined with antidepressant medication or psychotherapy, such as cognitive behaviour therapy, interpersonal psychotherapy or problem solving treatment. The European project MasterMind, started in March 2014 and, with 36 months duration, is an observational study aiming to implement collaborative care services and cCBT treatment in order to improve the care of people suffering from depression. The services will be tested in 15 European regions, for a total target of over 5280 patients and 141 professionals involved.**

Keywords - Integrated care, Depression, General Praticitioner, Collaborative Care, cCBT.

I. INTRODUCTION

Reducing quality of life and impairing social and personal relationships, a depressive disorder may start early in life and the course is often recurrent [1][2][3]. The World Heath Organization also said that depression is a treatable pathology [4], but most people with depression do not receive the care and support they need. Frequently, depression is not recognized and therefore not treated, and this exposes those affected to various negative consequences. In Italy, only the 29% of patients affected by depression receive treatment the same year in which the pathology appears [5].

The Local Health Authority n° 9 Treviso is one of the pilot of the MasterMind Project [6]. The European project MasterMind, started in March 2014 and, with 36 months duration, has the aim to implement collaborative care services and cCBT (computerized Cognitive Behaviour Therapy) treatment in order to improve the care of people suffering from depression.

MasterMind is an observational study and organizational improvement of the services, which aims to implement and disseminate care activities via tele psychiatry. The MasterMind Project was created to promote the development of guidelines for the application of tele psychiatry services in Europe in a safe, effective and efficient way. The project includes:

- The organization according to the model of collaborative care of the relationship between mental health services and primary care or General Practitioners (GPs) based on the use of tele psychiatry with data sharing, recruitment procedures agreed upon and use of videoconferencing;
- A service of Cognitive Behavioural Therapy available online (computerized Cognitive Behavioural Therapy - cCBT).

The services will be tested in 15 European regions and the trial evaluation will be developed using a rigorous method (Model for Assessment of Telemedicine MAST [7]) that follows the Health Technology Assessment (HTA) rationale.

In adherence to the project, the Local Health Authority n. 9 has the objectives to:

- Define an informative system (clinical database) shared between specialists services (Mental Health Department) and General Practitioners
- Supply to the General Practitioners diagnostic elements and guidelines for diagnosis and treatment
- Use computerized assessment tools (questionnaires and/or interviews) whose results can be accessible to all professionists involved in the network
- Provide possibility of advice, also with a videoconferencing system, for the cases discussion and for the evaluation of therapeutic plans.

The rest of the paper is structured as follows. In Section II, we mention the theories and methods used. Section III presents the results and discussion, and we conclude proposing next activities in Section IV.

II. THEORY/METHODS

Improving the collaborative care between the primary and secondary care and giving new tools to the patients to support them in the management of their disease is the core of the Veneto's pilot initiative. The collaborative care between General Practitioners and Mental Health Professionals has the scope to allow the progressive education of General Practitioners in the identification of first symptoms of depression and give to the primary care the support of highly skilled specialists when required during the treatment of a patient.

On the other hand, giving the specific tools to the patients, the system has the aim to support them in the management of their disease putting them in contact with their clinician, when necessary.

The project is on going and currently the videoconference service and the cCBT treatment have been implemented and are in use. At the end of the study (December 2016), the results of the implementation and introduction of the use of these services will be evaluated through an HTA method: the MAST will be applied with a particular focus on the organizational and economic aspects, in order to measure the impact on the organization involved, on the patients treated and on the operating clinical actors.

III. RESULTS/DISCUSSION

In accordance with the project objectives, the Local Health Authority n.9 of Treviso has created a network between General Practitioners and specialists through the territorial information service and videoconferencing (CCVC), for the early identification of depressive disorder and an effective treatment planning from the first access; and the implementation of a computerized cognitive behavioural therapy (cCBT) for patients with depression.

The Local Health Authority n. 9 – Treviso, in the Veneto Region, has implemented a new model of integrated care, improving the collaborative care between the primary and secondary care and giving to the patients the necessary support in the management of their disease. The actors involved are General Practitioners from two social district areas, psychologists and psychiatrists from the Mental Health Department of Treviso.

The new model includes two new services for what concerns the collaborative care: the videoconference tool and a new computerized therapy, the cCBT (computerized Cognitive Behavioural Therapy).

The videoconference tool is used by the General Practitioner to ask for support from the specialists. The General Practitioner can use the videoconference sessions to share the symptoms and the situation of his patients and define together the best follow up. Furthermore to permit the real integration of care, the Territorial Information System has been integrated with the primary care, giving to the General Practitioners the possibility to know the history of the patient, in every moment, and to have the necessary background to decide the right clinical path. Thanks to the new integrated relationship, the General Practitioners and the Mental Health Professionals share the same information about symptoms, drugs and actions taken and decide together the better way to take care of patients. Once a month, General Practitioners, care managers, and psychiatrists have video consultations discussing each individual patients' cases. This creates the opportunity for health professionals closely related to the case to discuss any problems with a psychiatrist whose role is more specialized and distanced from the case. Without videoconference, this level of cooperation and support is not feasible, because it would be too time consuming and not as efficient and focused.

The second service, the cCBT (computerized Cognitive Behaviour Therapy), is mostly a therapeutic service delivered through online sessions with a secure web-based online treatment platform that provides:

- Self-help modules that explain the situation the patient is living in and the relationship between his emotions and his daily life;
- Worksheets that actively involve patients regarding their moods, experiences, quality of sleep, planning for the future;

The duration of each module is about 30-45 minutes and the patient should complete one module per week. This tool supports the patient to deal with his disease, providing the method to recognize and change thought patterns, dysfunctional behaviours and perceived feelings, related to the disease of depression. The activities carried out through the modules and worksheets are intended to increase the capacity of people with depression to prevent relapse of depressive symptoms.

The new care model will be applied to a target of 200 patients followed for three or more months, to monitor the impact of their improved management and care, in term of organizational efficiency and clinical integration between different settings.

The clinician through a structured computerized questionnaire evaluates the patient, who refers to the General Practitioner to report a likely depression, that indicates if the patient will be recruited in the study. This assessment is shared, through the information system with the Mental Health Department who, subsequently, agrees on a time for a joint evaluation (by videoconference or at least through a phone discussion). If the two professionals agree on the diagnosis, they decide if the patient needs an intervention and/or a treatment plan. In the first case, the patient is paced in charge of the Mental Health Department, otherwise the General Practitioner supports the patient in the management of the disease. In the event of significant emergency, the General Practitioner has always the possibility to transfer the patient to the specialists' services.

In order to improve the collaborative care between the primary and secondary care and give new tools to the patients supporting them in management of their disease, the services proposed have the scope to allow the progressive education of General Practitioners in the identification of first symptoms of depression and give to the primary care the support of highly skilled specialists when required during the treatment of a patient.

Assessing the impact of the collaborative care model with videoconference and sharing of clinical data for patients with depression, the intervention aims to provide patients with high quality treatment in their immediate environment (General Practitioners' clinic), also extending the points of access for first assessment (General Practitioners, Mental Health departments, social health districts).

Currently, 30 clinicians and 70 patients are enrolled and the services presented are implemented and used by all.

During the study, data of the enrolled patients and professionals are collected in a central database; therefore, at the end of the study, some qualitative analysis are going to be done related to the organizations that provide the services. These data will be the basis for an HTA analysis (Health Technology Assessment) that will be made at the end of the project with the objective of assessing the impact of organizational, economic and social services proposed, with a view to large-scale deployment services.

The four objectives proposed from the LHA n. 9 and reporting at the beginning of the article, are all achieved: at today a specific informative system has been put in place and has been shared between specialists services (Mental Health Department) and General Practitioners involved in the project. In addition, the system has been used to collect data (questionnaires and/or interviews) that are accessible to all professionals involved in the network.

All the General Practitioners involved in the project have been trained through the videoconference sections and meetings done during the study, enhancing their expertise in diagnosis of depression and evaluation of efficient therapeutic plans.

IV. CONCLUSION AND FUTURE WORKS

Depression is a common mental disorder that can be long lasting or recurrent, substantially impairing an individual's ability to function at work or school or cope with daily life. At a most severe level, depression can lead to suicide. When mild, people can be treated without medication but when depression is moderate or severe they may need medication and professional treatments. Depression is a disorder that can be reliably diagnosed and treated by non-specialists as part of primary health care. Specialist care is needed for the proportion of individuals with complicated depression or those who do not respond to first-line treatments [8].

Through the continuous sharing of data, the course of treatment is carried out as cooperation between the GP, the psychiatrist and the patient. By increasing cooperation between the different healthcare actors, the new integrated and collaborative care model aims to delivery of treatment, care and learning.

REFERENCES

[1] L.J. Barney, et al., Stigma about depression and its impact on help-seeking intentions, Aust. N. Z. J. Psychiatry, 40 (1) (2006), pp. 51–54

[2] R.V. Bijl, A. Ravelli, Current and residual functional disability associated with psychopathology: findings from The Netherlands Mental Health Survey and Incidence Study (NEMESIS), Psychol. Med., 30 (3) (2000), pp. 657–668

[3] N. Titov, Internet-delivered psychotherapy for depression in adults, Curr. Opin. Psychiatry, 24 (1) (2011), pp. 18–23

[4] www.who.int/mental_health/en/ [retrieved in April 2016]

[5] www.salute.gov.it [retrieved in April 2016]

[6] www.MasterMind-project.eu/ [retrieved in April 2016]

[7] www.mast-model.info/ [retrieved in April 2016]

[8] www.who.int/topics/depression/en/ [retrieved in April 2016]

Barriers to Deploying Diabetes Self-management mHealth Services in the Chinese Market

Ying Li
Key Laboratory of Biorcheological Science and Technology
Bioengineering College, Chongqing University
Chongqing, China
e-mail: liyingberry@hotmail.com

Daidi Zhong
Key Laboratory of Biorcheological Science and Technology
Bioengineering College, Chongqing University
Chongqing, China
e-mail: daidi.zhong@cqu.edu.cn

Xiaolian Duan
Chongqing Academy of Science and Technology
Chongqing, China
e-mail: dxl@cast.gov.cn

Jianning Qu
the Second Affiliated Hospital of Chongqing Medical University
Chongqing, China
e-mail: 360113522@qq.com

Abstract—**Recently, a large number of mobile applications have been developed to address the needs of patients for diabetes self-management mobile health (mHealth) services. Given the explosion of mHealth technology enablers, increasing investment, and more favorable government policies, diabetes self-management mHealth services are generally seen as a promising domain by Chinese investors. However, the meta-analysis described here showed that even large numbers of cases and environmental factors did not result in a massive adoption or deep market penetration. The root causes of this were analyzed and there were five barriers, i.e., human resources, trust from clinical service providers, policies, functionality, and business models, preventing the adoption of this technology. This finding may serve as a reference for future decision-making with regard to the research, development, use, and policy-making related to mobile health services.**

Keywords-mHealth; diabetes; self-management; mobile applications.

I. INTRODUCTION

Diabetes mellitus is one of the most common chronic disorders in the world, and insufficient self-management can involve both considerable personal suffering and enormous costs. In 2013, diabetes caused 114 million deaths in the adult population of China. The estimated cost of managing people diagnosed with diabetes in 2014 was $612 billion [1]. Improving self-management is an important part of improving cost-effective patient-centered care and dealing with the growing health care challenge posed by diabetes [2][3].

Self-monitoring of blood glucose (SMBG) has been shown to be a useful tool in improving glycemic control in type 2 diabetes, helping patients make informed decisions in managing blood glucose [4]. Mobile health (mHealth) technology renders SMBG more flexible and efficient for the treatment of patients. Figure 1 presents a general architecture of such mHealth-based SMBG system.

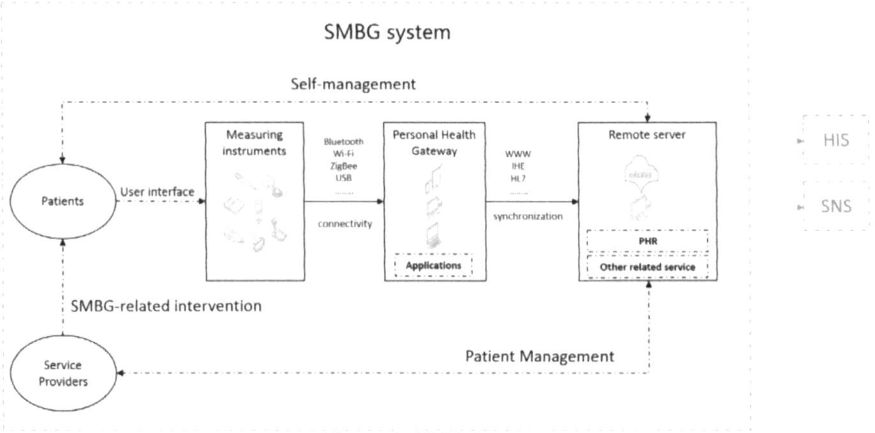

Figure 1. SMBG system

When practicing SMBG, patients normally monitor and manage their glucose level by themselves at home. Such activity may benefit from a SMBG system, which allows patients transmitting their data to service provider via an end-to-end data channel. Such a system is often constituted with one or more measuring instruments (e.g., a glucose meter), a gateway device (e.g., a mobile device) and a remote server. Patients can collect their glucose measurements and other context health data with glucose meter and other instruments. The common devices employed here are blood glucose monitor (BGM) and continuous glucose monitor (CGM). The remote server contains Personal Health Record (PHR) and other related services.

The established connectivity between measuring instruments and personal health gateway, and the connectivity between gateways to remote server, together populate an end-to-end data channel. Via this channel, the collected data can be transmitted from patients to service providers via uplink, and the instructions from service providers can be sent to patients via downlink. This allows patients and service providers to access the health data at any time. Service providers can further give appropriate interventions to patients based on certain data-driven strategy. Moreover, the external partners contain the Hospital Information System (HIS), the Social Network Site (SNS), etc.

The study [5] about the SMBG applications in EU (European Union) or foreign countries shows that (1) perceived barriers to use or continuous use, (2) perceived benefits of desired features of diabetes self-management, (3) facilitators to motivate use, and (4) information sharing with family, friends, and health professionals. The result shows that there is a problem about the usage of the SMBG applications in EU or foreign countries.

In recent years, the government of China has issued several policies that affect mHealth [6]–[8]. Top policymakers are targeting the development of portable health data collection devices, integration with the Internet and mobile Internet, and improvement of the level of automatic and intelligent health information services.

Under the support of government policy and the catalysis of the market demand, more and more companies and investment organizations have swarmed into the diabetes self- management application market. Such stakeholders are the key actors of the operation of online diabetes self-management.

In order to assess the status of adoption of diabetes self-management mHealth services in China, we conducted a meta-analysis [9], the meta-analysis was about the diabetes self-management related applications in China, and some aspects are compared, including the innovative functionalities, defects, prospects, conformance of standards, etc. And the following facts were observed: (1) the number of SMBG applications in China is low; there are only 78; (2) there is only 1 application whose download count exceeds 5000, and only 3 iOS applications have been scored by users; (3) the most common features of applications that have been studied include recording of blood health data, notification, and decision support. Obviously, the rate of adoption of diabetes self-management tools in the market is low, which seems to be contradicting to the continuously increasing investment interest in this domain.

In order to fully understand the root causes of this paradox, as well as the gains and gaps of the current Chinese mHealth-based SMBG industry, a thorough analysis was provided in Section 2, with regards to the barriers of its development and deployment.

II. BARRIERS TO THE DEVELOPMENT AND DEPLOYMENT OF DIABETES SELF-MANAGEMENT APPLICATIONS

A. *Lack of human resources to support diabetes self-management*

A sufficient number of service providers is needed to provide remote diabetes management services. This is a prerequisite to any good deployment of diabetes self-management applications. However, China lacks proper human resources for such purposes. This country is especially short of primary care service providers (e.g. GP, family doctors, family nurses). It has been reported that the amount of community-level service providers in 2010 was only 4.04% of the total amount of healthcare service professionals [10][11]. The same statistic in America, Britain, and Canada is 30–50%. There are 2.466 million registered primary care providers in China, which translates to 1.82 providers per thousand civilians. This ratio places China 80th among 193 countries, according the WHO. Until 2011, there were no dedicated personnel training system or official promotion channels for this group of providers. To solve this issue, the State Council issued guidance to improve the training of general practitioners (GP) [12]. It sets the goal of establishing a rigorous training system by 2020 to satisfy people's basic health service needs. It will be a long time before any substantial change will become observable.

Another reason for the shortage of service providers is that the current management policy for clinical practice does not fully allow clinicians to practice at multiple sites. In 1995, the Chinese government enforced the law establishing the national clinician registration system [13][14]. According to that law, clinicians can only practice at the point-of-care to which they were originally registered, and practicing at any other location is considered illegal. This situation is now changing because the State Council has recently issued a policy promoting multiple-site practicing [15][16]. However, the mobile Internet, where diabetes self-management applications are used, has not yet been acknowledged by this guideline as a permissible place to practice medicine. For this reason, further extensions to this policy are expected.

B. *Lack of trust from service providers*

The level of support from professional service providers is critical to accelerating the massive adoption of diabetes self-management mHealth services. When providing SMBG-related services, the service providers have to base their interventions on existing clinical guidelines [17][18]. These are generally considered the best practices in the field. However, the fact is that the currently available clinical guidelines were established mainly based on the medical studies conducted over controlled data within controlled environment, which is quite different from patients' self-generated data from ambient environments (including home and traveling). Those guidelines describe a SMBG process that is driven and managed solely by service providers (although the patient is the main actor); this is somewhat different from purely patient-dominant activities.

Theories underlying evidence-based medicine must be established based on clinical evidence. However, due to the lack of interoperability, regulation, and business impetus, no large-scale patient-oriented SMBG system has been established or used in the field. For this reason, the clinical researchers have no way to acquire useful research data, generated from patients' personal health devices in their personal environments in sufficient quality and quantity. Correspondingly, there has been no foundation upon which to create any clinical guidelines for SMBG activities. Without such guidelines, the service providers have generally hesitated to enter this market or to provide meaningful support for these SMBG-practicing patients. This can be seen as a classic chicken-egg puzzle: No deployment means no data, which means no research, no trust, no service, and no deployment. Solving this puzzle has become a core topic in today's SMBG industry.

Also, many users have the interest to use the apps, they said that "the app has made the SMBG more efficiently, it pontificated me to have medicine and record the glucose data".

There has recently been a rapidly increase in interest and investment in the Chinese mHealth market. Some personal connected health platforms are being established and deployed in many Chinese cities [19][20]. Many of them are designed for general health management services, but some of them do have the capability to support SMBG services. Domestic investors fully understand that mHealth is an emerging market, so they are patient enough to tolerate long revenue turn-around. Given the large population of China, this may provide the industry with an opportunity to collect enough data and to conduct proper medical studies before the true clinical efficiency of these applications can be proven and the dedicated best practices can be established. Hopefully, the puzzle will be solved in a few years.

C. Domestic regulatory policy regarding to the use of diabetes self-management applications

The diabetes self-management applications fall into a multidisciplinary domain between medical device and IT industry. The Food and Drug Administration of China (CFDA) and the Ministry of Industry and Information Technology (MIIT) are the corresponding regulatory bodies. However, none of them has published regulation policy dedicatedly for this multidisciplinary domain. The regulations issued by CFDA only cover situations inside hospitals, and they do not include the diabetes self-management applications outside hospitals. The regulations issued by MIIT only address Internet-based information services and promotes the development of the IT industry. None of them address the specific needs of the remote clinics and Personal Health Record (PHR) data utilization, which can be seen as another gap of this industry.

Currently, there is no clear legal definition of diabetes self-management applications. When using them, patients are unable to protect their rights in cases in which anything goes wrong (e.g., receiving the wrong prescription). Furthermore, due to the legal requirement that clinicians register at a certain physical point-of-care for all clinical work, the legal status of medical care provided online (such as via diabetes self- management applications) has remained murky. There is a true need to develop a proper regulatory policy for this type of technology and service, and there are currently no such policies in China.

The U.S. diabetes management platform developed by WellDoc [21] received FDA clearance. It is the first application approved by the FDA for the optimization of doctors' prescriptions. Unlike the U.S. FDA, the CFDA has not issued any dedicated regulation guidelines for mHealth. However, this does not necessarily mean there is no chance that it will get approval. A glucose meter device, Dnurse and its associated Internet-based service, both of which were developed by Beijing Dnurse Technology Ltd. [22], did get clearance from the CFDA. It operates in manner similar to WellDoc.

D. Functionality of diabetes self-management applications in China

In applications related to the diabetes self-management, most functionalities have conformed to those listed in the international standards and clinical guidelines. However, for some other types of functionalities, the situation is different. The missing key functionalities may lead to the low rate adoption.

Most applications provide no connectivity to measuring instruments. One possible reason for this is the obsolete mindset of domestic application developers and instrument manufacturers. When building their own products, their design logic is purely application-centric or device-centric rather than a holistic vision established over an interconnected infrastructure. In contrast, the leading tech company Google recently released Google Fit and Apple released Health Kit. Both of these applications leverage modern connectivity technologies to simplify or automate the heath data collection process. A similar trend is likely to appear in China in the future. It is only a question of when and how.

Despite of the benefits of PHR synchronization reported in some studies [23][24], the rate of adoption of such applications has small, partially because of end-to-end usability issues. Our interpretation of the possible reason is: gateway device vendors are quite dominant in current Chinese mHealth market (including the ecosystem of diabetes self- management mHealth services), the vendors have focused mainly on data synchronization with gateway devices, rather than on the cloud platform. Improving the aforementioned usability issue will require joint effort from all the stakeholders (operators, integrators, medical device manufacturers, application developers, healthcare providers, etc.) to establish a data synchronization channel.

E. *Business models of diabetes self-management applications in China*

According to the results of our meta-analysis, the business models of diabetes self-management applications in China were found to be less diverse. We have observed three categories of business models in China: (1) Consumers as payers; (2) device vendor providers as the payers; (3) service providers as payers.

In contrast, the variety of business models in developed countries is much better. For example, the Zocdoc [25] is free to patients but not to doctors. The Welldoc [21] charges insurance companies and cooperates with pharmaceutical companies in the sale of applications to hospitals. Vocera [26] charges hospitals.

The current business models in China have not incorporated hospitals, pharmaceutical companies, commercial insurance companies, and doctors. The current business models have not fully engaged all the stakeholders of the ecosystem of the diabetes self-management mHealth services. No operational infrastructure for diabetes self-management mHealth services has yet been established. All of the observed business models are purely technology-driven or product-driven, rather than built from the perspective of the stakeholders. Correspondingly, such business models cannot properly satisfy the needs of all potential stakeholders, thus cannot adapt to the quickly evolving business environment. There is an immediate need in the Chinese market for business models with enough comprehensiveness and sustainability for diabetes self-management applications. It is not practical to directly copy business models from developed countries. A local process must be developed to suit the actual situation in the domestic market. One of the future trends is to acquire large quantities of user data and to trigger innovative business projects based on this big data.

III. CONCLUSION

We observed that the low adoption rate of the diabetes self-management mHealth services was not consistent with the high level of interest in the domain. In this paper, a root cause analysis is performed to investigate barriers to technology adoption. As a result, five barriers to the development and deploying of applications in China have been identified and elaborated. There is a shortage of primary care service providers and those who are available are not allowed to practice at multiple sites. This leaves diabetes self- management applications without the support of doctors. The lack of validation from clinical data and large-scale data support leaves doctors reluctant to recommend diabetes self- management applications to patients. Without the protection of relevant regulations, patients cannot protect their rights in cases of harmful error (e.g., receiving the wrong prescription). Undeveloped functionalities leave doctors and patients unable to use the applications or realize actual diabetes self- management. The current business models are not diverse enough and do not incorporate hospitals, pharmaceutical companies, commercial insurance companies, and the doctors, leaving patients and doctors unable to reap the benefits of the system.

These findings may serve as a reference for future decision-making regarding the research, development, deployment, and policy-making related to these applications. Stakeholders may take action to reduce barriers to the development and deployment of diabetes self-management applications for public consumption.

ACKNOWLEDGMENT

Partial funding was provided by Chongqing University (Project No. CSTC2013JCSF10034 and No. 2015BAI01 B00). As the sponsor of this research, Chongqing University was not involved in the review or approval of the manuscript for publication.

REFERENCES

[1] IDF. http://www.idf.org/worlddiabetesday/toolkit/gp/facts-figures. [retrieved: 2.2016]

[2] S Norris, M Engelgau, and K Narayan. “Effectiveness of Self-Management Training in Type 2 Diabetes A systematic review of randomized controlled trials”. Diabetes Care, vol. 24, no. 3, 2001, pp. 561-587.

[3] S.L. Norris, J Lau, S.J. Smith, C.H. Schmid, and M.M. Engelgau. “Self- management education for adults with type 2 diabetes: a meta-analysis of the effect on glycemic control”. Diabetes Care, vol. 25, no. 7, 2002, pp. 1159-1171.

[4] J Tran, R Tran, and J.R. White. “Smartphone-Based Glucose Monitors and Applications in the Management of Diabetes: An Overview of 10 Salient "Apps" and a Novel Smartphone-Connected Blood Glucose Monitor”. Clinical Diabetes, 2012.

[5] W Peng, S.P. Yuan, and B.E. Holtz. “Exploring the Challenges and Opportunities of Health Mobile Apps for Individuals with Type 2 Diabetes Living in Rural Communities. Telemedicine and e-Health. doi:10.1089/tmj.2015.0180.

[6] The State Council. “"Twelfth Five (2011-2015)" National Strategic Emerging Industry Development Plan” .http://www.gov.cn/xxgk/pub/govpublic/mrlm/201207/t20120720_65368.html. [retrieved: 2.2016].

[7] The State Council. “The Issuance of Biomedical Development Plan”. http://www.gov.cn/zwgk/2013-01/06/content_2305639.htm. [retrieved: 2.2016].

[8] The State Council. “Opinions on Promoting the Development of Health Services Industry”. http://www.gov.cn/zwgk/2013-0/14/content_2506399.htm. [retrieved: 2.2016].

[9] Y Li, J Tan, B.Z. Shi, D.D. Zhong, X.L. Duan, X.L. Li, and J.N. Qu. “A Study of ICT-Powered Diabetes Self-management Systems in China”. JMIR Diabetes (forthcoming).doi:10.2196/diabetes.4475 (in press)

[10] Chinese Ministry of Health. “2011 China Health Statistics Yearbook”. http://www.moh.gov.cn/htmlfiles/zwgkzt/ptjnj/year2011/index2011.html. [retrieved: 2.2016].

[11] Chinese Ministry of Health. “2011 Statistical Bulletin of health development”. http://www.moh.gov.cn/zwgkzt/pnb/201204/54532.shtml. [retrieved: 2.2016].

[12] The State Council. "Guidance on the Establishment of General Practitioner System". http://www.gov.cn/zwgk/2011-7/07/content_1901099.htm.[retrieved: 2.2016].

[13] The State Council. "The Medical Practitioners Law in China" (in Chinese) .http://www.gov.cn/banshi/2005-05/25/content_973.htm. [retrieved: 2.2016].

[14] Chinese Ministry of Health. "The Interim doctors practicing registered Measures" (in Chinese) .http://www.gov.cn/banshi/2005-08/02/content_19342.htm. [retrieved: 2.2016].

[15] The State Council. "The Focused Task in Deepening Health System Reform in 2014". http://www.gov.cn/zhengce/content/2014-05/28/content_8832.htm. [retrieved: 2.2016].

[16] The Ministry of Public Health of China. "The notice on expanding the scope of practicing pilot" (in Chinese).http://www.moh.gov.cn/mohyzs/s3578/201107/52434.shtml. [retrieved: 2.2016].

[17] W.P. Jia. "Chinese Glucose Monitoring Clinical Guideline (2011 edition)" (in Chinese). Chinese Journal of the Frontiers of Medical Science (Electronic Version), vol. 04, 2011, pp. 62-72.

[18] W.P. Jia. "Chinese Continuous Glucose Monitoring Clinical Guideline (2012 edition)" (in Chinese) Chinese Journal of Diabetes, vol. 4, no.10, 2012, pp. 582-590.

[19] Baidu Jiankang. http://jiankang.baidu.com/. [retrieved: 2.2016].

[20] Ali Jiankang. http://www.alijk.com/. [retrieved: 2.2016].

[21] WellDoc. http://www.welldoc.com/. [retrieved: 2.2016].

[22] Dnurse. http://www.dnurse.com/. [retrieved: 2.2016].

[23] D Wiljer, S Urowitz, E Apatu, C DeLenardo, G Eysenbach, and T Harth. "Canadian Committee for Patient Accessible Health Records. Patient accessible electronic health records: exploring recommendations for successful implementation strategies". J Med Internet Res vol. 10, no. 4, 2008, pp. e34.

[24] C.P. Tang, et al. "Personal health records: definitions, benefits, and strategies for overcoming barriers to adoption". Journal of the American Medical Informatics Association, vol. 13, no. 2, 2006, pp. 121-126.

[25] Zocdoc. https://www.zocdoc.com/. [retrieved: 2.2016].

[26] Vocera. http://www.vocera.com/. [retrieved: 2.2016].

Results of the Australian CSIRO National Multi-site Trial of At-home Telemonitoring for the Management of Chronic Disease

Branko G. Celler, Marlien Varnfield, Jane Li, Rajiv Jayasena, Ross Sparks, Surya Nepal, Leila Alem, Julian Jang-Jaccard and Simon J. McBride

CSIRO eHealth Research Program,
Australian eHealth Research Centre
Brisbane, Queensland, Australia
email: b.celler@unsw.edu.au

Abstract—**A clinical trial of telehealth services was carried out along the Eastern seaboard of Australia from Townsville in the North to Launceston in Tasmania over a period of 18 months from mid-2013 to end-2014. Patients were selected based on their history of hospitalization for their chronic conditions. A Test group of 114 patients was supplied with at home telemonitoring equipment and 173 patients were enrolled as a matched control group who were subjected to normal care. The impact of telemonitoring over one year following the start of telemonitoring, was analyzed using linear regression and analysis of covariance (ANCOVA). Mortality was reduced by >40%, the rate of hospitalization was reduced by 53% and length of stay was reduced by 7.5 days. The rate of expenditure on medical services was reduced by 46.3% and that for Pharmaceutical expenditure by 25.5%. There were significant differences in results for patients with Heart Failure, those with Lung Disease and those with Diabetes. The effective introduction of telemonitoring was dependent on local factors such as workplace culture and capacity for organizational change management.**

Keywords- telehealth; home telehealth; home telemonitoring; chronic disease management; clinical trial protocol; BACI design; case matched control design.

I. INTRODUCTION

In industrialized nations, approximately 70-78% of healthcare budgets are spent on the management of chronic disease or its exacerbation. As the population ages the burden of chronic disease will increase and place healthcare budgets under increasing strain. Telehealth services have been demonstrated as an effective innovation in international contexts, but there are low levels of evidence from Australian studies.

This study analyzed whether the introduction of at-home telemonitoring services for the management of chronic disease in the community reduces patient use of the health system and improves mortality, healthcare outcomes and patient quality of life. We also explored the issues and challenges in deploying telemonitoring services in the community.

This trial was designed to create a robust evidence base for these key success factors and to demonstrate an effective and scalable model for Internet-enabled telehealth services in Australia. Armed with the insights provided by this evidence base, policy makers may have much of the data they require to implement funding models and create a sustainable Telehealth services sector in Australia.

II. AIMS

The project bbjectives were to demonstrate and document how telehealth services can be successfully deployed across Australia, by piloting services in five different settings across five states, with a range of health service providers, including Local Health Districts, Medicare Locals and not for profit community organisations. This was carried out by deploying and demonstrating the operation of Telehealth monitoring in a multi-site multi-state case matched control trial (Before-After-Control-Impact (BACI) design) of chronically ill patients living in their own homes in the community. This has never previously been attempted in Australia. Specific aims included;

- Provide evidence that at home telemonitoring has the potential to reduce unscheduled admissions to hospital compared to the control group.
- Provide evidence for an impact on hospital admissions, mortality, clinical events and symptoms and improvements in functional measures and patients' and carers' experiences with care.
- Evaluate health economic benefits
- Evaluate impact on clinical work force availability and deployment
- Evaluate impact of human factors (acceptability, usability by patients, carers, nurses, General Practitioners (GPs) and administrators)
- Evaluate impact of workplace culture and capacity for organizational change management.
- Derive clinical and health economic evidence on how Telehealth services can be scaled up nationally to provide an alternative cost effective health service for the management of chronic disease in the community.

III. METHODS

The clinical trial design has been previously reported in the literature [1] and will only briefly be summarized here.

The trial was carried out at five different sites representing two different models for the management of chronic disease in the community, one Hospital based and the other Community based. This allowed the analysis of site specific differences in workplace culture, organizational change management and staff and management capabilities, that contribute to differences in measured health, social and economic outcomes.

The trial design required 25 test patients and 50 case matched control patients at six sites in five states and Territories. Test patients were supplied with state of the art Telehealth technology in the home for the monitoring of vital signs, delivery of clinical questionnaires and messaging between patients and carers. Test and Control patients were closely matched using a range of clinical, demographic and socio-economic criteria. Before and after data was available from national data bases on medical expenses (MBS data) and pharmaceutical expenses (PBS Data), as well as hospitals admissions and length of stay (LOS) from National Hospital RoundTable data.

A preliminary graphical analysis of both PBS and MBS data, using the MATLAB function *normplot* as well as the Chi-square goodness of fit test indicated that the data were not normal. Both lognormal and sqrt transformations were found to be effective in normalizing the data. The sqrt transformation was chosen as a little better, and applied to data before linear regression analysis was carried out.

MBS data were summed over 30-day intervals back approximately three years (36x30-day intervals) from the start of intervention and forward by almost one year (12x30-day intervals). When a Test patient had two controls, the data for both controls were averaged.

Ignoring missing points, data were then averaged and the 95% confidence limits calculated across all rows for each time interval. Before and after data were generally varying over time, and were analyzed using analysis of covariance (ANCOVA) a general linear model, which blends ANOVA and linear regression.

IV. Results

Two sites in the Nepean Blue Mountains Area were ultimately merged into a single site for logistical reasons. Following dropouts, withdrawal of consent and patients rejected because of poor or missing data, 100 Test patients and 114 Control patients were analyzed in detail. Out of these, 67 were males and 33 were females.

The average age of patients was 71.9±9.4 years. There was no significant difference in age between males and females or Test and Control patients. Test patients were monitored for an average of 276 days with 75% monitored for more than 6 months.

TABLE I: LINEAR REGRESSION AND ANCOVA ANALYSIS OF SQRT(MBS) DATA

	BEFORE	AFTER		BEFORE	AFTER
	Slope	Slope	Sig	Intercept	Intercept
CONTROL	0.05098	-0.03953	0.1	12.58	12.98
	(0.0293, 0.0727)	(-0.1305, 0.0515)		(12.13, 13.02)	(12.29, 13.66)
TEST	0.0919	-0.2729	<0.001**	14.06	14.44
	(0.0625, 0.1213)	(-0.4236, -0.1222)		(13.47, 14.66)	(13.33, 15.55)
P	0.0268*	0.009**			
DIFF (Control - Test)	-0.9446	3.916	0.1025	-55.38	-30.91
	(-2.073, 0.1839)	(-3.251, 11.08)		(-78.71, -32.05)	(-83.66, 21.84)

In the period of 100 days, prior to the onset of telemonitoring, there were only minor differences in the number of GP visits (P=0.04) and no significant differences in number of visits to specialists, number and cost of medications prescribed or number and cost of laboratory tests carried out.

ANCOVA analysis was carried out on Test and Control patient data and control-test differences. A matrix of results with slopes, intercepts and P values is shown in Table 1 for all Test patients. Graphical data of sqrt(MBS Costs) against time is shown in Figure 1.

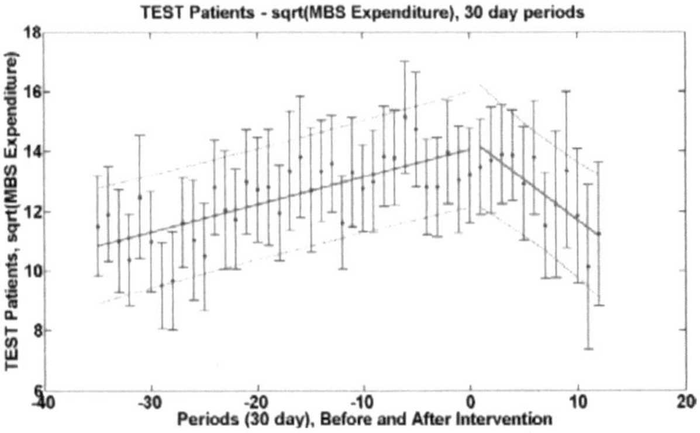

Figure 1. sqrt(MBS Costs) plotted for Test patients before and after intervention with 95% Prediction Intervals shown as red dotted lines

Similar analysis was carried out for PBS data as well as for different patient cohorts and different disease conditions. MBS data showed a 46.3% reduction in the rate of expenditure after one year and a 23.5% saving of $611. PBS savings were more modest with rates of expenditure falling by 25.5% and savings of 11.5% or $354 over one year.

Detailed hospital data was available for 53 Test patients and 64 Control patients. A similar ANCOVA analysis carried out on data averaged over 100 day intervals, revealed a 53.2% reduction in the rate of admission and a saving of approximately 23.8% or 0.67 admissions over the year following the intervention. Rates of LOS were reduced by almost 68% resulting in a 33.8% saving over the year of 7.5 days. Mortality for Test patients over the trial period was reduced by >40% relative to their Controls.

V. Conclusion

Results of the CSIRO National Telehealth Trial have been briefly summarized. Telemonitoring of patients with a range of chronic conditions has been shown to reduce expenditure on medical services and pharmaceutical items as well as significantly reduce the rate of hospitalization and length of stay. Mortality was also significantly reduced. Patients found the telemonitoring equipment easy to use and compliance rates >50% were achieved for daily measurement of vital signs. There were significant differences observed between different patient cohorts and different chronic conditions.

References

[1] B. G. Celler et al. "Design of a multi-site multi-state clinical trial of home monitoring of chronic disease in the community in Australia." BMC Public Health, 2014, 14(1), pp.1270-1285.

Distributed Case Management in the Public Health Area

Octavian Purcarea, Philip Cummings, Doina Patrubani, Cristian Taslitchi
European Commission- DG Employment, Social Affairs & Inclusion
Brussels, Belgium
E-mails: octavian.purcarea@ec.europa.eu, phil@vmns.co.uk,
dpatrubani@gmail.com, cristian.taslitchi@gmail.com

Abstract **– The free movement and mobility of citizens is a fundamental principle of European Union (EU). Currently, the data exchanges between social security institutions, including healthcare institutions are mostly paper-based. As a consequence, the process is time consuming, open to error and lengthen the resolution time for the citizens. Efficient and effective administrative cooperation between the institutions is, as a consequence, critical. Therefore, EU regulations triggered the European Exchange of Social Security Information (EESSI) project for the European Commission (EC) to provide the common secure framework that will facilitate electronic exchanges between relevant institutions. The previous attempts of implementing a similar system failed mostly because the clerks are accustomed to work with paper documents and the complex paperless solutions were perceived by the non-technical end-users as a major chance. The clerks need a system able to handle a partial snapshot of the European case, developed around their current social, communication and professional context, able to manage operational processes and, at the same time, to provide an efficient decision support. The present article describes the EESSI response to the main technical challenges encountered: the decomposition of business processes in order to provide a distributed case management solution, composition of business processes in order to provide standard functions across all case types, formalization of EESSI Business Messaging Protocol (BMP) in order to provide semantic and syntactic interoperability and case management visualization.**

Keywords-Public health data visualization; guidelines; timeline; case management; Business Process Modelling Notation (BPMN).

I. INTRODUCTION

Organisations have always been using the case paradigm to review and resolve various workflows like investigations, service requests and more. Typically, there are documents and artefacts associated with a case that are reviewed by authorised people in order to reach a resolution and close the case.

The case management model can be applied to diverse processes across a wide spectrum of industries and government agencies. The normative practice that leads to resolving and closing the case used to be encapsulated in process and procedure handbooks, as well as “stored” in the clerks’ community experience and best practice knowledge wealth.

The next wave of case management, after the paper-based one and often in parallel with it, was workflows that sprang from last-generation software.

In many industries, and most particularly in the healthcare one, many organisations still find themselves mired in IT systems that support only limited, if any, coordination across programs. Dating back to the 1980s, organisations were tethered to a legacy of large, disconnected transfer systems that took years to build, were outdated at implementation, and could not interoperate with one another. In the past, categorical funding requirements effectively dictated separate infrastructures.

New coordinated models ask healthcare organisations to work together and help connect people to services more efficiently. In the past several years, as technology has rapidly evolved, policymakers have begun to endorse the operating models that support a vision of distributed dynamic case management. These architectural guidelines recognize the value of a service-oriented approach to IT, in which technology components can be reused across organisations and programs that have similar business processes. By using dynamic case management applications, agencies can reduce the cost and complexity of both acquiring and supporting business applications.

Across the health care industry, information systems have much to offer in managing costs and in improving the quality of care. In addition to the embedded role of information technology in clinical and diagnostics equipment, the systems are uniquely positioned to capture, store, process, and communicate timely information to decision makers for better coordination of healthcare at both the individual and population levels.

At the most general level, a striking feature of the healthcare industry is the level of diversity that characterizes patients (e.g., physical traits and medical history), professional disciplines (e.g., doctors, nurses, administrators, and insurers), treatment options, healthcare delivery processes, and interests of various stakeholder groups (patients, providers, payers and regulators).

The healthcare delivery setting is characterized by a tension between the need for orderly routines and the need for sensitivity to variation in local conditions. As such, the current market offering in healthcare in its diverse organizational and regulatory settings covers several orientations, among which:

- ***Clinical Information Systems*** (CISs), which convert the medical data in relevant information about the patient’s health status. The current CISs' market covers most of the healthcare provider operational needs in regard to clinical services with some degree of support for point of care clinical decisions. The Graphical User Interface for documenting clinical cases is centred usually on patient and patient banner in order to help the medical personnel to easily identify the current patient clinical context. This ensures a comprehensive standard-based approach in regards to CISs functionalities, which is regulated by the world's leading medical informatics organizations: HL7 EHRS [1], CCHIT [2], Eurorec [3], etc. In this context, in order to analyse the current state of the art, we need to structure the existing knowledge in three domains: clinical pathway visualization, relevant medical data visualization (medical and administrative) and the combination of these two - medical data in clinical context.

Clinical pathway visualization should be considered from two points of view: from the point of view of clinical pathway encoder or designer, and from the point of view of the one who is executing the patient current clinical pathway requested actions. Projects/solutions like Protégé [4], Tallis Toolset [5], GUIDE [6], GLARE [7], VisiGuide [8], AsbruView [9], etc., are suitable for encoding and/or execution of a clinical pathway, but with limited adoption by the healthcare providers.

Not necessary in relation with the clinical context, projects like Graphical Summary of Patient Status [10], Time Lines and LifeLines [11], PatternFinder [12], KNAVE and KNAVE-II [13], VISITORS [14], VIE-VISU [15], Interactive Parallel Bar Charts (IPBC) [16], Gravi++ [17], and others moved in the direction of *visualizing the medical relevant data*.

There are also very few combined approaches of medical data in clinical context: Guideline Overview Tool (GOT) [18], Midgaard [19], CareVis [20], NHS Common User Interface [21], Visual-D [22], but most of them failed to be widely adopted.

- ***Patient Case Management Software***, such as FAM Care Human Service [23] by Global Vision Technologies, ClientTrack [24] by ClientTrack, Ahshay [25] by DataCare (the latter focusing on compensation industry better management of medical treatment and billing), Allscripts Care Management [26] by Allscripts, Penelope [27] by Athena Software, PracticePal [28] by PracticePal, etc. Such software are customised for the healthcare industry and cover all or a specific mix of features like: Activity Tracking, Assessment Notes, Billing & Invoicing, Calendar Management, Candidate Identification, Case List Management, Medical History Records, Patient Records, Referral Management, Treatment Planning, etc.
- ***Dynamic Case Management generic applications***, which need customisation to match healthcare industry specific requirements and context. Some of the most significant software providers are: Pegasystems with Pega Dynamic Case Management [29], Be Informed with Be Informed Business Process Platform [30], Kana Software with Kana Enterprise [31], IBM with IBM Case Manager [32], Isis Papyrus with Papyrus Platform [33] (including Framework Solution for ACM), Appian with Appian BPM Suite [34], OpenText with OpenText Cordys [35], OpenText BPM Everywhere [36], OpenText Process Intelligence [37], OpenText Cordys and Process Component Library [38], EMC with EMC Documentum xCP [39], Kofax with Kofax TotalAgility [40], Whitestein Technologies with Living Systems Process Suite [41], DST Systems with AWD10 [42], Oracle with Oracle BPM Suite 12c [43], and Hyland Software with OnBase [44].

The applications listed are characterised by various degrees of strong design time case management combined with strong runtime case management support use cases. There are two variations of Dynamic Case Management generic applications, variations described below.

Strong design time case management – capability that assumes that 90% or more of what the user will do is developed, tested, and deployed prior to user getting started. Case workers have less flexibility, e.g., adding new tasks or involving other users, and the overall process flow is well defined and more repeatable. These use cases tend to be more production oriented; for example, managing exceptions for financial transactions.

Strong runtime case management support – use cases where work is highly variable. The way in which the case unfolds over time is far less predictable. The case view is altered by user actions and system events, with users who are able to add tasks, processes and participants on the fly at the point of need.

The requirements that we had to meet for the EESSI project were focused on cross-European Member State cooperation between social security institutions through an electronic platform capable of supporting the current and future ability of all social security institutions to connect and fulfil their legal obligations of social security coordination through electronic exchange. Subsequent to market offer analysis, it was determined that the optimal approach was for the solution to be designed and developed in-house. The challenges and the way responses were elaborated to meet them are described in the next sections.

The paper structure, section by section, is presented below: Section II describes the business area that we are focusing on, public health and social security data exchange, Section III describes the system principles and high level architecture, Section IV describes the main challenges in the project implementation, Section V presents the response to the previously described challenges, Section VI refer to the project adoption, Section VII brings up the conclusions and future work and Appendix 1 depicts the system main features.

II. PUBLIC HEALTH DATA EXCHANGE IN THE CONTEXT OF EESSI

Better cooperation between social security institutions is a necessity in an increasingly mobile society in order for EU citizens to exercise their right to free movement and secure their social security rights.

The EU provides common rules to protect social security rights of citizens when moving within Europe. The European rules of coordination make sure that social security institutions of the EU plus Iceland, Norway, Liechtenstein and Switzerland, all communicate with each other to ensure social security rights are addressed correctly.

Currently, these data exchanges between the social security institutions are mostly paper-based and as a consequence are time consuming, open to error and lengthen the resolution time for the citizens, partly due to the method of exchange.

Efficient and effective administrative cooperation between the institutions is therefore critical and as a consequence the revisions of the European rules of coordination that came into force on May 1st, 2010 stated that "*The transmission of data between the institutions or the liaison bodies shall be carried out by electronic means either directly or indirectly through the access points under a common secure framework that can guarantee the confidentiality and protection of exchanges of data.*"[45].

Therefore, the aforementioned requirement of the EU regulations triggered the EESSI project for the EC to provide the common secure framework that will facilitate electronic exchanges between relevant institutions.

The overall vision for the EESSI project is to deliver an electronic platform that will support the current and future ability of all social security institutions to connect and fulfil their legal obligations of social security coordination through electronic exchange.

The EESSI platform has to enable secure exchanges for all relevant business messages that guarantee confidentiality, integrity and availability, and to have a sufficient level of validation. It should ensure wherever possible the right message is sent to the correct recipient on the first occasion at a time that is suitable to all institutions and contributes to the optimization of social security coordination.

EESSI is a cross-sectorial platform with competences in the area of public health. The main EESSI use cases in the area of public health are the following:

- Entitlement for short or long term healthcare related benefits;
- Validation of the person's right to healthcare related benefits during his/hers temporary stays in another Member State;
- Establishing the reimbursement rates for healthcare services;
- Cost reimbursement based and fixed amounts reimbursement or healthcare services;
- Certificate the incapacity of work and cost reimbursement for incapacity of work;
- Request for medical examinations or administrative checks, etc.

III. PRINCIPLES AND HIGH LEVEL ARCHITECTURE

Figure 1 below, provides a high level view of the conceptual architecture. This section will provide an explanation of the application domains in the conceptual architecture.

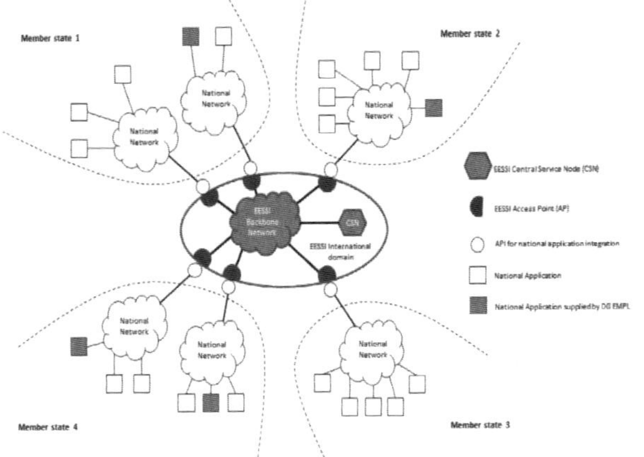

Figure 1. Conceptual architecture diagram

The diagram depicts the conceptual principles on how the inter-connection is achieved:

- A data exchange network interconnecting national administrations;
- National institutions are linked via national networks to the EESSI Access Points;
  -AP establishes the border between the national and international domains of EESSI end-to-end network. The APs are the gateways that enforce the stateless messaging protocol; they check that the structure and the semantic of the EESSI business messages are correct. A stateless protocol is a communications protocol that treats each request as an independent transaction that is unrelated to any previous request so that the communication consists of independent pairs of request and response. A stateless protocol does not require the server to retain session information or status about each communications partner for the duration of multiple requests. In contrast, a protocol, which requires keeping of the internal state on the server is known as a stateful protocol.

The main principle in building the EESSI platform is *"smart endpoints and dumb pipes"* [46] [59], which means that in most of the cases, when a National Application receives a request, it will apply the logic as appropriate and will produce a response and the AP will perform a minimal stateless validation. The diagram depicts the two fundamental domains of the EESSI network:

- The International Domain that hosts components, which are common to all participant countries. It is itself divided in two sub-domains: the Central Service Node (CSN) – designating the components that will be hosted centrally (e.g., hosted by EC) and the AP – a domain that holds components common to all countries, developed centrally and assumed to be hosted within each participant country. The components under the CSN and the APs, connected electronically as one system environment, constitute the EESSI Platform, a secure, reliable, pan-European data exchange platform;

- The National Domain that hosts national specific elements of the network can also be divided in two sub-domains: the National Application (NA) and National Gateways (NG).
  The NG are the National components that specifically integrate NA with the international domain and the Institutions' Domain – where the NA of the social security institutions resides.
  The NAs are the actual "client"; being themselves direct instruments of the "end-users", the clerks.

For a faster adoption of the EESSI platform and in order to help the Member States (MS) to provide better services for the citizens, the EC decided to provide an open source Distributed Case Management Solution called Reference Implementation for National Application (RINA). RINA consists in a collection of infrastructure and communication services, foundation, repository and publishing services, business, integration and user interface services, which will provide for clerks and their organizations, the tools to implement the EESSI data exchange protocol based on Structured Electronic Documents (SED).

IV. CHALLENGES

The main challenges are generated by the fact that EESSI is a peer-to-peer network where the International Domain of the platform, especially the Apps, are just the enablers of the communication between NAs, completely transparent from the business perspective.

Formalization of EESSI vision started about 10 years ago, and part of the project challenges at that time are still present in the nowadays industry:

- *Formalization of EESSI Business Messaging Protocol* in order to provide semantic and syntactic interoperability.
  One of the main outcomes of EESSI is the interoperability standard. The magnitude of the project is emphasized by the following facts: multi-sectorial data-exchange, around 110 businesses use cases (BUC) with at least two application roles involved, and around 320 business documents/SEDs.
- *Decomposition of business processes* in order to provide a distributed case management solution and composition of business processes in order to provide standard functions across all case types.
  All the EESSI business processes/case types are foreseen to be distributed and they involve from the applicative software perspective multiple application roles (e.g., case owner, counterparty, liaison body, etc.). Being a distributed system, the decoupling of the application roles in EESSI needs to be aligned to the EC guidelines in terms of messaging, more precisely, web services and ebMS3 [47] - AS4 profile [48]. Most of the Business Process Management Systems [49] (BPMS) are able to decompose the processes but they are not natively able to decouple the application roles through web services. The composition of business processes in order to provide standard functions across all case types is available in the BPMS tools build on top of Business Process Modelling Notation (BPMN) [50] [2] but the complexity and the multitude of the administrative processes in EESSI makes BPMN as it is difficult to use.
- *Case visualization* and case visualization in distributed environment.
  It is obvious that the level of adoption of case management solutions in the institutions participating in EESSI is quite limited and the main barrier in adopting it comes from the fact that implementation does not take into account the end user practices and their context.
  The clerks are accustomed to work with paper documents and previous attempts to create a distributed paperless solution have failed due to the magnitude of change perceived by the often non-technical end-users mainly because of Graphic User Interface (GUI) complexity.
  Therefore, the clerks need a Distributed Case Management System able to handle a partial snapshot of the international case, developed around their current social, communication and professional context, able to manage operational processes and, at the same time, to provide an efficient decision support.

V. EESSI RESPONSE TO DISTRIBUTED CASE MANAGEMENT CHALLENGES

This section presents the EESSI response to the previously mentioned challenges.

A. Formalising the EESSI Business Messaging Protocol

The data exchange protocol of EESSI consists in a collection of separate specifications that can be grouped in the areas of technical messaging and business messaging.

The technical messaging is aligned with EC messaging guidelines and consists in the ebMS3.0/AS4 EESSI profile and is physically implemented through XML Schema Definitions [51] (XSDs) and additional EESSI specific technical messaging validation rules.

The business messaging consists of three separate specifications:

- SEDs - physically implemented in around 320 XSDs,
- Standard Business Document Header [52] (SBDH) - specified through EESSI SBDH Implementation Guide and physically implemented through an EESSI constrained SBDH XSDs,
- The BUCs, which consist in around 110 descriptive documents including BPMN representation of the BUCs.

It is important to understand that the SED and SBDH schemas validate many of the business conformance requirements of EESSI, but are too general to enforce the data exchange in the context of Business Use Cases (BUC) [53]. This introduces genuine interoperability risks into the business domain. To address this, EESSI created a business messaging standard known as the EESSI BMP.

The EESSI BMP is the minimum standard by which all NA must adhere to produce business messages that can be accepted as fully validated transactions within the EESSI domain.

The EESSI BMP is a specification able to define constraints for business and data validation, the common denominators of EESSI through the integration of the key aspects of BUCs, SBDH and SED physical models.

The chosen approach for implementing the BMP is made through XSD constraints. XML technology is already an inherent part of the EESSI domain and continuing its use for the BMP does not introduce unnecessary additional technologies.

In EESSI, this technology is used by two application domains with two different scopes:

- The National Applications, the messages' authoring systems, by directly using the constrained XSD schemas of each defined transaction for producing valid messages and
- AP, the common denominator of EESSI, by validating any received message against a single constrained XSD schema for each transaction.

The following elements of the SBDH along with the SED itself will be constrained though the BMP: BUC type, BUC version, participant role (the role of the sender and receiver participants), number of participants (unilateral vs multilateral case types), case action, SED type, SED XSD version, SED version (if multiple SED versions are allowed) and attachments allowance (if the attachments are permitted for the transaction).

The BMP is a transaction oriented specification that defines specific authoring rules for a specific SED transaction, for a specific participant role, in the context of a specific BUC.

Each identified transaction will constrain the SBDH and SED schemas through a single XSD schema definition that will import and redefine the aforementioned schemas.

Business messages that validate against this constrained version of the transaction are, by definition, also valid SED instances and valid SBDH and therefore, a fully validated transaction within the EESSI domain.

In Figure 2, a BPMN collaboration diagram, illustrates how the transactions are identified, for a specific participant role, in the context of a BUC.

Within the EESSI domain there are four main business level exchange patterns that can occur between the participant roles. These are:

- A Case Owner sends a SED to a Counterparty;
- A Counterparty sends a SED to the Case Owner;
- A Counterparty sends a SED to another Counterparty
- A Case Owner forwards a SED to a Case Owner.

The BMP ensures these patterns are enforced within the context of BUCs though constraining the Sender and Receiver Roles in line with the BUC specifications.

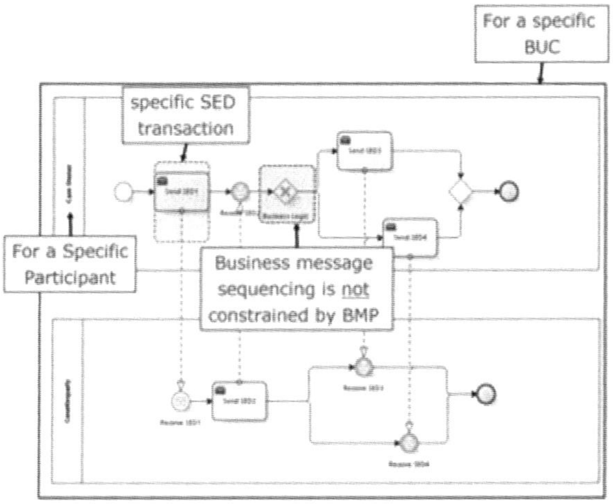

Figure 2. **BPMN** collaboration diagram

The syntax that is issued to uniquely identify each transaction with in the EESSI domain is the following: [BUC Short Name]-[BUC Version]-[Sender Role]-[Receiver Role]-[Case Action]-[SED Type] (e.g., S_BUC_19-1.0-CaseOwner-Counterparty-Start-S080.xsd).

For exemplification of the concept, a Public Heath case type will be briefly introduced. The illustrative example consists in a Healthcare Services Reimbursement Based on Actual Cost (S_BUC_19). The sequence diagram in Figure 3 bellow illustrates the interactions between the Case Owner and the corresponding Counterparty.

This case deals with the business transactions of a reimbursement based on actual costs whereby the Member State of Residence or Stay claims the reimbursement to the Competent Member State on behalf a Creditor Institution.

This case is used where the Institution in the Member State of Residence or Stay provides benefits when the treatment was necessary due to an accident at work or an occupational disease.

The Creditor Institution's Liaison Body (Case Owner) acting on behalf of a Creditor Institution (Claimant) or for itself sends a Reimbursement Claim for Benefits in Kind to the corresponding Debtor Institution's Liaison Body (Counterparty).

The Debtor Institution's Liaison Body (Counterparty) accepts, disputes or rejects the claim and notices it to the Creditor Institution's Liaison Body (Case Owner).

In this aforementioned exemplification, the Case Owner is the Creditor Institution's Liaison Body of the Member State of Residence or Stay that notifies the claim of a reimbursement payment on the basis of actual costs on behalf of a Creditor Institution and the Counterparty is the Debtor Institution's Liaison Body of the Competent Member State that replies for the claim of a reimbursement payment on the basis of actual cost on behalf of a Debtor Institution.

Figure 3 represents the sequence diagram that illustrates few transactions allowed between two the participant roles, in the context of S_BUC_19:

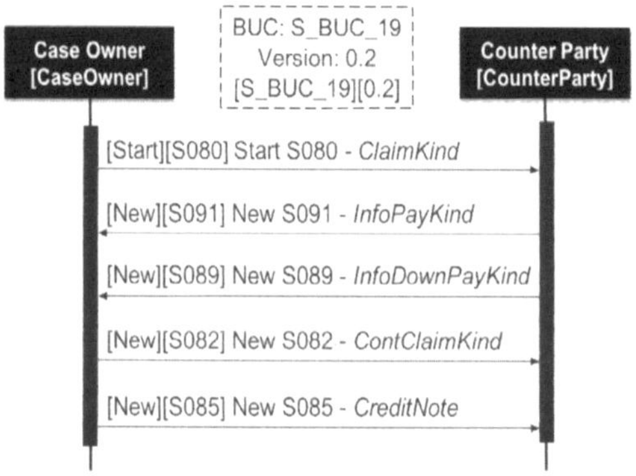

Figure 3. Sequence diagram

B. *EESSI Standard Business Document Header*

The Standard Business Document Header (SBDH) is an UN/CEFACT [54] standard that enables business applications to exchange documents using a consistent interface.

In the EESSI context, the SBDH enables the integration of Access Points with National Applications, with National Gateways or any other business-to-business infrastructure by providing consistent correlation data about a specific business document to be used across the EESSI ecosystem. It also enables any National Application to determine the logical routing and the logical processing of a SED/business document.

The EESSI implementation of SBDH standardizes the data presentation, the data elements within the SBDH that can be easily located and leveraged by multiple applications. The SBDH is created before the transport routing header is applied to the business document and is retained after the transport header is removed. SBDH data can be used also by transport applications like AP to determine the routing header since it does contain the sender, receivers and relevant document metadata. It can also be used by the national applications to determine the appropriate process instance to which the business document needs to be attached.

This EESSI SBDH Implementation Guide clarifies the function, design and implementation considerations of the SBDH in the context of EESSI.

The EESSI SBDH implementation guide deals with the sender and receivers identification together with their role in the case processing (process owner and counterparty), the unique case identifier, the case type, the version of the case definition, the sensitivity flag of the case (medical information case or protected person), the document identification attributes like SED type, schema version, the unique id of the SED, the SED instance version, the creation date, the attached files' metadata and the case action to be performed (start case, new/update document to an existing case, forward SED).

C. *RINA Distributed Case Management Visualization*

RINA is a Distributed Case Management Solution and the purpose of case visualization is to convert the case related data, the SEDs, in relevant information about the case status according to the main goal, which is case resolution. The process of conversion from disparate data into useful information should be analysed at least from three perspectives:

- Communication;
- Graphical user interface;
- Business use case or case type definition.

The SEDs-provided information used by the case management visualization process could be grouped in two categories: the social, demographic and administrative information and case specific information.

The case management view should represent the operational decision support system for the clerk. It is essential to visualize the current case and its corresponding case definition; also, to know what the existing exchanged SEDs are, what the contextual details (like medical history) are, what the social, demographic and administrative details are, etc. The progress observed in the case execution, the definition of the case with specific actions, stages, conditions and so on and nevertheless the best practices of the specific case type, should be accessible.

Figure 4. Case management - Timeline view

The primary goal of the Timeline-based visualization [58] into the RINA Case Management System, as illustrated in Figure 4, is to offer a comprehensive picture of the case, both real-time as well as historically, using textual and graphical means, and to sustain, in a task-oriented approach and based on decision support algorithms, the case progress.

Another important characteristic of this approach is the fast development and reusability of administrative SEDs, achieved through an external component, part of a Business Process Management Solution (BPMS) also used to execute the cases that encode and organize the decisions and action tasks for clerks.

A BPM-based case management solution in social security and public health offers the benefits of configurable workflows for both the medical areas as well as the administrative sub-processes.

Figure 5. RINA notification module

The RINA notifications, presented in Figure 5, are also part of the process definition by the fact that the process can be started based on a received message from a counterparty institution. Being a Distributed Case Management Solution, RINA notifications are focusing mostly on the collaboration with partner institutions. The application is able to notify the case assigned users when a new case is received, when a SED is received or updated, when a case is closed by the case initiator, but also when a user is assigned to a case by his supervisor or when a case defined alarm is triggered.

RINA overcomes the problem raised from the unavailability of easily interpretable guides within existing applications and takes the recording of clinical information beyond historical and statistical reasons, by being a proactive solution that provides to the clerk strong decisional support. RINA targets directly any clerk or medical personnel involved in public health cases.

The central piece of this approach is the Timeline, representing a historical view of SEDs disposed along a vertical time axis as thumbnails, which offers a clear and actionable insight into the case history, as presented in Figure 5.

As such, one of the main benefits of this approach, when compared with the tried and proven case management interfaces, is represented by the possibility of dynamically loading a large volume of data that is time-sorted, meaning that most recent documents will be shown first, allowing clerks to have a quick overview of the case status. Moreover, the timeline provides a unified browsing experience for a volume of heterogeneous data that was collected at different points in time.

The pervasiveness of computing platforms and their wide-scale adoption has led to the emergence of several new methodologies for user interface design that have been widely embraced by the public.

One of the relatively new means of data presentation is the timeline. In order to provide a gentle learning curve and natural grouping of information, RINA GUI is developed around familiarity gained from social networking services like Facebook and suitability of this approach to the social security and public health domain, and therefore, embraced the established graphical presentation patterns.

While not a new idea with regards to the presentation of public health information, timelines have been embraced with the advent of widely used social networks that popularized them. As such, they can be considered an already mature and well-known means for data presentation, significantly reducing the steepness of the learning curve.

VI. EESSI ADOPTION – CASE STUDY AND EXPERIMENTS

To ensure adoption, a **collaborative** and **incremental** approach has been selected for the design and development of EESSI.

By adopting a project approach based on successive incremental iterations, as well as close collaboration with stakeholders that have well defined roles in all project phases, EESSI ensures that Member States and designated sectorial experts actively contribute to the application design and development.

Healthcare experts and Member States representatives (clerks) have the opportunity to provide input to BUCs through the Business Playground, which is a central web-based platform, to be used as a "playground" by the relevant involved stakeholders in order to fulfil three major objectives:

- To review and confirm how the business use cases are to be implemented in the EESSI system;
- To facilitate the re-validation and prioritisation of the Business Layer (BL) requirements. Using the playground environment Member States representatives are able to review and provide feedback about how the business use case should be implemented. The BL requirements are revalidated and prioritised based on the feedback and requests from the user representatives;
- To provide a platform for execution of the Dry-Run activities based on voluntary involvement of Member States representatives (clerks). The Dry-Run activities are primarily aimed at confirming the business processes with real cases, and the corresponding SEDs that are to be used within the business processes.

The BUCs are validated and agreed on as they will provide the playground business content to be modelled and validated. They are also the main vehicle to be used (together with the data modelling efforts) on agreeing on the approach to tackle the points on which there are different opinions.

Bonita BPM [55] is used to implement, in stages, the various processes that are being used in the BUCs, to validate if there is some scenario/branch that cannot be executed in a workflow engine. During this phase agreement is reached over the level of IT implementation of the case (what is implemented in the workflow engine and what is left for the clerk to execute manually).

The Business Playground appears as an essential element taking into consideration the complexity level implied by performing business processes in a distributed environment of 32 countries, 8 social security sectors, with over 10.000 institutions to be connected. Having a Business Playground in place is an appropriate measure to adopt in the context of challenges raised by the integration of national-level applications with the AP.

A strong collaboration between the EC and the Member States as well as significant involvement of the end users in the development of the solution is a key factor in making the development of EESSI a success.

The playground work stream is being delivered in three phases – each phase delivering a number of BUCs (as illustrated in the below diagram) and providing an increasingly functionally rich NA. Additional details are presented in Figure 6.

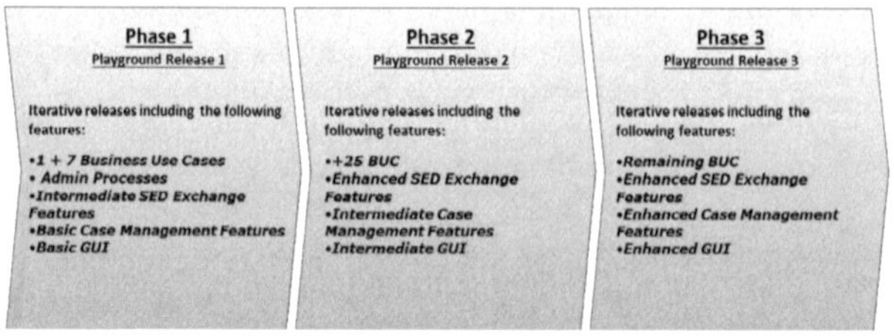

Figure 6. EESSI Playground phases

RINA is in the last year of industrial implementation as a multitenant/cloud solution and the underlying development technologies are: Bonita BPM [55], Elasticsearch [56] and AngularJS [57].

The sizing of Playground and Dry Run iterations is listed below:

Play Ground Environment:

- Number of RINA environments: 6
- Number of users: 103
- Total number of cases performed by participating institutions during the first two Play Ground phases: 2.830.

Dry Run Environment:

- Number of RINA environments: 16 (each RINA institution represents one country)
- Number of users: 240
- Total number of cases performed by participating institutions during the first two Dry Run phases: 13.130

More than 95% of the started cases in the Dry Run environments were successfully closed fully electronically.

The two mentioned environments: playground and dry run, are also involved in collecting feedback on RINA features using the EC change management process.

VII. Conclusion and Future Work

RINA overcomes the problem raised from the unavailability of easily interpretable guides within existing applications and takes the recording of medical information beyond historical and statistical reasons, by being a proactive solution that is giving a strong decisional support to the clerk. RINA targets directly any clerk or medical personnel involved in public health cases.

The central piece of this approach is the Timeline, representing a historical view of SEDs disposed along a vertical time axis as thumbnails, which offers a clear and actionable insight into the case history.

As such, one of the main benefits of this approach, when compared with the tried and proven case management interfaces, is represented by the possibility of dynamically loading a large volume of data that is time-sorted. Moreover, the timeline provides a unified browsing experience for a volume of heterogeneous data that was collected at different points in time.

In order to provide a gentle learning curve and natural grouping of information, RINA GUI is developed around familiarity gained from social networking services like Facebook and suitability of this approach to the social security and public health domain, and therefore, embraced the established graphical presentation patterns.

The tests conducted so far show a high level of acceptance from users (healthcare domain experts and clerks from a significant number of countries and institutions). Most Member States concerned are currently considering the options available for integration and EC surveys show that a large majority to use RINA, or at least one of its layers.

The next period will be dedicated to supporting the decision making at the Member State and institution level so that integration is planned on the most efficient option for each institution, taking into account on the one hand applicable RINA features and, on the other hand, specific environment characteristics, such as the existence of a centralised e-government infrastructure, the existence of an identity management solution, the technology used for existing national applications, specific national requirements regarding security and usability, specific national requirements for case routing and/or related to the SED content.

REFERENCES

[1] HL7 Electronic Health Record-System (EHR-S) Functional Model, retrieved: March, 2016 (http://www.hl7.org).

[2] Certification Commission for Healthcare Information Technology, retrieved: March, 2016 (http://www.cchit.org).

[3] European Institute for Health Records, retrieved: March, 2016 (http://www.eurorec.org).

[4] Protégé - open-source ontology editor and framework for building intelligent systems, retrieved: March, 2016 (http://protege.stanford.edu/).

[5] Tallis PROforma Primer, Advanced Computation Laboratory part of Cancer Research UK, retrieved: March, 2016 (http://www.cossac.org/tallis).

[6] GUIDE, Laboratorio di Informatica Medica Università di Pavia, retrieved: March, 2016 (http://www.openclinical.org/gmm_guide.html).

[7] GLARE, Università del Piemonte Orientale Amedeo Avogadro, retrieved: March, 2016 (http://www.openclinical.org/gmm_glare.html).

[8] VisiGuide - multi ontology guidelines browser, BGU Medical Informatics research center, Ben-Gurion University, retrieved: March, 2016 (http://medinfo.ise.bgu.ac.il/medlab/ResearchProjects/RP_visiGuide.htm#).

[9] The ASGAARD Project, AsbruView, Vienna University of Technology, Faculty of Informatics, Institute of Software Technology and Interactive Systems, retrieved: March, 2016 (http://www.asgaard.tuwien.ac.at/asbruview/index.html).

[10] S. Powsner and E. Tufte, "Graphical Summary of Patient Status", 1994, retrieved: March, 2016 (http://www.edwardtufte.com/tufte/lancet_p1).

[11] T. D. Wang, C. Plaisant, A. Quinn, R. Stanchak, B. Shneiderman, and S. Murphy, "Aligning Temporal Data by Sentinel Events: Discovering Patterns in Electronic Health Records. Time Lines and LifeLines", 2008, retrieved: March, 2016 (http://z3.aq.gs/papers/Aligning%20Temporal%20Data%20by%20Sentinel%20Events,%20Discovering%20Patterns%20in%20Electronic%20Health%20Records%20(2008).pdf).

[12] C. Plaisant et al., "Searching electronic health records for temporal patterns in patient histories: A case study with Microsoft Amalga", Technical Report HCIL-2008-13. College Park, MD: University of Maryland, 2008, retrieved: March, 2016 (http://www.cs.umd.edu/hcil/patternFinderInAmalga/UMD-AMIA08-v17finalsubmitted-forweb.pdf).

[13] KNAVE and KNAVE-II, BGU Medical Informatics research center, Ben-Gurion University, retrieved: March, 2016 (http://medinfo.ise.bgu.ac.il/medLab/ResearchProjects/RP_KNAVE.htm).E

[14] D. Klimov, Y. Shahar, and M. Taieb-Maimon, "VISITORS, Intelligent visualization and exploration of time-oriented data of multiple Patients", 2008, retrieved: March, 2016 (http://cs.uwaterloo.ca/~jchampai/papers/5235880526386162235.pdf)

[15] W. Horn, C. Popow, and L. Unterasinger, "Metaphor Graphics to Visualize ICU Data over Time. Intelligent Data Analysis in Medicine and Pharmacology, VIE-VISU", 1998, retrieved: March, 2016 (http://www.ifs.tuwien.ac.at/~silvia/wien/vu-infovis/articles/Horn-idamap98.pdf).

[16] L. Chittaro, C. Combi, and G. Trapasso, " Data mining on temporal data: A visual approach and its clinical application to hemodialysis", Journal of Visual Languages and Computing, Interactive Parallel Bar Charts (IPBC), 2003, retrieved: March, 2016 (http://www.cieffeweb.com/franz/hcilab/media/k2/attachments/DataMining_JournalVisualLanguages03.pdf).

[17] K. Hinum et al., "Gravi++: Interactive Information Visualization of Highly Structured Temporal Data", Journal of Universal Computer Science, Special Issue on Visual Data Mining, Gravi++, 2005, retrieved: March, 2016 (http://publik.tuwien.ac.at/files/pub-inf_2884.pdf).

[18] W. Aigner, "Guideline Overview Tool (GOT)", Vienna University of Technology, Institute of Software Technology and Interactive Systems, 2001, retrieved: March, 2016 (http://ieg.ifs.tuwien.ac.at/techreports/Asgaard-TR-2001-4.pdf).

[19] R. Bade, S. Miksch, and S. Schlechtweg, The MIDGAARD Project, "Connecting Time-Oriented Data and Information to a Coherent Interactive Visualization ", retrieved: March, 2016 (http://ieg.ifs.tuwien.ac.at/projects/midgaard.html).

[20] W. Aigner and S. Miksch, "The CareVis Project, Interactive Visualization Methods to Support Protocol-Based Care", retrieved: March, 2016 (http://ieg.ifs.tuwien.ac.at/projects/carevis/).

[21] Microsoft Corporation, Microsoft Health Common User Interface, retrieved: March, 2016 (http://www.mscui.net/Default.aspx).

[22] Patient data visualization for facile medical assistance process management, retrieved: March, 2016 (http://www.visual-d.ro/).

[23] FAM Care Human Services & Social Services Software, retrieved: January 2016 (http://www.famcare.net/).

[24] ClientTrack – software solution for health and human services, retrieved January 2016 (http://clienttrack.com/).

[25] Ahshay platform, retrieved January 2016 (http://www.datacare.com/ahshay-platform/).

[26] Allscripts Care Management, retrieved January 2016 (http://www.allscripts.com/products-services/products/care-management/).

[27] Penelope, retrieved: January 2016 (http://www.athenasoftware.net/)

[28] PracticePal, retrieved: January 2016 (http://www.practicepal.co.uk/)

[29] Pega Dynamic Case Management, retrieved: January 2016 (http://www.pega.com/insights/resources/dynamic-case-management)

[30] Be Informed Business Process, retrieved: January 2016 (http://www.beinformed.com/BeInformed/website/en/EN/BusinessProcessPlatform?init=true).

[31] Kana Enterprise, retrieved: January 2016 (http://www.kana.com/customer-service-experience-management-software).

[32] IBM Case manager, retrieved: January 2016 (www.ibm.com/software/products/en/casemana).

[33] ISIS Papyrus Platform, retrieved: January 2016 (http://www.isis-papyrus.com/e15/pages/software/platform-concept.html).

[34] Appian BPM Software, retrieved: January 2016 (http://www.appian.com/bpm-software/).

[35] OpenText Cordys, retrieved: January 2016 (http://www.opentext.com/).

[36] OpenText BPM Everywhere, retrieved: January 2016 (http://www.opentext.com/what-we-do/products/business-process-management).

[37] OpenText Process Intelligence, retrieved: January 2016 (http://www.opentext.com/what-we-do/products/business-process-management/process-suite-platform/opentext-process-intelligence).

[38] OpenText Process Component Library, retrieved: January 2016 (http://www.opentext.com/what-we-do/products/business-process-management/process-suite-platform/process-component-library).

[39] EMC Documentum xCP, retrieved: January 2016 (http://www.emc.com/enterprise-content-management/documentum-xcp.htm).

[40] Kofax TotalAgility, retrieved: January 2016 (http://www.kofax.com/smart-process-application-platform).

[41] Living Systems Process Suite, retrieved: January 2016 (https://www.whitestein.com/lsps-platform/lsps-overview).

[42] DST Systems - AWD10, retrieved: January 2016 (http://www.dstsystems.com/solutions/bpm/awd/).

[43] Oracle BPM Suite 12c, retrieved: January 2016 (http://www.oracle.com/us/technologies/bpm/suite/overview/index.html).

[44] Hyland OnBase, retrieved: January 2016 (https://www.onbase.com/en/product/onbase).

[45] Regulation (ec) no. 987/2009 of the European Parliament and of the Council of 16 September 2009 laying down the procedure for implementing Regulation (EC) No 883/2004 on the coordination of social security systems, retrieved January 2016 (http://eur-lex.europa.eu/legal-content/EN/TXT/PDF/?uri=CELEX:32009R0987&from=EN).

[46] Martin Fowler, Microservices, 2014, retrieved: January 2016 (http://martinfowler.com/articles/microservices.html#SmartEndpointsAndDumbPipes).

[47] OASIS ebXML Messaging Services Version 3.0: Part 1, Core Features (ebMS3), retrieved: January 2016 (http://docs.oasis-open.org/ebxml-msg/ebms/v3.0/core/ebms_core-3.0-spec.html)

[48] AS4 Profile of ebMS 3.0 Version 1.0, retrieved: January 2016 (http://docs.oasis-open.org/ebxml-msg/ebms/v3.0/profiles/AS4-profile/v1.0/AS4-profile-v1.0.html).

[49] Business Process Management, retrieved: March, 2016 (http://en.wikipedia.org/wiki/Business_process_management).

[50] Business Process Model and Notation, retrieved: March, 2016 (http://www.bpmn.org/).

[51] XML Schema, retrieved: January 2016 (http://www.w3.org/standards/xml/schema).

[52] Standard Business Document Header standard, retrieved: January 2016 (http://www.gs1.org/standard-business-document-header-sbdh).

[53] Business Use Cases - Rational Unified Process, retrieved: January 2016 (http://www.ibm.com/developerworks/rational/library/dec04/behrens)

[54] UN/CEFACT, retrieved: January 2016 (http://www.unece.org/cefact).

[55] Bonita BPM, retrieved: April, 2016 (http://www.bonitasoft.com).

[56] Elasticsearch, retrieved: March, 2016 (https://www.elastic.co/products/ elasticsearch).

[57] AngularJS, retrieved: February, 2016 (http://angularjs.org).

[58] C. Taslitchi, F. Moldoveanu and A.Moldoveanu "Timeline-based Clinical Case Manager", Faculty of Automatic Control and Computers, University Politehnica of Bucharest, 2014, retrieved: January, 2016

[59] Martin Fowler, "Patterns of Enterprise Application Architecture", Addison – Wesley Professional, 2002, retrieved January, 2016

VIII. APPENDIX 1 - RINA MAIN FEATURES

Below, we selected a number of illustrative core features implemented by RINA:

- Search management features
    - Free text search -
      (free-text search in fuzzy or exact manner on case metadata in order to identify a case or a group of cases)
    - Structured search
      (RINA allows users to define, modify or delete structured searches using visual search expressions and to persist users searches as pre-defined searches)
    - Applying a predefined search
      (Users are able to quickly select and apply a predefined search in order to minimize searching effort.)
    - Refine predefined search results
      (After the system has filtered the cases based on a user's predefined search request, the User can read the list of filtered cases. The User can also use the free text search to further refine the returned results)
    - Configure Search Result
      (Users have the possibility to configure for each case type: the list of columns that will be returned by the search engine, and the order of the columns and of the results.)
- Case management features
    - Create case
      (RINA allows a User to create a case through accessing a hierarchical structure of case types. RINA retains (per session) the last case type instantiated to enable to quickly creation of another case.)
    - Determine Exchange Partners
      (RINA persists the case participant(s) for each case instance RINA allows the User to view the current case participant(s) and any historical changes.)
    - Case available actions
      RINA provides Users with a list of available actions (tasks) related to a specific case instance depending on the state of the case - e.g., forward, create, update - and for any case action executed by a User RINA updates the case state. It also display the available action list upon the change of case state to ensure that Users can only perform actions that are correct based the case state.
    - Execute action
      (A User has the possibility to select one action from the action list presented to him.)
    - Group Case Actions
      (RINA provides logical grouping of case actions at case level or SED level and could provide further logical grouping of case actions as case level. RINA it also offer the possibility to filter case actions by grouping)
    - Case Importance/ Criticality
      (A User is able to specific the level of Importance and Criticality of a case and can change the importance and criticality of a case at time during the case. This can be done for more than one case at a time.)
    - Case Alarm
      (A User is able to set or delete an alarm by which they expect an action to have occurred. The user is notified when an alarm expires)
    - Create SED
      (RINA allows the persistence of documents in multiple draft states)
    - Create Portable Documents

(RINA can create Portable Documents (PD) were the content of the PD can be fully derived from SED content)

- SED Validation
  (RINA is able to display document validation errors in a clear and logical fashion, allowing the User to navigate directly to the source of the error from the error description.)
- Manage Document Views
  (RINA offers multiple views for Users to view sent and received documents, these can be filtered by: direction, partner, type, status. Users are able to clearly visualise draft documents from sent/received documents and documents that are in different document states. RINA it also offers access to view all document versions)
- Manage document attachments
  (Users are able to manage the attachments through attaching and detaching files. The attachments can be added at case level or at SED level. RINA can restrict the type of attachments allowed and can also restrict the access to attachments of a certain business type where necessary. RINA allows Users the access to read/open attachments directly)
- Print Documents
  (RINA allows a user to choose an action that will render a document as a printable form (PSED))
- Manage comments
  (The User is able to create, read or delete comments (case notes) at document level or case level. RINA persists and display the author and time of each comment and it also ensure that only the author of a comment (or a user with special permissions) can delete a comment)
- Case assignment
  (New Cases (both created locally & received) are automatically assigned to Users using a configurable case assignment rules (via the administration console). A User (with permission) is able to assign a case to users or groups of users that are part of the local organisation and also is able to unassigned/ reassign a case to users or groups of users that are part of the local organisation. A User (without permission) is able to request to a User (with permission) to assign a specific case to them. A User is able to assign/ unassigned/ reassign more than one case at a time. RINA ensures that only assigned users can actively work on/progress a case and is restricting access to certain case types where necessary (e.g., sensitive cases)
- Document Translation
  (A user is able to translate documents content (any free text elements) into any of the EU official languages. Translated document content is not persisted)
- User Settings
  (RINA provides users with possibility to save (persist) user preferences (such as view style, order style etc.))

- Notification management features
  - Generate notifications
    (The notification management requirements cover all aspects of case notification in RINA.)
  - View Notifications
    (Every time an event condition is fulfilled, RINA notifies the assigned Users of the case. Each notification has associated a type that is either: Error, Alert or Information.)
  - Notification Actions
    (A User should be able to take logical actions direct from the notification)
  - Filter Notifications
    (RINA offers the possibility to filter notifications based on type and status and to navigate the notifications' timeline.)
  - Notification Summary
    (RINA provides count of notification by types in the notification panel and the case level views. The counts be automatically re-calculated when User take actions on the notifications. A user is also able to receive anytime in every module of the application the delivery of a new notification.)
  - Suppress/Unsuppressed notification
    (The user is able to suppress a notification so that it is not taken into account anymore and also to unsuppressed a previously suppressed notification so that it is taken into account anymore.)
  - Mark notification as read/unread
    (The user can mark a notification as read or unread so that the next time the notifications are displayed the read/unread state is preserved)
- Administration features
  - Administration Console
    (All administrator tasks are provided through a dedicated administrator console)
  - User Management and organization structure
    (An administrator is able to define organization units, together with departments having any level of nesting and also able to define users or to refer them from an external identity management repository. An administrator is also able to configure the default user settings)
  - Authorisation policies
    (For each case type, the administrator is able to configure what users or groups are allowed to create, to execute, to administer or to audit new cases. These are corresponding to the regular user roles: Clerk, Supervisor and Auditor)
  - Audit
    (The administrator is able to configure what events the system needs to audit. The audit is also reflected to all main resources exposed by the functional modules, for all possible operations: create, read, update, delete and execute.)
  - Technical log

(All the modules of RINA populates a centralised log available for RINA Administrators. The administrator is able to configure the level of logging: trace, debug, info, warn, error and fatal.)

- Notification Management
  (An administrator is able to suspend notification types, and to set notification behaviour (i.e., how a notification is presented to a user))
- RINA updates and versioning of physical artefacts
  (The physical artefacts (forms, case behaviour, vocabularies, etc.) distributed by CSN through AP can be updated by a RINA administrator. These physical models are accompanied with a minimum set of metadata like: version, date of release, name)
- Messaging configuration
  (The messaging configuration is fully available for RINA administrators through the administration console.)
- Counter management
  (The administrator is able to define counters for national case ID. The policies for national case identification can be particular to a department and have a period of availability.)
- Retention policy (case, audit, technical log and notifications)
  (RINA is able to archive the closed cases, audit, technical log and notifications. The retention policy for all of this is configurable)
- Dead letter Queue
  (RINA provides an administrator with access to received documents that it could not logically processed (i.e., unrelated documents, documents with errors) and allows the administrator to return a business exception error automatically for these documents. The sending of business exception error could be a bulk operation)

- General non-functional features
  - User login
    (Any User having the right credentials is able to access RINA. The credentials can be user name and password or smartcards. RINA will logout a user if the application is left idle for a pre-configurable period of time.)
  - Multi-tenancy
    (One RINA deployment is able to host multiple institutions completely isolated between them and with independent capabilities of administration)
  - Localization
    (The User can decide any-time what is the language of its user interface. The user interface acts accordingly to this setting by translating all the screens together with the codified fields (vocabularies)).
  - Accessibility
    (RINA is WCAG 2.0 compliant to minimum AA standard.)
  - Browser Compatibility
    (RINA is fully compatible with the following browsers: Chrome (v.40 +), Firefox (v.32+), Internet Explorer (v.11+))
  - Error Handling
    (Application Errors are delivered to users in an unobtrusive manner, while providing sufficient information to understand the problem)
  - Help
    (RINA provides context aware help for users at SED level (Explanatory Notes) and Application level (general usage guidelines))

MoodLine and MoodMap
Designing a Mood Function for a Mobile Application with and for Young Patients

Maja van der Velden and Margaret Machniak Sommervold
Department of Informatics
University of Oslo
Oslo, Norway
majava@ifi.uio.no and magarma@ifi.uio.no

Abstract—**Tracking mood or emotional experiences over time is a popular function found in mobile health applications. In this study, young patients with chronic health challenges consider this also an important function of a multifunctional app supporting them in the transition to adult care. At the same time they expressed the need to be seen as a young person, not a diagnosed body. A lifeworld-led design approach, based on a Participatory Design methodology, resulted in a mood tracking and a mood mapping design, which was meaningful to the young persons' everyday experiences. Photos tagged with colors representing different emotional states were chosen as the best way to represent their moods. An overview of moods, by day as well as by color, gives an understanding of the wider context in which these moods appear and can play a motivational role in dealing with a difficult day or episode in their lives.**

Keywords-Mobile health application; lifeworld; Participatory Design; teenagers; transition

I. INTRODUCTION

The KULU research and design project[1] focuses on the design and use of interactive technologies with and for young people (15-25 years old) with chronic health challenges. Mobile applications are popular among young people, but a systemic review showed that there is no empirical evidence for their beneficial use in the personal health management of young patients [1]. The review did emphasize the value of involving young patients in designing these apps [1].

This paper addresses the design of the mood function in a multifunctional mobile application (app). The aim of the app is to support young patients in the transition from pediatric to adult healthcare. Mood is often differentiated from acute emotional states, such as being angry, sad or happy. They last longer and are often not related to an immediate trigger: "mood state appears to be an integrative function of the organism's acute emotional experiences over time" [2]. In the field of IT health, e-health, and m-health (health IT), this differentiation disappears when describing or designing mood technology. This becomes for example clear in [3], which categorizes mood technologies into *Technology that measures mood*; *Technology that expresses user mood*; *Technology that adapts to user mood*; and *Technology that influences user mood*. Another categorization of mood technology is diagnosis-based versus general mood. For example, mood apps can address specific mood disorders, such as bipolar disorder [4], anxiety disorders [5], and depression [6], or have a more general approach, such as happy apps [7].

Research focusing on apps for adolescent mental health report that the participation and adherence rate to treatment was higher for mobile phone apps than on paper [8][9], rating mood was seen as most useful [10], and "the ability of mobile phones to offer personal space is also considered to increase levels of perceived autonomy, control, and self-esteem in young users" [11]. Young people with chronic physical health challenges are more likely to have mood-related issues, ranging from emotional problems to mood disorders [12].

On the question about the preferred functions of an app that would support them in their transition to adult health care, a group of young chronic patients participating in KULU research proposed, among others, a mood tracking function. This paper presents and discusses the design of an app-based mood tracking functionality with and for young patients. The aim of the paper is to explore how we can design with and for young patients, while taking their whole being, as a teenager or young adult and as a patient, into consideration.

The contribution of this paper is threefold. It contributes to designing interactive technologies from the perspective of young people with health challenges. A *lifeworld*-led design approach, combined with a participatory design methodology, enables patients to be met as co-designers and experts of their own lives and facilitates patient contributions to the design of health IT. Secondly, it contributes to understanding young patients' lifeworlds. It shows how they value moods in their lives and the importance of taking a holistic perspective when tracking moods. Lastly, it contributes to understanding the role of technology in the lives of young people in general. Teenagers and young adults' technology preferences and use are often very different from those of the researchers. Designing with the future users of a technology increases the chance that the technology reflects the values and needs of that particular user group.

In Section II, we will first explore some concepts that support a holistic patient perspective in the design process of a lifeworld-led design approach. This is followed, in Section III, by a brief presentation of SHARM (Situation-based learning; Having a say; Adaptability; Respect; Mutual

[1] KULU is a Norwegian acronym for Cool Technologies for Youth with Long-term Health Challenges (www.kulu.no/en)

learning), our Participatory Design approach. In Section IV, we present the design process, the methods used, the different stages in the development of the mood tracking function, and the final results. In Section V, we discuss the design process and its results through the lens of the lifeworld concept and we evaluate the SHARM approach. In the final section, we present our concluding remarks and outline future research.

II. Conceptual Framework

Teenagers and young adults with chronic health challenges have made clear that they want to be met as young people, not as patients. Their wish to push their 'patientness', the quality of being a patient, to the background is also confirmed in the literature [14]–[16]. Young patients use the terms *normal* and *regular* to express how they want to be perceived and treated by the world around them [14][16]. They acknowledge their illness, but want to have lives like their peers and they do not want their caregivers to see only their diagnosed bodies: "the doctor should be interested in me, all of me, not just my diagnosis" [17].

This particular positioning by the young patients can be explained with the notion of the *lived body*, the body as experienced by the self and as being-in-the-world, as described in phenomenology [18]. The concept that encompasses both the lived body and its experiences in the world is the notion of lifeworld. Lifeworld can be described as "the world of lived experience or the beginning pace-flow from which we divide up our experiences into more abstract categories and names" [19]. Lifeworld theory describes five intertwined dimensions in which these experiences become meaningful: temporality, spatiality, intersubjectivity, embodiment, and mood [19]. Mood, in this context, is described as a "messenger of the meaning of our situation" or our being-in-the-world, "mood is complex and often more than words can say" [19].

Lifeworld-led care is a particular perspective on healthcare, which focuses on the wellbeing of the whole person, not just the illness or diagnosis [19][20]. This perspective is both a deepening of the understanding of patient-centered care and a critique on the dehumanisation and depersonalization of care, not the least through the use of technology [19][20]. The aim of a lifeworld-led design approach is to let the young patients' *lived experiences* of everyday life, diagnosis, and technology use, guide the design of new technology that supports them in living their everyday life with their health challenges [20].

In order to provide an enabling environment in which young patients can build forth on these experiences, KULU implements its design activities within a participatory methodology called SHARM, which is based on Participatory Design [20]. The participation of young people as co-designers of their own healthcare technologies enables a design space in which the young participants can position themselves in the way they perceive themselves and how they want to be perceived by others. The SHARM approach is based on five principles [22]: 1) *Situation-based action* locates the design activities in the lifeworld and relationships of the participants; 2) *Having a say* is about creating real opportunities for participants to share the decision-making power; 3) *Adaptability* is about applying tools and methods in the design activities that can easily adapt to the participants' changing physical or emotional state; 4) *Respect* is about treating the young participants as experts on their own life and body; and 5) *Mutual learning* refers to choosing methods and tools that enable the participants to lean as much from us as we do from them.

III. Results from the Design Process

The design of a mood tracking functionality was part of a larger design project with the Youth Council of the Akershus University Hospital (AHUS) in Norway. The Youth Council had made a wish list of issues and technologies they wanted to address in the design project with KULU. The transition to adult healthcare was one of the main concerns of the Council and they wanted to explore how a mobile phone application (app) could support them in the transition process.

We will report here on the design of the mood tracker functionality for a multifunctional transition app. The design process consisted of four workshops and an online prototype evaluation. The workshops took place in two large meeting rooms and were attended on average by seven Council members. In total, ten young patients participated, five male and five female participants, who were between 14 and 21 years old. They had a variety of chronic diagnoses. The project was evaluated and approved by the data authority for universities in Norway and the privacy officer of AHUS. All participants had given their consent to participate. Additional consent was sought from the legal guardians of the participants that were younger than 16 years old. Further details of the design process can be found in [22].

A. Identifying functionality

During the first workshop, the functionality of the transition app was explored with a brainstorming technique [23], resulting in a two lists: *Cool-to-have* and *Must-have* [24]. The *Must-have* list consisted of functionalities that the app needs in order to be used, such as calendar for doctor appointments, alarm for taking medicines, checklists, and *registration of the general state* of the user (e.g., mood, energy level), but also attributes, such as *colors* and *privacy*. Color preferences were perceived as very personal and one of the participants proposed that colors could be used to personalize the app. A password or pass code was proposed to keep the content of the app separate from other apps [24]. The *Cool-to-have* list mentioned aspects and functionalities that made the app extra attractive for young people, such as an 'Instagram'-like environment, music play-list (similar to 'Spotify'), and film and tv tips [24]. An analysis of the group discussion of all proposals resulted in the identification of three categories: to have an *overview* (medicines, appointments, routines); strengthen *autonomy* (registration of general state, checklists); and *entertainment* (music; tips).

B. General State

During the second workshop, the *registration of the general state*, was one of several functions further explored.

In a collaborative prototyping session, three design proposals were presented in the form of both paper-based and digital prototypes. Collaborative prototyping enables the translation of values and needs into design requirements [25][26]. The three proposals reflected different ways of mapping their energy levels and mood. The registration of the general state through images was perceived as more creative and personal. In the discussion that followed, the difference between taking your own photos and finding images on the net was explored, with one participant expressing the concern that finding and uploading images from the net needed focus and energy, which was not always available. Another participant mentioned "when you are really down, you can go back and look at the photos and see that there is one that makes you happy. For example, when you are admitted to the hospital, you can *go back* and look for what gives you energy, and look at the photos" [24].

C. Mood

The discussions of the three prototypes evolved around the use of photos, colors, and mood. The next iteration of the function for mapping the user's general state focused on these three aspects and consisted of three low-fidelity digital prototypes, which were also produced as plasticized paper printouts. In order to enable the user to "go back", we used the concept of timeline as an organizing principle for the photos. Each photo could be tagged with a colored frame, which would be an expression of the mood associated with the photo. During the third workshop the different prototypes were explored and discussed (Fig. 1). The participants preferred the option to scroll up and down through the list of photos. Secondly, they preferred photos of the same size to the option to have different sizes, because this gave a better overview of the photos in the timeline. The timeline itself should be based on the date, not on photos, so it would be clear to them on which days they were too tired or sick to add a photo to the timeline. Colored small dots on the side, based on the colors selected for the images, would give a quick overview of mood over time (Fig. 2).

Figure 1. Prototypes of MoodLines

Figure 2. MoodLine with dots

The timeline of photos with different color-tags inspired a discussion of what they could do with the colors. One participant proposed to add a new option to the mood tracker, namely the possibility to see only photos tagged with one color, similar to Instagram. For example, on a sad day, the user could scroll through photos tagged with the color-tag happy, in order to get through the day and inspire or motivate oneself with photos that presented happier times or moods. We also explored different options for personalization through colors [27]. The option to allow the user to configure the colors and their associated meanings was chosen over option to use a default set of colors with associated moods (See Fig. 3). The combination of images and colors enabled a focus on tracking their mood, not on taking pretty pictures. This option also expresses the wide variety of color associations found among the participants, which were the result of age, gender, and personal preferences [27].

D. Final iteration

The final iteration of the mood function was produced in InVision, an online prototyping software for clickable, high-resolution prototypes. During the last workshop, our co-designers were invited to access what was now named the KOOLO app on Invision, in order to click through the different options, such tagging colors with a mood, adding a photo, color-tagging the photo, scrolling through photos, and accessing the mood map to select a collection of photos

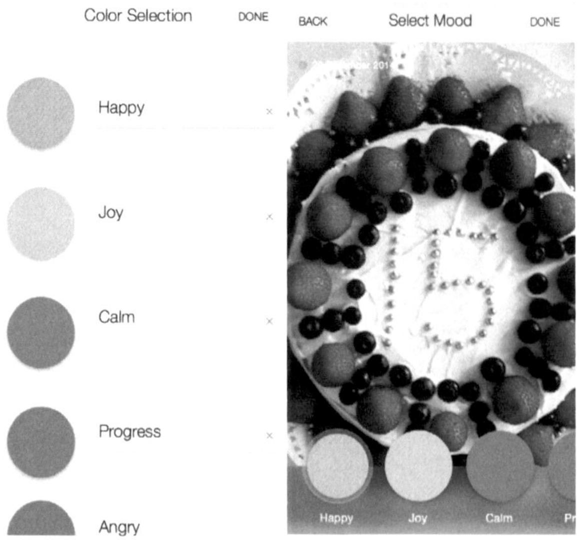

Figure 3. Color tags

tagged with the same color. All the participants received information on how to access the online prototype and were invited to use and evaluate the prototype.

IV. Discussion

"Lived experience is coloured by mood" [18]

The KOOLO app, including its mood function, has now been fully developed for both the Android and iOS platform. The co-designers' preferences and needs form the core of the mood function (see summary in Table 1).

TABLE I. IMPLEMENTATION OF DESIGN SPECIFICATIONS

Requirement	**Implementation**
1. Registration of the general state	Photos in an 'Instagram'-like environment
2. Time as organizing principle	A *MoodLine* organized by dates, including dates without photos
3. Color as organising principle	No default settings for tags
	Can be used for personalization of the whole app
	The photos can be organized by color via the *MoodMap* option
4. Privacy	Photos are stored in the app, which is password/code protected

In the discussions on the *registration of the general state*, the co-designers included their existing experiences with apps. This function therefore evolved very early in the design process into an image-based mood function. It could use the existing functionality of the mobile phone (the photo app) and was perceived as an intuitive, easy, and personal way to track one's mood. Also, the shape of the images reflected the participants' existing app use: square shape of the images was preferred over round-shaped ones, because of its similarity with square-shaped images of the popular Instagram app (see Fig. 1 and 2).

Figure 4. Calendar with color tag

Our co-designers also made clear that they wanted to track both positive and negative moods and did not want to favor one type of mood over the other by presenting them in different formats or styles. As an example they mentioned that a day with a negative mood could be a very important day, but that this could get lost in a design that would present positive mood images larger than negative mood images.

This example makes clear that the co-designers were able to find and explore connections between their *lifeworlds* and the specifications of the mood function. That they wanted to track their mood in relation to their lived experiences became also clear in the design of the timeline and the application of mood colors to other functions in the app. They preferred the *MoodLine* to track all days, not only the days in which a user added a photo. This way, a day without a photo has meaning as well, by evoking reflection on the reason for not adding a photo to the *MoodLine*, such as being too tired or too sick. Inspired by the color tags of the *MoodLine*, they proposed to use color tags in other functions of the app, such as the dates in the calendar function of the app. A date tagged with a color thus became a meaningful way to highlight days with doctor appointments or test results (Fig. 4).

The idea for the *MoodMap* (Fig. 5) came up in a discussion about keeping an *overview* of things. The larger the MoodLine would become over time, the more difficult it would be to find patterns that were meaningful in their lives. The proposal for a *MoodMap* was inspired by the Instagram photomap, which geographically maps where a user has taken a photo and shows all photos taken on that same location. All participants perceived this as a fun and intuitive way of organizing their mood images. The *MoodMap* gives an indication of how many photos are tagged by a particular color. Selecting one color in the *MoodMap* (Fig. 5) results in a *MoodLine* (Fig. 2) with only images tagged with the selected color.

The young patients can use the mood function to keep an overview of their moods over time as well as per mood. In addition, they can use mood colors to in other functions of the app, such as calendar and date function. This can give them an understanding of the context in which their moods appear. Keeping an overview and looking for meaning are related to mastery, the experience of emerging stronger from a very stressful health condition [28]. The experience of mastery increases when young patients can participate in a meaningful way in decisions that affect their lives. The *MoodMap* allows the user to focus on one particular emotional state, which may effect motivation, inspiration, learning, and change.

Lastly, the *privacy* specification: the co-designers proposed a strict division on their mobile phone between general apps and an app focusing on their diagnosis or health challenges. The design of the mood function, and the app as a whole, are designed according to *Privacy by Design* principles [29]. Privacy is default, as well as integrated to the system, without diminishing functionality. There is no communication between this app and the other apps stored on the mobile phone - data produced in the app, calendar events, photos, and checklist entries, are stored within the app – and there is no communication with a website or with third parties. The source code is open and available for investigation.

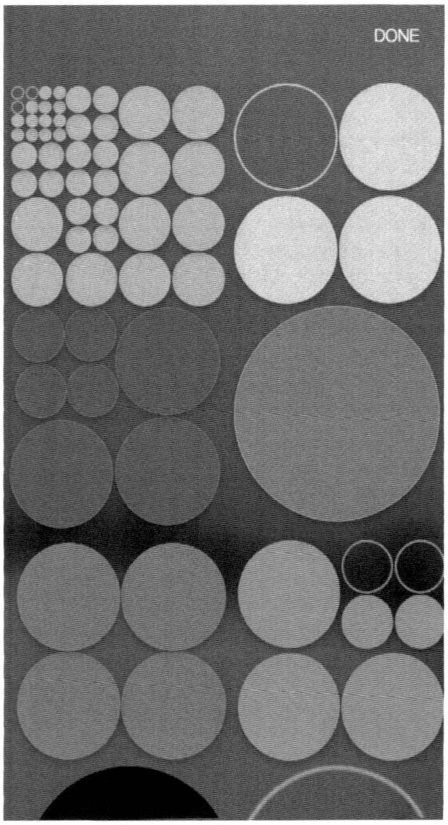

Figure 5. MoodMap – each dot presents one photo tagged with that color

The five principles of the SHARM approach played a central role in creating a *lifeworld-led design process*. The design workshops took place in the hospital, providing them a safe place for reflecting on their experiences and needs as young people with serious health conditions (situation-based action). The iterative approach, in which the design preferences, ideas, and results from the last workshop were presented in a more designed and developed manner, providing real opportunities for the co-designers to take part in decision-making based on their expertise of their own life and body. The methods we used in the workshops enabled a mutual learning process that was at the one hand explorative and inspiring, and on the other hand based on research and experience.

V. CONCLUSIONS

A systematic review of apps supporting adolescents' personal management of chronic and long-term physical conditions [1] shows that the lack of large-scale studies makes it difficult to find empirical evidence of their benefit, but that engaging the adolescents contributes to changes in the mobile intervention's design. Our study confirms that the participation of young people with health challenges in the design of their own interactive technologies can result in creative and important contributions to the design process. A participatory and *lifeworld*-led design process, based on collaborative methods and an iterative approach, allows young patients to explore mood-related needs and values in a more holistic and relational manner. This resulted in very specific design requirements that were closely related to the young persons' everyday experiences with technology. The popular Instagram app was an important inspiration in the design process. Secondly, it resulted in a more meaningful mood tracking and mapping practice, such as personalizing the use of self-selected colors and self-produced images; the equal importance of positive and negative moods and days with and without images (*MoodLine*); and the organization of photos by mood (*MoodMap*).

The young patients were met as co-designers and experts of their own lives, which enabled them to be heard as well as to have a say in the design process. This allowed them to make important contributions to the design of health IT. In turn, this enabled the researchers to learn more about the young patients' lived bodies, their *lifeworlds*, and the role of technology in their lives.

The design of the mood tracking function resulted in two new design concepts for organizing mood data, the *MoodLine* and the *MoodMap*. The use of photos, framed with colors representing emotional states, enabled a meaningful registration of the young persons' mood in a visually pleasant manner (*MoodLine*) and in a meaningfully organized way (*MoodMap*).

Future work consists of making the transition app, including the mood function, available to a group of test users, with and without a diagnosis, followed by qualitative interviews with the users as well as health personnel.

ACKNOWLEDGMENT

We thank the Youth Council of the Akershus University Hospital for sharing their expertise and time. We are grateful to the Hospital's staff of for supporting our research. We also acknowledge the contribution of former master student Nora Svarverud Aasen.

REFERENCES

[1] R. Majeed-Ariss et al., "Apps and Adolescents: A Systematic Review of Adolescents' Use of Mobile Phone and Tablet Apps That Support Personal Management of Their Chronic or Long-Term Physical Conditions," J. Med. Internet Res., vol. 17, no. 12, Dec. 2015, pp. e287.

[2] D. Nettle and M. Bateson, "The Evolutionary Origins of Mood and Its Disorders," Curr. Biol., vol. 22, no. 17, Sep. 2012, pp. R712–R721.

[3] P.M. A. Desmet, "Design for Mood: Twenty Activity-Based Opportunities to Design for Mood Regulation," International Journal of Design. [Online]. Available: http://www.ijdesign.org/ojs/index.php/IJDesign/article/view/2167/691. [Retrieved: 18-Jan-2016].

[4] J. Alvarez-Lozano et al., "Tell Me Your Apps and I Will Tell You Your Mood: Correlation of Apps Usage with Bipolar Disorder State," Proceedings of the 7th International Conference on PErvasive Technologies Related to Assistive Environments, New York, NY, USA, 2014, pp. 191–197.

[5] A. Miloff, A. Marklund, and P. Carlbring, "The challenger app for social anxiety disorder: New advances in mobile psychological treatment," Internet Interventions, vol. 2, no. 4, 2015, pp. 382-392.

[6] N. Shen et al., "Finding a Depression App: A Review and Content Analysis of the Depression App Marketplace," JMIR MHealth UHealth, vol. 3, no. 1, Feb. 2015, pp. e16.

[7] J. McNamara, "Get Happy iPhone App Gets You Happy One Tip At A Time," TheAppReviewZone.com, 28-Dec-2012.

[8] Y. Seko, S. Kidd, and D. Wiljer, "Apps for those who help themselves: Mobile self-guided interventions for adolescent mental health," Sel. Pap. Internet Res., vol. 3, 2013, pp. 1-4.

[9] M. Matthews, G. Doherty, J. Sharry, and C. Fitzpatrick, "Mobile phone mood charting for adolescents," Br. J. Guid. Couns., vol. 36, no. 2, May 2008, pp. 113–129.

[10] R. Kenny, B. Dooley, and A. Fitzgerald, "Feasibility of 'CopeSmart': A Telemental Health App for Adolescents," JMIR Ment. Health, vol. 2, no. 3, Aug. 2015, pp. e22.

[11] Y. Seko, S. Kidd, D. Wiljer, and K. McKenzie, "Youth Mental Health Interventions via Mobile Phones: A Scoping Review," Cyberpsychology Behav. Soc. Netw., vol. 17, no. 9, Jul. 2014, pp. 591–602.

[12] S. Turkel and M. Pao, "Late Consequences of Pediatric Chronic Illness," Psychiatr. Clin. North Am., vol. 30, no. 4, Dec. 2007, pp. 819–835.

[13] M. van der Velden and K. El Emam, "'Not all my friends need to know': a qualitative study of teenage patients, privacy, and social media," J. Am. Med. Inform. Assoc. JAMIA, vol. 20, no. 1, Jan. 2013, pp. 16–24.

[14] A. M. Kanstrup, P. Bertelsen, and C. Nøhr, "Patient innovation: an analysis of patients' designs of digital technology support for everyday living with diabetes," Health Inf. Manag. J., pp. 1–9, 2014.

[15] L. S. Liu, K. M. Inkpen, and W. Pratt, "'I'm Not Like My Friends': Understanding How Children with a Chronic Illness Use Technology to Maintain Normalcy," Proceedings of the 18th ACM Conference on Computer Supported Cooperative Work & Social Computing (CSCW15), ACM, 2015, pp. 1527–1539, ISBN: 978-1-4503-2922-4.

[16] Akershus University Hospital, Youth Council at HPH international conference. 2015 [Online]. Available https://www.youtube.com/watch?v=izWTk5lqzbM [Retrieved: 18-Jan-2016].

[17] J. A. Aho and K. Aho, Body Matters: A Phenomenology of Sickness, Disease, and Illness. Lexington Books, 2008.

[18] L. Todres, K. Galvin, and K. Dahlberg, "Lifeworld-led Healthcare: Revisiting a Humanising Philosophy that Integrates Emerging Trends," Med. Health Care Philos., vol. 10, no. 1, Jul. 2006, pp. 53–63.

[19] K. Dahlberg, L. Todres, and K. Galvin, "Lifeworld-led healthcare is more than patient-led care: an existential view of well-being," Med. Health Care Philos., vol. 12, no. 3, Aug. 2009, pp. 265–271.

[20] A. Pulman et al., "Empathy and Dignity through Technology: Using Lifeworld-Led Multimedia to Enhance Learning about the Head, Heart and Hand," Electronic Journal of e-Learning, vol. 10, no. 3, 2012, pp. 349–359.

[21] J. Simonsen and T. Robertson, Routledge international handbook of participatory design. New York: Routledge, 2012.

[22] M. van der Velden, M. M. Sommervold, and A. L. Culén, "Designing interactive technologies with teenagers in a hospital setting," in An HCI Perspective on Working with Teenagers in Research Projects, Springer, 2016, in press.

[23] V. P. Seidel and S. K. Fixson, "Adopting Design Thinking in Novice Multidisciplinary Teams: The Application and Limits of Design Methods and Reflexive Practices: Adopting Design Thinking in Novice Teams," J. Prod. Innov. Manag., vol. 30, Dec. 2013, pp. 19–33.

[24] N. Aasen, "Transitionapp – Responsibility for your own health," Master Thesis, University of Oslo, Oslo, 2014.

[25] S. Bødker and K. Grønbæk, "Cooperative prototyping: users and designers in mutual activity," Int. J. Man-Mach. Stud., vol. 34, no. 3, Mar. 1991, pp. 453–478.

[26] Y.-K. Lim, E. Stolterman, and J. Tenenberg, "The Anatomy of Prototypes: Prototypes As Filters, Prototypes As Manifestations of Design Ideas," ACM Trans Comput-Hum Interact, vol. 15, no. 2, Jul. 2008, pp. 71–727.

[27] M. van der Velden, M. M. Sommervold, and A. L. Culén, "Patient-Intitiated Personalisation: Privacy, Moods, and Colours," Proceedings of the International Conference on e-Health 2015 (E-Health), IADIS Press, 2015, pp. 95-103, ISBN 978-989-8533-42-5.

[28] J. B. Younger, "A theory of mastery," ANS Adv. Nurs. Sci., vol. 14, no. 1, Sep. 1991, pp. 76–89.

[29] A. Cavoukian, and M. Prosch, "The Roadmap for *Privacy by Design* in Mobile Communications: A Practical Tool for Developers, Service Providers, and Users," Inf. Priv. Comm. Ont. Can., 2010, pp. 1-27.

Re-establishing Interaction Through Design of Alternative Interfaces

Exploring new radio interfaces for elderly people with psychomotor disabilities

Suhas Govind Joshi
Department of Informatics
Faculty of Mathematics and Natural Sciences, University of Oslo
Oslo, Norway
joshi@ifi.uio.no

Abstract— **This paper explores new opportunities for elderly people who are no longer able to use their radio due to changes in psychomotor capacities in hands and fingers. We explore whether three alternative self-developed radio interfaces can provide these elderly people with new interaction mechanisms in the radio that allow them to re-establish use. The three radios are used to conduct measurements of psychomotor performance among 52 elderly participants. Our main findings indicate that providing elderly people suffering from reduced psychomotor capacities with appropriate interfaces cannot only re-establish interaction but also yield performance scores comparable to the scores of elderly people without any apparent reduction in psychomotor capacities. We further present our additional findings from a quantitative analysis of performance and discuss discovered opportunities.**

Keywords — psychomotor abilities; elderly; radio; assistive technology.

I. INTRODUCTION

One of the most appreciated and well-used devices for elderly people in Norway are radios. The radio has been with them throughout most of their long life, and in a world of rapid technological development, the radio as a piece of technology has withstood the test of time. Phones have become mobile, smart, small, and multi-purpose devices, and televisions have become bigger, flatter and expanded with secondary functions. In a local care home in Oslo (with the average resident age of 84 years), which was part of our empirical context, we observed that 91% of the 90 elderly residents had a radio device in the home that they would use on an average day. However, as much as the residents desire to hold on to their radios, the process of aging introduces a variety of cognitive and motor challenges that complicate the use of even familiar and simple technologies. Our prior studies have demonstrated how something seemingly simple as a radio is not considered as simple or functional when aging symptoms appear [1].

The purpose of this paper is to discuss the need for alternative radio interfaces for older people who, because of disabilities or other bodily challenges, are unable to operate normal radios. We investigate whether new interfaces can re-enable interaction between radios and users who are no longer able to operate them. We present three different functioning radios that are specifically designed for older people and discuss the psychomotor properties of these interfaces regarding the interaction opportunities they offer. This study involved 52 participants from 2013 through 2015 in a systematic testing of our proposed interfaces. We use Fleishman's taxonomy of psychomotor abilities and skills [2] to identify and measure the participants' ability to operate the three different radios. To support our discussion, we present a statistical analysis of the gathered data. The results are used to demonstrate how various participants preferred different interfaces based on their psychomotor capacities, and how participants with motor challenges in certain cases were able to match the performance of elderly people without these challenges.

The paper is structured as follows. We introduce the motivation for focusing on the radio in Section 2, and we present related work on psychomotor and age-related studies within HCI in Section 3. In Section 4, the taxonomy used to describe and measure psychomotor abilities is defined. The research methods and three developed radios are described in Section 5 followed by results and analysis in Section 6. The paper ends with a discussion of why we believe new radio interfaces can help re-establishing psychomotor interaction with radios.

II. BACKGROUND

According to statistics from Statistics Norway (SSB), the older part of the population (aged 67-79) remains stable in the national average of radio listening in Norway [3]. The red line in Figure 1 shows an overview of the average percentage of the population who listens to the radio while the blue line shows the corresponding percentage for people aged 67-79.

The number of minutes in average spent listening to the radio is illustrated in Figure 2, and as we can read from the graph, there is only one recorded case in the past 23 years (1997) where the elderly fraction of the population on average would listen less to the radio compared to the general population. We can also read from Figures 1 and 2 that even in years where the number of elderly radio listeners was lower than the national average (i.e., 1994, 1995, 2001, 2009, and 2010), the number of minutes spent in front of the radio was higher for the elderly radio listeners. The difference between the older generation and the rest of the population seems to have diverged over time, and the difference in a ten-year perspective is now greater than ever. The mean difference between the amount of time the elderly used for radio listening compared with the rest of the population in the period from 1991 to 2000 was 7.0 minutes while the corresponding difference in the period from 2005 to 2014 was over four times larger (33.1 minutes). This difference demonstrates an interesting phenomenon, namely that the radio as a piece of technology is not on its way to extinction. Quite the contrary, they are on the rise again regarding both share of the population that listens to the radio (Figure 1), and the number of minutes spent in front of the radio per day (Figure 2).

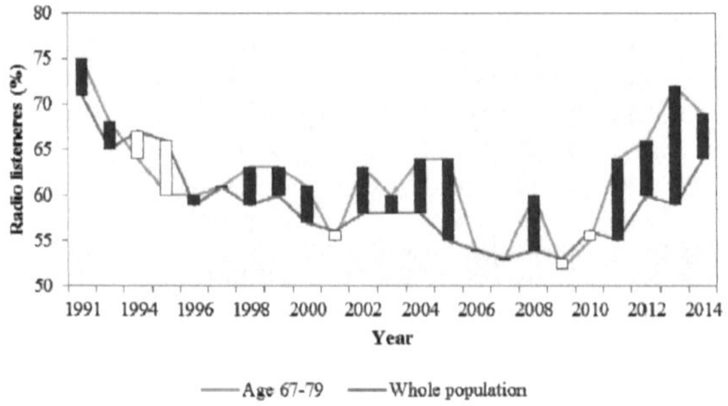

Figure 1. Percentage of population listening to the radio on an average day

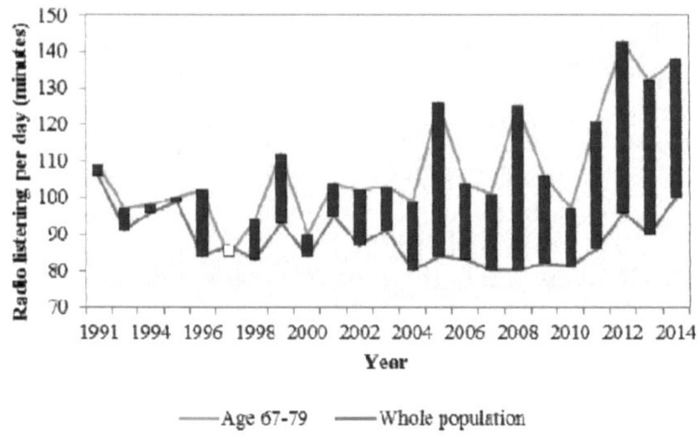

Figure 2. Number of minutes spent listening to the radio on an average day

In our prior research (e.g., [4] and [1]) we have discussed aspects of the role technology has in the lives of elderly people. We have touched upon related topics such as the social importance of being able to operate communicative technologies such as radios to stay in touch with the outside world [4]. We also explored deeper issues concerning the ability to operate such devices and the way such devices are presented, for instance, design that is oversimplified or stigmatizing [1]. These studies have concentrated on the experience of interacting with technology and would consequently be better suited to further discuss the social and contextual aspects of interaction with technologies, for instance, loneliness and boredom. However, in this paper, the focus remains on the psychomotor ability to interact with the radio, and more precisely re-establishing a lost relationship between old users and technology.

III. RELATED WORK

A long time has passed since researchers began systematically investigating the relationship between aging-related disabilities such as arthritis and the ability to interact with computers [5]. Morgan et al. [6] described significant differences in the execution of movement when comparing young adults with older adults, and more precisely the speed, sub-movements, and smoothness in movement. Similarly, Riviere & Thakor [7] use a comparative study between young, old, and motor-disabled subjects with regards to performance when operating tracking with a computer mouse. Their study claims that both aging and motor disability affect performance by increasing the inaccuracy and nonlinearity. Age has an apparent impact on our ability to interact and the extent to which we are able to adapt to new interaction mechanisms. This partly manifests itself through changes in psychomotor capacities. A recent study [8] claims the existence of age-related differences in the strategic repertoire, distribution, and execution within the sensorimotor domain. Regardless of the age of the intended user group, fine psychomotor abilities should be included in the determining of successful interactions [9].

One of the very few laws that attempt to descriptively explain the psychomotor role of human-computer interaction through mathematical formulas is Fitts' law. The original model was formulated over six decades ago and attempted described the linear relationship between movement time and index of difficulty. The model is still used today to quantify the difficulty of performing tasks and was in 2002 included in the ISO standard ISO 9241-9, which concerns non-keyboard input devices. However, since its conception, the model has undergone several modifications and refinements and does not pertain a universally accepted formulation today [10]. A shortcoming of Fitts' law is its ability to properly determine and evaluate differently observed result in the psychomotor performance when studying different task types, varying motor skills and differences in motor performance [11]. Others have argued that there are several factors affecting our endpoint performance that are not properly captured in the mathematical model [12].

In the context of aging, studies on how psychomotor abilities affect user performance with computer tasks within the field of HCI can also be traced back to at least the early 90s where researchers claimed and studied a relationship between the two [7]. Studies have been conducted within the field of HCI focusing on traditional interfaces, including WIMP and trackpads. For instance, psychomotor skills are an important part of the ability to operate a computer mouse, and several studies have investigated the relation between psychomotor abilities and use performance operating a mouse or trackpad [5][9][13]-[16]. Common for most of these studies is that they include several components that make up the list of psychomotor abilities described in Fleishman taxonomy, e.g., precision control, arm-hand steadiness, manual dexterity and wrist-finger speed [12].

IV. PSYCHOMOTOR ABILITIES

People undergo multiple reductions in both cognitive and motor skills as when entering later stages of life. In this study, we have chosen to focus on reduced psychomotor capacities in the hands and fingers, and how these changes affect the likeability to interact with radios. We have chosen not to describe this change as a limitation in the ability to interact since that would indicate an impossibility in an interaction between these individuals and the radio as technology. Instead, we believe that despite the undeniable changes in bodily capacities, our ability to interact with technology is not deprived, or necessarily not even reduced. We aim to demonstrate how adapting the technology to these changes in bodily capacities can prolong and re-establish interaction. Nevertheless, the focus of this study is older people with symptoms, illnesses, and diagnoses associated with reduced capabilities in the hands and fingers. This includes individual types of rheumatic disorder associated with hands and fingers, osteoarthritis, as well as more general motor system disabilities such as Parkinson's disease. Non-diagnosed elderly people showing symptoms affecting hands and fingers, such as trembling, involuntary movements, spasms were also included, as fine motor skills tend to decline with age [17]. We expanded our experimental group with elderly people claiming inability to operate radios, despite not being able to provide a medical record of a specific disability, as challenges associated with aging like inadequate blood flow and circulation to the muscles, injuries, stress, fatigue may also produce spasm in muscles that would reduce the psychomotor capacities. Several residents in our empirical context also reported similar symptoms of cramps from medical side effects, in particular from medication related to Parkinson's disease and Osteoporosis. Other types of developmental or genetic disorders that may have an impact on psychomotor capabilities, but that are not particularly prominent symptoms among the elderly people, were not included in this study (e.g., Down's syndrome, cerebral palsy and dystonia).

A. *Fleishman's taxonomy*

Based on cognitive, sensory, physical, and psychomotor factors, Fleishman derived 52 skills and abilities describing human performance. Although this model was initially developed for a job-related environment, the taxonomy of Fleishman describes abilities and skills that can be associated with performance in everyday tasks [18]. The taxonomy separates abilities from skills; abilities are defined as characteristics and traits shaped throughout the first phase of our lives while skills describe the degree to which we can effectively carry out an action directly related to a given task. Common for the two is that both skills and abilities related to psychomotor capacities involve complex movement patterns and require practice and maintenance in order to remain intact [19].

As the aging process does not follow a schematic or linear development, it is difficult to consider any abilities or skills as less relevant than others. For instance, the cognitive factor constitutes the biggest share of skills and abilities and is obviously relevant also in the discussion of aging-related reduction of interaction capacities. It is further apparent that some of the motoric challenges stem from changes in the cognitive capacities, e.g., ideomotor apraxia where changes in semantic memory capacity reduce the ability to plan or complete motor actions. Studying this category involves abilities and skills that fuse cognitive, perceptual and physical abilities [20].

Studies that focus on elderly people with motor challenges in their hands tend to carry an increased attention towards the abilities and skills that fall under the taxonomic category of physical factors. This is because the muscular restrictions and reduced bodily capabilities in the hands mainly tend to affect the abilities and skills covered by this category. Examples of abilities and skills included in this category are stamina, physical strength and flexibility, balance, and coordination. In previous studies of digital devices in the context of elderly people, we have been concerned with both stamina and physical strength (e.g., in [1]), but in this paper, we mainly focus on psychomotor factors. This is because most of the actions associated with the operation of a radio and other similar digital devices require movement and a configuration of hands and body that relies on the ability to combine physical movement with cognitive functions. Thus, psychomotor factors constitute our main interest, as this organically includes physical skills such as coordination, dexterity, reaction and manipulation. Unlike physical factors, psychomotor factors are also subjected to the influence of reduction in skills and abilities associated with secondary categories; psychomotor capacities often depend on a supportive capacity in addition to the physical. A reduction in other seemingly unrelated features (e.g., visual impairment) may, as Jacko & Vitense [20] point out, have an impact on psychomotor skills.

B. *Scope*

Our study is limited to psychomotor challenges of hands and fingers. Due to inadequate access to fully medically-assessed participants, as well as claimed expertise, we do not address the impact of the decline in cognitive abilities and skills in this paper (e.g., dementia, depression, and forgetfulness). Our scope does not allow us to identify the best interfaces for a given disease but instead let us study the relationship and possible correlation between motor challenges and performance when interacting with radio interfaces. Nor do we want to identify all skills and abilities that are included in the performance of work-related tasks; we aim to identify the specific abilities and skills that are involved in the operation of radios, and affected by reduced capacity in the hands and fingers. Abilities and skills in the taxonomy of Fleishmann are described as independent of each other [12], and it should consequently be possible only to study a selection of these. A similar approach has been conducted in prior research, more specifically in the research of [12, 14, 15, 16]. Table 1 gives an overview of the psychomotor abilities included in our study. A description is provided for each ability based on the original taxonomy of Fleishman [2] in the right column of the table.

TABLE I. OVERVIEW OF PSYCHOMOTOR ABILITIES

Psychomotor ability	Description
Precision control	Ability to move control and the degree to which they can be moved quickly and repeatedly to exact positions.
Arm-hand steadiness	Ability to keep the hand and arm steady, both when suspended in air and while moving. Independent of strength and speed.
Manual dexterity	Ability to make quick and skillful coordinated movements with arms and one or both hands, as well as the ability to assemble, grab and move objects.
Finger dexterity	Ability to make quick, skillful, and coordinated movements with fingers of one or both hands.
Wrist-finger speed	Ability to repeat fast movements with wrist and fingers.
Multi-limb coordination	Ability to use two or more limbs simultaneously to coordinate movements when the body is not in motion.

While all the abilities and skills described in the psychomotor category of the taxonomy are relevant in a broader scope, we have excluded certain abilities and skills from our test. These are abilities that are not relevant for our purposes, and the decision is taken by both the physical challenges we are focusing on, and the digital components and interfaces included in the study. Not all abilities are relevant for the operation of our radios; hence, measuring these abilities would be difficult with the radios. More precisely, rate control, reaction time, speed of limb movement, and response orientation have been excluded. The reason is that these four abilities are not directly determining the capacity to interact with our three radios, but instead, describe the degree to which we can interact with them, as well as the performance during use. Rate control is not appropriate in situations where speed and direction of an object are perfectly predictable [20] while the other three (reaction time, speed of limb movement, and orientation response) mainly concern efficiency of performance, rather than the distinctive ability to perform them. Also, both reaction time and response orientation are intended to capture our reaction to a given signal and our ability to quickly initiate the response routine, something which would be unnatural in a context where our participants are testing our radios. Thus, these four abilities have not been included in our tests.

V. RESEARCH METHOD

A. Radio #1

The first radio is the leftmost radio in Figure 3, and it was developed in 2013. The focus of the radio is to provide an interface that provides users with similar experiences and interaction mechanisms as they are used to from their traditional radios. The feedback one gets from operating radio is reminiscent of interaction found in traditional radios with a distinct response to actions. The focus has also been on finding the materials that provide the best grip and resistance during the interaction. We have explored the properties of various materials (wood, steel, plastic) to find the best functioning design for the knobs. The main interaction takes place by turning on a coarse switch that clearly snaps in place when selecting the channel. A second switch is used to adjust the volume.

Figure 3. The three radios included in the study

B. *Radio #2*

The second radio is the middle radio from Figure 3, and was developed in 2014. This radio depends on physical interaction and does not use traditional switches or buttons. As with the other two radios, this radio is also screenless. The user operates the radio with the use of wooden cubes with built-in Near Field Communication (NFC) chips. The NFC chips are preconfigured with a given radio channel, and by placing these physical cubes on top of the radio, one interacts with the interface. By placing a piece with a given channel on top of the radio starts playing. Removing the cube ends the playback. The focus has been on designing a radio that does not require fine motor skills in fingers. During the design process, material, weight, size and shape were explored in consultation with users to find the best objects for physical operation of the radio.

C. *Radio #3*

The third radio is the rightmost radio in Figure 3, and was developed in 2015. The purpose of this radio was to allow users with tremors, involuntary twitching, and reduced fine motor skills to operate it. The radio is made of oak and has an aluminum cylinder with a wooden knob that automatically snaps to predefined positions using magnets. One operates the radio by positioning the wooden knob at a predefined position. A secondary exploratory feature is that the wooden knob swivels around the cylinder. The design of the radio offers deliberate constraints that prevent users from making mistakes during the interaction. The wooden knob is locked to the pole and the magnet in the cylinder both guides and limits the positioning. This allows involuntary actions to have less impact on the accuracy.

D. *Empirical context and participants*

This study was conducted at three local care facilities in Oslo. Each care facility consists of a set of apartments, with the largest holding 90 apartments. The care facilities consist of senior residents residing in independent apartments, but with shared access to a range of facilities, e.g., cafeteria, lounge, fitness center, and 24-hour staffed reception. The limited access to participants with motor challenges in hands and fingers led to three years of data gathering in order to yield an appropriate set of data. The requirement for participation was that the participant suffered from reduced ability or no ability to operate a store-bought radio and thus needed a more customized interface. The three store-bought radios used for participant selection were Pinell Supersound DAB, Pop DAB Radio and Argon DAB Radio, three highly popular brands in Norway. The data for this study was collected in the period 2013-2015. The three radios used in the tests were also built during the same period. 39 participants (M = 82.1 years, SD = 6.31) participated in six tests. For each test, we recruited an independent control group consisting of 13 elderly people with no apparent motor disabilities (M = 80.4 years, SD = 5.29) who were asked to perform the same tasks as the experimental group.

The testing involved 52 participants in total. Most people had medical documentation to assess their motor disabilities. The documentation was provided to us by themselves or by the local care home administration with their consent. A few participants unable to operate store-bought radios and in the lack of proper medical documentation of disability were also invited to participate in the experimental group as they showed symptoms similar to those with proper diagnoses. Table 2 gives an overview of the participants and the documented or self-assessed disability or illness.

TABLE II. OVERVIEW OF PARTICIPANT GROUPS

Disability or illness	**N**
Cramps	8
Muscle stiffness	3
Osteoarthritis	8
Parkinson's disease	4
Rheumatoid arthritis	3
Tremor	13
Control group	13
Total	**52**

E. *Test procedure*

The participants in both groups were asked to interact with the three radios through a series of repeated tasks to measure their psychomotor performance. Three different tables and eight chairs were used to provide all participants with a setup that supported their preferred bodily configuration. Some participants were also sitting in their wheelchairs during the test, specifically three participants from the experimental group and one participant from the control group. For each of the radios, the participants were given a set of tasks that mimicked the context applicable parts of assignments given in standardized tests of psychomotor abilities, e.g., rotary pursuit test, steadiness tester, Minnesota manual dexterity test, Purdue pegboard, tapping board (as seen in [12], as well as O'Connor finger dexterity test, box and block test, Jebsen hand function test, and Moberg pick-up test. As we used our own set of tasks, the results are not meant to demonstrate the external validity and be directly comparable to other test results, but instead provide a set of tasks applicable to the three radios, thereby providing us with a measurement comparable within the study. To eliminate learning effects and bias due to unfamiliarity with novel interaction mechanisms, each participant was given a demonstration of the intended interaction of each radio, and each participant conducted ten trials for each radio (similar to [5]). The task order was randomized for each participant. We relied on randomized repeated measures to minimize bias due to interpersonal variations between tests. The task set consisted of 12 tasks: gripping, turning, positioning, re-positioning, and resetting the main and secondary interaction element, as well as lifting and moving the radio. Time (seconds), error (count) and precision (position and distance) were observed and measured for each task, and the performance was graded on a normalized scale from 1-10 to make the performance metrics comparable. The computationally-generated normalized score used the four metrics above (seconds, count, position, and distance) to calculate the final score. Thus, the performance scores are not intended to be comparable beyond the scope of our research. In Figure 4, we see two participants from the experimental group testing the positioning and re-positioning of the main interaction element for Radio #2.

Figure 4. Two residents participating in psychomotor measurements

VI. RESULTS AND ANALYSIS

The means and standard deviations for the psychomotor performance on a normalized scale from 1-10 across both groups are shown in Table 3. As expected, the control group had a better performance relatively compared to the experimental group for all three radios. The variation was larger for the control group, and we can read from the table that both groups demonstrated a similar within-group performance for each of the three radios. The average performance score was 7.43 (SD = 0.32) for the control group while it was 4.60 (SD = 0.22) for the experimental group. In Figure 5, we present the estimated marginal means for the control group vs. the regular group for all three radios.

TABLE III. PSYCHOMOTOR PERFORMANCE SCORE

Radio #	Group	Mean	Std. Deviation	Lower Bound	Upper Bound	N
1	Control	7.385	.7372	6.730	8.039	13
	Experimental	4.517	1.2824	6.471	7.658	39
2	Control	7.064	.5755	7.156	8.536	13
	Experimental	4.389	1.1787	4.139	4.895	39
3	Control	7.846	.4328	4.046	4.732	13
	Experimental	4.897	1.4000	4.499	5.296	39

A 2 (group: selection or control) x 3 (radio: #1, #2 or #3) between-subjects analysis of variance (ANOVA) was conducted to study the psychomotor performance between the three radios as a function of the performance. We registered significant main effects of group, $F(1,150) = 173.6$, $p < .005$, $n = .536$, and radio, $F(2,150) = 3.1$, $p = .048$, $n = .040$. The main effects were not qualified by an interaction between group and radio, $F(2,150) = 0.142$, $p = 0.867$, $\eta p^2 = .002$. The participants in the selection group (M = 4.601, SD = .107) had significantly lower performance than the participants in the control group (M = 7.432, SD = .186). The analysis also revealed a slightly lower performance difference between the three radios: (M = 5.951, SD = .186), (M = 5.726, SD = .186), and (M = 6.372, SD = .186). Levene's test for equality of variances was found to be violated for the present analysis ($p = .001$), and Bonferroni post-hoc analysis for the radios showed that Radio #2 had significantly lower performance than Radio #3 at the .05 level, while differences between Radios #1 and #2 and Radios #1 and #3 were not significant.

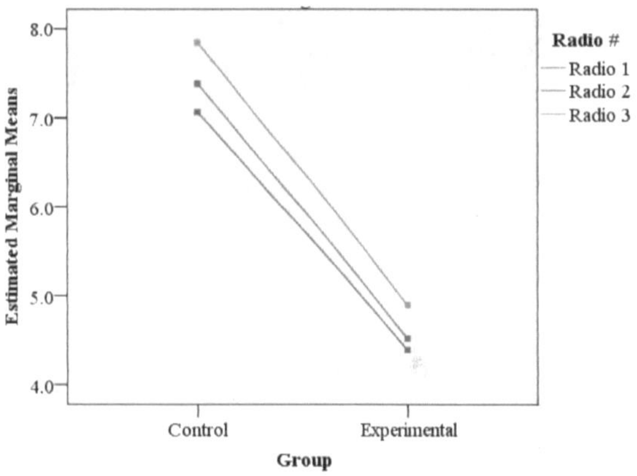

Figure 5. Estimated marginal means of performance for both groups

The results from Figure 5 only demonstrated how the estimated marginal means of the overall performance for all participants in the treatment group compared to the control group. For a post hoc evaluation of the performance within the experimental group, we performed a separate repeated measure analyzes for each level within the grouping factor to study the relationship between performance and psychomotor disability.

We analyzed the data with mixed-design ANOVA using a within-subjects factor of disability (cramp, muscle, osteoarthritis, Parkinson's disease, Rheumatoid Arthritis, Tremor) and a between-subject factor of radio (Radio #1, Radio #2, and Radio #3). Mauchly's test indicated that the assumption of sphericity had been violated ($\chi^2(2) = 2.681$, $p = .026$). Degrees of freedom were corrected using Huynh-Feldt estimates of sphericity ($\varepsilon = 1.000$) as Greenhouse-Geisser estimates reported an epsilon value above 0.75 ($\varepsilon = .926$) [21]. There were non-significant main effects of disability, $F(2, 66) = 5.566$, $p = .006$ and radio, $F = (1, 33) = 8.129$, $p = .007$. However, the main effects were qualified by a significant interaction between disability and radio, $F(10, 66) = 17.011$, $p < .001$. In Figure 6, we demonstrate how the interaction between disability and radio yielded a significant variation in the estimated marginal means of performance.

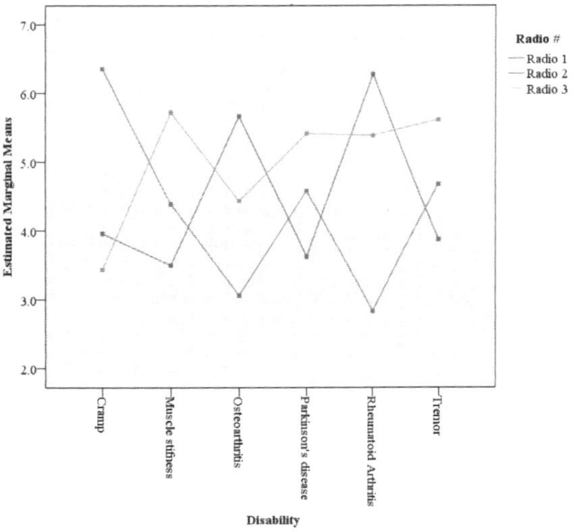

Figure 6. Performance for each disability group across all three radios

Again, the statistical results of this study do not attempt to provide a medical explanation for the performance but instead demonstrates a significant correlation in order to exemplify the need for various interfaces when addressing elderly people with psychomotor disabilities. The study only claims the presence of a significant difference in performance but does not provide any solutions.

VII. DISCUSSION

A. Psychomotor disabilities as a shift rather than a loss

The analysis presented in the previous section demonstrates some important findings. First and foremost, we see that grouping all elderly people in one common category cannot be considered scientifically justifiable when their needs, capacities, and performances are so different. To group the elderly in one common category is both stigmatizing and improper design practice as it neglects individual needs. Also, we have presented empirical data suggesting that even the specific group of older people suffering from motor deficit in the hands and fingers would highly benefit from designs that paid individual attention to their needs.

At first glance, it might look like Figure 5 illustrates a steady and consistent difference between the control group and the experimental group. However, this was not the case. Irregularities in performance resulted in statistically counteracting mean values, and glancing at Figure 5 one may wrongfully conclude that the older participants yielded a seemingly equal performance score for each radio regardless of their motor capacities. However, as presented in the secondary analysis of the relationship between disease and performance (illustrated in Figure 6), we see performance scores with high fluctuation within each group. We can confirm this by looking at the statistical analysis which indicated a significant interaction between disability and radio ($p <.001$).

One way of understanding this phenomenon is to look at average performance score for each group. The participants who suffered from trembling serve as a good illustration. This group, which accounted for a third of the participants in the experimental group, had the lowest performance score on Radio #2 ($M = 4.19$, $SD = 0.40$), an intermediate performance score on Radio #1 ($M = 05.07$, $SD = 1.57$), and the highest on Radio #3 ($M = 6.08$, $SD = 0.98$).These results can be explained by the different types and various symptoms of tremor. Participants reported issues with intention tremor that could affect their aim, specific tremor which influenced goal-oriented action), as well as general stressing tremor. As Radio #2 required participants to raise a cube in midair and place it within a designated area, it was difficult for several participants to operate this radio. With more degrees of freedom compared with the other two radios, there was more room for both intentional and deliberate errors. This group performed best on Radio # 3 as involuntary movements would not give adverse effect or hinder progress in solving the task.

A similar pattern can be seen in the group of participants who suffered from Rheumatoid Arthritis. They reported challenges with swelling, decreased sensitivity and reduced mobility, which resulted in problems with the interface of Radio# 1 ($M = 2.83$, $SD = 0.99$). The reduction in sensitivity in particular would mean that they struggled more with sensing moving, clicking, and snapping feedback from the radio. However, they delivered a good average performance score for Radio # 2 ($M = 6.28$, $SD = 1.74$), suggesting that they still had the capacity for interaction.

Thus, loss or reduction in motor capacities does not an automatically reduce or deprive our interaction opportunities; it mainly shifts them. All three of our radios were developed to allow people with motor impairments in their hands and fingers to still use these limbs for interaction. And our results suggest that they are highly capable of doing so if presented the right interface. In their studies of differences in pointing movements between older and younger users, [16] argues that older people maintain the use of residual sensory information (vision and proprioception) and can achieve similar precision levels as younger users. However, the radios in our study do not need to be operated by hands and fingers. There are also opportunities explore new bodily uses and configurations. In certain context, radios are naturally operated through different interaction mechanisms, for instance in cars. Prior studies have also demonstrated interaction opportunities for people with motor disabilities by the use of other bodily capacities. For instance, [22] use head gesture recognition for wheelchair control for elderly people suffering from Parkinson's disease and other restrictions in limb movement. The authors of [23] study wrist rotation as input mechanisms for mobile devices, and suggest that both hands-free and eyes-free interaction techniques would be feasible with further research. The research of [24] uses a voice-driven drawing application to include users with motor impairments.

We should never exclude any people as potential users just because their capacities prevent them from using a given interface. Incompetence or inability in use should not be tied with technologies, but instead, be a use dimension related to the specific interaction mechanisms that the technology provide. Radio might be considered one piece of technology, but there are limitless opportunities when it comes to the way it is presented to the user. The results in this paper have demonstrated that people can re-establish meaningful relationships with technology by shifting the way of presentation.

B. *Extending and re-establishing purposeful interactions*

It is important to note that none of the three radios was perceived as uniformly better than the rest. Each radio would yield good scores with one or more groups, but there was always another group that would struggle with the same interface. This supports our claim that radios designed for a specific group of people, and with features that may even fully compensate for the motor deficit, will still not necessarily work for everyone. Hence, the results of this study demonstrate not only the need but also the possibility, to make individual adjustments in the design of interfaces. Even though we developed three radios, they all utilized nearly identical hardware, and the basic electronic components are the same in all three radios. The back of the three radios and how their hardware is enclosed in similar casement taking up roughly the same size is demonstrated in Figure 7. They were developed in three independent processes focusing on various psychomotor challenges, yet we see that only the packaging, i.e., the "outer shell" enclosing technology, is changed. By designing three different interfaces, we have shown that it is possible to re-enable an entire group of elderly people who would otherwise have to abandon interaction with radios. And we achieved this while letting them continue to use their hands and fingers, something which is not a requirement for successful interaction. If we expand the design area to include all other bodily capacities, the potential to re-establish purposeful interaction would be even greater, and the chance is simultaneously greater for technology to remain meaningful longer, even when living through a decline in psychomotor capacities. To offer users a variety of interfaces on top of the technology also provides users the ability to customize the interaction to their capacity levels, even if they were to discover at some point that some of their motor skills develop in a positive sense. Adapting to skill levels is encouraged by [19]. It would also open up more room to address changes in movements, actions and bodily configurations as psychomotor skills among user group changed. The authors of [6] suggest that there is a natural discontinuation in slow movements among older people. This is also supported by [5] who suggest that older participants depend on interfaces that allow for more sub-movements in interaction. In general, it is considered reasonable to spend more time on interface adaptations since the majority of prior research only studied two-factor analysis of the interaction and psychomotor capacity in a context where other conditions such as frequency of use and expertise could have had an impact on the interaction [9].

Figure 7. The back of the three radios

In our empirical context, this idea of introducing multiple interfaces is particularly important as Norway is facing an infrastructural change where all radios are switching from FM broadcasting to digital audio broadcasting (DAB). This will render all current FM radios unusable as of 2017. People with older radios are forced to buy new devices where the interaction may depend on users properly learning and understanding new interfaces, new terminologies, new frequencies and new mechanisms. However, prior research simultaneously suggests that elderly people are less willing to modify current strategies or adapt new strategies [5, 25]. This forced transition gives us a golden opportunity to introduce a variety of interaction mechanisms that can be incorporated into routines and habits while people are relatively able-bodied and only shows early symptoms. By doing so, the technology could potentially remain with them even if they were to enter a downward phase with reduction of capacities. If someone should not develop symptoms consistent with the expectations, having incorporated these new interaction mechanisms may still have a positive effect as it is often the underlying factors that are to be blamed for reduction of psychomotor skills, e.g., in the performance of tasks aiming [12].

Another important factor is the degree of stigmatization associated with use. Technology tailor-made for a certain group of people often succumbs to design choices so distinctive that other people can interpret the intended users their weaknesses just from the design itself. Our participants claimed that all three radios, but, in particular, Radio #3, had looks that did not suggest being specifically designed for the target audience. The design did not emit the stigmatizing radiance often found in technology tailored for the elderly [4]. To early expose users interfaces that can have a secondary function ago, will also allow older staple acquaintances interaction mechanisms not yet been vital for their use. Thus, there is less chance that they will experience the design and interface as stigmatizing about it sometime in the future should be such that it was the specific interaction mechanism that allowed interaction; interaction is associated with routines and habits rather than to impose solutions. Early exposure to interfaces that can have a secondary function later will also allow older users to make acquaintances with interaction mechanisms that have not yet become vital for their use. This would mean fewer chances of experiencing the design and interface as stigmatizing, even though it sometime in the future may become the very interaction mechanism allowing interaction; the interaction is associated with routines and habits rather than to imposed solutions.

This discussion of avoiding stigmatizing design further aligns with the idea of universal design. Design tailored for specific disabilities or illnesses does not exclude people without disabilities from using them. On the contrary, we found that the design of our three radios, and, in particular, Radio #3, appealed to participants and stakeholders that were not in the user group such as family members, employees at the care home, and even our self as designers. In future research, it would be interesting to investigate this aspect of the design further. While our results does not provide any significant evidence of one radio fully re-establishing interaction for all types of psychomotor disabilities, we did see examples of radios elevating the interaction performance to the level of the control group for multiple types of disabilities and illness (as demonstrated with Radio #1 and Radio #2 in Figure 6). It is therefore not unreasonable for further research on this topic to generate designs that can reach even more people and help users achieve even better performance scores. Nevertheless, the aesthetics of the three radios demonstrate the important underlying idea that design tailored for a specific user group can very well be fully usable and appealing to everyone. There is no reason that design for elderly people cannot be design for all.

VIII. CONCLUSION

In this paper, we have discussed the opportunities to re-establish interaction through the use of alternative interfaces. We have studied the case of elderly people suffering from a reduction in psychomotor capacities in hands and fingers to demonstrate how three different radio interfaces allowed them to return to the use of radios. Six rounds of tests were conducted with 52 participants in 2013-2015. We demonstrate how none of the three radios were universally acclaimed, and how extending and re-establishing purposeful actions require a more nuanced and adapted interface, even for a specific demographic as the one included in this study. Our main findings suggest that providing interfaces that acknowledge and compensate for psychomotor disabilities can re-enable interaction comparable to the level of fully-functional users.

To investigate our findings further, we aim to conduct a long-term testing with these proposed models that will yield qualitative data better suited for an interpretive analysis; the addition of subjective opinions and experiences from elderly users living alongside our proposed designs will help support the claims presented in this paper. The work presented in this paper encourages further research into opportunities for re-establishing interaction. Our particular case should also only be considered a starting point for similar research with a different or expanded scope. We have studied a limited set of challenges, i.e., psychomotor disabilities in hands and finger, but we would like to advocate further research on similar topics such as discussion of other types of psychomotor abilities or the role of cognitive capacities in re-establishing interaction.

REFERENCES

[1] S. G. Joshi, "Designing for Experienced Simplicity. Why Analytic and Imagined Simplicity Fail in Design of Assistive Technology." The International Journal on Advances in Intelligent Systems, vol. 8, no. 3&4, 2015, pp. 324-338.

[2] E. A. Fleishman and M. E. Reilly, "Handbook of human abilities: Definitions, measurements, and job task requirements." Consulting Psychologists Press, 1992.

[3] S. Norway, "Norwegian Media Barometer, 2014," SSB.

[4] S. G. Joshi, "Emerging ethical considerations from the perspectives of the elderly," in Ninth International Conference on Cultural Attitudes in computer-Human Interactions, M. Strano, H. Hrachovec, and S. Fragos, 2014, pp. 1-15.

[5] M. W. Smith, J. Sharit, and S. J. Czaja, "Aging, motor control, and the performance of computer mouse tasks." Human Factors: The Journal of the Human Factors and Ergonomics Society, vol. 41, no. 3, 1999, pp. 389-396.

[6] M. Morgan, J. G. Phillips, J. L. Bradshaw, J. B. Mattingley, R. Iansek, and J. A. Bradshaw, "Age-related motor slowness: simply strategic?." Journal of gerontology, vol 49, no. 3, 1994, pp. M133-M139.

[7] C. N. Riviere and N. V. Thakor, "Effects of age and disability on tracking tasks with a computer mouse: Accuracy and linearity." Journal of Rehabilitation Research and Development, vol. 33, 1996, pp. 6-15.

[8] C. Poletti, R. Sleimen-Malkoun, J.-J. Temprado, and P. Lemaire, "Older and younger adults' strategies in sensorimotor tasks: Insights from Fitts' pointing task." Journal of Experimental Psychology: Human Perception and Performance, vol. 41, no. 2, 1994, p. 542.

[9] C. Sutter and M. Ziefle. "Psychomotor user characteristics as predictors for a successful usage of small input devices." in Proceedings of the HCI International 2005, pp. 1-11.

[10] R. W. Soukoreff and I. S. Mackenzie, "Towards a standard for pointing device evaluation, perspectives on 27 years of

Fitts' law research in HCI." International journal of human-computer studies, vol. 61, no. 6, 2004, pp. 751-789.

[11] C. Sutter and M. Ziefle. "Psychomotor performance of input device users and optimized cursor control." in Proceedings of the Human Factors and Ergonomics Society Annual Meeting. SAGE Publications. 2006, pp. 742-746.

[12] Y. Cheong, R. L. Shehab, and C. Ling, "Effects of age and psychomotor ability on kinematics of mouse-mediated aiming movement." Ergonomics, vol. 56, no. 6, 2013, pp. 1006-1020.

[13] Y.-K. Cheong, "Performance and movement kinematics of mouse pointing task: Perspectives from age, psychomotor ability, and visual ability," University of Oklahoma, 2007.

[14] N. Đorđević, D. Rančić, and A. Dimitrijević. "Evaluation of User Cognitive Abilities for HCI." in Proceedings of the 11th WSEAS International Conference on COMPUTERS, Agios Nikolaos, Crete Island, Greece. Citeseer. 2007, pp. 468-473.

[15] J. Knight and G. Salvendy, "Psychomotor work capabilities." Handbook of industrial engineering, 2nd edn. Wiley-Interscience, New York, 1992, pp. 978-989.

[16] L. Zhang, J. Yang, Y. Inai, Q. Huang, and J. Wu, "Effects of aging on pointing movements under restricted visual feedback conditions." Human movement science, vol. 40, 2015, pp. 1-13.

[17] B. Makofske, "Manual Dexterity," in Encyclopedia of Clinical Neuropsychology. Springer, 2011, pp. 1522-1523.

[18] A. Furnham, S. Von Stumm, A. Makendrayogam, and T. Chamorro-Premuzic, "A Taxonomy of Self-Estimated Human Performance: The General Factor." Journal of Individual Differences, vol. 30, no. 4, 2009, pp. 188-193.

[19] M. H. Fogtmann. "Kinesthetic empathy interaction–exploring the possibilities of psychomotor abilities in interaction design." in Second International Workshop on Physicality, 2007, p. 37.

[20] J. A. Jacko and H. S. Vitense, "A review and reappraisal of information technologies within a conceptual framework for individuals with disabilities." Universal Access in the Information Society, vol. 1, no. 1, 2001, pp. 56-76.

[21] A. Field and G. Hole, "How to design and report experiments." Sage, 2002.

[22] J. Gray, P. Jia, H. H. Hu, T. Lu, and K. Yuan, "Head gesture recognition for hands-free control of an intelligent wheelchair." Industrial Robot: An International Journal, vol. 34, no. 1, 2007, pp. 60-68.

[23] A. Crossan, J. Williamson, S. Brewster, and R. Murray-Smith. "Wrist rotation for interaction in mobile contexts." in Proceedings of the 10th international conference on Human computer interaction with mobile devices and services. ACM. 2008, pp. 435-438.

[24] S. Harada, J. O. Wobbrock, and J. A. Landay. "Voicedraw: a hands-free voice-driven drawing application for people with motor impairments." in Proceedings of the 9th international ACM SIGACCESS conference on Computers and accessibility. ACM. 2007, pp. 27-34.

[25] N. Walker, D. A. Philbin, and A. D. Fisk, "Age-related differences in movement control: Adjusting submovement structure to optimize performance." The Journals of Gerontology Series B: Psychological Sciences and Social Sciences, vol. 52, no. 1, 1997, pp. 40-53.

ICT Platform for Cognitive Stimulation: Technological Description and Evaluation of Effectiveness and Impact of the Practice in Alzheimer's Disease Patients

Andrea Caroppo, Alessandro Leone, Pietro Siciliano
Institute for Microelectronics and Microsystems, Italian National Research Council (CNR), Via Monteroni, c/o Campus Università del Salento, Palazzina A3
Lecce, Italy
email:andrea.caroppo@le.imm.cnr.it
email:alessandro.leone@le.imm.cnr.it
email:pietro.siciliano@le.imm.cnr.it

Daniele Sancarlo, Grazia D'Onofrio, Antonio Greco
Geriatric Unit & Laboratory of Gerontology and Geriatrics, Department of Medical Sciences, IRCCS "Casa Sollievo della Sofferenza"
San Giovanni Rotondo, Foggia, Italy
email:d.sancarlo@operapadrepio.it
email:g.donofrio@operapadrepio.it
email:a.greco@operapadrepio.it

Abstract— **Cognitive impairments, such as memory problems or other thinking problems, are a defining feature of the early stages of Alzheimer's Disease (AD). Cognitive Stimulation (CS) is a relatively new approach to improve well-being for people with mild to moderate AD. At present only preliminary evidence regarding efficacy is available but it is enough to suggest that this CS has the potential to bring about changes in behavior and maintain involvement in daily life. The main contribution of this work is to evaluate the effectiveness and impact of CS in AD patients using an information and communication technologies (ICT) platform, implemented through a customized Virtual Personal Trainer (VPT) that can allow the patients to perform the CS practice directly at home. The whole system is made up of an embedded PC connected to a TV monitor with internet connection, a low-cost 3d sensor and an optional e-shirt with textile electrodes for clinical signs detection. Moreover, the system provides an audio/visual link with the medical center, so the physician can interact with the patient during the rehabilitation practice, increasing the compliance and the efficacy and making sure that type and intensity of treatment are appropriate. Results obtained after the experimentation stages demonstrates that the use of the aforementioned ICT platform can improve the cognitive and neuropsychiatric state of the patient and its quality of life, confirming that CS could be represent an attractive option in the support of caregivers and patients affected of AD.**

Keywords- Cognitive Stimulation; Alzheimer's Disease; ICT Platform; Quality of Life.

I. INTRODUCTION

In the recent years, there has been significant progress in using Information and Communication Technologies (ICT) in the field of healthcare; in particular, a great effort has been addressed by researchers in order to develop enabling solutions that are cost-effective. Alzheimer's Disease (AD) is the most common form of dementia, a neurologic disease characterized by loss of mental ability. This kind of deficit interferes frequently with normal activities of daily living (ADLs) and usually occurs in old age. Generally, it is marked by a decline in cognitive functions such as remembering, reasoning and planning.

The importance of Cognitive Stimulation (CS) in the treatment of patients with dementia is underlined by recent scientific publications [1]-[3]. Therefore, a recent study showed that integrated treatment with rivastigmine transdermal patch and CS for six-months significantly improves the cognitive, emotional, behavioral aspects and mortality risk of AD patient [4]. The development of a low-cost platform/home-care service with CS functionalities could be very useful in order to increase the chances of an appropriate medical therapy. Some preliminary studies show that ICT tools are well accepted by elderly people, although education and ICT skills level is often low. Moreover, it is scientifically proven that ICT technologies improving quality of life and increasing the permanence at home.

In the field of healthcare, different kinds of technologies have being developed and used for cognitive training and stimulation in the past years [5]-[7]. For example, virtual reality offers training environments in which human cognitive and functional performance can be accurately assessed and rehabilitated [8][9]. On the other hand, augmented reality provides safer and more intuitive interaction techniques allowing interaction with 3D objects in real world [10][11]. In this scenario, social communication channels (natural speech, para-language, etc.) are not blocked, breaking down mental barriers applying such a technology to specific problems or disabilities. New solutions for cognitive assistance based on touch system have been implemented: in the field of CS, commercial products (Nintendo's Brain Age, Big Brain Academy, etc.) have been tuned as educational tools helping to slow the decline of AD [12][13]. More recently, the large diffusion of interaction devices enabling body movements to control systems have been investigated, with specific focus on ICT technologies for natural interaction. Microsoft Kinect is the state-of-the-art [14] as 3d device for body movements acquisition and gesture recognition and the effects of this kind of technology for rehabilitation purposes is widely investigated [15][16].

In this work, a Natural User Interface (NUI) platform has been designed with the aim to support different kind of patients during the multi-domain stimulation practice without the presence of medical staff or caregiver. The remainder of the paper is organized as follows. Section II reports the overview of the platform with specific focus on the hardware devices used for the interaction with the system. In Section III, some details about stimulation practice and environment setup are described. Section IV reports the evaluation of effectiveness and impact of the ICT platform in AD patients. In the Section V the result obtained are reported and finally the conclusions of this paper are presented in Section VI.

II. OVERVIEW OF THE PLATFORM

The developed ICT platform (called AL.TR.U.I.S.M. - Alzheimer patient's home rehabilitation by a Virtual Personal Trainer-based Unique Information System Monitoring) provides a digital tool for CS through VPT allowing the patients to perform the rehabilitation practice at home. From this perspective, the process is a highly innovative compared to existing systems [17] as the caregiver/physician defines a specific sequence of exercises (the therapeutic session) according to the residual abilities of the patient. From the hardware point of view the platform is made up of a set-top-box (a commercial embedded PC) connected to a TV monitor with Internet connection, a Microsoft Kinect sensor for human body tracking and gesture recognition and a WWS system with textile electrodes for clinical signs monitoring (Figure 1).

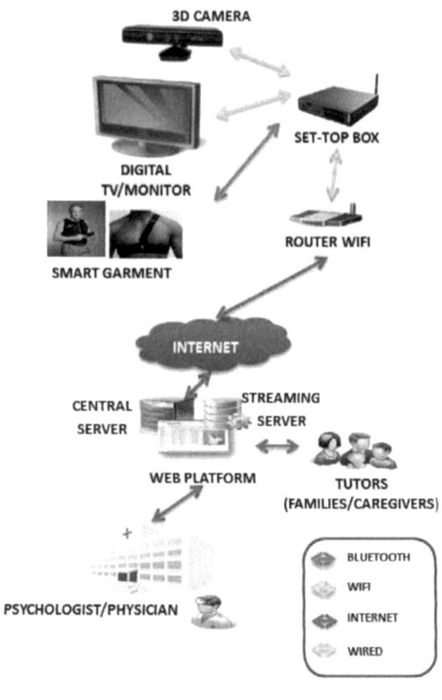

Figure 1. AL.TR.U.I.S.M. platform overview

Moreover, the platform integrates a software module able to stream visual and audio data acquired during the therapy. This module can record the video streaming for post-verification from a remote architecture by the caregiver or physician. From this perspective, the physician or the psychologist of the reference centre could communicate to the patients through a remote connection and then monitor the progress or trouble in the execution of the different required tasks.

At the end of the rehabilitation session, the central platform collects different kinds of data locally stored on the embedded PC. Finally, an ad-hoc multi-modal messaging procedure (e-mail, SMS, Mobile App for Android devices, etc.) is performed and relevant data are sent to the physician allowing instant verification of the performance through an easy-to-use Graphical User Interface (GUI). In the next sections some details regarding the hardware devices involved in the platform are reported.

A. *Microsoft Kinect Sensor*

In order to approach the CS practice, the end-user needs a specific hardware device (Microsoft Kinect) which allows to interact intuitively with a GUI using their bodies (Figure 2.a). The Kinect is a motion sensor that can measure three-dimensional motion of a person. From the functioning principle point of view, the Kinect device integrates both a high resolution RGB camera and an infrared depth sensor.

Microsoft's 'Kinect for Windows SDK'[18], was used to provide an Application Programmer's Interface (API) to the Kinect hardware. The API was used to interface with the Kinect sensor and its skeletal tracking software, providing an estimate for the position of 20 anatomical landmarks at a frequency of 30 Hz and spatial and depth resolution of 640 × 480 pixels.

The space resolution along the x and y axis is 3 mm at a depth of 2 meters, whereas the resolution of z-depth is 1 cm at the same depth. While increasing distance from the sensor, the accuracy decreases remaining within an acceptable range for people body part detection.

B. *Wearable Wellness System*

Another important feature of the platform is the continuous monitoring of physiological parameters during the execution of the therapy, for the evaluation of psycho-emotive stress of the patient involved in the CS practice. Physiological parameters are collected by a Wearable Wellness System (WWS) produced by Smartex [19]. The system includes a sensorized garment (Figure 2.b) and an electronic device (Figure 2.c). The sensorized garment is equipped with two textile electrodes directly connected to the device (named SEW). A jack connector links the sensorized garment with the electronic device. The e-shirt is available in both male and female version, with size ranging from S to XL. WWS is able to simultaneously acquire ECG, heart rate, breath rate and acceleration values along x-axis, y-axis and z-axis. Data acquired are sampled with different rates (breath-rate@25Hz, heart-rate@0.2Hz, 1-derived ECG channel@ 250Hz) and transmitted to the set-top-box via Bluetooth radio link.

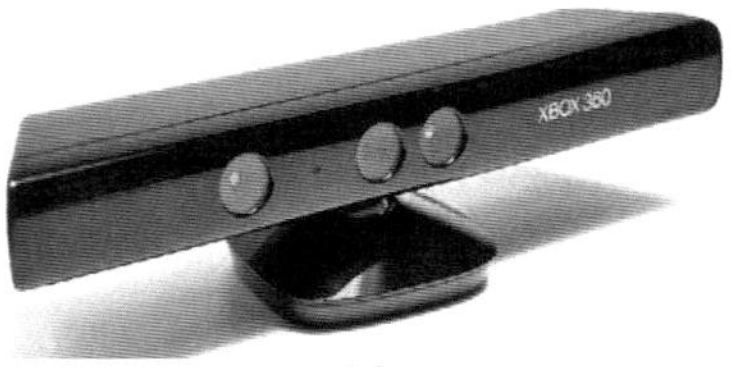

(a)

(b)

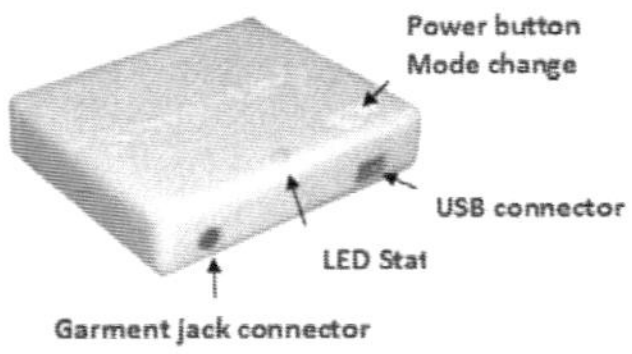

(c)

Figure 2. (a) Kinect 3d sensor device, (b) smart garment and (c) electronic device able to collect and elaborate physiological parameters

III. STIMULATION PRACTICE DETAILS AND ENVIRONMENT SETUP

A. *Multi-Domain CS Practice*

The CS program is composed by sequences of exercises appropriately tuned by the physician or psychologist. Each exercise belongs to a category, bringing out specific cognitive activities according to guidelines of the state-of-the-art international evaluation scales for AD (e.g., Mini Mental State Examination - MMSE [20]). An innovative feature of the platform deals with the opportunity to customize each exercise on the basis of the severity of cognitive impairment and residual skills of the target. For this purpose, during the setting procedure, few input parameters need to be defined a-priori (e.g., execution time, maximum numbers of allowed errors, movement sensitivity). From the taxonomic point of view, the following categories of exercises have been implemented: temporal orientation, personnel guidance, topographical memory, visual memory, hearing attention, visual attention, categorization and verbal fluency. Figure 3 shows the GUI of some CS exercises.

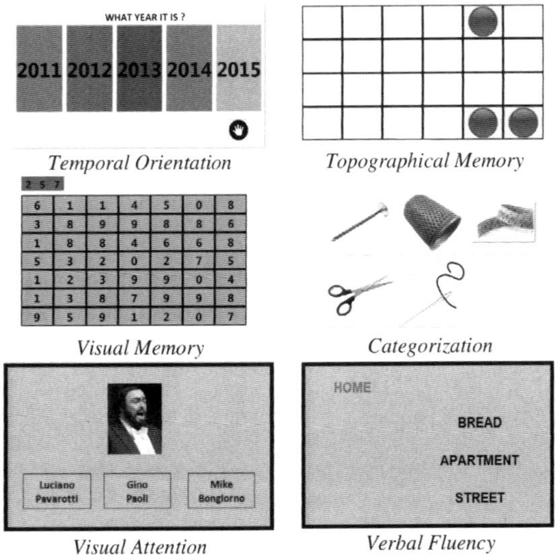

Figure 3. Examples of GUI for different categories of CS exercises.

The design of the therapy can be remotely performed, thanks to a web application that allows physician to configure all the exercises based on the patient's residual abilities and related performance. As a result, the interface of a specific exercise can be different for each patient. For example, the exercise "Topographical Memory" requires the following parameters: number of rows in the grid, number of columns in the grid, number of red dots (correct answers), whereas the exercise "Categorization" requires only the number of images to display. In addition to the GUI modelling, it is possible to establish the maximum length (in time unit) of every exercise and to set (only for a certain categories of exercises) the number of aids. The specific input required for each category of exercise with the information about the presence or not of aids are reported in the next table:

TABLE I. SPECIFIC INPUT REQUIRED FOR EACH CATEGORY OF EXERCISE

Category	*Specific Input*	*Help?*
Temporal Orientation	Year, month, day, day of the week	Yes
Personnel Guidance	Body part target	Yes
Topographical Memory	N° of answers, N° grid rows, N° grid colums	Yes
Visual Memory	N° of images	Yes
Hearing Attention	Text/Story, N° of words to identifiy	No
Visual Attention	N° of answers, N° grid rows, N° grid colums	No
Categorization	N° of images	Yes
Verbal Fluency	Target word, synonym/contrary word	Yes

An appropriate display of every exercise and the independence from the specific output device (digital monitor, HD TV, etc.) is obtained thanks to the implementation of a software module for the best video rendering. The definition of graphics objects displayed on the GUI has been designed according to the principles of ergonomics, usability as referred in ISO/IEC 2001a [21] and acceptability through an extensive literature search, user experience and expert opinion.

B. Environment Setup

The setup of the environment can be done without any specific help, but it is important to observe a few simple requirements for proper CS practice execution.

For the best performance during the practice, Kinect must be placed allowing to acquire the whole body (see Figure 4). Some tips on how to place the sensor are listed in the following:

- sensor must be place near the edge of a flat, stable surface;
- sensor should be within 15 cm above or below a TV monitor, and between 0.6 meters and 1.8 meters from the floor;
- avoid positioning the sensor in direct sunlight or within 0.3 meters of audio speakers.

The end-user must be far from Kinect sensor at least 1.5 meters and never more than 3 meters (Figure 5), assuring the proper functioning of 3d skeletonization procedure provided by Microsoft Kinect SDK. Some categories of exercises can be executed from a seated position; however the user must always respect the specific operative range of the platform.. In order to interact with the VPT, the patient must move the hand minimizing occlusion with other body parts. Hand tracking algorithms have been implemented providing a customized level of movement sensitivity which is manually tuned by the physician (three different level of sensitivity have been implemented). For specific exercise (e.g., Personnel Guidance) gesture recognition algorithms have been developed in order to verify the correctness of the hand movement with respect to a previously recorded template.

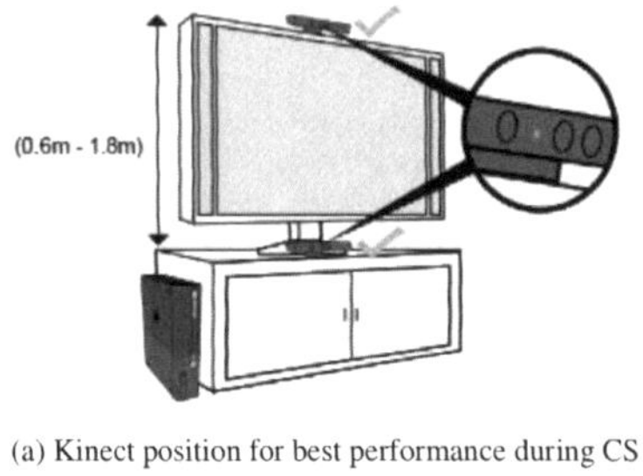

Figure 4. (a) Kinect position for best performance during CS practice

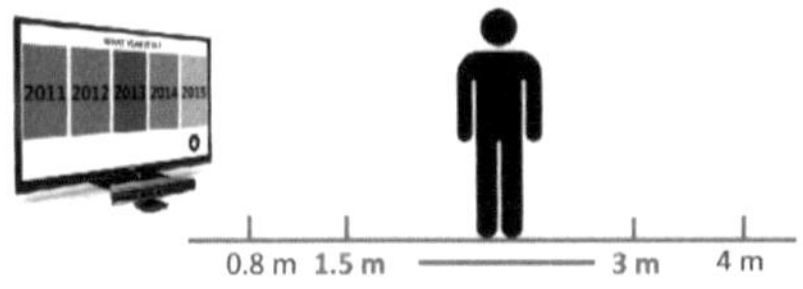

Figure 5. Spatial position of end-user for proper interaction with the platform

IV. EFFECTIVENESS AND IMPACT OF THE PRACTICE IN AD PATIENTS

The study was conducted according to the Declaration of Helsinki, the guidelines for Good Clinical Practice and following the CONSORT statement. Was approved by our local Ethic Committee. The experimentation stages have begun at September 2013 up to February 2014. Inclusion criteria were: 1) age ≥ 65 years; 2) diagnosis of Dementia according to the National Institute on Aging-Alzheimer's Association (NIAAA) criteria [22]; 3) ability to provide an informed consent or availability of a proxy for informed consent. Exclusion criteria were: presence of serious comorbidity, tumours and other diseases that could be causally related to cognitive impairment (ascertained blood infections, vitamin B12 deficiency, anaemia, disorders of the thyroid, kidneys, or liver), history of alcohol or drug abuse, head trauma, and psychoactive substance use; presence of severe cognitive impairment (MMSE< 10). At the baseline and at the follow-up, performed after experimental stage, the following parameters, explained in details in the text, were collected by a systematic interview, clinical evaluation, and review of records from the patients' general practitioners: demographic data, clinical and medication history and a complete multidimensional and cognitive-affective assessment.

A. Cognitive evaluation and diagnosis of dementia

In all patients, cognitive status was screened by means of the Mini-Mental State Examination (MMSE) Babcock Story Recall Test (BSRT) [23], Attentional Matrices (AM) [24], Verbal Fluency (VF) [25], and Copying of Geometric Figures (CGF) [26].

Dementia was diagnosed by the Diagnostic and Statistical Manual of Mental Disorders – 5 Edition (DMS 5) criteria [27]. Diagnoses of possible/probable AD were made according to the NIAAA criteria and supported by neuroimaging evidence (CT scan and/or NMR).

B. Neuropsychiatric and affective assessment

Neuropsychiatric symptoms was evaluated with the Neuropsychiatric Inventory (NPI) [28] including the following 12 domains: delusions, hallucinations, agitation/aggression, depression mood, anxiety, euphoria, apathy, disinhibition, irritability/lability, aberrant motor activity, sleep disturbance and eating disorder. Affective status was evaluated using the Hamilton Rating Scale for Depression (HDRS-21) [29].

C. *Comprehensive Geriatric Assessment (CGA)*

A CGA was carried out using assessment instruments widely employed in geriatric practice. Functional status was evaluated by activities of daily living (ADL) index [30], and by instrumental activities of daily living (IADL) scale [31]. Cognitive status was screened by the Short Portable Mental Status Questionnaire (SPMSQ) [32]. Comorbidity was examined using the Cumulative Illness Rating Scale (CIRS) [33]. Nutritional status was explored with the Mini Nutritional Assessment (MNA) [34]. The Exton-Smith Scale (ESS) was used to evaluate the risk of developing pressure sores [35]. Medication use was defined according to the Anatomical Therapeutics Chemical Classification code system, and the number of drugs used by patients was recorded. Social aspects included household composition, home service, and institutionalization.

D. *Quality of Life and Satisfaction assessment*

The instrument to be used to assess the quality of life and satisfaction. will be Quality of Life Enjoyment and Satisfaction Questionnaire (Q-LES-Q) [30]. It's a self-report measure designed to easily obtain sensitive measures of the degree of enjoyment and satisfaction experienced by subjects in various areas of daily functioning

V. RESULTS

A. *Questionnaire results before experimental period*

Between January and April 2013 a survey was conducted at the Alzheimer Evaluation Unit of the "Casa Sollievo della Sofferenza" Research Hospital in San Giovanni Rotondo (FG). The questionnaire included 17 items that explored different areas; the survey included 72 patients and their caregivers. The summary of the results are the following: 51.6% of patients need care in self-hygiene, 15.6% in moving at home, 12.5% in moving out of home and 10.9% in taking drugs. Among the patients with mild AD: 46.9% had a slight deficit in personal care, 37.5% had a mild motor disability, 6.3% sever loco motor disability, 4.7% present a serious lack in personal care, and 4.8% significant sensory deficits. The 48.4% of patients live with a spouse, 26.6% with a child or sibling, 23.4% alone and 1.6% with a private caregiver. Patients are assisted by family members (92.2%, with a mean age of 63.59 years ± 16.50) and private caregiver (7.8%, with a mean age of 48.20 ± 2.78 years). Family members take care for patients for an average of 6.22 ± 1.02 days a week and 6.94 ± 5.44 hours per day; private caregivers caring for patients for 6.00 ± 0.71 days a week and 12.0 ± 7.87 hours per day.

In the sample, 54.7% of patients like to watch TV and 26.6% watch preferably film. The 92.2% of patients are capable to turn on and off the TV, 85.9% is able to choose the programs in autonomy and use the remote control, the 67.2% is able to maintain a level of attention from the beginning to the end of a television program; the 79.7% of patients can use a PC only if helped and 50% cannot use the PC keyboard. Family members/caregivers have reported in their preferences that the AL.TR.U.I.S.M. system could be useful to facilitate communication with the medical center of reference (34.4%) and improve the quality of care through constant monitoring of the exercises performed (32.8%). Moreover, the system could be very useful to improve the quality of life (73.4%) and quality of care and assistance (73.4%). Finally, 62.5% of patients could accept the use of interactive TV and devices to monitor vital signs and gesture recognition at a distance.

B. *Pilot Results*

The pilot study has included 6 patients enrolled in three different sites. Each patients have an initial program of six exercises for session with parameters established on the basis of the first rehabilitation test. Analysing the user responses, the number of exercises and parameters were increased or reduced respectively. All the patients enrolled showed a similar good acceptability to the use of the platform as measured through the use of subjective feed-back. All patients have concluded the study and no drop-out were registered. The sensorized shirt was used to better set-up the system during the first rehabilitation test meanwhile at home these system weren't used.

After experimental period, the end users showed an improvement of 10.4% on the MMSE score, of 1.2% on the BSRT, 1.3% on the Rey-15, 12.64% on the AM, 2.5% on the VF and 1.3% on the CGF. In addition, the end users showed an improvement of 24.5% on the HDRS-21 score, 13.2% on the NPI score and 11.78% on the NPI-D (subscale of NPI that assesses the distress of the caregiver) score. They showed a significant improvement of 47.89% on the Q-LES-Q score. The data included the CGA domains showed no differences. Unfortunately the results were not significant different from a statistical point of view for the extremely small sample size but the trend are promising.

VI. CONCLUSION

For elderly people, home is a place of memories where they spend most of their time. Their demands on to stay home will increase and change with growing age, especially when their health status starts to worsen. An important aspect for all people having the need to be supported in their daily-life-activities is to remain integrated in social life, despite of their age and/or existing disabilities.

We demonstrated that the use of the VPT can improve the functional, nutritional, cognitive, affective and neuropsychiatric state. Furthermore, the VPT can improve the customer's satisfaction and quality of life.

The home cognitive rehabilitation could be represent an attractive option in the support of caregivers and patients affected of AD. VPT is being implemented into more and more homes of the elderly in order to maintain their independence and safety. These VPT allow the elderly to stay in their homes where they feel comfortable, instead of moving to a health care facility. The transition to a health care facility can cause a lot of anxiety and to stay home using the VPT can either prevent or delay this anxiety. Moreover, VPT can provide the elderly with many different types of emergency assistance systems, security features, automated timers, and alerts. These systems allow for the individual to feel secure in their homes knowing that help is only minutes away. VPT will make it possible for family members to monitor their loved ones from anywhere with an internet connection.

ACKNOWLEDGMENT

This work has been carried out within AL.TR.U.I.S.M. project, funded by Apulia Region under the Health-care call for the promotion of regional partnerships for innovation. Authors thank the staff of the Geriatric Unit of the Hospital in San Giovanni Rotondo (Italy) for the validation of the platform.

REFERENCES

[1] L. Clare and R. T. Woods, "Cognitive Rehabilitation in Dementia", in Neuropsychologistical Rehabilitation, Psychology Press Ltd, 2001, 193 – 196.

[2] M. Neal and M. Briggs, "Validation therapy for dementia", (Cochrane Review). The Cochrane Library, Issue 2. Chichester, UK, 2004.

[3] L. Clare, R. T. Woods, E. D. Moniz Cook, M. Orrel, A. Spector, "Cognitive Rehabilitation in Alzheimer's Disease", Aging Clinical and Experimental Research,2006, 18:141-143.

[4] G. D'Onofrio et al. "A pilot randomized controlled trial evaluating an integrated treatment of rivastigmine transdermal patch and cognitive stimulation in patients with Alzheimer's disease" Int J Geriatr Psychiatry, 2015, 30(9):965-975.

[5] E. Richard, V. Billaudeau, P. Richard, G. Gaudin, "Augmented Reality for Rehabilitation of Cognitive Disable Children: A preliminary Study" Virtual Rehabilitation, 2007, 102-08.

[6] P. Standen and D. Brown, "Virtual reality and its role in removing the barriers that turn cognitive impairments into intellectual disability" Virtual Reality, 2006, 10 (3), 241-52.

[7] M. J. D. Taylor, D. McCormick, R. Impson, T. Shawis, M. Griffin, "Activity-promoting gaming systems in exercise and rehabilitation" Journal of Rehabilitation Research, 2011, 48 (10), 1171-86.

[8] M. Wang and D. Reid. "Using the virtual reality-cognitive rehabilitation approach to improve contextual processing in children with autism." The Scientific World Journal 2013, 2013

[9] E. B. Larson, M. Feigon, P. Gagliardo, A. Y. Dvorkin, "Virtual reality and cognitive rehabilitation: A review of current outcome research." NeuroRehabilitation, 2014, 34.4:759-772.

[10] R. Azuma, Y. Baillot, R. Behringer, S. Feiner, S. Julier, B. MacIntyre, "Recent advances in augmented reality" IEEE Computer Graphics and Applications, 2001, 21(6): 34–47.

[11] R. Azuma, "A survey of augmented reality" Presence: Teleoperators and Virtual Environments, 1997,6(4): 355–385.

[12] L. E. Nacke, A. Nacke, C. A. Lindley, "Brain training for silver aged gamers: effects of age and game form on effectiveness, self-assessment, and gameplay" in Cyberpsychology & Behavior, 2009, 12(5), 493 – 499.

[13] F. Imbeault, B. Bouchard, A. Bouzouane, "Serious Games in Cognitive Training for Alzheimer's Patients" IEEE International Conference on Serious Games and Applications for Health, 2011, 122 – 129.

[14] http://www.microsoft.com/en-us/kinectforwindows, retrieved: March,2016

[15] A. Da Gama, T. Chaves, L. Figueiredo, V. Teichrieb, "Improving Motor Rehabilitation Process through a Natural Interaction Based System Using Kinect Sensor" Proceedings of IEEE Symposium on 3D User Interfaces, 145-46, 2012.

[16] B. Lange, C. Y. Chang, E. Suma, B. Newman, A. S. Rizzo, M. Bolas, "Development and evaluation of low cost game based balance rehabilitation tool using the Microsoft Kinect Sensor" 33rd IEEE International Conference on Engineering in Medicine and Biology Society, 2011.

[17] J. Solana et al., "PREVIRNEC, A new platform for cognitive tele-rehabilitation" Third International Conference on Advanced Cognitive Technologies and Applications, 2011

[18] https://dev.windows.com/en-us/kinect, retrieved: March,2016

[19] http://www.smartex.it , retrieved: March,2016

[20] M. F. Folstein, S. E. Folstein, P. R. McHugh, "Mini-mental state: a practical method for grading the cognitive state of patients for the clinician" Journal of psychiatric research, 1975, 12(3), 189-198.

[21] ISO/IEC. 2001. International Standard ISO/IEC 9126 – 1. Software engineering - Product quality - Part 1: Quality model. International Organization for Standardization/International Electrotechnical Commision, Geneva, 2001.

[22] G. M. McKhann, D. S. Knopman, H. Chertkow, et al. "The diagnosis of dementia due to Alzheimer's disease: recommendations from the National Institute on Aging-Alzheimer's Association workgroups on diagnostic guidelines for Alzheimer's disease Alzh. Dement, 2011, 7:263–269.

[23] Rey A., L'examen clinique en psychologie. Paris: Presses Universitaires de France, 1964.

[24] H. Spinnler and G. Tognoni, "Standardizzazione e taratura italiana di test neuropsicologici" Ital J Neurol Sci, 1987, 8(suppl):1–120.

[25] H. M. Lezak, D. Howieson, E. Bigler, D. Tranel "Neuropsychological Assessment" New York, NY: Oxford University Press, 2012.

[26] G. Arrigoni and E. De Renzi. "Constructional apraxia and hemispheric locus of lesion" Cortex, 1964, 1 pp. 170–197.

[27] J. L. Cummings et al. "The Neuropsychiatric Inventory: comprehensive assessment of psychopathology in dementia" Neurology,1994, 44:2308–2314.

[28] M. Hamilton, "A rating scale for depression" J Neurol Neurosurg Psychiatry, 1960, 23:56-62.

[29] S. Katz, T. D. Downs, H. R. Cash, et al. "Progress in the development of an index of ADL" Gerontologist, 1970, 10:20–30.

[30] M. P. Lawton and E. M. Brody, "Assessment of older people: self-maintaining and instrumental activities of daily living" Gerontologist, 1969, 9:179-186.

[31] E. Pfeiffer, "A short portable mental status questionnaire for the assessment of organic brain deficit in elderly patients" J Am Geriatr Soc, 1975, 23:433-441.

[32] P. A. Parmelee, P. D. Thuras, I. R. Katz, M. P. Lawton, "Validation of the Cumulative illness rating scale in a geriatric residential population" J Am Geriatr Soc, 1995, 43:130-137.

[33] B. Vellas et al. "The Mini Nutritional Assessment (MNA) and its use in grading the nutritional state of elderly patients" Nutrition, 1999, 15:116-122.

[34] M. R. Bliss, R. McLaren, A. N. Exton-Smith "Mattresses for preventing pressure sores in geriatric patients" Mon Bull Minist Health Public Health Lab Serv, 1966, 25:238-268.

[35] J. Endicott, J. Nee, W. Harrison, R. Blumenthal "Quality of life enjoyment and satisfaction questionnaire" Psychopharmacol Bull, 1993, 29(2), 321-326, 1993.

Sense-Making in Complex Healthcare Domains: The Role of Technology in Every Day Lives of Youth Suffering from ME/CFS

Jorun Børsting
Department of Informatics
University of Oslo
Oslo, Norway
jorubo@ifi.uio.no

Alma L. Culén
Department of Informatics
University of Oslo
Oslo, Norway
almira@ifi.uio.no

Abstract—**In this paper, we argue that the technology design needs to take a more holistic perspective, well beyond opportunities offered by the technology development alone. This is especially important when developing technologies for complex domains, such as healthcare, and with users who are at risk. We propose a sense-making process that helps human-computer interaction researchers/designers to develop a broader understanding of such complex domains in which they often have no expert knowledge. We exemplify our approach by exploring the complex domain related to the design of technologies that could support youth with Myalgic Encephalomyelitis to cope better with their situation. The approach is based on a combination of user research, Actor-Network Theory adapted to the design context, a complex systems design tool (Giga-mapping) and a literature review. We find that this approach leads researchers to reflect deeper over the domain complexities, and to avoid solving problems that do not untangle complexity or lead to real-life solutions. Our conclusion is that such initial sense-making processes are fundamental when considering the design of new technologies for healthcare, and with vulnerable users.**

Keywords-complex systems; sense-making; ME/CFS; ANT; healthcare; design thinking.

I. INTRODUCTION

Diverse self-management applications and technologies that aim to support people in improving and maintaining their quality of life, or helping them through an illness and a recovery process, are increasingly sought after within healthcare and addressed through Human-Computer Interaction (HCI) design and research. However, these applications are frequently developed without sufficient understanding of the complexities of the healthcare and problems within this domain that are often wicked [1]–[3]. Wicked problems have been discussed by Rittel and Webber already in 1973 [4], but have gained more attention lately through the use of design thinking, e.g., [5]. Wicked problems are incomplete, contradictory, changing and interdependent. Working with them is difficult, and is often relinquished too fast in favor of finding a good solution to some problem. Finding real design opportunities and problems to solve may then be lost. The design of healthcare technologies that support users and fit well the context of use require a broader and deeper inquiry into the nature of problems and their relation to the situated design space [6].

In this paper, we discuss the initial sense-making in complex domains, a process that enables this broader and deeper understanding of the context, and design options. As an illustrative case, we use the design of supportive technologies for a particular user group, adolescents suffering from Myalgic Encephalomyelitis (ME), often referred to as the Chronic Fatigue Syndrome (CFS), or ME/CFS. Almost no research results were to be found on this topic. Without a-priory knowing what kind of solution(s) to offer to these users, be it self-management tools, or other systems or services that could be useful in their everyday lives, we propose a framework for sense-making and finding problems that are meaningful to solve in this context.

In recent years, the increase of chronic diseases, in particular among youth, has been significant [7]. Among these diseases, ME/CFS has become a growing concern, not only for those suffering from it and their families, but also for the medical science, governmental health management, and society at large. Tweens and adolescents who are affected by the illness, would be, if healthy, considered to be willing adopters and users of technology [8]. However, even for their healthy peers, intensive use of technology may lead to negative health consequences [9]. Thus, for those with ME/CFS, it is crucial that technology does not introduce a new set of problems or worsens their health condition [10].

From this perspective, we consider youth with ME/CFS to be a group of vulnerable users [11]. Vulnerability is to be understood as a set of risks that these adolescents are exposed to. Understanding these risks, and how to reduce or eliminate them, is a step towards understanding design and technology design spaces when considering designing for and with this user group.

Furthermore, the understanding of the design space in complex contexts may require understanding of several different fields pertinent to the context. Understandings can be at the micro-level (e.g., how a specific solution affects the user) and understandings at the macro-level (such as social, cultural and governmental influences on the design space). In a development of supportive technologies for and with youth with ME/CFS, we have chosen to use Actor-Network Theory (ANT) to produce rich descriptions of the interrelated social and material processes [12][13], Giga-mapping to represent and communicate our thinking about design, and finally, a firm focus on users and use. We argue that such combined approach helps researchers to develop and communicate growing insights efficiently, become aware of the broader range of issues and their interconnectedness, engage in more innovative designerly practices and reflect critically on own position and values in the sense-making process. When using Design Thinking (DT), an HCI researcher typically adopts designerly practices when making prototypes. The difference between a research prototype and a product in use is significant [14], as are the implications for knowledge production [15]. When working with wicked problems in complex domains, an HCI researcher needs to work with his/her own mindset and develop an ongoing sensibility and sensitivity to problems inherent to the context.

The paper is structured as follows: we provide the literature review in Section II, followed by the section on ME/CFS and technology design. Section IV combines our findings and discussion around technology for adolescents. Finally, Section V presents our conclusions.

II. The Literature Review

For the case presented in this paper, we started by mapping out our understanding of the context, see Fig. 1. The purple area in the center of the figure represents users for and with whom we aim to find and design solutions that could support them. To do so successfully, understanding what it means to be an adolescent with ME/CFS is crucial. This understanding is often referred to in the literature as an empathy with users [16]. We, however, believe that going beyond empathy, and venturing into the understanding of scientific, in our case medical, results related to the condition is also necessary, even when outside of our field of competence. These medical results, for example, could help to find a broader set of risks those adolescents with ME/CFS experience. Naturally, HCI and designerly thinking, being our own and familiar practices, are a fundamental ingredient in this process. ANT, as it applies to design and designers [13], is chosen to represent cultural, governmental and educational perspectives.

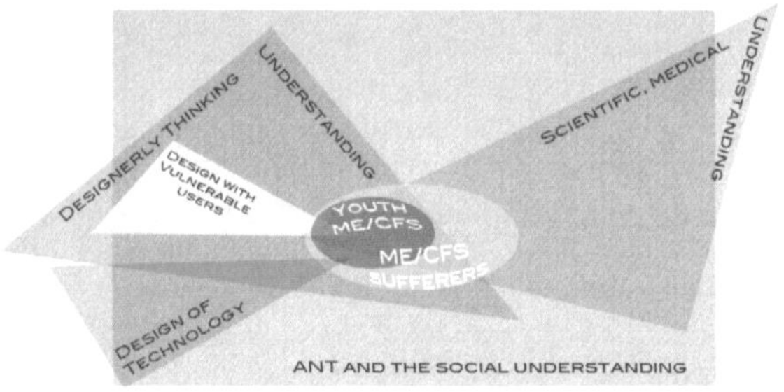

Figure 1. Giga-mapping fields of concern for design related to ME/CFS.

In what follows, each subsection represents a short summary of literature findings related to areas depicted in Fig. 1 that became the starting point for our sense-making.

A. A medical understanding of adult and paediatric ME/ CFS and its effects: the challenging nature of the disease

ME/CFS, is a debilitating multisystem illness resulting in a plethora of symptoms that include severe physical and cognitive exhaustion, confusion, difficulties with memory, concentration, sensitivity to light and noise among others, see [17], [18]. What causes the illness remains an enigma, and the condition is presently researched on a broad scale, e.g., [19]. Many theories, ranging from viral infections to psychological stress have been proposed [17]. Some studies suggest that the immune system may be chronically active in CFS sufferers. This might relate to a theory that ME/CFS is caused by an abnormal reaction to common infectious agents. This theory could link ME/CFS to autoimmune diseases such as the Lyme disease and the Epstein-Barr virus.

A review of thirty-four qualitative studies on ME/CFS, see [20], was done and the authors find that *"For sufferers, illness development influenced identity, reductions in functioning, and coping. Physician-specific themes described lack of awareness about ME/CFS and recommended improvement in educational resources. Intersecting themes expressed issues with diagnosis creating tensions and fueling the stigmatization of ME/CFS"*. Also, some research points in the direction where sufferers themselves help to perpetuate their condition. For example, Afari and Buchwald state that *"current knowledge about chronic fatigue syndrome suggests that genetic, physiological, and psychological factors work together to predispose an individual to the condition and to precipitate and perpetuate the illness"* and, further, that *"sufferers' perceptions, illness attributions, and coping skills may help to perpetuate the illness"* [17, p. 230]. Such results, unless verified carefully, may help to perpetuate stigmatization. Other research, such as that of Geelen et al. [21], or Winger et al. [22], offers insight into personality issues relevant for understanding the youth suffering from ME/CFS.

The treatment of those suffering from ME/CFS is highly individualized, frequently symptom-based, and includes both pharmacological [19] and behavioral approaches [23]. The most common form of help to those suffering from the illness in Norway [24] are self-management courses, offering guidance on how to stabilize the symptoms, find a balance between rest and activity, adjust to a life with ME, etc. This is in line with the trend to promote self-management wherever appropriate for sufferer care, and especially so for sufferers with chronic diseases [25]. Self-management often involves the management of medical conditions, behavior or emotions [26], [27]. For users with ME/CFS, self-management also includes the management of so-called *energy balance*. An ME/CFS sufferer is considered to have a limited amount of energy available for use per day. If an activity takes too much energy, the overuse results in extra exhaustion over a period of time lasting from several hours to several days, leading to increased pain levels, sensitivity and overall worsened physical and cognitive condition. The problem is often that a person with ME/CFS does not know what amount of energy is required for an ordinary activity such as, for example, meeting a friend for a cup of coffee. Therefore, the energy balance management is difficult.

A user study carried out by the Norwegian ME/CFS association [18] found that 40% of people suffering from ME/ CFS did not receive medical help with relieving symptoms caused by the illness. When sufferers were asked what kind of help they would like to receive from the healthcare sector, the results indicated professional assistance with sleep problems, stress management, stomach problems and general pain relief.

For children and youth, the situation is additionally difficult because it is harder to make their voice on the matter heard:

"Do children and adolescents suffer from ME/CFS? Simple common sense tells most parents, teachers and doctors that they do and often more severely than adults. Yet there remain a sizeable proportion of professionals in Health Care, Education and Social Services who are still prepared to ascribe the numerous, disabling but seemingly unconnected symptoms of this illness in young people to anorexia, depression, school phobia or a dysfunctional family background. All are, at least, agreed that the illness presents a considerable economic, educational and social problem" [28].

Resonating with these findings, and in part, because the illness does not leave visible marks on a person suffering from it, they are often dismissed as suffering from psychological problems, or worse, stigmatized by general population [20].

The sufferers from ME/CFS have to bear the medical uncertainty of what causes the disease and, as a consequence, absence of an effective diagnostic tests and treatment for the illness [29]. Furthermore, ME/CFS social and political uncertainties related to the disease are of undetermined duration, depending most strongly on scientific, medical findings around the condition. The youth is affected by nearly the same rates, from about the age of 11, as adults [30].

For someone to be diagnosed with ME/CFS, the symptoms need to be present to such a degree that they clearly limit a person's ability to carry out ordinary daily activities [31]. Sufferers are classified into four groups: mild (an approximate 50% reduction in pre-illness activity level), moderate (mostly housebound), severe (mostly bedridden) or very severe (totally bedridden and in need of help with basic functions) [32].

B. *Previous research on ME/CFS in adolescence and technology*

As mentioned in the introduction, there is very little in the literature on assisting people, including adolescents, suffering from ME/CFS through the use of technology. ME/CFS sufferers, as mentioned above, have multiple challenges, at the physical and cognitive level, coupled with reduced tolerance to light and noise. They, thus, may not be the prime candidates for the use of screens for receiving information, whether they are smartphone screens, tablets or personal computers. All the solutions mentioned here are screen based. While screens may be a good option for adolescents in general, they may have limitations for those with ME/CFS. Thus, before deciding on any new solutions for these adolescents, one needs to make sure that proposed solutions do not introduce new problems and health risks.

From the literature survey, we found that social isolation and access to education are important for these users. Several research efforts point in these directions, studying either the social media or platforms for education. General use of social media in medical care was described in [33], where authors have analyzed and synthesized 76 articles, 44 websites, and 11 policies/reports and presented findings according to 10 different categories of social media such as blogs, micro blogs, social networks, professional networks, thematic networks, wikis, sharing sites and others (represented by the Second Life). Findings, in particular those related to the second life are relevant for adolescents with ME/CFS and have been taken further in the work of Best and Butler [34], [35]. The paper [35] describes how a virtual support center was constructed in the Second Life, featuring meeting areas, relaxation areas, library resources and a gallery of art by and for people with ME/CFS. However, the results of the investigation gave mixed conclusions, possibly reflecting the fact that certain level of mastery of the virtual environment was needed, and that was not possible for all users to accomplish. Also, a physical and cognitive condition of ME/CFS sufferers may affect their willingness to dedicate energy to learning about new virtual environments. The Second life application, in addition to providing information related to the condition, was aiming to address loneliness and social isolation.

Considering the educational platforms for ME/CFS sufferers, the paper on e-learning, [36], can be brought forward. In this study, e-learning platforms were explored as an opportunity for children with ME/CFS to participate in classroom activities even though they are staying at home. As the authors state, this is especially relevant to Scotland, where many live far away from schools. However, the access to school and educational materials is only the first hurdle in receiving regular education and does not address other problems that youth with ME/CFS often have.

C. *Design Thinking and Design with and for Vulnerable users*

Design Thinking may be defined in many different ways. For example, it may be defined as a process that fosters innovative and creative thinking [37], or as an approach to mitigate complex problems through design [5], [38]. In [39], the authors also include the important discussion on the role of design research and designerly practice, with core concepts that include reflexive practice, meaning making and designerly ways of knowing. When working with vulnerable users, it is important to take into consideration ethical concerns as part of the reflection, and meaning making that ideally includes real users or those who represent them well.

While the screen-based technology can play a supportive role in providing care for many users, the ME/CFS sufferers may have a very limited ability to concentrate, read, or even just look at a screen. ME/CFS adolescent sufferers are perceived as vulnerable also on being at risk of not using technology because of how their diverse illness symptoms limit their capabilities (physical or cognitive). In [10], authors argue that the term 'vulnerable' imply a set of risks.

When identified, the risks could be helpful in defining design goals that aim to reduce or eliminate risks. They further argue *"the awareness of risks/vulnerabilities in a design situation may be helpful in designing better products for vulnerable people"* [10, p. 3]. These and other design concerns in design for and with vulnerable users need to come forth through a sense-making process.

We find Giga-mapping [40], to be a particularly useful tool for sense-making in complex contexts. The Giga-mapping aims to facilitate thinking and communication, at the same time inviting (basically by its large physical space) participation and collective production of understanding of the design context. Giga-mapping can be used to visualize fields of knowledge, as was done in Fig. 1. Furthermore, the complexity of a problem space can be mapped out, for example, in layers (e.g., understandings on micro, mezzo, and macro levels), fostering further understanding of complexity and relatedness of problems.

D. Theorizing technology and Actor-Network Theory (ANT)

As this research aims to understand a larger, holistic picture around ME/CFS sufferers and technologies that could support them in their everyday lives, we propose the use of a pragmatic ANT inspired analysis of design opportunities in the initial, sense-making phases of a design process. To that end, a short overview of ANT is provided.

ANT can be used as a framework to understand a heterogeneous network consisting of diverse human and non-human actors. ANT is different from other social theories applied in information systems research, in how it helps to theorize the technology artefact, by focusing on both the social and the technical [41]. In ANT all the social-technical elements are included in networks of actors/actants. Latour describes the networks by stating *"Behind the actors, others appear; behind one set of intentions there are others; between the (variable) goals and the (variable) desires, intermediate goals and implications proliferate, and they all demand to be taken into account"* [42, p. 100]. The social-technical focus draws attention to analyzing both the visible (the technical/objects) and the invisible (the social) and identifying the relationships between these. These relationships can be both material and semiotic simultaneously and combined into a network of actors that acts as a whole [43]. Tatnall and Gilding (2005) argue that ANT can be particularly useful for studies in areas that involve a consideration of some of the social and political issues in information systems. They further mention interface design, usability testing, and the use of distributed systems within organizations as examples of areas where ANT could be beneficial [44].

There is an increasing focus on how ANT could be fruitful in a design of information systems [14], [45], [46]. Stuedahl and Smørdal found that *"involving ANT concepts in co-design does help to frame co-design processes within the wider context and consequences of emerging knowledge development"* [47, p. 204]. Others, focused on ANT in relation to designing visualizations that makes things public, in order to interest and engage people in participatory processes [48]. Storni explored how ANT can be used to offer an alternative perspective to co-designers, focusing on *"a convention from Latour's call for risky descriptions to a call to design things together"* [14, p.167]. He further argues that ANT suggests three general turns to rethinking co-design and participatory design practices. The first turn addresses the question of *what* to design? It also involves the idea to first *design actor networks,* and then look for ways to map them. He further describes the resulting maps as not only descriptive, but also as supporting participation in the design process. The second turn concerns the question of how to co-design and suggests the idea of designing as actor-networking (in public). The last turn, described as epistemological, involves what Storni describes as *"moving from the idea of the designer as a network prince to the idea of the designer as an agnostic Prometheus"* [14, p. 167].

III. ME/CFS AND DESIGN OF TECHNOLOGY SESSION

As mentioned in the introduction, this study uses a combination of user research, literature review and analysis inspired by ANT and Giga-mapping in order to tease out real problems and design opportunities related to assistive technologies for young people with ME/CFS.

In terms of user research, we have organized a sense-making session with experts on ME/CFS. These experts have a deep knowledge of ME/CFS. They have daily contact with children, youth and adults with ME, as well as their families, and therefore have a broad, personal knowledge of ME and experiences with ME challenges in lives of the sufferers and their families. These people came from 1) the Norwegian ME organization [18], 2) an organization that provides support to families of ME/CFS sufferers, 3) ME/CFS youth organization and 4) a medical doctor. The session lasted for close to three hours. The first part of the meeting was dedicated to getting to know what these organizations do. The results of insights gained are incorporated in Fig. 2.

The second half of the session was used to discuss future technologies that could be useful for ME sufferers. Given the aforementioned deep insight available within the group, many design opportunities for future technologies, as well as novel uses of existing technology, were considered. Some of the suggestions that do not exist today, and are viewed as helpful, are presented bellow.

The first suggestion was related to education, an important area for adolescents. It was an online e-learning solution with streamed lectures, accessible any time. It was brought up that most young people with ME/CFS want to follow up on their studies, but they have to be able to take a break when needed. This addresses the need for a flexible educational system.

The second suggestion was an App designed with a critical design approach with special focus on how the energy levels are different for people with ME/CFS. The App could simulate the amount of energy needed for simple everyday tasks in order to reduce prejudice and increase general understanding of the disease.

The third suggestion was another App that could be designed for self-management of the disease. The App could provide reminders to rest. It could also give short advice on things that have worked for others, which they could try to reduce illness symptoms. The App could also measure sound levels and give warnings when the levels are too high. Since exposure to high sound levels is known to increase illness symptoms, the App could also help visualize this, by showing sound measurements and giving reminders to use earplugs in spaces with high sound levels, e.g., when using public transportation. General personalized everyday reminders were also seen as valuable, as increase in disease symptoms may lead to memory problems and forgetfulness.

The fourth suggestion was an electronic diary that could be used to register symptoms daily. The knowledge and awareness gained through daily activities, and their effects on illness symptoms were seen as valuable. This diary could also be a tool to explain better the current health status to healthcare professionals and social workers. For this diary to be used in periods with increased illness symptoms, it would be of outmost importance that the self-rating process is easy to conduct.

The fifth suggestion was related to wearable devices, such as the Fitbit. ME/CFS sufferers experienced the possibility of tracking activity, sleep patterns and energy expenditure as desirable. It was suggested that a similar wearable technologies, tailored to the needs of people with ME/CFS, could be valuable.

The sixth suggestion is a Social technology specially designed for people with ME/CFS, to meet and seek support from other people with the illness.

The last one considered novel discussion platforms.

IV. ADOLESCENTS WITH ME AND ANT

Instead of considering any one of mentioned suggestions as a design opportunity, and in isolation, we performed a broad analysis, focusing on all data collected from the sense-making session with experts. At first, the traditional ANT analysis was used, but then, a more pragmatic, design and technology oriented approach was taken, in line with the literature such as [13] and [46]. The result is shown in Fig. 2. The image is a visual representation, a Giga-map at the macro level. All the relations shown are significant for the initial sense-making process. Each relation explicitly or implicitly contributes to discussion around technology that could aid adolescents suffering from ME/CFS in some way. Three large connected areas of the network emerge, related to what we named Personal/family, Educational and Governmental design spaces. At present, these spaces are not strongly connected, and connections within each one of them are sparse. As mentioned earlier, also in the literature, there were very few results to mention.

In Fig. 2, a circle represents an actor and lines represent relations between actors. The number of actors shown is limited to the actors identified in the data analysis from the sense-making session with experts on ME/CFS. Actors in the network are both human (such as the ME/CFS sufferer, family members, hospital doctors and school teachers) and non-human (such as politics, prejudice, economics, and social media). They all have a role in and influence on the network. These roles and influences are not predefined but continuously evolve. For example, a family member may consider himself or herself to be without prejudice in other situations, but in the network representing the youth with ME/CFS they still can think that the adolescent is lazy rather than sick, and so contribute to maintaining the prejudice around the disease.

Adolescents, different people that they interact with, as well as non-human entities, obviously form different heterogeneous networks. This directly relates to the wickedness of problems related to defining technologies that aim to help youth to have a better life, given their condition. Not only adolescents suffer from the condition to varying degrees, but their condition changes from day to day and period-wise as well. This changeability needs to be addressed as part of the sense-making.

Furthermore, the analysis revealed two mediators. Mediators are non-human entities that increase difference among actors and, thus, should be carefully studied [49]. Mediators that we have identified impact how our society currently meets ME/CFS sufferers. Referring to Fig. 2, and the actors (human and non-human) in the network, the two mediators are marked with red circles. They are *Understanding of ME/CFS* and *Prejudice*. The first one is seen as a contributing factor that defines relations between ME/CFS sufferers and many other actors in the network. It influences, for example, how the social service caseworkers or physicians handle youth with ME/CFS. The *Prejudice* seeks to impose its version of reality on ME/CFS sufferers. *Prejudice* may determine what people see and in turn how they act towards people suffering from ME/CFS. As one person from our expert session said *"If one could only eliminate the burden of the shame that many people with ME/CFS live with, and shift the focus to acceptance and to living their lives as best as possible, the quality of their lives would be so much better."* The red edges drawn in Fig. 2 between the *Understanding of ME/CFS* and other nodes are also relevant for *Prejudice* as an actor in the network. However, the two are co-dependent: as the knowledge about, and understanding of, the ME/CFS increase, prejudice decreases.

Other concepts from ANT, such as agency and delegation, are also very relevant but will not be further described here.

V. CONCLUSION

The numerous challenges related to ME/CFS are not easy to solve and clearly relate to the main characteristics of wicked problems. We have proposed a holistic approach to sense-making when considering technology design for adolescents with ME/CFS. Using a literature review, a sense-making session with experts, ANT and Giga-mapping, we gained insight in what ME/CFS is, developed the empathy with youth suffering from it. ANT provided rich descriptions of the complex design context and Giga-mapping visual representations of relations that emerged as relevant. They also helped to define possible areas of interest for design (Personal/family, Educational and Governmental). Our findings align well with the literature related to Personal/family and Educational space. Another possible design focus could be relate to the mediators found - understanding of ME/CFS and helping to reduce prejudice and stigmatization through design.

All relations among actors are significant for the initial sense-making process. They all, explicitly or implicitly, contribute to discussion around technology that could aid adolescents suffering from ME/CFS. Jointly, they provide a tool for critical reflection in the early sense-making phase of trying to understand adolescents with ME, and the situated design space related to technology that could be useful in their lives.

ACKNOWLEDGMENT

Profound thanks are due to all participants in the expert session.

REFERENCES

[1] B. W. Head and J. Alford, "Wicked Problems Implications for Public Policy and Management," Adm. Soc., vol. 47, no. 6, pp. 711–739, Aug. 2015.

[2] S. Waddock, "The Wicked Problems of Global Sustainability Need Wicked (Good) Leaders and Wicked (Good) Collaborative Solutions," J. Manag. Glob. Sustain., vol. 1, no. 1, pp. 91–111, Jun. 2013.

[3] R. Buchanan, "Wicked Problems in Design Thinking," Des. Issues, vol. 8, no. 2, pp. 5–21, Apr. 1992.

[4] H. W. J. Rittel and M. M. Webber, "Dilemmas in a general theory of planning," Policy Sci., vol. 4, no. 2, pp. 155–169, Jun. 1973.

[5] J. Kolko, Austin Center for Design, Wicked problems problems worth solving. Austin, Texas: Ac4d, 2012.

[6] J. Redström, "Towards user design? On the shift from object to user as the subject of design," Des. Stud., vol. 27, no. 2, pp. 123–139, Mar. 2006.

[7] J. M. Perrin, L. E. Anderson, and J. V. Cleave, "The Rise In Chronic Conditions Among Infants, Children, And Youth Can Be Met With Continued Health System Innovations," Health Aff. (Millwood), vol. 33, no. 12, pp. 2099–2105, Dec. 2014.

[8] E. A. Witt, A. J. Massman, and L. A. Jackson, "Trends in youth's videogame playing, overall computer use, and communication technology use: The impact of self-esteem and the Big Five personality factors," Comput. Hum. Behav., vol. 27, no. 2, pp. 763–769, Mar. 2011.

[9] R.-L. Punamäki, M. Wallenius, C.-H. Nygård, L. Saarni, and A. Rimpelä, "Use of information and communication technology (ICT) and perceived health in adolescence: The role of sleeping habits and waking-time tiredness," J. Adolesc., vol. 30, no. 4, pp. 569–585, Aug. 2007.

[10] A. L. Culén and A. Karpova, "Designing with Vulnerable Children: a Researchers Perspective," in Human-Computer Interfaces and Interactivity: Emergent Research and Applications Book, IGI Global, 2014, pp. 118–136.

[11] A. L. Culén and M. van der Velden, "The Digital Life of Vulnerable Users: Designing with Children, Patients, and Elderly," in Nordic Contributions in IS Research, Springer Berlin Heidelberg, 2013, pp. 53–71.

[12] B. Latour, "On using ANT for studying information systems: a (somewhat) Socratic dialogue," Soc. Study Inf. Commun. Technol., pp. 62–76, 2004.

[13] C. Storni, "Notes on ANT for designers: ontological, methodological and epistemological turn in collaborative design," CoDesign, vol. 11, no. 3–4, pp. 166–178, Oct. 2015.

[14] P. K. Chilana, A. J. Ko and J. Wobbrock, "From User-Centered to Adoption-Centered Design: A Case Study of an HCI Research Innovation Becoming a Product," in Proceedings of the 33rd Annual ACM Conference on Human Factors in Computing Systems, New York, NY, USA, 2015, pp. 1749–1758.

[15] A. L. Culén and A. Følstad, "Innovation in HCI: What Can We Learn from Design Thinking?," In Proceedings of the 8th Nordic Conference on Human-Computer Interaction: Fun, Fast, Foundational, NY, USA, 2014, pp. 849–852.

[16] A. A. Gasparini, "Perspective and Use of Empathy in Design Thinking," in ACHI 2015, 2015.

[17] N. Afari and D. Buchwald, "Chronic Fatigue Syndrome: A Review," Am. J. Psychiatry, vol. 160, no. 2, pp. 221–236, Feb. 2003.

[18] G. J. Bringsli, A. Gilje, and B. K. G. Wold, "The Norwegian ME Association National Survey Abridged English Version." The Norwegian ME Association, 2014.

[19] Ø. Fluge et al., "Benefit from B-Lymphocyte Depletion Using the Anti-CD20 Antibody Rituximab in Chronic Fatigue Syndrome. A Double-Blind and Placebo-Controlled Study," PLoS ONE, vol. 6, no. 10, p. e26358, Oct. 2011.

[20] V. R. Anderson, L. A. Jason, L. E. Hlavaty, N. Porter, and J. Cudia, "A review and meta-synthesis of qualitative studies on Myalgic Encephalomyelitis/chronic fatigue syndrome," Patient Educ. Couns., vol. 86, no. 2, pp. 147–155, Feb. 2012.

[21] S. M. van Geelen, G. Sinnema, H. J. M. Hermans, and W. Kuis, "Personality and chronic fatigue syndrome: Methodological and conceptual issues," Clin. Psychol. Rev., vol. 27, no. 8, pp. 885–903, Dec. 2007.

[22] A. Winger, M. Ekstedt, V. B. Wyller, and S. Helseth, "'Sometimes it feels as if the world goes on without me': adolescents' experiences of living with chronic fatigue syndrome," J. Clin. Nurs., vol. 23, no. 17–18, pp. 2649–2657, Sep. 2014.

[23] P. White et al., "Comparison of adaptive pacing therapy, cognitive behaviour therapy, graded exercise therapy, and specialist medical care for chronic fatigue syndrome (PACE): a randomised trial," The Lancet, vol. 377, no. 9768, pp. 823–836, Mar. 2011.

[24] Norwegian ME Association (in Norwegian only): me-foreningen.com/. Accessed on April 12th, 2016.

[25] H. Holman and K. Lorig, "Patient self-management: a key to effectiveness and efficiency in care of chronic disease.," Public Health Rep., vol. 119, no. 3, pp. 239–243, 2004.

[26] J. M. Corbin and A. Strauss, Unending work and care: Managing chronic illness at home, vol. xviii. San Francisco, CA, US: Jossey-Bass, 1988.

[27] K. R. Lorig and H. Holman, "Self-management education: history, definition, outcomes, and mechanisms," Ann. Behav. Med. Publ. Soc. Behav. Med., vol. 26, no. 1, pp. 1–7, Aug. 2003.

[28] E. G. Dowsett and J. Colby, "Long-Term Sickness Absence Due to ME/CFS in UK Schools," J. Chronic Fatigue Syndr., vol. 3, no. 2, pp. 29–42, Jan. 1997.

[29] M. E. B. Smith et al., "Treatment of Myalgic Encephalomyelitis/Chronic Fatigue Syndrome: A Systematic Review for a National Institutes of Health Pathways to Prevention WorkshopTreatment of Myalgic Encephalomyelitis/Chronic Fatigue Syndrome," Ann. Intern. Med., vol. 162, no. 12, pp. 841–850, Jun. 2015.

[30] A. Farmer, T. Fowler, J. Scourfield, and A. Thapar, "Prevalence of chronic disabling fatigue in children and adolescents," Br. J. Psychiatry, vol. 184, no. 6, pp. 477–481, Jun. 2004.

[31] National Collaborating Centre for Primary Care (UK), Chronic Fatigue Syndrome/Myalgic Encephalomyelitis (or Encephalopathy): Diagnosis and Management of Chronic Fatigue Syndrome/Myalgic Encephalomyelitis (or

Encephalopathy) in Adults and Children. London: Royal College of General Practitioners (UK), 2007.

[32] B. M. Carruthers et al., "Myalgic Encephalomyelitis: International Consensus Criteria," J. Intern. Med., vol. 270, no. 4, pp. 327–338, Oct. 2011.

[33] "JMIR-Social Media: A Review and Tutorial of Applications in Medicine and Health Care | Grajales III | Journal of Medical Internet Research." [Online]. Available: http://www.jmir.org/2014/2/e13/. [Accessed: 18-Jan-2016].

[34] K. Best and S. Butler, "Second life avatars as extensions of social and physical bodies in people with Myalgic Encephalomyelitis/Chronic Fatigue Syndrome," Continuum, vol. 27, no. 6, pp. 837–849, Dec. 2013.

[35] K. Best and S. Butler, "The Necessity of Control for Computer and Internet Users with Myalgic Encephalomyelitis/Chronic Fatigue Syndrome," Inf. Commun. Soc., vol. 16, no. 7, pp. 1042–1060, Sep. 2013.

[36] A. K. Sheridan and and members of the C. P. G.-E. of C. with M. (Scotland), "Exploring E-learning provision for Children with ME in Scotland," Educ., vol. 2, no. 1, pp. 78–80, Jan. 2013.

[37] T. Brown, "Design Thinking," Harv. Bus. Rev., vol. 86, no. 6, p. 84, 2008.

[38] T. Brown and J. Wyatt, "Design Thinking for Social Innovation (SSIR)," Stanford Social Innovation Review, 2010.

[39] U. Johansson-Sköldberg, J. Woodilla, and M. Çetinkaya, "Design Thinking: Past, Present and Possible Futures," Creat. Innov. Manag., vol. 22, no. 2, pp. 121–146, 2013.

[40] B. Sevaldson, "Giga-Mapping: Visualisation for complexity and systems thinking in design," Nordes, vol. 4, Mar. 2011.

[41] O. Hanseth, M. Aanestad, and M. Berg, "Guest editors' introduction: Actor-network theory and information systems. What's so special?," Inf. Technol. People, vol. 17, no. 2, pp. 116–123, 2004.

[42] B. Latour, Aramis, or, The love of technology. Cambridge, Mass: Harvard University Press, 1996.

[43] J. Law, "Notes on the theory of the actor-network: Ordering, strategy, and heterogeneity," Syst. Pr., vol. 5, no. 4, pp. 379–393, Aug. 1992.

[44] A. Tatnall and A. Gilding, Actor-Network Theory in Information Systems Research. 2005.

[45] Y. Akama, "Being awake to Ma: designing in between-ness as a way of becoming with," CoDesign, vol. 11, no. 3–4, pp. 262–274, Oct. 2015.

[46] A. Yaneva, "Making the Social Hold: Towards an Actor-Network Theory of Design," Des. Cult., vol. 1, no. 3, pp. 273–288, Nov. 2009.

[47] D. Stuedahl and O. Smørdal, "Matters of becoming, experimental zones for making museums public with social media," CoDesign, vol. 11, no. 3–4, pp. 193–207, Oct. 2015.

[48] J. Schoffelen et al, "Visualising things. Perspectives on how to make things public through visualisation," CoDesign, vol. 11, no. 3–4, pp. 179–192, Oct. 2015.

[49] E.Sayes. "Actor–Network Theory and methodology: Just what does it mean to say that nonhumans have agency?" Social Studies of Science, vol. 44, no. 1, pp. 134-147, Feb. 2014.

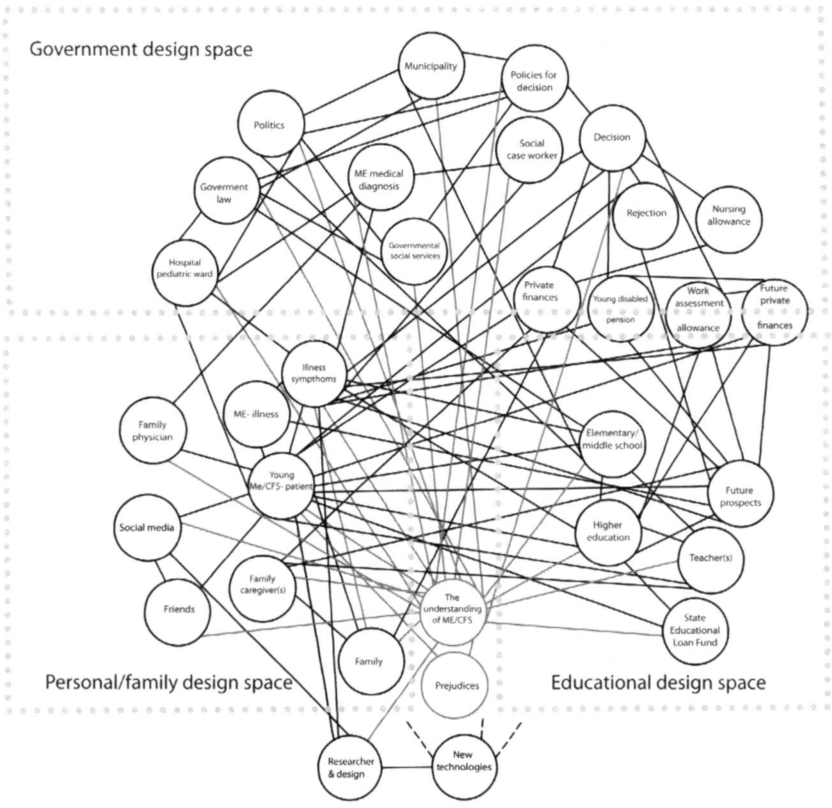

Figure 2. Using a combination of Giga-mapping and ANT to map out design spaces and consider design opportunities.

Do-It-Yourself Health Care:
A Three-Step Approach to Supporting Patient Self-Management in Clinical Practice

Veronica Janssen
Health, Medical and Neuropsychology
Leiden University
Leiden, the Netherlands
e-mail: VJanssen@FSW.leidenuniv.nl

Roderik Kraaijenhagen
Hearts4People Foundation
Amsterdam, The Netherlands
e-mail: roderik.kraaijenhagen@cardiovitaal.nl

Abstract— **Implementation issues plague the uptake and effectiveness of self-management programs in clinical practice. Using a participatory design, we developed an innovative three-step approach to support patient self-management in the field. This approach is based on existing evidence-based techniques and is supported by a Web-based patient portal. The focus of this online patient portal is threefold: firstly, patients are taught the skills and helped to build the confidence to adequately self-manage their disease in the home-environment. Secondly, professionals are provided with the tools and techniques to guide and monitor this process. Thirdly, the portal allows for data sharing, long-term progress monitoring and efficient communication between different health care providers. The authors are currently in the process of evaluating this approach and assessing the uptake, usability and usefulness in clinical practice.**

Keywords-self-management; health care; e-health; implementation; participatory design; patient portal

I. INTRODUCTION

Support is growing for a new conceptualization of health that views health as 'the ability to adapt and self-manage' [1]. This definition is process-oriented and encourages people to take an active role in managing their own health. Self-management programs aim to support patients in coping pro-actively with being (chronically) ill. Such programs focus on building the skills and confidence necessary to make informed decisions, engage in health promoting activities and manage the impact of the illness on life. Research shows that supporting self-management has beneficial effects on people's health behaviors, quality of life, clinical symptoms and use of health care resources [2]. However, the wide dissemination and uptake of self-management initiatives in clinical practice lags behind [3]-[5]. Next to financial, managerial and technical problems, there are two important reasons for the problematic uptake and implementation of self-management in practice. Firstly, adequate self-management requires behavior change on multiple levels: patients need the skills and confidence to learn how to 'adapt and self-manage' their illness, and professionals need the motivation and tools to build a collaborative partnership with their patients and families. Most self-management programs, however, focus on providing skills training for patients, but do not provide any such training for professionals [6]. Secondly, most self-management programs are offered in isolation from the chronic care system, meaning that self-management skills, activities and (electronic) tools for support are not integrated by the multiple partners involved in long-term disease management or embedded in the day-to-day routines of care [6].

In an attempt to address these issues, we have developed an innovative approach to aid self-management in practice that focuses on (a) providing patients with a set of tools and techniques to help them build their confidence, manage the impact of their illness on their lives and engage in healthy behaviors, (b) providing health care professionals with a set of tools and techniques to help them transform the patient-caregiver relationship and (c) aiding collaboration between primary and secondary health care providers using an overarching electronic self-management support system. This approach joins several theories of health behavior change [7]-[9] and is based on existing evidence-based techniques and interventions [7][10]-[12] and is supported by online self-management tools.

This paper is organized as follows. Section II gives the theoretical and conceptual background of the three-step approach to self-management. Part A in Section II goes into finer detail with respect to the development of the Web-based patient portal. Part B addresses the use of the self-management approach and patient portal in clinical practice. Part C describes the evaluation of the approach. The conclusions close the article.

II. A THREE-STEP APPROACH TO SELF-MANAGEMENT

Central to our approach is the idea that self-management is an interactive, 'staged' process – rather than a set of skills used in isolation [7]-[12]. Each stage, or step, commands patients to obtain specific knowledge, learn specific skills and form adaptive cognitions. It supports patients to move through these series of stages and requires health care professionals to employ different methods for behavioral change, tailor the mode and level of intensity of communication, and offer varying settings in which to practice self-management skills. We propose that full self-management may be achieved in three distinct steps, the first of which is termed 'self-confidence'. High levels of self-efficacy are a prerequisite for behavior change [8]. Thus, in this first phase, self-management support focuses on empowering patients and enhancing self-control by means of education, skills training and by offering patients a safe and controlled environment in which to explore and extend their boundaries. In this stage patients will communicate frequently with health professionals, and practice their newly learned skills mostly in a controlled, face-to-face setting (e.g., rehabilitation or health (care) center). The second phase, which is termed 'self-regulation', focuses on guiding patients in self-regulating their (new) behaviors and learning them to monitor relevant risk factors and parameters for the management of their disease [10]-[12]. Support for behavior change in this phase focuses on learning patients how to set salient and achievable goals, practice the skills they need to achieve these goals in real-life situations, and overcome barriers to change [10]-[12]. Patients will still communicate regularly with health professionals in this stage, but face-to-face contact will be alternated with online and/or telephone contact (e.g., e-coaching). The third and final phase is the 'self-management' phase in which patients draw upon their skills and experiences to adequately manage their risk factors and embed their healthy behavior in day-to-day life. Self-management support in this stage focuses on leveraging the skills developed and integrating new behaviors and skills in the home-environment [7]. This means that communication with health care professionals in this stage will be almost solely internet-or telephone-based and that the frequency of contact will gradually fade-out.

A. Development of a Web-Based Patient Portal

As we have a strong background in cardiology and since there is a clear need to help cardiac patients in maintaining adequate self-management over time, we have developed and pilot-tested our three-step approach in close dialogue with patients in cardiac rehabilitation. Using both qualitative and quantitative methods, we first assessed cardiac patients' as well as health care professionals' needs and preferences regarding self-management. Such participatory design of (online) self-management programs is thought to increase implementation success and uptake rates in practice [13]. Structured interviews were carried out with 13 patients and 5 health care professionals [14]. Both patients and professionals confirmed the need for a more systematic approach to aiding self-management in practice. Patients indicated that they would appreciate online access to their personal health information in addition to face-to-face contact. Health care professionals valued efficient communication and stressed the importance of being able to track patients' progress over time. Moreover, they indicated a need for tools and techniques to help them ask motivational questions to aid patients in the process of behavior change. On the basis of this input a mock-up version of a Web-based patient portal to support self-management (MyHealthePortal) was developed. Using a questionnaire, patients' needs and preferences as well as factors associated with intention to use this Web-based patient portal were assessed [15]. The questionnaire was filled out by 113 cardiac rehabilitation patients (34% females and 66% males, mean age 63 years). Patients especially valued being able to objectively monitor their progress, and being able to obtain adequate feedback from health care professionals on progress, lifestyle behaviors and relevant risk factors. Furthermore, they indicated a need for low-threshold communication with professionals [15]. The vast majority (97%) of patients used the internet several times a week and 69% indicated that they strongly intended to use the Web-based patient portal [15]. On the basis of this input, the mock-up version was improved. The look and feel and the different functionalities of this updated version were assessed in three patients by means of a qualitative thinking aloud study. The Web-based patient portal was rated as useful, but several design-related problems (e.g., consistency with other systems, error management, visibility of system status) appeared to impact ease-of-use. These problems were addressed and the design was updated accordingly; screenshots of 'MyHealthePortal' are displayed in Fig. 1.

B. Do-It-Yourself Health Care

MyHealthePortal is currently used in 'blended' (i.e., using a combination of online support and face-to-face contact) forms of cardiac rehabilitation. It allows both patients and health care professionals to set salient goals that are linked to patients' life goals, choose suitable face-to-face treatment modalities that will help support these goals, and monitor progress on both objective (e.g., blood pressure, cholesterol, weight, physical activity etc.) and subjective (e.g., well-being, self-reported goal progress etc.) outcome measures. The patient portal dashboard visually displays the journey through the three steps of behavior change and the actual progress on outcomes. The program is inherently empowering, as the patients themselves – as opposed to the clinician – determine their own curriculum and (learn how to) indicate to health care professionals what it is they need to move from one stage to the next. Progress towards self-set goals is rewarded by bonus points, which can be exchanged for lifestyle-related gadgets. Such elements have been shown to increase adherence to and effectiveness of interventions [16]. The Web-based portal also allows patients to continue with self-monitoring their dietary habits, exercise and smoking behavior, blood pressure, body weight and cholesterol after termination of the program. Not only does continuous monitoring and feedback prevent relapse into old lifestyles [7], the portal also allows electronic input of data from various devices (such as blood pressure monitors, weight scales and activity trackers) and sharing of this data between different levels of the health care system. Thus, the portal smooths the transition from primary to secondary care (and vice-versa), and aids the interoperability of health care providers and systems.

The Web-based portal also acts as a powerful tool to help health care professionals shape the patient-caregiver relationship. The three-step approach to full self-management is reflected in a behavior-change 'ladder': a series of small behavior change steps that ultimately lead to full self-management of the behavior. A series of questions based on motivational interviewing techniques guide health care professionals in their communication with the patient; thus, professionals have the tools to assess the stage the patient is in with regards to a specific behavior, discuss motivation to change with the patient and help the patient determine what their next step might be and what they need in order to take this step. Moreover, the Web-based portal allows health care providers involved in different levels of care to guide and monitor this process in the long-term, share data and communicate efficiently about the patient's progress.

C. Practice-Based Implementation-Evaluation

The authors are currently in the process of implementing and evaluating our approach in a number of cardiac rehabilitation centers in the Netherlands. We feel that our three-step approach to full self-management and the do-it-yourself patient portal are applicable to other areas in which lifestyle modification plays an important role, such as cardiovascular risk management, diabetes and chronic obstructive pulmonary disease management.

In order to evaluate actual use in the field and to obtain feedback on both the user experience and on clinically relevant outcomes measures, we aim to include several approaches. First of all, Rapid Application Development Processes, such as the Agile methodology, can be used for iterative usability evaluations. Thus, the patient portal can be continuously updated and the user experience improved.

Secondly, in order to evaluate which components of the self-management approach impact upon important outcomes, such as cardiac risk factors and health behaviors, we suggest an alternative to the randomized controlled trial (RCT) design. RCT designs typically evaluate the efficacy of an intervention - which can be defined as the effectiveness of an intervention under 'ideal conditions'. When a (behavioral) intervention is subsequently implemented in clinical practice, the residual effect may be much smaller as a result of delivery and compliance issues [17]. Therefore, alternatives to RCT-designs have been suggested that have greater ecological validity and greater generalizability as they are intertwined with the process of implementation [18]. So-called 'multivariate testing' (or 'split testing') methods are widely used to compare which of several versions of a Website performs better in terms of conversion rates [19]. We suggest that the concept of A/B testing may also be applied to comparing different versions of a behavioral intervention. For example, by comparing automated versus human coaching, text-message reminders versus online reminders, automatic registration of behavior versus self-monitoring etc., the content of the intervention can be optimized. Performance indicators would be progress in terms of the three self-management steps (e.g., move from self-confidence to self-regulation), and actual behavior shown (e.g., physical activity, smoking, diet etc.). When carried out in multiple settings and across a large group of participants, A/B testing will be able to project the impact of our approach in real-life situations, as well as allow for ongoing development and innovation. Thus, the 'optimal intervention' (for a specific setting) can be determined from a range of evidence-based behavior-change methods and supportive tools. Furthermore, uptake and implementation issues can be tackled whilst they are occurring.

III. CONCLUSIONS

This paper argues that adequate self-management requires (1) gaining the self-confidence to change behavior(s), (2) monitoring and self-regulating the new behavior(s), ultimately leading to (3) embedding the new behavior(s) in daily life. In order to support this process, a Web-based patient portal has been developed, which allows progress and outcome monitoring, and personal coaching. The portal provides tools and techniques for the professional to monitor patient-progress and give motivational feedback. The authors are currently implementing and evaluating this approach in cardiac rehabilitation.

REFERENCES

[1] M. Huber et al., "How should we define health?" BMJ, vol. 343, 2011, pp. d4163-4166.

[2] D. de Silva, "Helping people help themselves." Long Acre, London: Health Foundation, 2011.

[3] R. E. Glasgow, D. Kurz, J. M. Dickman, D. Osuna, L. Strycker, and D. K. King. "Linking internet-based diabetes self-management to primary care: Lessons learned and implications for research translation and practice implementation." Transl. Behav. Med., vol. 2, 2012, pp. 313–321.

[4] B. Appiah et al. "Challenges and opportunities for implementing diabetes self-management guidelines." J. Am. Board. Fam. Med., vol 26, no. 1, 2013, pp. 90–92.

[5] M. F. Harris, A. M. Williams, S. M. Dennis, N. A. Zwar, and G. P. Davies. "Chronic disease self-management: Implementation with and within Australian general practice." Med. J. Aust., vol. 189, no. 10, 2008, pp. 17-20.

[6] A. Kennedy et al. "Implementation of a self-management support approach (WISE) across a health system: a process evaluation explaining what did and did not work for organisations, clinicians and patients." Implement. Sci., vol. 9, 2014, pp. 1–16.

[7] V. Janssen, V. de Gucht, H. van Exel, and S. Maes. "A self-regulation lifestyle program for post-cardiac rehabilitation patients has long-term effects on exercise adherence." J. Behav. Med., vol. 37, no. 2, 2014, pp. 308–321.

[8] A. Bandura. "Self-efficacy: Toward a unifying theory of behavior." *Psychol. Rev.*, vol. 84, no. 2, 1977, pp. 191–215.

[9] C. S. Carver and M. F. Scheier. "On the Self-Regulation of Behavior." Cambridge: Cambridge University Press, 2001.

[10] V. Janssen, V. de Gucht, E. Dusseldorp, and S. Maes. "Lifestyle modification programmes for patients with coronary heart disease: a systematic review and meta-analysis of randomized controlled trials." Eur. J. Prev. Cardiol., vol. 20, no. 4, 2013, pp. 620–640.

[11] V. Janssen, V. de Gucht V, H. van Exel, and S. Maes. "Beyond resolutions? A randomized controlled trial of a self-regulation lifestyle programme for post-cardiac rehabilitation patients." Eur J Prev Cardiol., vol. 20, no. 3, 2013, pp. 431-441.

[12] S. Michie, C. Abraham, C. Whittington, J. McAteer, and S. Gupta. "Effective techniques in healthy eating and physical activity interventions: a meta-regression." Health Psychology., vol. 28, no. 6, 2009, pp. 690–701.

[13] G. Demiris et al. "Patient-centered Applications: Use of Information Technology to Promote Disease Management and Wellness." A

White Paper by the AMIA Knowledge in Motion Working Group. J. Am. Med. Informatics. Assoc. vol. 15, no. 1, 2008, pp. 8–13.

[14] L. Lam, "Acceptatie en waardering van eHealth onder patienten van het Cardiovitaal programma," unpublished.

[15] L. van Wieringen, "Development of a self-management portal within cardiac rehabilitation," unpublished.

[16] J. Hamari, J. Koivisto, and H. Sarsa. "Does gamification work? - A literature review of empirical studies on gamification." Proc. Annu. Hawaii. Int. Conf. Syst. Sci., 2014, pp. 3025–3034, doi:10.1109/HICSS.2014.377.

[17] C. G. Victora, J. P. Habicht, and J. Bryce. "Evidence-Based Public Health: Moving Beyond Randomized Trials." Am. J. Public Health., vol. 94, no. 3, 2004, pp. 400–405.

[18] C. P. Bonell et al. "Alternatives to randomisation in the evaluation of public health interventions: design challenges and solutions." J. Epidemiol. Community Health., vol. 65, no. 7, 2011, pp. 582–587.

[19] R. Kohavi, R. Longbotham, D. Sommerfield, and R. M. Henne. "Controlled experiments on the Web: Survey and practical guide." Data. Min. Knowl. *Discov.*, vol. 18, no. 1, 2009, pp. 140–181.

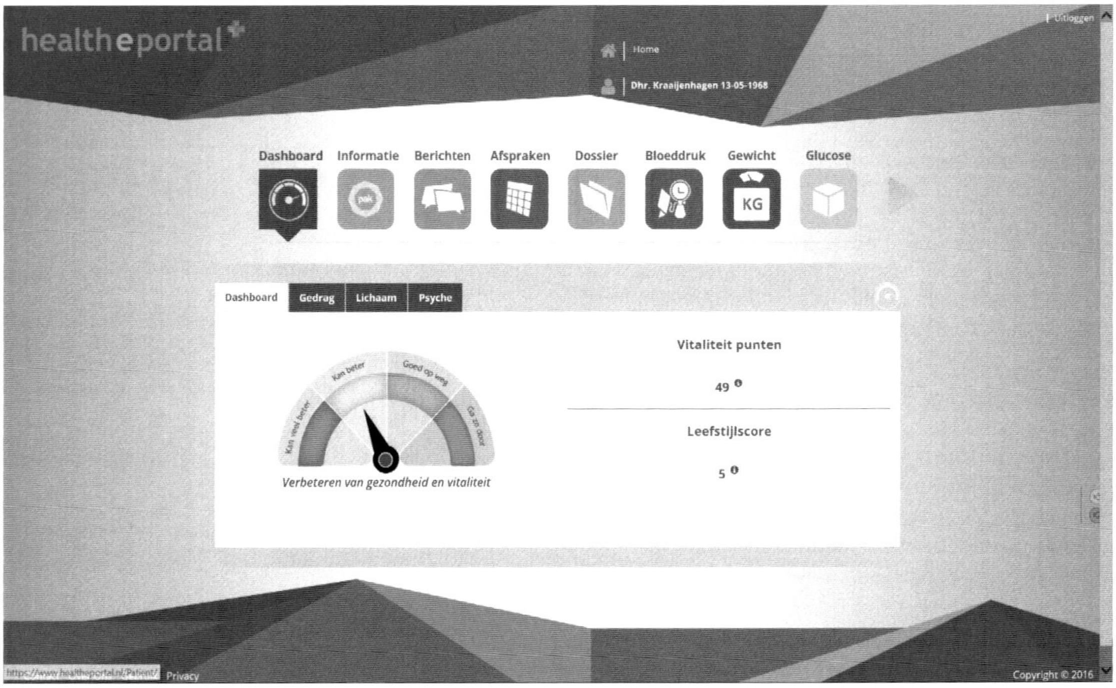

Figure 1. Screenshot of 'MyHealthePortal'

Performing Telecare: Recognizing New Nursing Care Practices

Annemarie van Hout, Ruud Janssen, Marike Hettinga
Research Group ICT-innovations in Health Care
Windesheim University of Applied Sciences
Zwolle, The Netherlands
Email: {a.van.hout, tjwm.janssen, m.hettinga}@windesheim.nl

Abstract—**Telecare is increasingly becoming a part of nursing care, and also in mental health care. In this paper, we study the effects telecare has on nursing care, by showing the first results of research on care by webcam for SMI (severely mentally ill) patients who live at home. Based on ethnographic fieldwork, we show how the nursing practice is altered by the use of technology. New nursing care practices emerge when care is given at a distance, using a webcam. The changes in care practice are noticed by nurses, but not fully recognized. Even more importantly, the new practices are not shared and discussed. Discussing changing practices among nurses is essential, in order to support nurses to name and purposely use the opportunities technology brings.**

Keywords: nursing telecare; mental health care; ethnography.

I. Introduction

Nursing care is substantially changing, as it is increasingly performed at a distance with the use of technology. To care at a distance is already commonplace in mental health care; especially Internet therapy is customary for groups of patients [1,2]. The use of webcams is rising as well; for example, for SMI (severely mentally ill) patients who live at home. Where nurses normally visit patients at home [3], the use of the webcam changes routines significantly. Some changes are obvious, as care at a distance first means that patient and care professional are not in the same room. Physical absence means that there is no room for touching, smelling, walking around or observing anything that technology does not cover [4,5]. Some changes might not be so obvious though and alter care into new forms.

When indicating care as changed when technology is used, we draw from Science and Technology Studies, by analyzing the role of the technology. Technology is comprehended as an active participant in the care relationship, besides nurses and patients. Being active means that the technology is not neutral [6,7], but that its design [8] and its (unforeseen) usability [9] influences its environment. In healthcare, this means that the relationships between care professionals and patients changes. Basic examples are technology that takes over the actions of professionals, like a hoist. Other examples show how practices are altered by transferring tasks and responsibilities. Blood pressure can be self-measured for example, by which a patient performs tasks which originally belonged to care professionals [10,11]. Healthcare is redefined by the use of technology [12] as roles, tasks, functions, relationships, places and spaces change.

Last year, in preliminary research [13], focus groups were held in order to find out what experienced telecare users, nurses that is, thought about this new practice. One of the nurses spoke about an experience she had with one of the patients: *One client is actually care-avoiding and with him I am text messaging via an iPad. Like last Friday, I worked all day and he was sending me messages all day and then I responded. And at the end of the day he spilled the beans: he texted that he was very stressed! And I thought: well done!*

In this case, the use of text messaging was an incidental feature, as the aim of the use of the project's iPads was telecare by webcam. It turned out to be of extra value as this client was able, by having frequent, but brief contacts during the day through a messaging app, to gain self-insight into his anxiety, supported by the nurse. In this example the use of technology would seem to have added to new forms of care. The technology is intended for one thing, but actually fulfills another function as well. The nurse and patient seem to have uncovered a new, unforeseen care option, which enables new forms of care. We therefore ask the question whether or not healthcare is not only being redefined but even reinvented.

This paper is based on a research project that aims to improve the use of telecare for care professionals in mental health care. In order to do so this 'Videoconferencing in mental health care' project will develop several online instruments. With these online instruments, professionals can get familiar with telecare and practice specific situations, in order to bring their professional knowledge and experience in line with the new care practice. Seven mental health care organizations, two universities of applied science, a research institution, a hospital and a university are working together in this project. The project aims to add to knowledge on how telecare changes nursing care by analyzing how frontrunners are using nursing telecare. These insights can support nurses in the use of telecare.

In this paper, we want to answer the following questions: What new nursing care practices emerge when using a webcam? Are professionals aware of these new practices and are they purposefully performed?

II. Method

A. The project teams

The mental care organizations mainly participate with FACT teams. FACT stands for Flexible Assertive Community Treatment; its teams are multidisciplinary and consist of case managers (mostly nurses), psychiatrists, psychologist and sometimes social workers [14]. All have their own caseload of patients, but they share responsibility for the total case load. Whenever patients need extra care, for example because of an imminent crisis, they are listed on the so-called FACT board. This board is a tool to follow the patient more closely, a task that is the shared responsibility of the team. All the FACT teams that participate in the project are involved in using telecare to care at a distance with a webcam. The technical systems vary, just as the implementation phases of the teams. Some teams have embraced the new technology fully, in some teams just a few nurses have and there are teams that are still discussing if and how they want to use telecare.

For this article, we drew from material from two teams: team E and team H. Nurses and patients of team E use a dedicated computer with a screen and webcam and in team H an iPad is used. In team H three case managers are forerunners and use a webcam to care at a distance regularly. The web contacts they have are part of planned care. Team E consists of appointed members from several FACT teams. They take turns manning a health care post where from 8 am till 8 pm they are able to receive all unplanned webcam contacts from clients throughout the region.

B. Ethographic research

In order to understand the daily practice of telecare we, a multi-disciplinary research group (ethics, nursing and user-centered design) use ethnographic research techniques. We assume that new care is in the making as telecare is performed. We use techniques that allow us to become part of the care practice, or at least approach the practice closely, so we can recognize and understand these new practices. We observed and talked with nurses engaged in telecare whilst we were in the room. We took field notes while observing webcam contacts with patients. In between, informal interviews were held, subsequent to the webcam contacts, which were taped and transcribed. We have joined the different teams 32 times, during two to four hours each time. We have observed and talked to 18 nurses, who were in contact with 41 patients, some multiple times. The research group has jointly performed the data analysis. The observations and analysis are led by sensitizing concepts, guiding the notes and the coding process [15]. This is not a neutral process, as neither is the observer. Our observations and analyses were shaped by the theoretical notions used by the researchers. The researchers discussed and articulated these notions during the analytical process. In that way we specified what we have discussed they specify what they have seen and discussed with the nurses, in order to describe the new practices [16]. Patients were informed of our presence beforehand and if they did not consent, the researcher left the room. A letter with extra information for patients was available. The independent ethics committee judged this project to be exempt from review.

III. Results

This section presents our results. We identified three different themes which we will discuss here. In each theme we use examples from the data to clarify our points. These examples may be described or quoted from the data; so called thick descriptions, to support understanding of the results.

A. Several roads to Rome?

We found a large variety of ways in which telecare is used. Even more interesting is the variety in how nurses define this usage of telecare. To illustrate this use, we use three quotes, all of which are about patient Bob, who has an anxiety disorder. Bob 'calls in' regularly to the team E. As he contacts them frequently, he sees a lot of different nurses. In the next three quotes, nurses Mary, Daniel and Rudi reflect on the contacts with Bob:

1) Interviewer: What do you think the purpose of telecare is for people like Bob?
Mary: Well, when people get stuck for example. People who are unable to start the day by themselves and then they start calling their case manager every five minutes. With the screen, I feel they can learn to give themselves a signal, like: I get stuck, I have to do five things and I do not know how to start. Structuring your day, that is a perfect way of using it. *(20150501OtE)*

For nurse Mary, telecare is an excellent way to enable Bob to structure his day. Instead of having to face his doubt all day, he can call in and ask for support when he needs it. Mary feels this helps Bob on a crucial moment and will give room for a better day.

2) Bob calls in. He says: I want to talk a bit.
Daniel: Why?
Bob: I want to get rid of my tension.
Daniel: You always do, but you have to talk to your psychiatrist, I cannot help you.
Bob: I want to know what I can do about it.
Daniel: What do you think?
Bob: I think I will go for a ride on my bike.
Daniel: Good idea!
Bob: I hang up now ... and he immediately terminates the call.

Daniel says that all conversations with Bob take this course. He even thinks that whenever Bob calls in and sees that Daniel is there, he quickly comes up with his own solution, like riding his bike. I try to find out if Daniel intents to give Bob little room to show this behavior, but Daniel doesn't know, it is just how it goes. (20150423OtE)

Whenever Bob has contact with Daniel, conversations take another course than in the first example. Daniel, unconsciously or so it seems, steers Bob into the direction of a self-found solution. The conversations do not have the atmosphere of a quick relief, even though they still might have that effect on Bob.

3) Bob already called in once this morning. He knows what to do, but needs confirmation. Rudi thinks that Bob should be taught how to handle his thoughts himself, without the continuing intervention of others, for example with the help of cognitive behavior therapy. He does not know if that would be an option for Bob or if anything like that has been tried yet. (20150430OtE)

Rudi, finally, reflects upon why Bob would need such quick reliefs and whether he should be able to do this himself, without the telecare. Mary, Daniel and Rudi differ clearly in their opinion on how Bob should use telecare and if and how he would benefit from it. Different lines of thought on care would probably be the case before telecare as well, but this dispute is on the use of telecare, which makes these differences particular to telecare.

B. *Part of the plan?*

In team E, where most of the care is unplanned, nurses are aware of changes in their practice, even though they found them difficult to name:

Nurse Taco talks about patient Tobias, who told very dark stories after his last admittance. One example was on how Tobias claimed that one of the nurses at the clinic had instructed him to 'go grab that borderline bitch'. Taco tells how much he is affected by such contacts and how difficult these conversations are by webcam.
We discuss this for a bit, but do not seem to get to the heart of the matter. Taco says these contacts seem a kind of stopover, like it is not part of the process. I ask him if it would have been different had he been Tobias' case manager. Taco ponders on the treatment, on how telecare is a part of the care offer and how this is part of the treatment, but that does not seem right to him after all.... It seems clear though that conversations like these with Tobias have a larger effect on Taco because they are by webcam. (20150422OtE)

Nurse Taco is aware of his reaction to intrusive contacts, which he links to the fact that he is not part of the care process of most patients he has contact with by webcam. He finds it difficult to pinpoint the differences with regular care, but he sees it in the context of the treatment plan.

What happens here is that the webcam creates a new form of contact, contact without the context of a treatment plan. And even though the nurses mostly report the contacts they had in the patient's file, especially the ones they (on a professional basis) diagnose as difficult or important, the fact remains the contacts are incidental instead of part of a professional plan. In one way this fits the mental health care paradigm of 'recovery' very well. Patients are the center of care and are encouraged to take charge of their lives and care. Unplanned care fits this seamlessly, but from a professional point of view a new form of care is in the making.

C. *Telecare: the intertwining of aid and aim*

The cases of how nursing care changes in the above paragraphs are just a few examples of the very rich data we have gathered. Within the data there are many different forms of telecare and many different opinions of nurses on telecare. Apart from the new forms of care that were shown before, another interesting phenomenon has come from the data: the intertwining of aid and aim within care. Nurses talk about telecare as a way of sparing time because they do not have to travel to patients. They also mention how helpful it can be in contact with patients, both care-avoiding and excessively consuming patients, as patients do not have to come to the office for regular appointments anymore:

C [team H]: Well, in that case, I would tell him: we have a deal. I visit you once a week and I could contact you an extra time by telecare at the beginning of the week. This way I try to direct him so to speak and to restrict him in order to make him realize that he has to wait for our planned contacts so he will not keep calling in between.

So, telecare can be very supportive in alleviating logistic obstacles. On the other hand, the nurses mention how telecare enables more frequent, businesslike and concise contacts, which leads to more substantial guidance or supervision of some patients, again both care-avoiding and excessively consuming patients. So sometimes telecare can be used as an aid to solve logistic obstacles and sometimes it is the aim: more frequent contact with patients.

IV. DISCUSSION

We started this paper with the following questions: What new nursing care practices emerge when using a webcam? Are professionals aware of these new practices and are they purposeful performed? We have seen how nursing practice changes when telecare is introduced and we identified three themes within these changes. In the first theme, we saw the different functions of telecare, even for the same patient. Different practices might be common in the everyday practice of mental health care, but these different practices are established with the webcam and not recognized or discussed. The differences we saw between the three nurses is on a new form of care: in what way can nurses offer options for contact to patients with anxiety disorders? And even more important: what do nurses, in their professional opinion, think these options for patient contact should be? The nurses of patient Bob have not discussed his case. They belong to different teams and in Team E, only set up to man the webcam, no care-meetings are held. So technology changes practices without any necessary extra adjustments are made.

In the second theme, the webcam created a new form of contact, namely contact that was not part of the treatment plan. Nurses find it difficult founding these contacts, as they occur in a different manner and context than regular contacts. It seems as if the new dynamic and opportunities of these kind of contacts are not explored yet.

And thirdly, we saw how an intertwining of aids and aims in care occurs when webcams come in. This might lead to all kind of hindrances, especially when the logistic profits of patients not coming to the office conflict with a care goal that aims at supporting patients intensively, which may require bodily presences of the nurse.

The technology has brought new opportunities and different ways of working. What is remarkable is that the changes the technology brings about in the care practice are not discussed among nurses. This lack of discussion on the content of telecare might also be the reason that telecare does not seem to be a component of the treatment process. Even when telecare is included in the treatment plan as a tool for contact with the patient, the unplanned, incidental contacts can leave nurses with a distinct idea of the purpose of the contact. In the analyses, both these themes are linked to the problem that aim and aid are intertwined within the telecare practice, which makes it quite difficult for nurses to purposefully use telecare.

So, we have seen so far that new nursing care practices emerge when using a webcam. Nurses are aware of changes in their care practice, but do not have the opportunity to fully understand the changes in order to put them to use in care. We therefore conclude that it is very important to facilitate nurses to discuss these changes in order to name the new practices and purposely use them in care.

As our project is still running, at this moment this research has logically led to first results. Over time, and with possible extra field work, new insights can be gained on other new practices. Other perspectives will be used as well, especially from nursing and care theory. Finally, to name the unsaid is a part of ethnographic research. Observations and reflections on practices do not always cover all that is happening or the unsaid [17]. We will continue to pay attention to this issue.

ACKNOWLEDGMENT

We would like to thank the professionals who participated in our study and their respective organizations: Dimence, GGZ Drenthe, Lentis and GGZ Noord-Holland-Noord. The research described in this paper was made possible by a research grant from Nationaal Regieorgaan Praktijkgericht Onderzoek SIA (grant number 2014-01-52P).

REFERENCES

[1] F. A. Moreno, J. Chong, J. Dumbauld, M. Humke and S. Byreddy. "Use of StandardWebcam and Internet Equipment for Telepsychiatry Treatment of Depression Among Underserved Hispanics". Psychaitric Services. 2012 12: 1213-1217.

[2] L. van der Krieke, L. Wunderink, A.C. Emerencia , P. de Jonge, S. Sytema. "E-mental health self-management for psychotic disorders: state of the art and future perspectives". Psychiatric Services. 2014 1:33-49.

[3] J.R. van Veldhuizen and M. Bähler. "Manual Flexible Assertive Community Treatment". Groningen, 2013.

[4] L. Barnes and T. Rudge. "Virtual reality or real virtuality: the space of flows and nursing practice". Nursing Inquiry.2005 12: 306-315.

[5] K. England and I. Dyck . "Managing the body work of home care". Sociology of Health and Illness. 2011 33: 206-219.

[6] B. Latour. Science in Action. Cambridge (MS): Harvard University Press, 1987.

[7] J. Law. "Actor-network theory and material semiotics". In: The New Blackwell Companion to Social Theory. BS Turner Ed. Oxford 2009: Blackwell, 141-158.

[8] M. Akrich. " The De-Scription of technical objects", in Shaping technology / building society. Studies in sociotechnical change. W.E. Bijker and J. Law, Eds. Cambridge (MS), London (GB) 1992: The MIT Press, 225-258.

[9] B. Latour. "Where are the missing masses? The sociology of a few mundane artefacts". In: Shaping technology / building society. Studies in sociotechnical change. W.E. Bijker and J. Law, Eds. Cambridge (MS), London (GB) 1992: The MIT Press, 225-258.

[10] M. Mort, T. and C. May. "Making and unmaking telepatients: identity and governance in new health technologies". Science Technology and Human Values. 2009 34 -9: 9-33.

[11] D.Willems. "Varities of goodness in high-tech home care". In: Care in practice. On tinkering in clinics, homes and farmes. A.Mol, I. Moser and J. Pols Eds. Bielefeld 2010: Transcript Verlag257-277.

[12] N. Oudshoorn. Telecare technologies and the transformation of healthcare. New York: Palgrave Macmillan, 2011.

[13] R. Janssen, H. Prins, A. van Hout, J. Nauta, L. van der Krieke, S. Sytema and M. Hettinga, "Videoconferencing in Mental Health Care", paper for eTelemed 2015.

[14] J.R. van Veldhuizen, "FACT: a Dutch version of ACT", Community Ment Health J, vol. 43, no. 4, pp. 421-433, August 2007.

[15] H. Boeije. Analysis in qualitative research. London: Sage, 2010.

[16] J. Pols, Care at a Distance: On the Closeness of Technology. Amsterdam, The Netherlands: Amsterdam University Press, 2012.

[17] S. Hirschauer. "Putting Things into Words. Ethnographic Description and the Silence of the Social". Human Studies. 2006 29-4:413-441.

Smartphone-based Optical System for Blood Coagulation Self-monitoring

Ramji Sitaraman Lakshmanan and Vitaly Efremov
Biomedical Diagnostics Institute
Dublin City University
Dublin, Ireland
e-mail: ramji.lakshmanan@dcu.ie; vitaly.efremov@dcu.ie

Abstract— **Blood coagulation self-monitoring is vital in modern day healthcare, in particular supporting patients on anticoagulant therapy. Commercially-available coagulation self-monitoring devices are typically based on conventional end-point-based haemostatic tests and, measuring clotting time within a narrow range of values being connected to specific aspects of clot formation. In the present study, novel optical method combined with a smartphone was applied to measure the clotting time. The method was utilised for analysis of normal whole blood (WB) sample coagulation, activated by various tissue factor (TF) concentrations. The results demonstrated the ability of such a system to measure a wide range of clotting time values with appropriate level of accuracy and precision similar to standard thromboelastography (TEG). Significantly, the use of a smartphone as an optical detector and signal processor potentiates system miniaturisation, pertinent to high levels of functionality, cost-efficiency and user-friendliness.**

Keywords-blood coagulation; nephelometry; clotting time; light scattering; smartphone.

I. INTRODUCTION

Oral anticoagulation therapy is prescribed for both prophylactic and therapeutic use for patients at increased risk of thromboembolism including ones with mechanical heart valves [1]. In current practice, the efficacy of anticoagulation therapy is routinely determined by blood coagulation monitoring. This monitoring is typically based on a clotting time value measurement, specifically the time (in seconds) taken by method of blood sample activation to commence the coagulation process. Clotting time has a simple unambiguous interpretation. Its lower value corresponds to high blood coagulability and hence, increased risk of thromboembolism whereas higher clotting time values indicate increased bleeding risk. Furthermore, clotting time depends not only on patients' blood coagulability but also on the chosen method of coagulation activation. The oral anticoagulation drugs such as warfarin, heparin and many others became commercially available more than 60 years ago and their dosage monitoring was initially carried out in stationary clinical laboratories. More specifically, coagulation monitoring during a therapeutic regime is of critical importance for dosage control, since it provides key information about actual drug efficiency and reduces the risk of potential bleeding as a result of therapeutic overdose. In current practice, compact coagulometry devices for home-based self-monitoring are available [2]. The self-monitoring makes it possible to reduce the number of visits to clinics and optimise the resources of a local health-care system. Here, the patient collects a small volume of a blood sample with a finger-prick at the time agreed with by a clinical specialist, loads it on the device using an appropriate mechanism of transfer and records the clotting time value detected by the device, which is presented as a measurement output. Results are typically obtained in relatively short times, ultimately improving patient care and empowering patients as it allows them to take more responsibility for their own health [3].

There are several blood coagulation self-monitoring devices available on the market [2][4][5]. The absolute majority of these are designed to adapt some of the traditional laboratory assay of coagulation for use in self-monitoring. These conventional assays measure clotting time via the intrinsic and extrinsic pathways of the coagulation cascade, and selected methods of activation include prothrombin time (PT) and activated partial thromboplastin time (aPTT). Each assay measures a different part of the coagulation cascade. For example, PT uses TF as an activator of coagulation via the extrinsic pathway, and is sensitive to the effect of many widely-used anticoagulation drugs such as warfarin. Moreover, the PT is the only coagulation assay which to-date has been effectively standardised via the International Normalisation Ratio (INR), which explains why the PT assay is the most commonly-used in coagulation self-monitoring systems. In contrast, the aPTT assay uses negatively-charged contact activation material such as glass or kaolin in the presence of phospholipids as an activator of coagulation via the intrinsic pathway. This test is sensitive to the effect of anti-coagulants such as heparin. A panel of self-monitoring devices have been developed to essentially automate these assays through the integration of mechanical, optical, and electrochemical detection methodologies. One of the most recognisable device is the Roche CoaguCheck [2][4][5], which combines the PT/INR assay with microfluidics and a sample warming module. In the original device, a blood sample is applied to a disposable test strip containing activation reagents and iron filings. Next, an electromagnetic field moves the fillings, which is detected optically. As clot forms it prevents this movement, and the corresponding time where optical signal decays are returned to the end-user as the clotting time values. The Alere INRatio [5] also uses PT/INR assay and employs electrochemical impedance measurements to determine clotting time. However, both systems are not compatible with mobile devices, such as smartphones.

Although blood coagulation self-monitoring devices take a quite strong position on the market, there are at least two critical aspects that need to be addressed. Firstly, current devices typically measure clotting time only within narrow range of values since they are specifically adapted to be used with certain conventional coagulation assays. This is a quite serious limitation for the monitoring of the effect of specific and/or atypical anticoagulation drugs. For instance, a PT/INR-based device is usually tuned to provide effective resolution of clotting time values not exceeding 60 seconds, whereas many current therapeutic strategies require the rate of coagulation monitoring to be decreased to facilitate measurement (see, e.g., [6] where diluted TF activation is used and clotting time exceeds normally 4-5 minutes). Moreover, Siemens Healthcare has recently published a report where the applicability of the standard laboratory assays for the monitoring of the effect of novel drugs is analyzed [7]. The results set forth in the report show that PT, APTT, and other conventional coagulation assays are less sensitive or not sensitive to a number of novel effective anticoagulants such as thrombin inhibitors (e.g., dabigatran, argatroban) and direct factor FXa inhibitors (e.g., rivaroxaban). These novel drugs offer greater advantages over the traditional anticoagulants such as warfarin or unfractionated heparin which present the aforementioned well-documented drawbacks. Therefore, the extension of the dynamic range of measured clotting time values seems to be crucial for the ongoing evolution of self-monitoring devices. In an ideal case scenario, the device should have almost unlimited dynamic range, such as that afforded through thromboelastograph (TEG). TEG offers assays similar to the conventional assays that investigate the intrinsic and the extrinsic pathway of coagulation [8] and is notwithstanding its deficits, is considered as one of the gold standard in coagulation monitoring in clinical laboratories. TEG is not a self-monitor since it is too expensive, is not compact and has a level of operational complexity. Hence, reaching the TEG's level of performance and functionality but in the form of simple and compact device is an actual challenge for the biomedical diagnostic industry in developing innovative assays monitoring the onset of coagulation.

Other imperfections of the current coagulation self-monitoring devices include their poor compatibility with the modern telecommunication technologies. A simple connection to a smartphone with a specific application (App) on-board may deliver a new level of functionality, communicability and user-friendliness.

A novel optical method for blood coagulation monitoring is presented in this work. The method is based on recording of an optical signal acquired from a whole blood (WB) sample. The physical principle is optical nephelometry [9] where intensity of the light scattered by red blood cells (RBC) during their sedimentation is recorded. The time at which the fibrin clot becomes strong enough to stop the RBC sedimentation process is returned as the clotting time. The method has a wide dynamic range of measured clotting time values and is successfully implemented in the form of a compact device where a smartphone is used as the optical detector and the signal processor.

The paper is organized in four sections. In Section 2, the assay methodology, optical system operation principle and software algorithm are presented. In Section 3, the results are presented and compared to TEG. Finally, in Section 4, the conclusions are evidenced.

II. MATERIALS AND METHODS

A. *Preparation of blood samples*

Normal healthy controls were recruited at Dublin City University and patients with cardiovascular disease or individuals taking anti-platelet medication (e.g. Aspirin) were excluded from the study. Venous blood was collected from the antecubital vein with minimum stasis, and was citrated to provide a final sodium citrate concentration of 10.5mmol/L. Blood was stored at 37^0C for a period no longer than 60 min post-collection. Coagulation was activated by diluted tissue factor reagent (Dade Innovin®) which was obtained from Sysmex. Three types of TF reagent dilution (0.29, 2.9 or 29.0 pM) were used. A 10uL aliquot of TF and 10uL of saline were added to a vial containing 100 uL of whole blood (WB) sample, maintained at 37°C, and incubated for 5 minutes. Pre-warmed (37°C) 10uL of $CaCl_2$ (100 mM) was added to this vial and 20uL of this mixture was introduced into measurement cuvette and immediately transferred to the measurement chamber of the optical system.

B. *Optical System*

The system's optical layout is shown on Figure 1. The measurement chamber consists of a rectangular hole in a black nylon panel body sealed with a transparent plastic window from below. The polycarbonate rectangular cuvette containing the WB sample is loaded vertically to the measurement chamber. The black nylon panel body also contains a rectangular-shape adit with a mirror on its distal end to collect the light scattered by the sample via one of the four cuvette walls. The proximal end of the adit is sealed by a transparent light collection window being in direct optical contact with the lower half of measurement chamber volume. The system also comprises a source of collimated light, a warming module and an optical focusing module. The source of collimated light is a diode laser module with focusing elements (wavelength:650nm, power: 3mW, spot diameter: ~1mm) located directly above the measurement chamber. The focusing module comprises a mirror, apertures and lenses to deliver the light from the adit's mirror to an optical detector pupil. Finally, a Samsung Galaxy S3 smartphone was used as the optical signal processor with its main camera being used as the optical detector.

The operational principle of the system is very similar to nephelometry. Here, a laser light beam passes through the liquid sample introduced into the measurement chamber. If the chamber is empty or the sample is transparent, the light passes through the chamber with no scattering and minimal loss of intensity. However, when the analytical sample is turbid, this effects light scattering in a wide range of spatial angles. In this analysis, the scattering indicatrix depends on sample thickness, chamber geometry, and the concentration and shape of the light-scattering particles in the sample of interest. There are numerous examples of technologies with diagnostic applications which are based on interactions of red light with blood cells, and one may find typical scattering indicatrices elsewhere (see, e.g., [10]). The part of the scattered light leaving the measurement chamber through the light collection window is delivered to the smartphone camera pupil. The advanced design of the system could also incorporate a panel of microfluidic channels and additional chambers for blood sample preparation, incubation, mixing with specific reagents. However, the current design does not integrate these features.

The customized software (Android App) driving the camera, recording and processing the optical signal was written in Processing© 2 Java-style language with the use of Android® Development Tools and Ketai© library. The App takes the 640x480 pixel image every 0.25 sec and crops it in order to operate with its central region (100x100 pixel) where the light spot is located. The red colour channel is used only. When 3 images are taken and cropped, the camera signal value, *CS*, is calculated as follows:

$$CS = \{\Sigma RPI^i_1+\Sigma RPI^i_2+\Sigma RPI^i_3\}/7650000 \qquad (1)$$

where $RPI^i_{1,2,3}$ is the red-channel pixel intensity in a format of 8-bit number (0-255 scale) for the first, second and third images taken, respectively. Summation is set for the entire 100x100 pixel array, i.e., for a total of 10,000 pixels. The result of summation is then divided by 7650000=255x3x100x100, i.e., by a product of maximal single pixel intensity value, the number of images taken and the size of the pixel array.

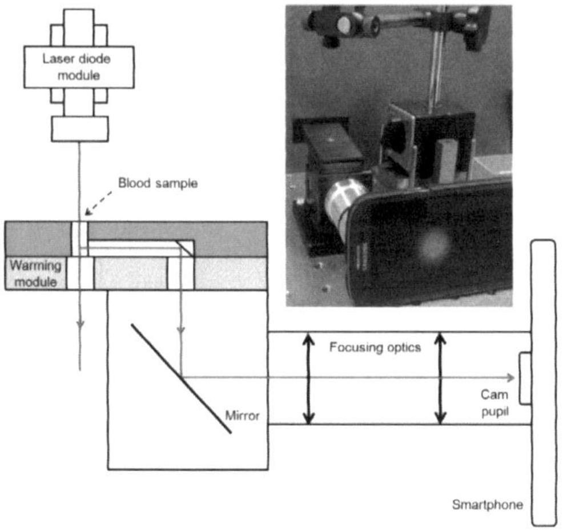

Figure 1. Optical schematics of the blood coagulation monitoring system and its actual implementation based on Samsung Galaxy S3 smartphone.

Thus, *CS* is proposed to be a normalised value characterising a luminosity level in the range from 0 to 1, i.e., from absolute darkness to a camera saturation level respectively. The App generates *CS* values every 1.0 sec and stores it in a phone memory. Further, the signal time-course, *CS* = *CS(t)*, alongside with the key diagnostic indices characterising coagulation can be directly seen on the smartphone screen or transferred to a PC via a USB port connection. The App also comprises an interface for the camera adjustment procedure, which is recommended to be completed before every signal recording session in order to provide reproducible results and maximise camera sensitivity. The user-friendly adjustment routine involves firstly switching the laser module off, and subsequently adjusting the camera to its "auto-settings" mode with the focus being set to "infinity". Auto-adaptation of the camera to absolute darkness makes its sensitivity intensely high, and close to the limit afforded by Android hardware drivers. Next, the "auto-settings" mode is locked following which the laser module can be switched on. At this stage, the camera is ready for the sensitive measurement and reproducible signal recording.

III. RESULTS AND DISCUSSION

Typical time-courses of the camera signal recorded by the system for normal WB samples are presented in Figure 2. For these initial measurements, coagulation was not activated and the observed monotonous increase in camera signal can be explained by the continuous sedimentation of RBCs onto the sensor surface, with haematocrit concentrations directly influencing red light scattering efficiency. During the initial stages of signal recording, WB samples can be identified as a suspension of homogenously-distributed cells in plasma. As RBCs sediment, there is a proportional increase in the concentration of cells at the bottom of the chamber, and an inverse decrease in cell concentration at the top of the chamber.

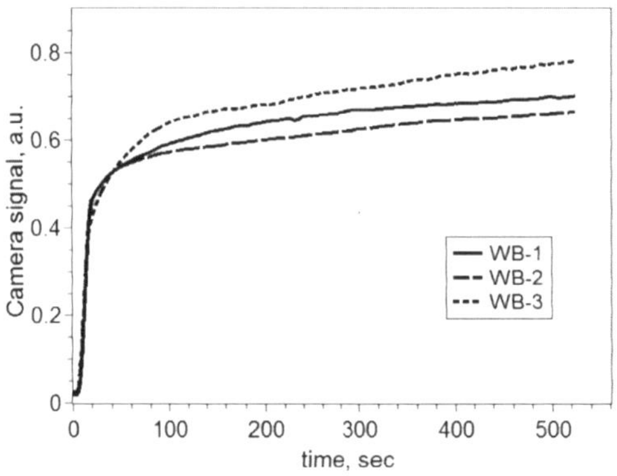

Figure 2. Camera signal recorded for not coagulating blood samples collected from 3 normal healthy individuals.

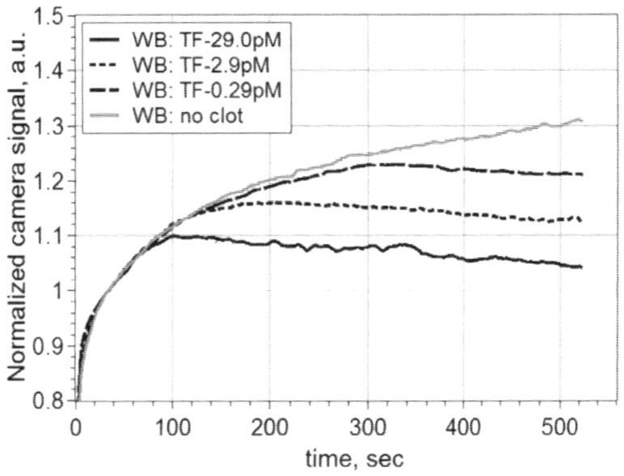

Figure 3. Camera signal recorded for coagulating blood sample collected from 1 normal healthy individual and activated by different TF concentrations.

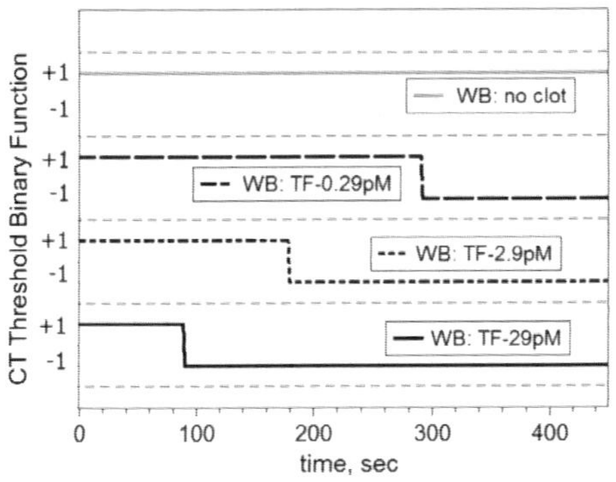

Figure 4. Clotting time determination by the threshold binary function. The results correspond to the camera signal time courses presented on Figure 3.

Concomitantly, the camera signal also increases since the light-receiving window is located at the lower half of the measurement chamber volume, and is in proximity to the sedimenting RBCs. The suspension of cells tends to reach its final state where the lower part of the volume is occupied by RBC mass, with the almost transparent plasma layer settling on top of this. It is important to note that the camera signal doesn't stop increasing during at least 1000 sec of the test when non-coagulating blood samples are selected for analysis. RBC count and sedimentation rate are the physiological parameters which can directly affect the scattered light intensity. The variable normal range is one contributing factor to the differences observed in the signal. A second contributing the factor the observed differences in the signal is due to the fact that the liquid sample is housed in the open-top cuvette and, hence, its upper layer forms a meniscus playing the role of a lens with optical characteristics which are challenging to predict. In order to perform an adequate comparative analysis, it seems reasonable to normalise time-course by the signal value corresponding to some point in the beginning of the reading.

Figure 3 represents the camera signal time-courses in the normalised form recorded for a WB sample collected from the same healthy individual and activated by 4 different concentrations of TF, namely 29pM, 2.9pM, 0,29pM and 0pM (non-coagulating). Every time-course is normalized by its value corresponding to 40sec after start of the test. One can see that the profile of the RBC sedimentation curve has deviated (at approximately 80s) from its normal course during coagulation. This point of deviation can be interpreted as the state of a sample where the fibrin clot becomes strong enough to prevent the RBC sedimentation process, and the corresponding time can be measured as the clotting time. Moreover, if TF concentration is higher the deviation is observed earlier, strongly suggesting that it is indeed indicative of the start of the coagulation process. Moreover, the time-course tends to decrease after the break-point since the growing fibrin network influences the scattering characteristics of the sample.

Indeed, single fibrin fibres can be presented as thin and long cylinders with a diameter from 50 to 200nm [11], i.e., comparable with the quarter of the light wavelength. Hence, the fibrin network is a quite effective scattering object preventing straight light propagation through the sample and reducing its intensity at the lower half of the measurement chamber where the light collection window is located.

In order to measure actual clotting time based on the above measurement outputs, a simple data processing algorithm was developed. Here, $TBF(t)$ is a threshold binary function taking two possible values of -1 or +1 indicating two possible states of a blood sample, namely "coagulating" or "non-coagulating" respectively. The CS value is being updated every 1.0 sec during the test. Let's consider the time series, $CS^{50}(t)$, containing 50 CS values following the current $CS=CS(t)$ value. Let $A(t)$ and $SE_A(t)$ be the first coefficient (slope) of linear regression for $CS^{50}(t)$ and its standard error respectively.

$$TBF(t) = sign(A(t) - 2SE_A(t)). \qquad (2)$$

One can see that the $TBF(t)$ function indicates "non-coagulating state" ($TFB(t)=+1$) only where the slope of the camera signal time-course takes the statistically significant positive value. If the slope cannot be determined by linear regression or its value is negative the $TBF(t)$ function indicates "coagulating state" ($TFB(t)=-1$). Thus, the time where threshold binary function changes its value can be returned as the clotting time (see Figure 4). One obvious weakness of the algorithm is that it requires > 50 seconds to make a decision since the clotting time is detected retrospectively. On the contrary, a strong point of the algorithm is the low risk of false clot detection insofar as $TFB(t)$ function value is always equal to +1 for a non-coagulating blood sample. Other strong point is that $TFB(t)$ function changes its sign only once and there are no multiple shifts for all three coagulating samples. Therefore, the algorithm provides unambiguous clotting time determination.

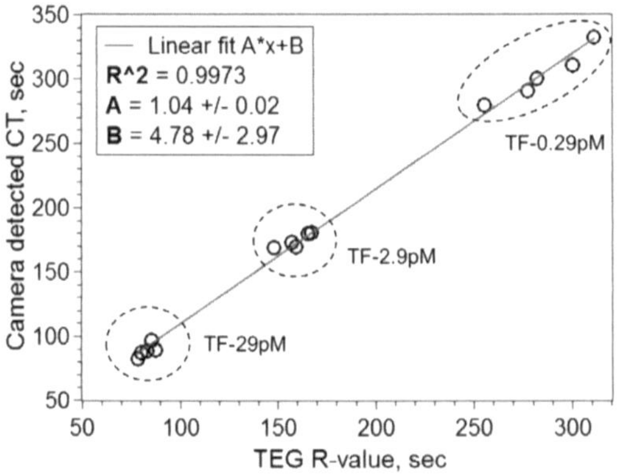

Figure 5. Comparison of clotting time values measured by the presented optical system and TEG for 5 healthy individuals and 3 TF levels.

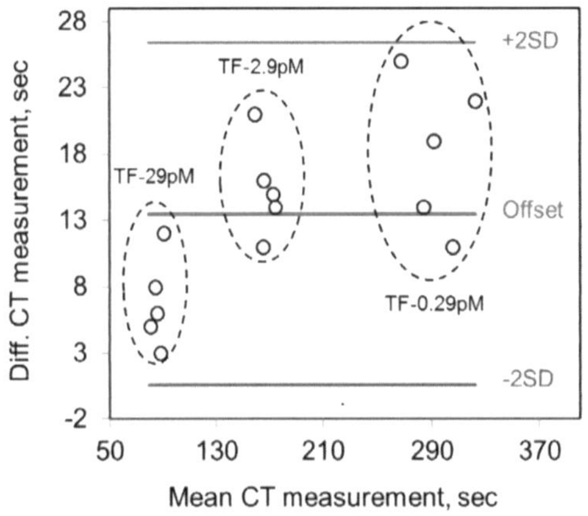

Figure 6. Bland-Altman plot for the presented optical system and TEG for 5 healthy individuals and 3 different TF levels.

As previously mentioned, clotting time depends not only on the blood sample coagulability but also on concentration and type of activation reagent. Coagulation self-monitoring device is supposed to use some standardized assay. If activation assay is based on TF, its concentration should be fixed in order to capture and recognise actual coagulability status of a sample.

In the present study, the real patients with abnormal coagulability and/or taking anticoagulation drugs were not available for examination. Thus, the inverse methodology was applied where 3 different TF concentrations were used for coagulation activation for conventionally normal blood coagulability status. In order to approve the method relevance and estimate the dynamic range of detectable clotting time values 5 normal healthy individuals were examined. The results then were compared to TEG. Scatter-plots for the clotting times evaluated by the presented system versus TEG R-values (clotting time equivalent in terms of TEG) are presented in Figure 5. Total number of data points is equal to 3x5=15, and we demonstrate a strong correlation between the two methods.

The fitting line intercept parameter equal to approximately 5 sec indicates the existence of an offset in the clotting time values measured by the two methods. Figure 6 represents the offset value along the measured clotting times. One may note that the system accuracy tends to be lower for the slower blood coagulation process. The delay in clot detection can be explained in terms of the optical system measurement principles. Here, a clot is detected only when the fibrin network is strong (or mature) enough to stop the RBC sedimentation, i.e., at some time after the actual physical appearance of the first fibrin fibres in the sample volume.

Nonetheless, the agreement between the system presented herein and the gold standard TEG is appropriate, and the offset can be corrected by more advanced data processing algorithms in the future system.

IV. CONCLUSION

Here, a novel optical coagulation monitoring method was developed and implemented in a form of compact system where a commercially-available smartphone is used as the optical detector and the signal processor. The customized Android App was developed to calibrate and drive the smartphone camera, record the optical signal during the test and process the data in order to return clotting time value being the main output parameter of the monitoring. The data processing algorithm provides low risk of false clot detection and precise clotting time measurement.

The method was approved by the series of clotting time measurements with tissue factor activation assay for 5 normal healthy individuals and 3 tissue factor concentrations. The results correlate well with the gold standard thromboelastography method, and demonstrate the ability of the method to measure a wide range of clotting time values with high levels of precision. The use of a smartphone as a part of future self-monitoring system can provide compatibility with modern telecommunication technologies, high level of functionality, cost-efficiency and user-friendliness.

ACKNOWLEDGMENT

The work was supported by the Science Foundation Ireland under Grant No. 14/TIDA/2443.

REFERENCES

[1] T. Christensen, "Self-management of oral anticoagulation therapy - Methodological and clinical aspects," Dannish Med. Bull., 58(5), B4284, 2011.

[2] L. Harris, V. Castro-Lopez, and A. Killard, "Coagulation monitoring devices: Past, present, and future at the point of care," Trends Anal. Chem., 50, 2013, pp. 85–95.

[3] M. Bissell and F. Sanfilippo, "Empowering patients with point-of-care testing," Trends Biotechnol., 20, 2002, pp.269–270.

[4] A. van den Besselaar and M. Hoekstra, "Long-term stability and reproducibility of CoaguChek test strips," Thromb. Haemost., 93(6), 2005, pp. 1189-1192.

[5] A. van den Besselaar, F. van der Meer, C. Abdoel, and E. Witteveen, "Analytical accuracy and precision of two novel Point-of-Care systems for INR determination," Thrombosis Research, 135(3), 2015, pp. 526-531.

[6] S. Butenas, C. van't Veer, and K. Mann, ""Normal" Thrombin Generation," Blood, 94(7), 1999, pp. 2169-2178.

[7] D. Funk, " Coagulation assays and anticoagulant monitoring," Hematol. Am. Soc. Hematol. Educ. Program. 2012, pp. 460–465.

[8] M. Thakur and A. Ahmed, "A Review of Thromboelastography," Int. J. Perioperative Ultrasound and Appl. Techn., 1(1), 2012, pp. 25-29.

[9] H. Oishi, Y. Hatayama, H. Shiraishi, Y. Miyashita, and Y. Sakata, "Development of a multifunctional laser nephelometer using backward light scattering detection for clinical diagnosis of blood coagulation and fibrinolysis," Industrial Metrology, 1(1), 1990, pp. 55-58.

[10] M. Yurkin et al., "Experimental and theoretical study of light scattering by individual mature red blood cells by use of scanning flow cytometry and a discrete dipole approximation," Appl. Opt., 44(25), 2005, pp. 5249-5256.

[11] E. Ryan, L. Mockros, J. Weisel, and L. Lorand, "Structural Origins of Fibrin Clot Rheology," Biophys. J., 77, 1999, pp. 2813-2826.

Technology for Transition

Needs and preferences of young patients

Margaret Machniak Sommervold and Maja van der Velden
Department of Informatics
University of Oslo
Oslo, Norway
margarma@ifi.uio.no and majava@ifi.uio.no

Abstract—**This paper presents a study of the technology needs of young patients with Irritable Bowel Disease, who are in the process of transitioning from pediatrics to adult-centered healthcare. The study is part of a Participatory Design process and is based on the assumption that information and communications technologies can potentially support young patients in achieving increasing independence from parental support and in engaging with their own healthcare. We argue in favor of designing for context-awareness and appropriation in technologies for young patients in transition.**

Keywords- Participatory Design; IBD; ICT needs; context-aware; appropriation

I. Introduction

Today, more than 90 % of children born with chronic or disabling conditions will survive well beyond their 20th birthday due to the advancements in medicine and treatment [1]. As they mature, young patients' personal and medical needs change and in their late teens they transition to the adult clinic or ward in order to receive age-appropriate care [2]. Transition in healthcare is being defined as "the purposeful, planned movement of adolescents and young adults with chronic physical and medical conditions from child-centered to adult-oriented healthcare systems" [3] and is triggered by the age of the patient. Before the actual transition, "adolescents are prepared to assume responsibility for their care" [4] because there are important changes in the way care is provided [5]. While pediatric providers rely on input from the young patient's family and have a more holistic approach to care, adult-centered healthcare providers communicate exclusively with the adult patient and present a more fragmented healthcare system. Applebaum et al. [6] argue that a seamless transition process is critical to the future health of the young patients; flaws in this process may lead to worse health outcomes. Adolescents with medical needs seldom make a smooth transition to adult healthcare [1]. Hence, transition has gained much needed attention during the past decade and has been described as "one of the greatest challenges facing peadiatricians" [7].

In a period already characterized by major physical, psychosocial, and social changes, young patients face additional challenges having to deal with changes in the way healthcare is provided [5] and "loose respected and loved carers and are forced to trust new and unknown ones" [8].

Because of their vulnerable position, young patients face a greater risk of dropping out of the healthcare system, which may have adverse health consequences. As a major life stressor, transition may also have direct negative health outcomes especially in case of diagnoses where psychological stress can increase the likelihood of relapse, such as Inflammatory Bowel Disease (hereafter, IBD) [2].

A growing body of evidence suggests that information and communication technology (hereafter, ICT) and technology-mediated methods hold the potential to improve outcomes for people with chronic conditions [9]. ICTs are especially promising in case of young patients who tend to be technology savvy and enjoy using new technologies. In terms of altering behavior, it is especially important to target this population, as it is during adolescence that adult behaviors are established and hence represent an opportunity to promote health oriented behavior and influence the healthcare needs of tomorrow's adults.

As we will discuss later in this paper, downloading a mobile application is not equivalent to using it. While not problematic in the context of regular apps, this issue presents a considerable challenge in health-oriented solutions, where positive health outcomes are dependent on consistent use.

This paper presents research on mHealth use and preferences for transition-oriented ICTs among a group of young people with IBD. The aim of the paper is to explore young patients' perspectives on *what*, *when*, and *how* technology can play a supporting role in transition.

The remainder of this paper is organized as follows. In the next section, we briefly present the existing trends in improving the transition process. Section III describes our methodology, methods, and population sample used to obtain the results, which we present in Section IV. In Section V, the main part of this paper, we discuss our results and the implications they might have for the design of technology for transition by using the concepts of context-awareness and appropriation. Finally, in Section V, we present some concluding remarks and opportunities for future work.

II. Background

Existing initiatives surrounding transition can be divided into three categories: i) *guidelines and recommendations*, ii) *paper- and form-based tools*, and iii) *mHealth and eHealth initiatives*. Guidelines form a large body of literature, proposing ways to help patients gain the necessary knowledge and skills to become competent in adult services and make decisions about their care. Important elements of successful transitions are a) cooperation between pediatrics and adult medicine, b) enabling the young patient to be part of the decision-making and management of their healthcare as early as possible, c) balancing parental support with the need for independence, d) continuity of care, and e) health education [6].

The recommendations stress that the barriers to achieving good transitions are not adolescents but the limitations of the healthcare itself. Acknowledging transition as a process rather than a singular event or transfer, Blum [3] was one of the first to suggest that the process should be initiated when the patient is 13 years old. Aiding the patient in planning and initiating the work towards independency in adult medicine, a range of paper-based tools has been developed. A transition checklist has been designed and implemented by The Royal Children's Hospital Melbourne and has been adapted by several Norwegian hospitals. Similar tools, such as health passes and 3 questions (Hospital for Sick Children, SickKids, Toronto) encourage young patients to learn more about their conditions and prepare for doctor's appointments.

There is a unified consensus that transition is an important challenge for both pediatrics and adult medicine. However, little has been done to implement the abovementioned guidelines and recommendations for transition. The growing demands are met with scarce resources in capacity, budgets, and healthcare professionals. This has contributed to the recent growth in the number of mHealth initiatives or healthcare supported by mobile technologies. Teens and young adults are an especially relevant group of potential users as they are tech savvy and enjoy using mobile technologies, making mHealth suitable and promising in terms of improving healthcare. TransitionMate which offers reminders, self-assessment of health and emotional state, and suggestions for health promoting activities [10], and Healthy Transitions which consists of video training sessions and quizzes [11] are two mobile applications targeting medical transition. As part of their ongoing research, [12] have presented various platform-based considerations for designing a transition app. There is a considerable number of ICTs focusing on IBD. Among these we find myIBD, developed specifically for *young patients* by SickKids, which focuses on logging and managing symptoms.

III. METHODOLOGY

Young patients, especially during their teenage years, tend to separate their identity as a patient from their identity as a teenager [13]. In order to meet them in a context where they felt comfortable to talk from a patient perspective, we conducted our research while they received treatment at the Children and Youth Clinic of Akershus University Hospital (AHUS) and the Medical Gastroenterology Clinic of the Central Hospital in Vestfold (SiV), respectively. The focus of the study was to understand transition from a patient's perspective and investigate *whether*, *where*, and *when* technology could play a supporting role in this process.

A. Participatory Design

There exists a lack of understanding of the teenage agenda and the challenges they face in transition – both as patients and adolescents [1], [14]. This study has therefore taken a Participatory Design (hereafter, PD) approach as it "takes into consideration the needs, interests and abilities of the youngsters, but also includes a more profound interest in their hopes, fears dreams, and opportunities to express themselves as someone of importance" [15]. With a focus on understanding *knowledge by doing*, the study was based upon PD's principles of *having a say, mutual learning* between the participants and researchers, and *situated action* [16].

B. Methods

For the purpose of this study, we used a closed sort [17], card-based method that we have described extensively in [18]. The method, called Transition Cards (TCs), is a qualitative card sorting method that enables participants to express their expectations and experiences surrounding transition. The TCs were used during individual interviews, in which we asked the participants to sort 70 cards representing *people*, *things*, *skills*, and *feelings* into categories representing *pediatric ward* ('BUK'), *adult ward* ('Voksen'), and *both* ('Begge'), which represented things relevant across these wards (see Figure 1). The cards referred both to the participants' identity as a patient, as well as to their life outside the hospital, thus providing a holistic approach to understanding transition [18].

Figure 1. A participant's use of the Transition Cards representing *things*.

C. Recruitment and participants

Recruitment of the participants was undertaken by the head of research at AHUS and by the staff at the Medical Gastroenterology Clinic at SiV. The participants were recruited while receiving treatment at the clinic and the medical staff decided whether the patients were well enough to participate. The patients received an information sheet, as well as a consent form, which they read before meeting the researcher. Once they agreed to participate, the researcher would meet them and both the researcher and the participant signed the forms. In the case of patients under the age of 16, a legal guardian also had to sign the consent form.

The research was registered and approved by the ethical board at both hospitals, and the Norwegian Social Science Data Services (NSD).

D. Sample

Participatory Design calls for involvement of future users based on the motivation that they are the sole experts on their needs and practices that the design should support. However, designing for a process that the future users have not undergone yet or know much about, required the involvement of patients who had transitioned. Therefore, the sample consisted of pre- and post transition patients.

An acceptable sample size in qualitative research studies involving participants with impairments or other vulnerabilities is between 5 and 10 [19]. In this study, fifteen young patients participated in this study – 8 pre-transition (ages 13-17) and 7 post-transition (ages 18-25) out of whom six were male and 9 female.

IV. Results

The interviews provided important information for mapping the young patients' expectations and experiences with transition and the role of ICTs in supporting them in any of the challenges that they presented and discussed using the Transition Cards. It was also important to understand which technologies they were using the most and what they used them for. We therefore started all interviews with asking whether the participants had access to mobile phones, computers/laptops, and tablets. All of the interviewed patients owned smart phones and only one out of fifteen did not own a computer. Majority of the participants had access to tablets, but reported on using mostly laptops and mobile phones. The participants reported that they used their phones for social media, such as Instagram, Facebook, Snapchat, Vine, and mobile applications offering instant messaging. The participants used their computers for schoolwork, gaming, and social media.

When mapping their use of health applications, several participants conveyed that they were advised by their doctors to download the 'IBD app' developed by a Norwegian company. Once downloaded, few of them tried to use the app but stopped due to the problem of too many notifications and reminders to log their symptoms or to enter a symptom-free period. Other health apps used by the participants were Endomondo and other training apps, and menstrual calendars. One of the participants reported on using an app to log migraine episodes and headaches.

In addition to ideas for more youth-friendly adult wards, the participants also proposed several areas where they thought that new technology could play a supporting role in their transition and life as a patient. These areas and needs can be categorized as *information, medical dictionary, reminders,* and *health rights and financial support.*

A. Paperless Information

Only seven of the patients received information about transition at least a year prior to their transfer. Two of the interviewed teens received the paper-based checklists during what was their last visit at the Child and Youth clinic. Majority of the interviewed patients suggested sites or apps presenting information about the new ward and an explanation of the differences in routines and practices of their old clinic and the new ward. Getting information about what they could expect at the adult wards, such as different specializations, having to engage with different doctors as opposed to their own pediatrician at the pediatric wards, could reduce some of the uncertainty. The pre-transition patients, who were informed about transition, were told that the adult health services were managed by the nurses and not by the doctors. They mentioned that they did not fully comprehend what it meant.

The patients suggested that the information about the adult wards should not only inform the patients about what would be different with regards to the above-mentioned changes, but also with regards to the changes in routines specific to their diagnosis. Conditions related to the digestive system often require endoscopic examinations, which in pediatric wards are performed under anesthesia. Anesthesia is not offered to adult patients, but they can receive temporary sedatives. The participants felt that they should be informed about this prior to the transfer in order to avoid stress and fear. The expressed need for diagnosis-specific information on transition is not the same as need for information about the diagnosis, and in particular the outlooks and life expectancy. Two of the participants expressed that it was too scary to think about the severity of their diagnosis, and that they did not read about the diagnosis online to avoid confrontation with information on this matter.

In addition to information about the different practices, the patients wished to be able to see the wards – ideally in person and accompanied by the personnel from the pediatric ward, but pictures or videos were considered helpful as well. Because the treatment and examinations were offered at different wards and clinics, the patients also thought that a map they could have on their mobile phones would help them in finding their way around the hospital.

We asked the patients about how they wanted to be informed about new health-oriented ICTs. Posters in the pediatric and adult medicine wards were suggested as a good way of spreading the initiatives. Another important factor was confirmation of the legitimacy and origin of the technology by the healthcare professionals.

B. Medical Dictionary-app

Two of the younger participants suggested medical dictionaries containing explanations for difficult words, acronyms, and treatments. When asked whether they would like to learn difficult terms through games, quizzes or textual explanations over time, the patients expressed that they needed information when difficult terms came up during their consultations and stay at the hospital, indicating that the information should be easy to find – even without knowing the correct spelling and be short enough so they could deal with the information also during the consultations with the medical staff.

One of the post-transition patients shared her experiences with not understanding what a stoma bag was, which affected her treatment decisions. She discussed a possibility of offering images or videos of treatments to show patients what they really looked like in order to reduce fear and uncertainty. As one of the younger participants explained, *When they call it 'blah blah', you don't get any of it. It sounds like gibberish or turns out to be something else than you thought. (...) So that it said what it was. It would be easier to understand what they put in your body. (...) I think it's weird that you get stuff put in your body that you don't know what it is* (girl 13).

C. Reminders for mobile devices

A large part of becoming an adult and being responsible for one's own healthcare is taking charge of appointments and medications. Today, Norwegian hospitals routinely send out text messages reminding the patients about their appointments at the hospital in addition to a letter. Remembering appointments with general practitioners are still the patients' responsibility. Several of the participants suggested reminders as desired functions on their mobile devices. The reminders would help them to remember their appointments and to take their medications. Reminders during the holidays and while travelling were highlighted as especially useful as different routines and time zones would often cause the participants to forget their medications especially if they had to be taken at specific times of the day.

D. Information on Health Rights and Financial Support

The Norwegian healthcare system offers free healthcare to persons under the age of 16. Upon turning 16, patients start to pay a deductible when receiving healthcare services, have the right to confidentiality, and can make decisions surrounding their treatment. When patients turn 18, they lose their right to free appointments with a child and youth psychologist. On the other hand, the Norwegian Welfare system offers benefits and financial subsidy for adults with chronic conditions.

Two the interviewed patients voiced a need for ICTs offering information on health rights targeting pre-transition youth. In addition, several post-transition patients reported on suddenly having to learn about health rights during the course of their transition.

E. Platforms

Both websites and apps were suggested as good platforms for information. One of the participants explained that: *You can take an app or cell anywhere, but a website is better for information. So both are important* (boy 17). The choice of platforms for the suggested tools was closely coupled with the purpose and context of use. While websites were explained as great for learning and reading up on things, apps were preferred where information was supposed to be offered immediately - without the need of internet connection. The participants suggesting the medical dictionary as a technology for transition argued that it would have to necessarily be an app so that they wouldn't have to rely on an Internet connection, which was often scarce at the hospital. Only one of the participants suggested paper-based brochures containing information, and none of the participants perceived the checklists offered at the hospital as helpful. One of the participants explained: *"it sucks more when it's on paper. I don't know why but I feel that it more serious and stuff like that when it's on paper, that you have to read through everything that will happen. It's certainly easier on a website"* (boy, 17).

F. Social networking

To our surprise, none of the participating patients suggested a patient-social network or other means of connecting with other patients in a similar life situation. However, they proposed that especially information sites should have comment fields so that others could discuss the topics or share their experiences. Several participants reported on reading blogs by other patients, but shared their concern about the reliability of the information, and relevance for their own situation – especially in cases of rare or multiple diagnoses, the participants were not able to relate or gain new knowledge from experience-based information found online.

V. Discussion

Transition is situated between the different contexts and stages in life of adolescents. The process involves changes in how healthcare services are provided, different legal rights, and growing economic responsibilities. It also involves responsibilities related to growing up and becoming an adult and, above all – being a young person – not a patient. [20] argues that despite the dominant role of their condition, they do not see themselves as patients. When designing and implementing measures to aid them in aspects connected to their patient identity, it is important to understand their daily life and how technology can fit there.

Powell et al. [21] imagine medical practitioners prescribing apps alongside medications. The patients interviewed in this study reported on their doctors recommending them to download and use the "IBD-app" developed by a Norwegian company. The "IBD-app" can remind the patients to take their medications and recommend booking appointments on the basis of their logged symptoms. The participants downloaded the app, but reported that they didn't use it because the app sent out too many notifications and urged them to log their symptoms when they were in a symptom-free period. Although they expressed the need for a reminder app, the IBD-app required them to log their symptoms in order to access the rest of the functionalities. The youth participating in this study did not adopt and use the IBD-app over time and hence did not experience the full functionality and benefits of the app. This stands in contrast to the research suggesting that combination of tracking tools, with mobile technologies makes it easier for users to keep track of the different aspects of their health. Based on the participants' explanations, the prescribed app did not offer them the experience they were looking for and became intrusive in periods when they did not have the need for its functionality. In other words, the app did not react to the patients' dynamic life-situation and lost its instrumental value when interfering with their non-patient identity. Dey [22] argues that today's technologies lack the ability to use and adapt to the situational information, or *context*, in the same way humans do. Context is defined as "any information that can be used to characterize the situation of an entity" [22, pp.5]. Further, [22] argues that having the ability to access and adapt to context will lead to better usability and usefulness of ICTs. In relation to technologies for transition and managing one's health, the understanding of context may help in determining *what, when,* and *how* we want to support.

A. When to support

The timing, or chronological age context, for introducing different ICTs and functions is crucial for adoption. The majority of patients suggested the timing for introduction of the information regarding transition to be close to the transfer. This is in contrast with what is suggested in the literature, which proposes the introduction of information between the ages of 12 and 14. As one of the teens explained: *When you're 14, it's still four years until you'll move, so it's not so important to know so much about it then* (girl, 17).

This aversion toward planning and preferences toward instantaneous and relevant ITs can be explained from a biological perspective. The ongoing re-structuring of the brain and development of cognitive abilities is not completed before reaching the mid-twenties, causes notable differences between adults and adolescents in their ability to control impulses, make rational decisions, and affects long-term planning [23]. This bio-developmental perspective combined with the findings in this study affects the overarching context of technologies for transition and indicate a need for technologies that are either context-aware, as defined by [22], and provide relevant information based on for example the user's age, location, and condition; or technologies developed specially for a relevant period during transition, which excludes the possibility and motivation for persistent use. Developing multiple technologies specifically for relevant periods of transition poses challenges with regards to sustainability and maintenance of these technologies and adds to the already high number of available apps and sites, making the task to find legitimate and relevant technologies harder for the users.

B. How to support

Another approach to this issue is to focus on appropriation. Within the field of PD, appropriation is being described as "the way that users 'take possession' of a technology innovation over time" [24]. As PD is concerned with giving users a say, appropriation becomes a way for users to "control and shape technologies to their own ends" [25]. Appropriation becomes especially relevant in the context of ICTs directed toward promoting health and supporting patients in managing their condition and care. This is due to the claim that appropriation involves *mutual adaptation* [14], where the users shape technologies and technology shapes the users' practices. We propose that the process of *adaptation* holds the potential to produce context-aware technologies through users directly shaping and controlling the technology to fit with their current context, whether it is chronological age, physical and emotional condition, or physical or social environment. The majority of the health-oriented ICTs focus on the condition of the patient. However, when focusing on the context, our design interest should shift towards inclusion of health and wellbeing alongside the chronic condition. When meeting a healthy person, humans don't tend to inquire about their symptoms – doing so would be unnatural. Similarly, context-aware technologies should be designed to stop collecting data when patients are symptom-free or are within physical or social contexts where they are detached from their identity and responsibilities as a patient.

One of the patients used an app to log migraine episodes and headaches. She reported that she found the app very useful to make sense of what triggered her migraines. Why did then the "IBD-app", which was recommended by the doctors - a source preferred by the participants, not receive the same response? We believe that the motivation for adaptation and the perceived usefulness affected the non-adaptation of the IBD-app. Combined with the intrusive reminders to log their symptoms, the IBD-app was not perceived as useful even though it offered reminders, which the patients interviewed in this study showed an interest for.

C. What to support

The spaces for initiatives and designs proposed by our participants – the medical dictionary, information on transition and their legal rights, signaled that there is a need for information on the *practicalities* of transition on mobile and web platforms. When trying to support the informational needs of patients, we need to balance our literature-based knowledge on transition, which focuses on transition as a continuous process [6], with the patients' perspective on transition as a series of fragmented events with specific and practical implications – such as turning 16 and having to pay a deductible. An example of practical information could be to *(One could) write what the routines were in relation to meals and visiting hours* (girl 21).

Based on our results, we need to support the information needs of the patients aware of the context in which *they* want to acquire this information. Turning toward the patients' perspective allows us to discover that there is also a need to inform young patients about the benefits of transitioning; *Treatment is more efficient here* (boy, 21). *Actually I think that the whole arrangement was more complicated when I was downstairs at the children's ward because there were so many people around me that I had to deal with – here things are a bit simpler* (girl 22).

The requirement for contextual and developmental awareness needs to be balanced with the requirement for privacy. The interviewed patients, as well as the findings from our earlier studies [13], suggest that young patients value their privacy and any technology aimed at health or extending the reach of healthcare institutions needs to be based on the values of confidentiality, privacy, and patient-centeredness.

VI. Concluding Remarks

There are no diagnoses with the exact same course for all patients – all diseases affect unique bodies at different points in life, and are triggered by different events in patients' lives. There is therefore a need for technologies that are aware of and will adapt to these lives and contexts. By designing for appropriation, the technologies may achieve context-awareness and hopefully shape the practices of the users by themselves allowing to be shaped.

In our future research, we aim to work towards understanding how we can implement context-awareness and facilitate appropriation in already existing technologies, as well as in the transition app, which we are currently developing in our ongoing research. We hope that this study will inspire designers, as well as healthcare professionals to work toward understanding the needs and context of young patients and design solutions that support them on their way toward independency.

Acknowledgment

We thank the Children and Youth Clinic of Akershus University Hospital and the Medical Gastroenterology Clinic of the Central Hospital in Vestfold for their support and cooperation during the planning and implementation of the study and their staff for facilitating the interviews. Our warmest thanks go out to all the participants for their willingness to share their knowledge and experiences.

References

[1] D. L. Patterson and C. Lanier, "Adolescent health transitions: focus group study of teens and young adults with special health care needs," Fam. Community Health, vol. 22, no. 2, 1999, pp. 43–58.

[2] R. Baldassano, G. Ferry, A. Griffiths, D. Mack, J. Markowitz, and H. Winter, "Transition of the Patient With Inflammatory Bowel Disease From Pediatric to Adult Care: Recommendations of the North American Society for Pediatric Gastroenterology, Hepatology and Nutrition," J. Pediatr. Gastroenterol. Nutr., vol. 34, no. 3, 2002, pp. 245-248.

[3] R. W. Blum, D. Garell, C. H. Hodgman, T. W. Jorissen, N. A. Okinow, D. P. Orr, and G. B. Slap, "Transition from child-centered to adult health-care systems for adolescents with chronic conditions: A position paper of the Society for Adolescent Medicine," J. Adolesc. Health, vol. 14, no. 7, Nov. 1993, pp. 570–576.

[4] M. E. Bowen, J. A. Henske, and A. Potter, "Health Care Transition in Adolescents and Young Adults With Diabetes," Clin. Diabetes, vol. 28, no. 3, Jul. 2010, pp. 99–106.

[5] F. Campbell, P. M. O'Neill, A. While, and J. McDonagh, "Interventions to improve transition of care for adolescents from paediatric services to adult services," in Cochrane Database of Systematic Reviews, John Wiley & Sons, Ltd, 1996.

[6] M. A. Applebaum, E. F. Lawson, and E. von Scheven, "Perception of transition readiness and preferences for use of technology in transition programs: teens' ideas for the future," Int. J. Adolesc. Med. Health, vol. 25, no. 2, Jan. 2013, pp. 119–125.

[7] R. Remorino and J. Taylor, "Smoothing things over: the transition from pediatric to adult care for kidney transplant recipients," Prog. Transplant., vol. 16, no. 4, Dec. 2006, pp. 303–308.

[8] R. Viner, "Transition from paediatric to adult care. Bridging the gaps or passing the buck?," Arch. Dis. Child., vol. 81, no. 3, 1999, pp. 271–275.

[9] S. M. Sawyer, S. Drew, M. S. Yeo, and M. T. Britto, "Adolescents with a chronic condition: challenges living, challenges treating," Lancet, vol. 369, no. 9571, pp. 1481–1489, 2007.

[10] R. A. Calvo, A. Pardo, Y. Zhao, E. Klineberg, M. Lam, and K. Steinbeck, "TransitionMate: a mobile application for chronic illness transition support," in OzCHI 2013, Adelaide, Australia, 2013, pp. 1–4.

[11] "Healthy Transitions", App Store. [Online]. Available from: https://itunes.apple.com/us/app/healthy-transitions/id953246958?mt=8 [Retrieved: March, 2016].

[12] J. Dixon, J. Dehlinger, and S. D. Dixon, "Designing, Implementing and Testing a Mobile Application to Assist with Pediatric-to-Adult Health Care Transition," in Human-Computer Interaction. Applications and Services, M. Kurosu, Ed. Springer Berlin Heidelberg, 2013, pp. 66–75.

[13] M. van der Velden and K. El Emam, "'Not all my friends need to know': a qualitative study of teenage patients, privacy, and social media," J. Am. Med. Inform. Assoc. JAMIA, vol. 20, no. 1, Jan. 2013, pp. 16–24.

[14] L. Jacobson, G. Richardson, N. Parry-Langdon, and C. Donovan, "How do teenagers and primary healthcare providers view each other? An overview of key themes.," Br. J. Gen. Pract., vol. 51, no. 471, Oct. 2001, pp. 811–816.

[15] O. S. Iversen and R. C. Smith, "Scandinavian Participatory Design: Dialogic Curation with Teenagers," in Proceedings of the 11th International Conference on Interaction Design and Children, New York, NY, USA, 2012, pp. 106–115.

[16] F. Kensing and J. Blomberg, "Participatory Design: Issues and Concerns," Comput. Support. Coop. Work CSCW, vol. 7, no. 3–4, Sep. 1998, pp. 167–185.

[17] J. Brucker, "Playing with a Bad Deck: The Caveats of Card Sorting as a Web Site Redesign Tool," J. Hosp. Librariansh., vol. 10, no. 1, Jan. 2010, pp. 41–53.

[18] M. M. Sommervold and M. van der Velden, "Transition Cards: Designing a method with and for young patients," IADIS Int. J. Comput. Sci. Inf. Syst., vol. 10, no. 2, 2015, pp. 79–94.

[19] D. J. Lazar, D. J. H. Feng, and D. H. Hochheiser, Research Methods in Human-Computer Interaction. John Wiley & Sons, 2010.

[20] P. Jones, Design for Care: Innovating Healthcare Experience. New York, NY, USA: Rosenfeld, 2013.

[21] Powell AC, Landman AB, and Bates DW, "In search of a few good apps," JAMA, vol. 311, no. 18, , May 2014, pp. 1851–1852.

[22] A. K. Dey, "Understanding and using context," Pers. Ubiquitous Comput., vol. 5, no. 1, 2001, pp. 4–7.

[23] B. Bower, "Teen Brains on Trial," Sci. News, vol. 165, no. 19, 2004, pp. 299–301.

[24] J. Carroll, "Completing design in use: closing the appropriation cycle," ECIS 2004 Proc., 2004, pp. 44.

[25] L. J. Bannon and P. Ehn, "Design matters in participatory design," *Routledge Int. Handb. Particip. Des.*, 2012, pp. 37.

Textile Performance Assessment for Smart T-Shirt Development

Mechanical and eletrical study for conductive yarn

Paolo Perego, Giuseppe Andreoni
Department of Design
Politecnico di Milano
Milan, Italy
{paolo.perego, giuseppe.andreoni}@polimi.it

Marco Tarabini
Department of Mechanical Engineering
Politecnico di Milano
Milan, Italy
marco.tarabini@polimi.it

Abstract— **This paper discusses a methodological approach for the metrological characterization of features in conductive textile. One of the outcomes of this paper is an integrated testing protocol aiming to assess the mechanical and electrical specification of conductive fabric. The protocol has been developed focusing on smart garments; this allows for evaluating the conductive features over elasticity, moisture, temperature and washing. The fine characterization of the fabrics allows to use them both as electrodes, sensors or wired connections. In particular, the paper studies the metrological aspects of conductive textile in order to optimize and verify their usage in sensorized t-shirt (as ECG- Electro-CardioGram electrodes, strain gauge, etc.) for continuous non-invasive monitoring in hospital settings, activity of daily living (ADL), sports and fitness. This study looks at optimizing the conductivity of the yarn that follows in minimizing the movement artifacts; this could increase the general performances and reliability of smart garments.**

Keywords-component; conductive yarn characterization, textile sensors, smart t-shirt.

I. INTRODUCTION

The future of healthcare system is based on the development of new and innovative technologies with the purpose of a more accurate, personalized, comfortable and widespread diagnosis and treatment of pathologies [1].

The 2015-2025 decade has been elected the "Wearable Era" for the diffusion and incidence of new miniaturized and wearable products and related services in our life. Most of these systems are related to health and lifestyle. Wearable Biomedical System (WBS) and Wearable Health System (WHS) are probably the most important among these emerging technologies.

The most effective feature of these systems is the non-intrusiveness that is more than non-invasiveness; they can offer solutions for continuous monitoring by measuring non-invasive bio-signal and biomedical parameters without awareness of the user [4]. The non-invasiveness quality of a Wearable Health and Biomedical System is often due to garment embedded sensors. These sensors usually are built on conductive textile sensor. One of the best example for these systems are the smart t-shirts. Smart t-shirts are garments with embedded textile sensors which can record different biomedical signals: ECG (Electro-CardioGram), Heart-rate, Breath-rate, skin temperature and other parameters [5]. All the signals are recorded by means of conductive textiles allowing for several measurements based on different metrological characteristics.

Moreover, this kind of conductive yarns, due to their intrinsic features, could be also used as light wire, antenna or electromagnetic shield. For this reason, the characterization of the yarn and textile becomes mandatory in order to increase the capability of this fabric.

In the last two years, different companies launched different types of sensorized garment, from socks (Sensoria®) [8] to Bra (OM-Signal®) [9]. Despite this, the main purposes of these products are focused on sport monitoring, which is mainly a one-spot measure. These devices/garments are not developed for medical continuous monitoring, which requires to match and respect strict specifications for the medical applications. In order to improve the performance of these smart garments and increase the application field of these conductive tissues, this paper reports a study on the performance and reliability of different fabrics. The metrological characterization and the study of stability over usage, time and washes, allows for developing new sensors, applications and approaches which can be used for sensorized garment. The rest of the paper is structures as follow. In Section II, we defined the system underling all the features that need to be validated; Section III describes the metrological issues defining the proper measuring. Section IV proposes the experimental setup for each measures. Finally, we conclude with Sessions V and VI which show the experimental results and some considerations on these.

II. MATERIALS

Electrocardiogram (ECG) is one of the most used exams for assessing the health status of the cardiovascular system. Moreover, a few parameters extracted from ECG could also be very useful in sport monitoring to detect the fatigue stages and the athletes' performances. Usually, these studies are conducted by means of ambulatory ECG with standard Ag-AgCl electrodes or chest belt based heart rate monitors. In the first case, the use of electrodes and conductive gels could modify the results due to the invasiveness of the procedure. In the second case, the use of the chest belt requires the electrodes to be well moistened and the belt fits snugly around the chest; this could cause discomfort to the users. Moreover, in some sports, the use of these system is prevented by continuous impacts and displacements, caused by the athletes' gestures, which can cause noise and errors on signals detected by the belt.

The study proposed in this paper starts from an already validated sensorized t-shirt for sport [6]. One of the goals of this study is the characterization of conductive textile in order to optimize the capability and the reliability of this fabric developing a new concept of sensorized t-shirt which could be applied for non-invasive continuous medical monitoring. The sensing t-shirt for medical application consists of I-lead embedded ECG electrodes (Figure 1) and a conductive strain gauge for respiratory activity monitoring. Both ECG and respiratory signal are recorded by an ad-hoc device which is connected to the sensing part by means of snap buttons. As visible in Figure 1, snap buttons receive the signal from ECG electrodes via conductive fabric. According to the hereof description, studying the conductivity, mechanical features and reliability of the yarn is mandatory.

Performances of the smart t-shirt depend also on mechanical characteristics of the yarn and the design of the t-shirt [7] and not only on conductive fabric. T-shirt usually is designed as a technical t-shirt developed starting from special yarn which is not only able to constrain electrodes and sensors in the right positions, but it also inserts a certain level of compression and thermoregulation, shaping the body while maintaining the same comfort of a normal t-shirt. The presence of two or more different yarns in the same cloth requires a mechanical study of textures, especially for elasticity which can compromise the stability and the contact of the embedded textile sensors.

Textile electrodes consist into textile containing silver yarn, mixed with cotton, lycra, or other fabric. The composite mixture of conductive and non-conductive yarn changes the features of the textile in elasticity, conductivity, difference of elasticity over length and elongation, etc.

Due to the different composition of the conductive textile, the quality of signals recorded by the t-shirt may differ significantly. For this reason, this work aims to investigate the performances of the different types of conductive textiles in terms of:

- Conductivity;
- Elasticity;
- Measurement repeatability and reproducibility;
- Sensitivity versus disturbances;
- Resistance to washing.

III. Metrological Issues

There are several works in the literature focused on the identification of performances of textile strain gauges, but only few of them focus on the effect of disturbances on the measurement quality. The first step of the sensor characterization is the identification of the textile strain gauge sensitivity. These tests aim to identify the relationship existing between the electrical resistance and the sensor elongation, and the experimental setup usually pairs a displacement measurement system with a 4-wire resistance measurement circuit. In addition to the sensitivity versus the measurand, sensors should also be characterized in terms of sensitivity versus disturbances [2, 3].

The main issues deserving for investigation are:

1) Measurement repeatability, which provides for an indication of a lower boundary for the measurement uncertainty;
2) Measurement reproducibility, that should assess, at least, the change of electrical conductivity after several washing cycles;
3) Contact resistance: in the classical setup for the identification of the strain gauge sensitivity the fabric is usually clamped at its ends so that it can be stretched for the calibration; however, the electrical resistance measured in this way is the sum of the resistance of the fabric and the electrical resistance between the setup and the fabric. The latter has to be measured before the tests.

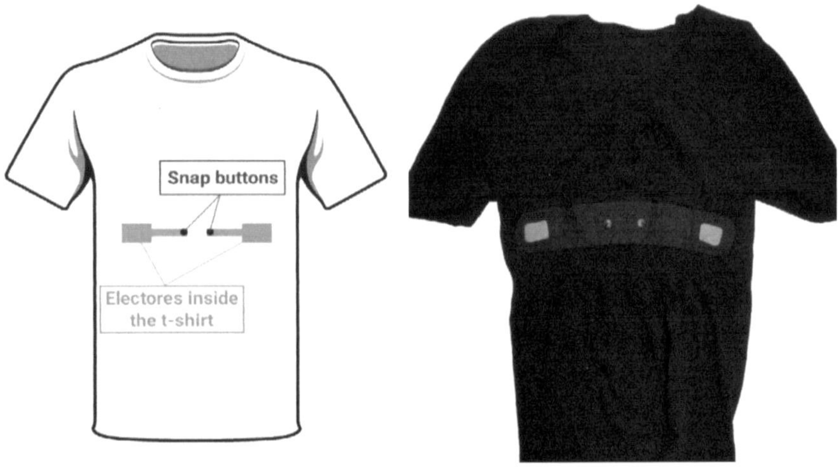

Figure 1. The smart t-shirt schema and a photo of the prototype.

4) Sensor creep: the creep results in a progressive and slow modification of the electrical resistance in presence of constant sensor deformation and should be quantified with proper tests if sensors are used for long-term monitoring of posture.
5) Sensors' dynamic behavior: often the frequency response function of the sensors is far from the ideal one, i.e. the sensitivity depends on the frequency of the stimulus. The dynamic calibration is therefore needed in order to identify the sensor frequency response.
6) Sensitivity versus temperature and humidity: the sensor behavior may depend on the temperature and on the humidity, thus resulting in a decrease in measurement performance in sports, where the sweating and thermoregulation strongly vary.

IV. EXPERIMENTAL SETUP

Different experimental setups for the complete sensor calibration are mandatory. The first setup is obviously the one for the static calibration, i.e. for the identification of the variation of electrical resistance versus the displacement. We have chosen to measure the sensor elongation with a linear variable differential transformer (LVDT) DC/DC; this sensor has a friction-free core and incorporate oscillator, demodulator and filter providing a self-contained unit accepting a DC input and providing a DC output relative to armature position. The electrical resistance of the textile sensor is measured by a four wire (volt-amperometric) circuit: a stabilized current generator creates a current of 100 mA (to reduce the self-heating). The voltage drop across the tensile sensor is measured by a National Instruments data acquisition board and the electrical resistance is derived using the Ohm law. The experimental setup is shown in Figure 2.

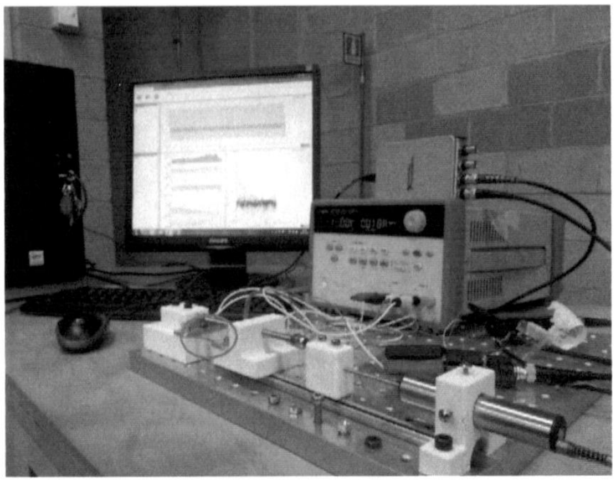

Figure 2. Experimental setup for the static calibration. The red circle shows the sensor

This setup cannot be used to impose quick displacements to the sensors; the textile strain gauge was therefore tested with an electrodynamic shaker, as shown in Figure 3. The upper extremity of the strain gauge and the lower one was moved by the shaker head. The tests to identify the effects of temperature and humidity were performed putting the setup of Figure 2 into a climatic chamber, as shown in Figure 4. The temperature range was between 10 and 40 °C and the relative humidity was not constant during the tests. Further tests were performed by spraying the sensor with water. Tests for resistance to washing were conducted comparing four different conductive fabrics. The composition of the textiles makes them different in conductivity, elasticity and on various mechanical aspects including the characteristics decay after washing. The four fabrics were cut in 2x9 cm patches and were washed for 20 times at 30° degrees with mild soap. The textiles were dried in open air at 22° for about 4/5 hours. The conductivity was measured with the setup used for the static tests.

Figure 3. Setup for dynamic tests. The green circle shows the sensor, mounted between the fixed support (upper part of the figure) and the shaker head.

Figure 4. Tests performed in the climatic chamber.

V. EXPERIMENTAL RESULTS

Results of the static calibration on one of the specimens that underwent our tests is shown in Figure 5. The specimen is 24% lycra with 74% silver and 2% polymer. As in many other fabrics of this kind, the sensitivity is not constant and therefore the resistance/length curve is not linear. The approach, in this case, can be the use of a variable sensitivity or the reduction of the useful range (45 to 65 mm) where the linearization is not critical.

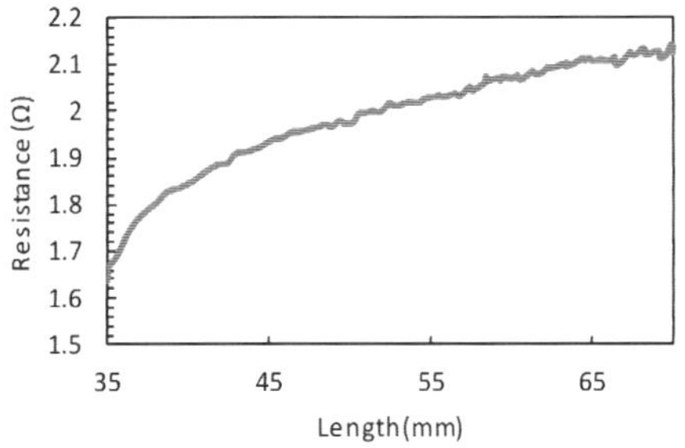

Figure 5. Example of results obtained with the static calibration.

The contact resistance was measured by testing two samples of different length; their electrical resistance is proportional to their length and the measured resistance is the sum of the sample resistance and the contact resistance. The contact resistance with our setup was 0.17 Ω; this value has to be subtracted from the raw measurements obtained with the experimental setup.

The creep of one of the sensors that underwent our tests is shown in Figure 6. In this particular case the displacement was constant but resistance of the sensor increased of approximately 6% after 20 hours. The creep may be critical in all the application where the sensor is used to identify the posture; DC reading is important and therefore a correction procedure similar to that described in ref. [3] should be adopted. In dynamic applications (e.g. breath monitoring) the creep might be not relevant and this issue could be ignored.

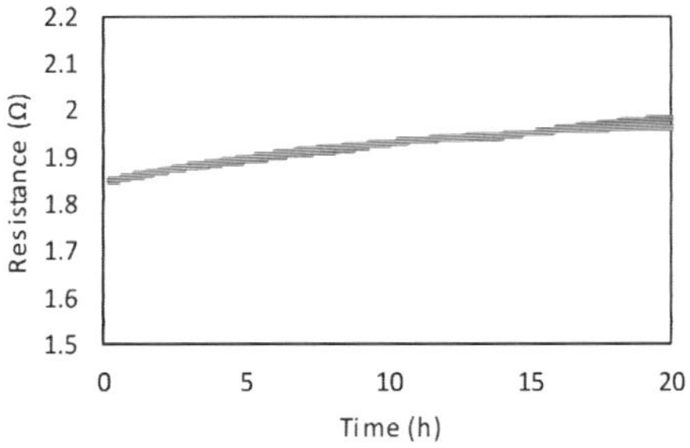

Figure 6. Creep tests of a specimen: resistance variation after 20 hours

Also the frequency response function may significantly differ from the ideal (flat) one. Our tests showed that the response to a sinusoidal excitation contains high order harmonics, as shown in Figure 7. This can be due to the lack of preload of the specimen and can lead to biased amplitude estimation in presence of quick subject movements.

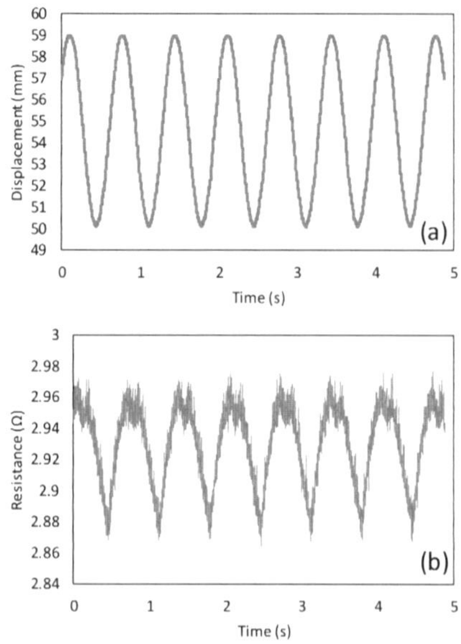

Figure 7. Results of dynamic tests at a frequency of 1.5 Hz. Imposed displacement (a) and specimen electrical resistance (b)

Also, the temperature and the humidity affect the measured electrical resistance: the dependence might be very complex, as shown in Figure 8. In this case, the resistance was influenced not only by the temperature, but also by the humidity, that was not constant during the tests. From this perspective, results were coherent with the ones obtained by spraying the specimens with water.

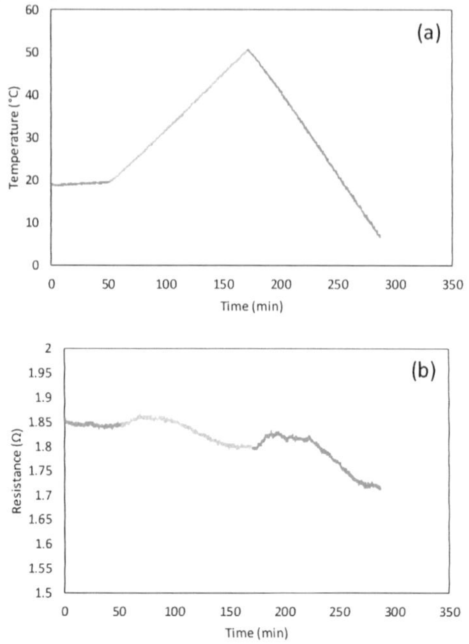

Figure 8. Variation of temperature (a) and of the specimen resistance (b) in environmental tests.

The last step to be investigated is the sensor aging: the setup for static tests was used to compare the aging after 20 washing cycles of four different conductive textiles:

- 3D conductive textile: the silver yarn is mixed with a 3d static filament which give to the fabric a thickness of 2mm. It consists in two different layers (one on the top and one on the bottom) connected in the middle by another more rigid conductive yarn which contributes to thickness.
- High elasticity textile: This fabric is composed by 76% of nylon covered silver and 24% of elastic yarn. Thanks to the nylon and elastic yarn, this textile is very elastic in one direction (65% elongation), and less elastic (30% elongation) in the perpendicular one.
- Low elasticity textile: It consists in 99,9% silver yarn coupled with a polymer. It is elastic in only one direction (about 20% elongation).
- Cotton/silver textile: This last textile is structured into two layers: the firs layer is composed by 100% cotton while the second layer consist in 50% silver and 50% cotton. It has an elasticity that is intermediate between the two previous ones.

The results are shown in Table 1 and Figure 9. The measures were not taken in a controlled environment in order to simulate the normal aging of an article of clothing. Results show that the largest resistance variation occurs for cotton and silver textile; this is due to the fact that this yarn is very fragile and rupture of conductive fibers causes a considerable increase of the resistance. Moreover, the result outlining that the percentage of silver yarn in the textile is not directly correlated to the high conductivity of the fabric.

TABLE I. VARIATION OF ELECTRICAL RESISTANCE OVER WASHING

ELECTRICAL RESISTENCE (OHM) OF FABRICS Vs WASHING NO.	3D TEXTILE	LOW ELASTIC TEXTILE	HIGH ELASTIC TEXTILE	COTTON AND SILVER TEXTILE
0	0.553	8.411	1.721	3.130
1	0.536	8.525	1.917	3.420
2	0.526	9.873	2.544	4.498
5	0.526	10.828	3.337	9.943
10	0.759	19.491	4.942	26.032
20	0.930	37.040	5.171	46.687

All the measure are in ohm

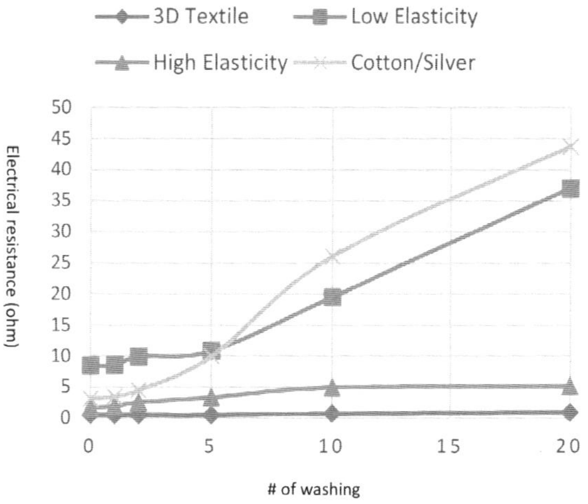

Figure 9. Variation of resistance over washing.

VI. CONCLUSION

In this work we have described the tests that should be performed for a complete metrological characterization of a textile sensor. The setup is based on a classic voltamperometric circuit but allows performing static and dynamic tests on the different specimens. The contact resistance was approximately 0.17 Ω; this value is small, but is of the same order of magnitude of the resistance of specimens that underwent our tests. The static calibration outlined that the behavior of some sensors is not linear, similarly with what was evidenced in studies already published. The creep was also relevant but could be compensated using a first order regression model. Dynamic tests are also mandatory in the sensors' characterization, given that on different specimen families the frequency response function decreased above 1 Hz, mainly because of the harmonic distortion due to the lack of elastic preload of the sensor. Also the temperature, the humidity and the washing affected the electrical resistance, thus showing that a proper calibration and a pre/washing are mandatory to obtain reliable results. Washing tests shows also that the conductivity is not directly related with the percentage of silver (conductive) yarn, but it is related with the knitting technique.

In order to optimize the performances and the reliability, all these aspects need to be taken into consideration when developing a smart t-shirt for medical purposes. The proposed set of measurements is one of the first proposals to build a "standardized" protocol in this directions. Future work will go in the direction of testing other fabrics and eventually to identify other parameters and a related testing methodology.

ACKNOWLEDGMENT

This work has been funded by the European Commission: Project PEGASO – Fit for Future, FP7-ICT-2013.5.1 - Grant Agreement n° 610727. The Authors would like to thank all the partners of the PEGASO consortium for their proactive collaboration in the project.

REFERENCES

[1] Bonato, Paolo. Wearable sensors/systems and their impact on biomedical engineering. *IEEE Engineering in Medicine and Biology Magazine* 22.3 (2003): 18-20.

[2] Saggin, B., Scaccabarozzi, D and Tarabini, M. Metrological Performances of a Plantar Pressure Measurement System. *Instrumentation and Measurement, IEEE Transactions on* 62.4 (2013): 766-776.

[3] Tarabini, M., Saggin, B., Scaccabarozzi, D., & Moschioni, G. (2012). The potential of micro-electro-mechanical accelerometers in human vibration measurements. *Journal of Sound and Vibration*, 331(2), 487-499.

[4] Lymberis A, Gatzoulis L., Wearable Health Systems: from smart technologies to real applications, inCo Proc 2006 IEEE Eng. Me BioI. Soc. pp. 6789-6792. 2006

[5] Karlsson, J. Stefan, et al. Wireless monitoring of heart rate and electromyographic signals using a smart T-shirt. *Proceedings of International Workshop on Wearable Micro and Nanosystems for Personalised Health*. 2008.

[6] Andreoni, Giuseppe, et al. Smart garments for performance and training assessment in sport. *Wireless Mobile Communication and Healthcare (Mobihealth), 2014 EAI 4th International Conference on. IEEE.* 2014.

[7] Andreoni, Giuseppe, Carlo Emilio Standoli, and Paolo Perego. Sensorized Garments for biomedical monitoring: Design issues. *2nd International Electronic Conference on Sensors and Applications. Multidisciplinary Digital Publishing Institute.* 2015.

[8] Sensoria Inc. - http://www.sensoriafitness.com/

[9] OM-Signal - http://omsignal.com/pages/omsignal-bra

A System for Real Time Monitoring of Users' Postural Attitudes

Alberto Cavallo, Alessio Robaldo, Flavio Ansovini and Alessandro De Gloria
University of Genoa – EliosLab
Genova – Italy
Email: Cavallo.Alberto.ca@gmail.com Robaldo.Alessio@gmail.com Flavio.Ansovini@unige.it
Alessandro.DeGloria@unige.it

Abstract— **This paper presents a system for real time monitoring of users' postural attitudes with particular focus on children. Maintaining for too long a wrong posture could cause serious problems to children and, in worst cases, they could receive permanent spinal deviations. The system is composed of two different elements: 1) a device equipped with inertial sensors (accelerometer, magnetometer and gyroscope), which are applied to the patient's spine and are able to measure his spinal curvature; 2) a mobile application for smartphone that provides a graphical user interface. Considering the age of the target of this project, we have decided to make use of the system similar to a game, giving the user a score based on his posture and creating a ranking. Our system can also give an alarm when the patient is maintaining a bad posture in order to educate him.**

Keywords- Posture; Inertial Measurement Unit; Serious Game for Health.

I. INTRODUCTION

Bad posture attitude is one of the main causes of backache and it can easily degenerate into real pathologies, especially when the patient is in his childhood and his growth is not concluded yet [1]. Thanks to some studies [1], we know that postural interventions, such as reminders about the good posture during a school lesson, significantly reduce the level of pain reported by children at the end of the class.

We also know that users are more willing to use whatever kind of product (in our case a medical device), also improving its effectiveness at maximum, if it is presented in form of a game, especially if it is aimed at children [2]. So that, this system would fall under the category of serious games and, in particular, games for health (games that are designed for a primary purpose other than pure entertainment) which are becoming largely adopted in various health problems for improving the people wellness [3].

The remainder of this paper is organized as follows: in Section 2 we briefly present a study on the state of the art related to posture problems, talking about what is needed to put our project into practice and considering some of the systems already existing focused on the same problem. In Section 3 we present the project itself, analyzing the very elements of which the system is composed and the algorithms we have developed. In Section 4 we examine the benefits deriving from the inclusion in our project of some typical aspects of gamification and we explain the idea we are developing. We conclude our paper in Section 5 by analyzing the results obtained with tests and proposing further experiments in order to verify the real impact of our device.

II. STATE OF THE ART

What is needed to put into practice our project is a system able to monitor postural attitudes suitable for an environment different from a laboratory (e.g., it could be used in a classroom); this should be non-invasive, and finally it should not either restrict or alter in any way user's motion. Nowadays, many systems have already been proposed but without achieving all our three goals at the same time. The main analysis methods of movement are based on videos and optoelectronics; they offer a good precision but they cannot be employed outside of an environment specifically studied for them [5]. Other interesting techniques perform the analysis by means of electromagnetic tracking systems and potentiometric goniometers, but the first one can be affected by the presence of metallic objects in the work environment and the second one restricts the patient's movements.

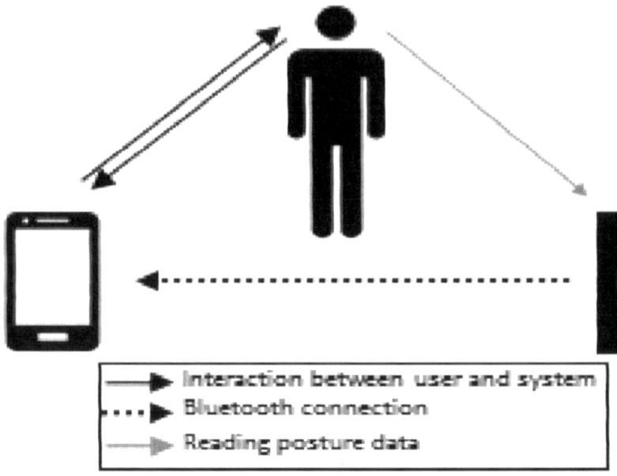

Figure 1. System components relation

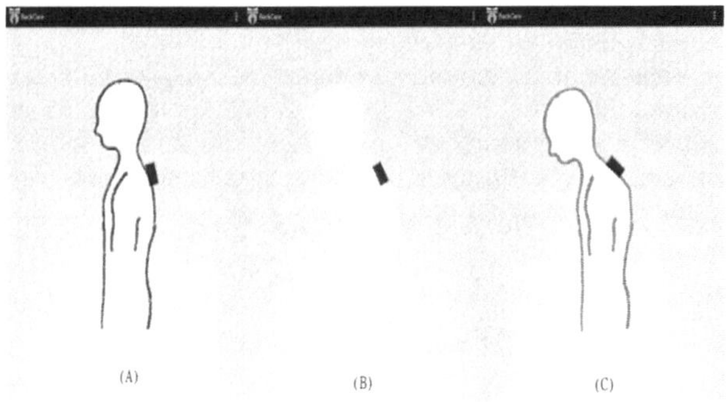

Figure 2. Graphic User Interface: a) Correct curve; b) Medium curve; c) Wrong curve

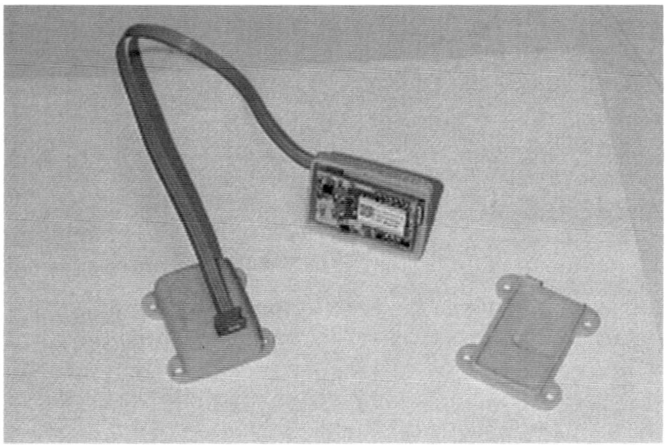

Figure 3. The system hardware prototype

III. THE PROJECT

Once we evaluated the above considerations, we decided to implement a system based on inertial sensors and in particular a three-axis gyroscope, a three-axis magnetometer and a three-axis accelerometer. These sensors are used to obtain the Eulero angles, three data that describe the orientation of a rigid body in the space. The use of accelerometers and gyroscopes to analyze the spine position of a human being has already been studied [4]-[7], but we have introduced in our system a magnetometer in order to avoid a typical problem of gyroscopes, called drift.

Our system is basically composed of two components: the first one is dedicated to the measuring of the patient posture, and the second one analyses the data and allows the user to interact with the system through a graphic interface. The two elements can communicate each other via Bluetooth connection (Figure 1).

We have designed and developed a device (Figure 2) equipped with an ST iNemo module [8]. With this module it is possible to receive data from a three-axis accelerometer, a three-axis gyroscope and a three-axis magnetometer, and filter them to compute Euler angles.

The measurement of the posture kept by the user occurs by placing one or more devices on different spots of his back in order to monitor the main curves of the spinal column. In particular, our sensors are used to measure back curves of the patient on the sagittal plane (to monitor kyphosis and lordosis) but they can also be used to monitor angles on the coronal plane (to monitor scoliosis).

The device we have developed is composed of two Inertial Measurement Units (IMUs): the first one, called master has to handle the data transmission to the Android application via Bluetooth, in addition to measurement task using the sensors; the second IMU, called slave, is only able to measure and elaborate data, working as master satellite. This structure allows us to perform simultaneously measurement in different areas of patient's back, giving a more accurate set of data.

The graphic user interface is implemented to be executed on a mobile device (smartphone or tablet) equipped with an Android operative system; in turn, the device is connected with the sensor component using a Bluetooth connection. When it receives data from the sensors, it gives back to the user a real time view of his position, emitting three different kinds of warning in case of a prolonged bad posture: a graphic one (Figure 3), an acoustic one, and a vibrating one. In order to avoid excessive noise from our system, we have set the alarm so that it will sound only after a certain span of time during which the patient is maintaining a wrong position; in this way the system will not measure patient's temporary movements/changes of position.

Another function performed by our app is the possibility to save various information concerning monitoring shifts within a database so that the user can examine them later.

Finally, the software has been designed to be used by more than one user on the same device; when the system is launched, a user authentication is required so each patient can access their own personal data and profile.

IV. POSTURE MONITORING AS A SOCIAL GAME

An important aspect we have considered in developing our system is the great success that social games have been recently achieved. Several studies show that the fact of sharing the results obtained between user and the possibility for them to leave a usually positive feedback, make the system more attractive. This is a feature common to every kind of activities, be they virtual or real, experienced by the user who is nowadays used to share many aspects of his life through social networks.

For all these reasons, while developing our system, we have thought about giving a score to the user based on the time he spends in maintaining the correct position. This score is then inserted within a rank to create a certain sense of competition among users/players and, most importantly, to push them to maintain the correct position. This idea would be of particular effectiveness in case of children or teenagers belonging to the same class.

The score is given accordingly to the framework shown below:

- +1 point every minute spent in the correct position;
- 0 points every minute spent in a position partially incorrect;
- -1 point every minute spent in the wrong position.

The system updates the results of all participants every thirty minutes to highlight children who have kept the position particularly wrong. Besides, we have thought of the possibility to form teams to increase interaction among children. Finally, as mentioned before, users' scores can be shared on social networks and inserted within a global rank to develop a challenge.

V. CONCLUSIONS

In order to evaluate our system, we have carried out experiments in lab. They showed the validity of the approach both for the precision in measuring the posture and for the feedback given to the user.

The experiments conducted so far left us quite optimistic about the results our system could give but, the real impact of social games have not been completely verified.

In order to verify the real impact of the device we have developed, shaping it as a social game, we suggest working in partnership with a school. In this way, we could test the system on a significant number of subjects. The general idea we propose is to split up the students in two group called *Group A* and *Group B*. The first group should use our system in "social game modality"; in the meanwhile Group B should be using the system only with "real time feedback modality". During the experiment we expect the members of Group A to maintain a correct position for a longer period of time than the one kept by members of Group B.

REFERENCES

[1] M. Robbins, I.P. Johnson, and C. Cunliffe, "Encouraging good posture in school children in using computers," Clinical Chiropratic, March 2009, pp. 35-44.

[2] "10 stats on the growth of gamification," http://www.gameifications.com/learning/10-numeri-sulla-gamification-nelleducazione/, [retrieved: March, 2016].

[3] "Health a fitness news," http://seriousgamessociety.org/index.php/2014-07-11-14-15-51/explore/137-health-a-fitness-news, [retrieved: March, 2016].

[4] W. Y. Wong, M. S. Wong, "Trunk posture monitoring with inertial sensors," Eur Spine J, January 2008, pp. 743-53.

[5] R. Y. W. Lee, J. Laprade, and E. H. K. Fung, "A real-time gyroscopic system for three-dimensional measurement of lumbar spine motion," Medical Engineering & Physics, December 2003, pp. 817-24.

[6] E. Lou, J. Raso, D. Hill, N. Durdle, and M. Moreau, "Spine-straight device for the treatment of kyphosis," Studies in Health Technology and Informatics, 2002, pp. 401-4.

[7] R. J. Nevins, N. G. Durdle, and V. J. Raso, "A posture monitoring system using accelerometers," Electrical and Computer Engineering, 2002, pp. 1087-92.

[8] "INEMO_M1 official documentation," www.st.com/web/en/catalog/sense_power/FM89/SC1448/PF253162. [retrieved: March, 2016]

[9] University of Pennsylvania Annenberg School for Communication, "Social networks can motivate people to exercise more: Compared to motivational messaging, the social influence of online peers proved more effective in improving exercise habits," www.sciencedaily.com/releases/2015/10/151007110738.htm, October 2007 [retrieved: March, 2016]

[10] J. Simoes, R. D. Redondo, and A. F. Vilas, "A social gamification framework for a K-6 learning platform," Computers in Human Behavior, March 2013, 345-53.

[11] J. Hamari, J. Koivisto, "Social motivation to use gamification: an empirical study of gamifying exercise," ECIS 2013 Completed Research, January 2013. Paper 105.

iCARE - Telematic Platform for Network Management of Pediatric Patients with Incurable Disease and High Complexity Care

The Ligurian Experimental Model

Luca Bianconi, Cristiana Degano, Matteo Toma
R&D Business Unit
Gruppo SIGLA S.r.l.
Genova, Italy
e-mail: luca.bianconi@grupposigla.it, cristiana.degano@grupposigla.it, matteo.toma@grupposigla.it

Riccardo Berta
Elios Lab, DITEN, Università di Genova
Genova, Italy
e-mail: berta@elios.unige.it

Luca Manfredini(1), Federica Penco(1), Sandro Dallorso(2), Paolo Petralia(2)
(1)Centro Regionale di Riferimento di Cure Palliative e Terapia del Dolore Pediatrico,
(2)Istituto Giannina Gaslini
Genova, Italia
e-mail: lucamanfredini@gaslini.org, federicapenco@gmail.com
sandrodallorso@ospedale-gaslini.ge.it, paolopetralia@ospedale-gaslini.it

Abstract— **The Information and Communication Technologies (ICT) support service to the activities of home care has, as its main objective, the creation of a technological tool aimed at improving the operation and the quality of home care services for pediatric palliative care. In particular, it will provide a tool to support the performance of the core activities required to provide assisted home care of excellent quality and, to clinicians, a whole process management system able to ensure simplification and efficiency.**

Keywords-Telematic; mHealth; IoT; Home Care.

I. INTRODUCTION

iCARE has been designed and is under development within Health@Home (H@H) project, funded by the Italian Ministry of University and Research (MIUR) as part of the Smart Cities and Communities Call, and whose goal is to create assistance services to citizens based on a network of integrated services in the health, territory, and society filed, through the use and implementation of interoperable devices and systems [1].

The development of information technology solutions for the collection, processing and use of critical data, can provide a very useful support to the young patient, family and clinicians. Furthermore, the telemedicine, so as the m-Health applied to the field of Paediatric Palliative home-care may allow the patient and his family to interact effectively and real-time with the medical team (local and the one of reference center) through a remote control of the different phases of the care process, from diagnosis to treatment.

In this context, iCARE will be a telematics platform for the management of pediatric patients with incurable disease and with high-complexity care needs.

Particularly, iCARE platform firstly aims at the realization of a system able to ensure adequate quality of services by improving and optimizing the situations described previously and the process that characterizes them. Secondly it aims to avoid unnecessary house calls, reducing the feeling of hospitalization and, while ensuring security in the continuity relief, facilitate the stay of patients at home and reduce improper hospitalizations.

Because the research project is currently in progress, we can here propose a short presentation of the iCARE platform dealing with the main end users' required functionalities and the first application implemented.

This paper is composed of Section II that describes the clinical context and functionality, and of Section III, in which the two main applications are described: *Home Care management system* and *Mobile Diary*.

II. CLINICAL CONTEXT AND FUNCTIONALITY

The context within iCARE platform will be developed for the home-care services and is for the palliative care at home of *Giannina Gaslini Institute* that treats pediatric patients at their residence or at care facilities (in the case of children coming from outside the region), administering first-line treatments and, if necessary, appropriate palliative care.

In this context, one of the main objectives of the research is the development of a tool based on the so-called *Family Centered Care* [2].

Since iCARE belongs to the so-called *Digital Medicine*, the system will provide the following functionalities:

- Communication among patient and Medical Doctor (MD)
  - Doctor's visit
  - Drug prescription and administration
  - Warning of complicated health status
  - Knowledge exchange about therapy
  - Visit recording with automatic generation of the report
  - Storing and sharing of visits performed
- Communication among MD and MD

    - Management of the patient's medical staff
    - Sharing of not solved problems and important information about
    - Countersign of drug prescription
- Patient's electronic diary (clinical events, therapeutic events, etc.).
- Patient's information management.
- Automatic and semi-automatic warning of MD actors relating with particular situations.

III. APPLICATIONS

It is mandatory that iCARE platform must satisfy the following requirements:

- Appropriate interface between the biomedical devices at home and the data transmission systems.
- User friendly system providing clinical information to the young patient e family (taking into account cultural and linguistic differences).
- IT protocols for the storage of data, coming from home devices, in the medical record system belonging to the *Reference Palliative Care Center*.
- Clinical and Information Technology (IT) protocols to analyze the collected data and to verify their accuracy and quality.

In order to provide the functionalities mentioned in Section II and to satisfy the abovementioned requirements, two applications have been developed: *the Home-Care management system* and *the Mobile Diary*.

A. Home-Care management system

The main functional requirements provided by the *Home-Care management system* are:

- Management and tracking of therapy.
- Management of clinical intervention at home.
- Noninvasive data collection about health status of the young patient.
- Efficient communication among the young patients (young patient and his family) and the medical staff (oncologist, pediatrician, nurse, healthcare assistance, etc.).
- Sharing of therapy data among the actors belonging to the medical personnel (MD of *Reference Palliative Care Center*, specialized MD, MD of palliative care, pediatrician, etc.).

The system is accessible through the family's device(s) such as personal computer, tablet and smart phone, while data will be processed and managed centrally via cloud technologies. Furthermore, the system is composed by a front-end, dedicated to users, and by an administrative back-end. Back-end access to the different functionalities is regulated through roles and authorizations based on the users' categories.

B. Mobile Diary

The *Mobile Diary is* the first application implemented. It is the user's (patient, family, medical personnel) interface providing the described functionalities. Particularly, it is an application that has been developed to record and monitor the daily clinical and psychophysical status of the patients.

The use of *Mobile Diary* allows the patient, family and medical staff to answer to ad-hoc questionnaires through a dedicated interface, and to fill information related to the clinical parameters and drug therapies associated with the pathology and the adopted treatment protocol. On the other hand, the *Mobile Diary* allows the MD staff to create and manage new kinds of questionnaires, to analyze collected data and to export them in different format through a back-end that is accessible by the MD staff only.

Then, the application is composed of two different interfaces: the interface devoted to the patient and the administrative back-end devoted to the MD staff.

The interface for the patient is available in two ways: as a Web application (thus accessible through a common web browser) or as a mobile application itself. Both client types provide the same functionality.

To the contrary, the administrative back-end takes the form of a Web portal for administration and management of both questionnaires and data collected. Its natural means of fruition is the Web browser.

The application is still accessible via the mobile device (smartphone and tablet) so as from any other device equipped with a browser and Internet access.

The data generated and processed by the system will be managed centrally via cloud technologies.

The system provides the following features:

- Interaction with questionnaire (creation, management, publication, response).
- Data analysis.
- Monitoring of physical/psychophysical status of critical events.

IV. CONCLUSIONS

The reduction of the residence times in the hospital and the costs associated with them will result in an improvement of the conditions of patient's safety and comfort.

Particularly, the home-care service will allow to:

- rationalize and improve the management of clinical and therapeutic data of the patient by ensuring appropriate access to people involved in the various phases and assistance situations.
- improve the availability of communication between assisted care facilities and patients, ensuring efficiency in communications and non-invasiveness of the same.
- improve the lifestyle of patients, each actor of the care process will help, according to the tools provided by the system, to provide additional data in support of care activities, sometimes also as a preventive perspective.

REFERENCES

[1] http://www.si4life.com/index.php?option=com_content&view=article&id=71:smart-cities-en&catid=20&lang=en

[2] Smith, T., & Conant Rees, H. (2000). Makinng family-centered care reality. Semin Nurse Manag.; 8 (3), 136-142.

Development of Smart Garments and Accessories to Support Behaviour Change in Teen-agers: Considerations on the Use of Interactive Virtual Prototyping

Maria Renata Guarneri
Department of Design
Politecnico di Milano
20158 Milano – Italy
e-mail: mariarenata.guarneri@polimi.it

Abstract—**This paper provides some considerations and preliminary results on research work for the development and experimentation of a behaviour change system to promote healthy lifestyle in teen-agers and prevent the development of lifestyle related disease. Lifestyle-related illnesses are among the top healthcare challenges in Europe. As an example, the rapidly increasing prevalence of overweight and obesity among children and adolescents reflects a global 'epidemic' worldwide. Due to the associated serious medical conditions, it is estimated that obesity already accounts for up to 7% of healthcare costs in the European Union (EU), as well as costs to the wider economy associated with lower productivity, lost output and premature death. Obesity in younger age groups has been recognized as an alarming key predictor for obesity in adulthood, but also entails a number of short-term health complications in juvenile age along with greater risk of social and psychological problems. The rapid development of Information and Communication Technologies (ICT), and in particular mobile technologies, together with their increasing diffusion among the EU populations, offers an important opportunity for facing these issues in an innovative manner introducing the possibility of a new technological framework to re-design the healthcare system model. Starting from and leveraging the work being performed in the course of the EU-funded project "PEGASO", the research focuses on developing solutions for self-managing life-style with specific target on the younger population. More specifically, an important part of the solution is a wearable sensor system for lifestyle monitoring and awareness development, composed of smart garments and bracelets. Due to the core relevance of the system for the overall solution, it is important that their design meets a set of requirements linked to the performance of the system, but also in terms of user requirements. Tapping into the potential of Interactive Virtual Prototyping (IVP) offers the possibility of designing and testing solutions that offer higher user acceptance and market opportunities. This paper briefly examines this potential approach, looking in particular at the fashion sector and how IVP is used, highlighting the additional issues posed by smart garments and accessories.**

Keywords- wearable sensors; smart garments, smart texiles; Interactive Virtual Protityping; lifestyle management, prevention in healthcare.

I. INTRODUCTION

The research leverages the work in progress of the EU Project PEGASO [1], coordinated by Politecnico di Milano. The project addresses the growing epidemic of obesity in the younger population, with focus on prevention. While there is a general agreement that prevention is the only viable strategy – also in economic terms – for the future of healthcare, prevention is also a vague concept, as the object of prevention is undefined. Strategies are therefore very difficult to define, and they must include the economic model for sustainable prevention. Attempts to put this burden on the public health system will remain unanswered, at least in the present economic situation. It is therefore important to develop a culture of prevention that empowers individuals to become co-creators of their healthcare and wellbeing.

The paper focuses on the issue of how to design a solution that offers a compelling user experience and on the tools that can support the design phase: the case of the PEGASO smart garments is considered.

In particular, Section II provides an overview of the current approach in PEGASO, while Section III illustrates how the use of Interactive Virtual Prototyping tools and techniques can support the design of solution that meet the user requirements and expectations. Section IV briefly highlights potential strands for future work and provides some initial conclusions.

II. OVERVIEW OF PEGASO APPROACH TO LIFESTYLE MANAGEMENT AND OBESITY PREVENTION

As mentioned in the Introduction, the project PEGASO Fit-for-Future aims at developing a comprehensive solution for lifestyle management in order to prevent the risk of obesity and related illnesses, targeting the teen-agers population. In order to address this complex issue, a multidisciplinary approach is required, covering:

- Research on current approaches to prevention (including studies on patients' empowerment strategies and approaches);
- Research on behaviour change theories and approaches;
- Research on how technologies (and more specifically wearable technologies) can be used to support behaviour change towards healthy lifestyles and develop a culture of prevention in young people.

Objective of the work is the development of a product/service system that supports young people (teen-agers) in becoming aware of their habits, while learning what healthy lifestyles are, so that healthier habits are developed and sustained in time.

Adopting a practice-centred research, the way to proceed is seemingly straightforward: mobile apps, games and gamification together with wearable sensors to support self-monitoring and development of self-awareness. However, the teen-agers, as target users, add some difficulty to the development of a successful solution. Indeed while wearable sensors (bracelets in particular) are meeting with market success with the adult population, their potential is still not fully expressed with the younger population. In order for teen-agers to understand and accept wearing sensors, they have to be positively engaged; they need to feel involved; the solution needs to make them feel part of a community and not set apart as technology nerds or people in need of care.

The current methodological approach is articulated along the following steps:

- Identification of scenarios: through story telling, teen-agers are involved in the development of different lifestyle scenarios, in order to understand which objects and services may exist or may be imagined that can help them develop self-awareness and improve their lifestyles (a scenario design game such as "The thing from the future" can be adapted to this aim [2]);
- User studies are conducted in parallel: these are based on ethnographic approach, to observe what young people do and like, in different socio-economic sectors (different socio-economic situation have significant influence in lifestyles and on the attention to health and well-being);
- Technology studies and development: focus is on mobile technologies and wearable sensors and actuators, including smart garments, bracelets, etc.;
- Development of case studies, with students from high schools, using also mock-ups.

III. USE OF INTERACTIVE VIRTUAL PROTOTYPING

Using Interactive Virtual Prototyping (IVP) has the potential to improve the overall approach to the research described above; in particular, with regard to the technology studies and development, IVP can be successfully applied to the design and development of smart garments and accessories.

Co-creation has already been applied as a potential approach involving teen-agers in focus groups and making use of questionnaires to understand their lifestyles, their approach to health and wellbeing and also their familiarity with technology and wearable sensors. However, during the first year of the project a full adoption of co-design has proven difficult, due to the limited contact with students in the schools and the consequent difficulty of creating a team able to follow the research, participating in a continuous manner in the design process and therefore championing the project with other students.

In particular with regard to the design on the smart T-shirt, it was clear from the start that "fashion" in a wide sense played an important role in user acceptability. Focus groups were conducted using drawings developed by the designers of the research group; few samples of T-shirts were developed based on the results. However, the samples produced were very costly due to having to set-up the production for such a small number of items. Further the wearability of the T-shirts was not exactly what was expected. The "technical" fabric - which had been selected as very suitable for sports apparels - turned out to be very little stretchable reducing significantly the wearability and comfort. Figure 1 below shows the first set of T-shirts developed within the PEGASO project and shown to the students.

Figure 1. Initial set of T-shirts developed in the PEGASO project

The further development of the model was abandoned and it became evident that drawing was not sufficient and that there was a need to be able to feel the fabric.

IVP is therefore an interesting approach and provides suitable tools for delivering a comprehensive user experience on the basis on which a product with high user acceptance and market potential can be developed and understood.

Figure 2 provides a representation of the IVP process, showing how all senses can be leveraged to develop a truly engaging user experience with products.

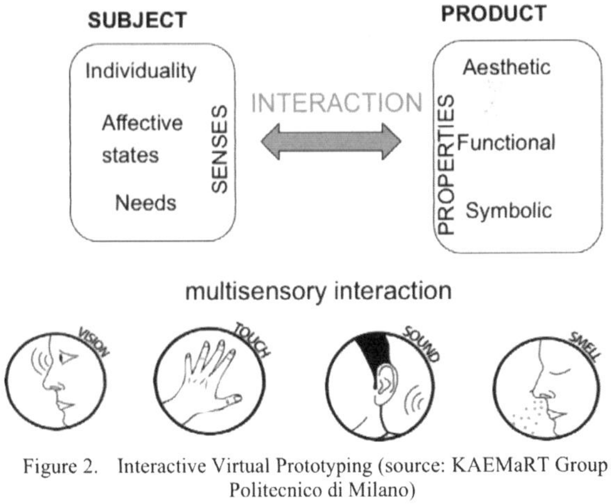

Figure 2. Interactive Virtual Prototyping (source: KAEMaRT Group – Politecnico di Milano)

Virtual prototyping is a method in the process of product development. A Virtual Prototype is an anticipation of a product Shape & Function that does not exist in reality yet, but that appears and behaves as it were real. Virtual Interactive Prototyping simulates how a product looks like, how it works, how users can manipulate and use it, etc.

Virtual Prototyping has been first developed and applied to the manufacturing sector due to pressure to reduce time to market. Products are developed in the form of virtual prototypes in which simulation software is used to predict performance prior to constructing physical prototypes. Different design alternatives can be evaluated leading to improvements in performance and design quality.

A. Interactive Virtual Prototyping and Fashion

The user of Interactive Virtual Prototyping in the fashion sector is relatively novel and still presents areas for further study and development. In the course of the project a small search has been performed to understand what is available and which features presents.

Interesting work on IVP for fashion has been carried out at MIRALab – University of Geneva, Switzerland – where many projects have been carried out covering different aspects such as cloth modelling, with the project Virtual Clothing where MIRACloth, a system for building and animating the garments on virtual actors, was developed (see Figure 3), or the projects Fashion Dream (1998) and Dreams of a Model (2004), where real models performed on stage with virtual models, in real time (see Figure 4).

The group has also conducted relevant works in modelling of the human body, as well as in haptic sensing of virtual textile (HAPTEX project, see Figure 5).

[3] and [4] summarise the evolution from modelling clothing to Interactive Fashion Design explaining the different modelling techniques used and how these have been combined.

Politecnico di Milano, with the KAEMaRT group, is also working on applying IVP to the fashion industry [5]. Cooperation with the lab is being investigated to evaluate whether there are tools that can easily be adopted in the project and experimented also with the teen-agers involved.

In the academic environment, many relevant results are available that can be used to develop a user experience in which teen-ager can be engaged in the design of the smart T-shirt, without having to develop costly real prototypes.

Figure 3. Virtual Clothing – use of MIRACloth tool (source MIRALab – University of Geneva)

Figure 4. Project "Dreams of a Model" (source MIRALab – University of Geneva)

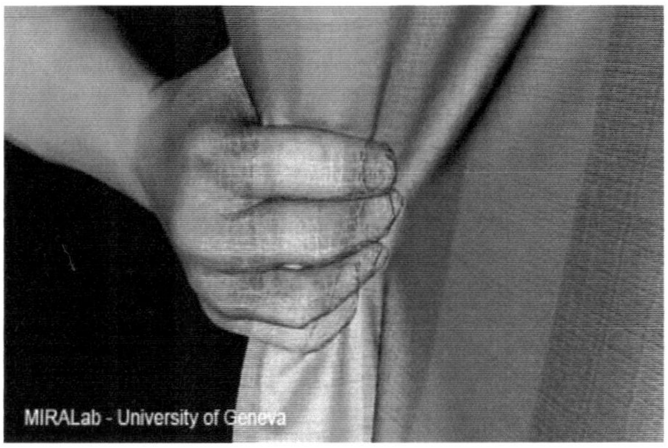

Figure 5. Project HAPTEX (source MIRALab – University of Geneva)

On the commercial side tools are available from OPTITEX [6] (Innovative & Easy-to-use 3D Virtual Prototyping + 2D CAD/CAM Pattern & Fashion Design Software), LECTRA [7], TUKATECH [8]. These products are being investigated to see under which conditions they can be used in the academic environment and within the PEGASO project, and its follow up activities, in particular.

With regard to the start garments and accessories, in addition to the issues and characteristics already taken into consideration by the studies and products described above, also the sensing aspect - which are an integral part of the product - needs to be modelled in order to provide the full functionality required for the development of a full user experience.

This means that ideally we would need to model different body shapes in order to position the sensors, so that the body parameters that need to be monitored are correctly acquired. Also the relative position of sensor and data logger as well and the transmission of the sensed signal to the data logger (WES – Wearable Electronic System) may need to be modelled, so that different positions in the clothing could be tested in order to measure how the quality of the signal is affected. This would allow to test different designs of the smart T-shirt, also changing the position of the data logger, creating different designs that are better suited to different the sports practised by the students at school and also in sports' teams.

Working with teen-agers the appearance is also very important. The smart T-shirts proposed for the tests performed in December (total black as other designs have proven difficult to get an overall agreement) are produced in stretchable material (which is needed to ensure good contact between the skin and the sensors) that is tight on the body. This makes them of difficult acceptance for people that are overweight and prefer loosely fitting apparels. Given the constraint that the sensors have to be in touch with the skin, using flexible tools to try different designs would help to ensure user acceptance by working with the teens to test different options and validate with them wearability and function.

While there are many open points, and also complex modelling issues, Interactive Virtual Prototyping is a very promising approach to use, also as an engaging technology for the target users that can be involved in co-creation session.

IV. CONCLUSIONS AND FUTURE WORK

The work in the project has involved the target users – students in high schools – since the beginning adopting a user-centred co-design approach. However, traditional methods have been used based in focus groups, trying to abstract general user requirements, which have guided the implementation.

It has to be noted that PEGASO is a wide-ranging project, aiming at a solution of which the wearable sensor system is only a subset. In consideration of the wide scope, the focus groups – though specific to the different elements of the overall solution – in the area of smart clothing and accessories have not been able to provide clear user requirements, taking into consideration the restrictions posed by the sensor system. And indeed the first feedbacks are not very positive, both on the design of the T-shirt (too small, and not useful in a hotter period; the suggestion was for a more summer design) and on the WES, which has to be considered an integral part of the garment – needed for the sensing function but otherwise to be considered as a decorative element (the students indicated that they did not like others to see them using the sensors, which is very visible when lights are on).

The need for a more flexible tool that allows trying different designs and shapes while taking into account the constraints of the sensing system has emerged very strongly from the last tests with the students. Interactive Virtual Prototyping is a promising approach and, although the time limits of the PEGASO project will not allow its full use for the development of the solution that will be tested by the students from June 2016 onward, it can be used in parallel to evaluate alternative designs in view of exploitation of project results.

Further studies in this area will consider fancier types of smart clothes, which are based on, e.g., smart fabrics [9] that can change colour or patterns based on sensed environment or body parameters, including reaction of the fabric to people's skin. Currently we are considering how emotions can be sensed and reflected in clothes appearance. Work has been already performed in this area [10] and can be used for further development that – with an approach based on the Internet of Things philosophy – allows different wearables to connect and communicate with each other. Smart accessories with embedded communication capabilities and smart fabrics are the future of fashion. Designers and virtual prototyping have a very significant role to play in this development. Together with the further evolution of computer graphics and technologies, the capability of modelling the different aspects is the key research issue to be addressed.

ACKNOWLEDGMENTS

This paper is partly based on the work performed within the context of the PEGASO project. The PEGASO project is co-funded by the European Commission under the 7th Framework Programme. PEGASO is part of the cluster of projects in the ICT for health area; it has started in December 2013 and will run for 42 months. The author wishes to thank all the project partners for their contribution to the discussion.

REFERENCES

[1] http://pegasof4f.eu/ [retrieved: March, 2016]

[2] http://situationlab.org/projects/the-thing-from-the-future/ [retrieved: March, 2016]

[3] N. Magnenat-Thalmann and P. Volino, "From early draping to haute couture models: 20 years of research", The Visual Computer, 2005, Volume 21, Number 8-10, Page 506

[4] P. Volino, F. Cordier, and N. Magnenat-Thalmann, "From early virtual garment simulation to interactive fashion design", Comput. Aided Des. 37, 6 (May 2005), 593-608.

[5] U. Cugini, M. Bordegoni, and R. Mana, "The role of virtual prototyping and simulation in the fashion sector", International Journal on Interactive Design and Manufacturing (IJIDeM), February 2008, Volume 2, Issue 1, pp 33-38

[6] http://www.optitex.com/en [retrieved: March, 2016]

[7] http://www.lectra.com/en/fashion-apparel/product-development-modaris [retrieved: March, 2016]

[8] http://www.tukatech.com/3D-fashion-design-software/TUKA3D [retrieved: March, 2016]

[9] R. Gaddis, What Is The Future Of Fabric? These Smart Textiles Will Blow Your Mind, Forbes / Lifestyle, May 7 2014, http://www.forbes.com/sites/forbesstylefile/2014/05/07/what-is-the-future-of-fabric-these-smart-textiles-will-blow-your-mind/#389142964914 [retrieved: March, 2016]

[10] L. Stead, P. Goulev, C. Evans, and E. Mamdani, "The Emotional Wardrobe", Personal Ubiquitous Computing (2004) 8: 282–290 DOI 10.1007/s00779-004-0289-

A Graph Framework for Multimodal Medical Information Processing

Georgios Drakopoulos

MDAKM Lab
University of Patras
Patras 26504 Hellas
Email: drakop@ceid.upatras.gr

Vasileios Megalooikonomou

MDAKM Lab
University of Patras
Patras 26504 Hellas
Email: vasilis@ceid.upatras.gr

Abstract—**Multimodal medical information processing is currently the epicenter of intense interdisciplinary research, as proper data fusion may lead to more accurate diagnoses. Moreover, multimodality may disambiguate cases of co-morbidity. This paper presents a framework for retrieving, analyzing, and storing medical information as a multilayer graph, an abstract format suitable for data fusion and further processing. At the same time, this paper addresses the need for reliable medical information through co-author graph ranking. A use case pertaining to frailty based on Python and Neo4j serves as an illustration of the proposed framework.**

Keywords–Frailty index; Co-morbidity; Neo4j; Tensor analysis; Multimodal data mining

I. Introduction

Multimodal medical information processing is a technique for providing improved diagnosis by fusing data from heterogenous sources. This is especially useful when clinical data are scarce or difficult to obtain. Moreover, multimodal processing may provide a solution to co-morbidity cases, namely in cases where individuals suffer from at least two diseases with overlapping symptoms. As a rule, co-morbidity cases are hard to treat since all symptoms can be mistakenly attributed to only a single disease, complicating thus the cure. For instance, a specific brain anomaly may be attributed by medical practitioners to lesions based on EEG readings but functional neuroimaging may reveal additional brain damage. Medical information processing systems are currently expected to handle a multitude of data forms including, among others, research papers, field reports with raw data, medical images, EEG waveforms, and MEG recordings.

In order to support multimodality, a versatile and generic data structure should be used. The framework proposed in this work relies extensively on multilayer graphs, namely graphs whose labeled edges belong to at least two distinct categories. With the advent of large scale graph processing systems such as Apache Giraph, of graph oriented machine learning tools such as GraphLab, and of graph databases such as Neo4j and GraphDB, there is a plethora of high quality, open source, software tools using graphs as the default storage and query format to select from.

The primary contribution of this work is Perseus. The name is a direct reference to the mythological Hellenic hero Perseus. His main achievement was the direct handling of Medusa, a creature whose hair were deadly snakes. In a very abstract representation, snakes can be drawn as simple edges connecting two vertices. Also, in a liberal interpretation, their venom corresponds to an exceedingly degree of complexity. Perseus is a conceptual framework for retrieving, maintaining, querying, and processing medical data from a number of sources. Perseus is strongly based on the explicit assumption that data as well as their interconnection patterns can be expressed as graphs. To demonstrate the potential of Perseus to be tailored to specific application needs, a use case for the frailty index, a significant health indicator for the elderly, is presented.

The remainder of this paper is organized as follows. In section II the scientific literature regarding multimodal medical information processing is overviewed. Subsequently, in section III Perseus is presented and analyzed. The frailty index use case of section IV illustrates the application of multilayer graphs to medical cases with potential co-morbidity. Finally, in section V future research directions are outlined.

II. Previous Work

Fully multimodal information processing systems are a relatively new breakthrough. However, there have been notable recent efforts towards a true multimodal system. In [17] document similarity functions are merged with document ranking. Text mining information from medical documents based on the grammatical structure as well as on field reports and possibly text metadata is outlined in [1] and in [3].

The advent of Semantic Web sparked additional interest in multimodality in gene ontology [5][4] as well as in reasoning wether specific genes are related to a given disease [2][14]. Further momentum was gained with the addition of retrieval methods which are based on medical images [19][17], which allowed more targeted and context aware database search. To the best of the authors knowledge, no medical multimodal data mining system currently exists that relies on a graph database at the back end, although [16] describes a graph database implementation for human gene ontology processing.

III. Framework

A. Architecture

In figure 1(a) the Perseus architecture is displayed. It is centered around the contoller module, which directs data acquistion from a number of heterogenous sources through the appropriate interfaces. Data collected can be stored either in their original form or after processing in various databases depending on the operational needs.

Multimodality is ensured by obtaining the appropriate APIs for at least two different types of data sources. Besides

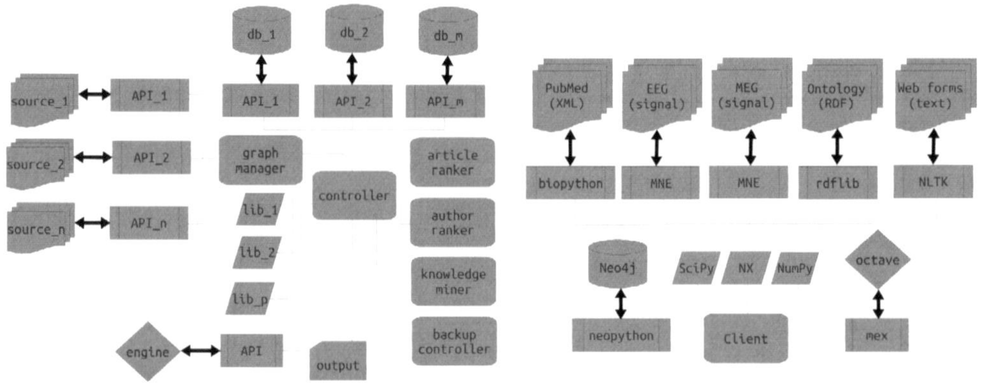

(a) Perseus generic architecture

(b) Frailty index specific architecture

locally available clinical results and articles from open access databases such as PubMed, common sources include medical ontologies, statistical aggregations from organizations like WHO, and biosignal repositories.

Notice that the star architecture, which has a sigle critical point of failure, combined with the nature of a medical processing system calls for a backup controller which is normally shadowing the main controller and replacing it, should the need arise.

Versatile storage is ensured through multilayer graphs. They have been selected as the primary data structure not only because many significant data sources provide data in a tree or graph format, such as PubMed, but also it facilitates information fusion in a space efficient manner [13]. Additionally, graphs allow the easy information localization as well as the easy composition of a summary based on various sources. Finally, the connectivity patterns of an author-to-author collaboration network and of a document-to-document citation network allow the ranking of both the authors and the articles, serving thus as a quality control mechanism.

IV. USE CASE: FRAILTY INDEX

A. Architecture

As a concete example of its true multimodal nature, Perseus can be tailored to integrate and maintain information about the frailty index, a major health indicator for the elderly. More formally

Definition 1: Frailty is a biological syndrome of decreased reserve and resistance to stressors resulting from cumulative declines across multiple physiological systems and causing vulnerability to adverse outcomes.

With the sharp ageing of population globally, frailty index has risen in significance as a factor of social coherence. Moreover, proper elderly monitoring through frailty index can prevent an array of fatal accidents.

The tailored architecture is illustrated in figure 1(b). It integrates information from a wide ragne of sources, namely field reports from Web forms, EEG and MEG biosignals, PubMed articles, and medical ontologies which have been explicitly developed for frailty. Notice that data source selection is indicative, as an alternative Perseus implementation might as well include sensor readings from mobile devices or smart homes or results from Rorsach tests. Notice that besides the frailty index, its confidence level can be determined from factors listed in table III.

Algorithm 1 shows the information flow for the frailty index case.

Algorithm 1 Frailty index information flow

Require: Quality criteria list T
Ensure: Frailty index I is computed
Ensure: Confidence level L is computed
Collect data from Web forms through NLTK
Extract a set of frailty related terms $\{\tau_k\}$
for all terms τ_k through Biopython **do**
  Retrieve PubMed documents $\{d_i\}$ containing τ_k
  if $\{d_i\}$ meets the criteria of T **then**
    Retrive PubMed popularity scores of $\{d_i\}$
  end if
end for
Retrieve EEG and MEG waveforms through MNE
while unexplained symptoms remain **do**
  Retrieve a frailty ontology O through rdflib
  if O matches symptoms **then**
    Store O in Neo4j
  end if
end while
return computed (I, L)

Python has been selected as the primary language because of its ability to generate glue code, namely code connecting two or more software components whithout modifying them. Moreover, there are Python interfaces for each of the other tools.

B. Biopython and Entrez

Entrez is a special purpose, federated search engine designed to provide seamless access across a vast array of medical databases. Currently Entrez APIs have been developed for various programming languages and problem solving environments, most notably the Matlab API, the BioGo package for Go, and the Java Entrez Eclipse API.

Entrez stores PubMed documents as XML trees, facilitating thus lexical analysis. At the same time it renders appealing the use of a graph database for local storage and analysis. Table

TABLE I. PUBMED XML DOCUMENT TAGS.

FileHeader	ArticleSet	Article
Journal	PublisherName	JournalTitle
Issn	Volume	Issue
PubDate	Year	Month
Season	Day	Replaces
ArticleTitle	VernacularTitle	FirstPage
ELocationID	Language	AuthorList
Author	FirstName	MiddleName
LastName	Suffix	CollectiveName
Affiliation	Identifier	GroupList
Group	GroupName	IndividualName
PublicationType	ArticleIdList	ArticleId
History	Abstract	OtherAbstract
CopyrightInfo	ObjectList	Object
Param	LastPage	

I lists the mandatory and optional tags from the Entrez XML schema.

Each API for Entrez should implement as a minimum functionality methods for searching medical documents based either on ID or on search terms, retrieving document abstracts and bodies, and retrieving relevant articles. Usually methods for text parsing are provided by a number of APIs.

Biopython is a Python open source Entrez API. The most common Biopython methods are listed in table II. Furthermore, Python is heavily employed in the NLP field, mainly because of NLTK, the natural language toolkit. Finally, even if certain Perseus components are written in another language, Python is strongly famed for its so called glue code, namely code that provides connection and communication between different applications or modules of the same application, offering thus a unified view of an otherwise segmented software.

TABLE II. BIOPYTHON FUNCTIONS.

Method	Description
efetch	Retrieves records from a list of ids
epost	Posts a file containing a list of ids in the user environment for future use
esearch	Searches and retrieves ids and term translations and optionally retains results
elink	Checks for the existence of an external or Related Articles link in a list of ids
einfo	Provides field index term counts, last update, and available links for each database
esummary	Retrieves document summaries
egquery	Provides Entrez database counts in XML for a single search using Global Query
espell	Retrieves spelling suggestions
read	Parses the XML results returned by any of the above functions
parse	Parses the XML results returned by functions. Appropriate for large files.

C. Neopython and Neo4j

Since the beginning of the current decade there has been a rising interest in four branches of non-relational databases collectively referred to as NoSQL databases. For a brief NoSQL review see [15].

Neo4j is a major graph database supporting queries to large graphs through either Cypher, an ASCII art high level query language, or through APIs for various programming languages, most notably Python, Scala, and Java. It is also possible to query graphs in Neo4j in SPARQL through suitable extensions. Neo4j is currently open source and it is written mostly in Scala.

Property 1: Neo4j is schemaless.

Property 2: The three operational Neo4j requirements are basic availability, soft state, and eventual consistency collectively known as BASE [15].

Compared to relational ACID requirements, the BASE ones are easier to implement and place less strain on the database system, including fewer data duplications and locks. The downside is that the system does not become immediately consistent.

Property 3: The primary conceptual data model supported by Neo4j and offered to a high level user through Cypher is the property graph [15].

For a review of the graph property model see [7][6].

Neopython is a Python interface for Neo4j which allows the formation of dynamic Cypher queries as Python strings. For this reason, Neopython has been included to the proposed implementation.

D. Document Ranking Criteria

But when a rule is extremely complex, what it is in conformity with it passes for irregular.
(Leibniz)

Building T requires care, as it can easily degrade to a nearly all-reject rule. In fact, as a safety valve in real life situations there should be a component evaluating the effectiveness of T by checking the Perseus null return rate. Significant criteria for T are listed in table III.

TABLE III. DOCUMENT RANKING CRITERIA.

Article	
PubMed popularity	age
citation graph ranking	number of authors
journal impact factor	acceptance rate
Content	
terminology frequency	keywords
attached data	reproducible research
Authors	
co-authorship graphs	research continuity

The ranking of the article depends on at least three group of factors. First, the article is ranking as a member of a document collection. This can be achieved by examining any scientific citation graph. Then, the actual article contents contribute to article ranking. Finally, the authors are ranked, for instance from any established research collaboration graph, and their reputation affects article ranking. Research continuity means that the authors should have written other articles on the same subject. Reproducible research means that the data are available along with the publication or in a public Web location, giving thus the opportunity to other research teams to process them.

E. Graph Analytics

Perseus relies heavily on graph analytics in order to perform computations. Specifically Perseus

- ranks authors in co-authorship graphs.
- ranks articles in citation graphs.
- parses PubMed XML documents.
- converts EEG and MEG waveforms as graphs.
- reasons on frailty ontologies.
- derives brain regions of interest.

Graph algorithms for performing the above tasks are listed in table IV.

TABLE IV. HIGHER ORDER GRAPH ANALYTICS.

Structure	
Density	Expander graph
Bridges	Articulations
Triangles	Squares
Euler path	Hamilton path
Connected components	Independent sets
Total connectivity	Cheeger number
Centrality (Structural)	
Degree	Betweeness
Delta	Egonet
Centrality (PageRank family)	
PageRank	Eigenvector
Gell point	Harmonic point
Centrality (Resolvent family)	
Resolvent series	Mercator series
Katz series	Neuman series
Community detection	
Newman-Girvan	Blondel
Walktrap	Fast Greedy
Shortest paths	
Dijkstra	Bellman-Ford
A^* algorithm	A^+ algorithm
Spanning tree	
Prim	Kruskal

Common metrics for ranking vertex centrality include PageRank, the eigenvector centrality [7], Gell point centrality [18], as well as resolvent centrality [11]. There have been also developed variants of these algorithms for fuzzy graphs [8]. Vertex set sizes can be estimated with techniques similar to those in [9]

NetworkX is an open access Python module for handling and storing in memory graphs and multilayer graphs. NumPy and SciPy are also Python modules implementing advanced mathematical functions for evaluating graph quantities among others.

F. Frailty Ontologies

Fraitly ontologies have been developed in order to describe many of the associated actors, events, and entities. Entities related to frailty include brain activity, psychological indicators, as well as basic physical status indicators such as blood pressure. Actors include the elderly, their familty, and medical experts and healthcare personel in general. Rules to validate the relationship between the entities or to discover new entities have been proposed.

G. MEG and EEG waveofmrs

MNE is an open access Python tool for processing EEG and MEG readings [12]. These readings can be used to represent and analyze brain activity, which is an important factor in assessing frailty. As both modes collect brain data over a time interval, their readings can be stored in a graph representing a brain map. Moreover, they can be used to build a second graph which represents events of interest. These events can be correlated with machine learning methods such as sparse neural networks [10] to events stored in frailty ontologies, facilitating brain activity abnormalities.

V. FUTURE WORK

This work can be expanded in a number of ways. A current trend in nearly every retrieval and data mining problem is that of multimodality, namely the combination of heterogeneous information in order to maximize the effectiveness of retrieval procedures. In the particular setting of medical retrieval multimodality translates to boosting text based retrieval with knowledge derived from other sources of medical data such as neuroimaging, EEG, or health statistics. In fact, since both neuroimaging and EEG data can be represented as graphs, they can be expected to be easily incorporated to Perseus.

Another research direction is medical multilingual information retrieval. There are many non-English public medical databases from where significant knowledge can be drawn, facilitating thus cooperation between researchers and allowing better coordination between field teams and practitioners. Moreover, multilingual retrieval allows quicker international response to severe disease outbreaks as incidents as well as potentially related symptoms recorded in non-English medical databases can be retrieved on behalf of an English speaking practitioner without prior translation. Additionally, advanced text semantic analysis can locate specific points of interest or data for such a user and even selectively translate part of the retrieved documents. As a caveat though, term polysemy can be aggravated within a multilingual context, as the same term can have different meanings in different languages.

ACKNOWLEDGEMENTS

This work was supported by the European Union's Horizon 2020 research and innovation programme under grant agreement No 690140.

REFERENCES

[1] A. R. Aronson. Effective mapping of biomedical text to the UMLS Metathesaurus: the MetaMap program. In *Proceedings of the AMIA Symposium*, page 17. American Medical Informatics Association, 2001.

[2] M. Bada, D. Turi, R. McEntire, and R. Stevens. Using reasoning to guide annotation with gene ontology terms in GOAT. *ACM SIGMOD*, 33(2):27–32, 2004.

[3] R. Baud, A.-M. Rassinoux, and J. R. Scherrer. Natural language processing and semantical representation of medical texts. *Methods of Information in Medicine*, 31(2):117–125, 1992.

[4] D. Botstein, J. Cherry, M. Ashburner, C. Ball, J. Blake, H. Butler, A. Davis, K. Dolinski, S. Dwight, J. Eppig, et al. Gene ontology: tool for the unification of biology. *Nature Genetics*, 25(1):25–29, 2000.

[5] G. O. Consortium et al. The gene ontology (go) database and informatics resource. *Nucleic acids research*, 32(suppl 1):D258–D261, 2004.

[6] G. Drakopoulos and A. Baroutiadi. Implementing centrality measures for Neo4j. In *Proceedings of the 8th ECESCON*, ECESCON 2015. ECESCON, April 2015.

[7] G. Drakopoulos, A. Baroutiadi, and V. Megalooikonomou. Higher order graph centrality metrics for Neo4j. In *Proceedings of the 6th International Conference on Information, Intelligence, Systemems, and Applications*, IISA 2015. IEEE, July 2015.

[8] G. Drakopoulos, A. Kanavos, C. Makris, and V. Megalooikonomou. On converting community detection algorithms for fuzzy graphs in Neo4j. In *Proceedings of the 5th International Workshop on Combinations of Intelligent Methods and Applications*, CIMA 2015. IEEE, November 2015.

[9] G. Drakopoulos, S. Kontopoulos, and C. Makris. Eventually consistent cardinality estimation with applications in biodata mining. In *Proceedings of the 31st Symposium on Applied Computing*, SAC 2016. ACM, April 2016.

[10] G. Drakopoulos and V. Megalooikonomou. On the weight sparsity of multilayer perceptrons. In *Proceedings of the 6th International Conference of Information, Intelligence, Systems, and Applications*, IISA 2015. IEEE, July 2015.

[11] E. Estrada and D. J. Higham. Network properties revealed through matrix functions. *SIAM Rev.*, 52(4):696–714, Nov. 2010.

[12] A. e. a. Gramfot. Meg and eeg data analysis with mne-python. *Frontiers in neuroscience*, 7(267), December 2013.

[13] S. Kontopoulos and G. Drakopoulos. A space efficient scheme for graph representation. In *Proceedings of the 26th International Conference on Tools with Artificial Intelligence*, ICTAI 2014, pages 299–303. IEEE, November 2014.

[14] S. Myhre, H. Tveit, T. Mollestad, and A. Lægreid. Additional gene ontology structure for improved biological reasoning. *Bioinformatics*, 22(16):2020–2027, 2006.

[15] I. Robinson, J. Webber, and E. Eifrem. *Graph Databases*. O'Reilly, June 2013.

[16] L. M. Schriml, C. Arze, S. Nadendla, Y.-W. W. Chang, M. Mazaitis, V. Felix, G. Feng, and W. A. Kibbe. Disease Ontology (DO): a backbone for disease semantic integration. *Nucleic acids research*, 40(D1):D940–D946, 2012.

[17] H. D. Tagare, C. C. Jaffe, and J. Duncan. Medical image databases: A content-based retrieval approach. *Journal of the American Medical Informatics Association*, 4(3):184–198, May 1997.

[18] H. Tong, B. A. Prakash, T. Eliassi-Rad, M. Faloutsos, and C. Faloutsos. Gelling and melting large graphs by edge manipulation. In *Proceedings of the 21st ACM international conference on Information and knowledge management*, CIKM12, pages 245–254, New York, NY, USA, 2012. ACM.

[19] C.-H. Wei, C.-T. Li, and R. Wilson. A content-based approach to medical image database retrieval. *Database Modeling for Industrial Data Management: Emerging Technologies and Applications*, pages 258–290, 2005.

Designing a Mobile Serious Game to Promote Healthy Lifestyles

Motivating teenagers to adopt healthy habits through play

Stefano Carrino, Maurizio Caon and Elena Mugellini
HumanTech Institute
University of Applied Sciences and Arts Western Switzerland (HES-SO)
Fribourg, Switzerland
e-mail: {firstname.lastname}@hes-so.ch

Ian Dunwell
Serious Games Institute
Coventry University
Coventry, United Kingdom
e-mail:idunwell@cad.coventry.ac.uk

Dalia Morosini
Imaginary SRL
Milan, Italy
e-mail:dalia.morosini@i-maginary.it

Abstract—**Behaviours such as diet, sedentariness and how much exercise we do play a major role in influencing health and wellbeing. As a medium, mobile serious games have the potential to educate and foster behaviour change whilst engaging and motivating users. This paper illustrates the user-centred design approach used to develop a mobile serious game to promote healthy lifestyles. In particular, this research, developed in the frame of the PEGASO European project, focuses on teenagers and healthy behaviours linked to the prevention of obesity. In this approach, teenagers from different European countries (Italy, Spain, and United Kingdom) participate to the co-design of the serious game. This activity is still ongoing but first findings show encouraging feedback concerning the generic game mechanics and the designed game.**

Keywords—Serious game; behaviour change techniques; health and lifestyles; mobile application.

I. INTRODUCTION

Making healthy lifestyle choices in terms of diet, physical activity levels and in avoiding unhealthy behaviours such as smoking have the potential to dramatically impact an individual's mental and physical well-being. For this reason, a number of interventions have already been designed and implemented to avoid the wide range of negative outcomes of poor habits and encourage individuals to maintain healthier behaviours. However, one of the main restrictions to the effectiveness of these interventions is a lack of motivation, alongside a lack of time and difficulties in accessing sufficient proportions of the target group [1][2]. These issues, along with a rise in the accessibility of technology, have encouraged the implementation of game-based interventions, which incorporate all the benefits of interventions delivered remotely but with the added advantages of greater enjoyment and in-built intrinsic motivation [3][4].

A number of serious games aimed at improving lifestyle choices in young people have been developed, many of which have reported significant improvements in behaviour change. While likely to be at least partially due to publication bias against studies reporting non-significant findings, it is still noteworthy that only very few studies report a lack of change following game-based interventions (e.g., [5][6]). This highlights the strong potential for positive outcomes in terms of health-related behaviour change.

The PEGASO European project aims at fostering healthy habits concerning physical activity, sedentariness and healthy nutrition in teenagers through a PEGASO ecosystem, composed of different elements and approaches, among which also a serious game for mobile platforms. In particular, PEGASO focuses its behaviour change approach on specifics target healthy behaviours. Target behaviours are those behaviours, pertaining to nutrition, physical activity, sedentary behaviours and sleeping habits, which have been recognised to have impact on overweight/obesity prevention, and are amenable to be changed. It is important to highlight, that each user of the PEGASO platform will focus on one specific target behaviour at a time.

In the next section, Section 2, we motivate the use of serious games as means for healthy lifestyle interventions presenting the state of the art in this field. In Section 3, we present and schematise the approach that we are following to design the PEGASO serious game. Section 4 discusses the first results that we have achieved testing the game with teenagers in three different countries in Europe (Italy, Spain, United Kingdom). Finally, Section 5 concludes this paper highlighting the future work.

II. RELATED WORK

One of the main arguments for the use of serious games over more traditional methods or interventions is how ubiquitous and familiar technology is for young people, making game-based delivery a comfortable and potentially enjoyable situation [7]. Learning through play encourages faster, more implicit learning [8], and digital games in particular have been argued to be especially motivating, encouraging a greater level of engagement than other activities [9]. Digital games are also enjoyed by a large number of young people, and this expectation of enjoyment can be capitalised in educational contexts; while children may not be interested in participating in traditional interventions, digital games are still seen as a fun activity, and they therefore enter the intervention expecting to enjoy it [10]. As a consequence, in the correct contexts, serious games can prove an ideal medium for the delivery of educational material.

Previous researches into game-based interventions in relation to healthy lifestyle adoption also demonstrates this tool to be largely successful (albeit in relation to different kinds of measures). While the effect sizes tended to be small, they are comparable to other non-game, computer-based interventions (e.g., [11][12]). Moreover, DeSmet et al. [13] argue that coupled with the increased intrinsic motivation inherent serious games compared to other computer based methods, as well as the greater levels of interactivity and visual interest, games may be a better medium. This is in line with other assertions that as long as the games present adequate challenge, participants are willing to spend more time on them than they would on traditional learning methods [14][15].

However, it also worth noting that all reviews on the topic of game based interventions and interventions for health-related behaviours agree that the existing literature base is incredibly diverse in terms of target demographics, game formats and outcome measures, making overall conclusions very difficult [16-28]. Furthermore, a lack of information in many of the articles reviewed regarding game characteristics make it difficult to pin-point what elements are especially effective or successful [29].

III. PEGASO GAME DESIGN

The design of the PEGASO serious game has been structured according to eight different iterative phases of design/development and test. These phases are fundamental steps of a unique iterative process that follows the user-centred design approach. Indeed, in user-centred design, "development proceeds with the user as the centre of focus" [30]. Every phase is performed involving the user many times. The analysis, design and development phases are performed rapidly in order to provide first prototypes to test with actual users in order to understand what is appreciated and what can be improved. Then, the whole process is performed again and again to converge towards the optimal solution. This approach allows responding better to the users' needs and desires in order to increase the acceptance of the final product.

In PEGASO, the test phases involving users are divided in three short time duration pre-pilots (from one day to two weeks) and one final pilot of about six months.

PEGASO Game design – Iterations:

- Phase 0: User requirements
- Phase 1: First design iteration
- Phase 2: Game acceptability and preliminary usability tests (first pre-pilot)
- Phase 3: Serious game first prototype
- Phase 4: Game mechanics and usability II tests (second pre-pilot)
- Phase 5: Design Iteration and Integration in the PEGASO Companion
- Phase 6: Integrated game test (third pre-pilot)
- Phase 7: Pilot version development
- Phase 8: Pilot study

The full process will take 3 and half years. At the time of writing this paper the "pre-pilot 2" phase is still on-going.

A. *Phase 0: User Requirements*

The first step before the actual development of the game was aimed at determining the game typologies that are attractive for the target population as long as their knowledge in the domains related to the PEGASO project (healthy habits concerning physical activity, sedentariness and nutrition). Therefore, we conducted a review of the current literature, examining existing game-based interventions to make a list of the evidences that can enable interventions for lifestyle change; a summary of this analysis can be found in [31].

B. *Phase 1: First Design Iteration*

Behaviour change interventions have been designed in PEGASO following the COM-B model and the Behaviour Change Wheel (BCW) framework [32]. The COM-B model of behaviour is a powerful model to understand behaviour. This model has the great capability of taking into account the context in which a behaviour is taking place and is particularly valuable for PEGASO because it can be used as a starting point for intervention design. On the other hand, the Behaviour Change Wheel (BCW) is a complementary framework (synthesis of 19 frameworks to classify interventions), based on COM-B, facilitating the design of interventions for behaviour change. BCW is a generic framework and can be adapted to different behaviours: the PEGASO target behaviours. Therefore, the goal of this first phase of design was to conceive key game mechanics starting from the requirements defined by the PEGASO objectives and the preliminary study phase to foster behaviour change through the guidelines of behaviour change models. In addition, the serious game has specific objectives within the PEGASO ecosystem: 1) Convey educational material; 2) Keep teenagers engaged in the whole PEGASO ecosystem; 3) Foster healthy habits.

These elements have brought to the design of several game mechanics. In this paper, we present two core game mechanics that are the most specific within the PEGASO project. The first game mechanic is called "research benefits". Since a major learning outcome of PEGASO is to provide teenagers knowledge about food and aliments we designed an *ad hoc* mini-game. In this game, food items found in the game have to be combined in order to benefit the character. However, in order to be effective for the game, the combination should be achieved taking into account the real nutritional parameters of the food item (which are shown to the user). In this way, the user is implicitly learning the composition of the different aliments. The second mechanic concerns the "energy" bar. Playing the game, the player makes use of and consumes "energy". Without energy some of the character abilities are weakened or disabled. Whilst the user will still be able to continue to play and complete tasks that allow her/him to gain nutritional knowledge. While the character energy is depleted within the game, it is replenished by the behaviours of the user in real world. In particular, more the user behaves close to her/his target behaviour more energy is available to the character in the game. It is important to highlight that the energy game mechanic overcomes the boundaries of the serious games since it needs the integration in the PEGASO ecosystem in order to assess the user activities.

C. Phase 2: Game acceptability and preliminary usability tests

This pre-pilot phase has taken place in three countries: Italy (Lombardy region), Spain (Catalonia region), and UK (England). The teenagers were provided of first mock-ups of the game and have to evaluate the game acceptability and usability. Users had to consider that the game will be developed for a mobile platform.

The results of this phase are reported and explained with more details in [33], in this article we highlight the more interesting findings that had an important impact on the game design and development.

- The teenagers liked the idea that activities in real life will have impact in the game (to spend energy in real life means to gain more energy in the game). They suggested that activities in real life should not be limited to physical activities and exercises but, for instances, calculating their resting and sleeping hours may be reflected in the energy bar as well.
- Both male and females teenagers stated that the zombie theme is appealing. Demonstrating that there is not no clear gender differences for game themes that are typically associated to a male audience.
- Nevertheless, a cartoonish version is slightly preferred if compared to a realistic one (Spanish teenagers are a little exception on that). This allows the game to have a funnier and more casual look to balance the darker atmosphere suggested by the theme.

D. Phase 3:Serious Game First Prototype

The goal of this development and more technical phase is to transform the mock-ups in first functioning and playable applications for mobile platforms (Android OS) taking into account the remarks of the teenagers obtained during phase 2. Figure 1 shows some screenshots of the serious game as developed during this phase: the start-up screen (top); the main character seeing a food item container to be used in the "research" mini-game (middle); during the night the character is attacked by zombies (bottom).

Main outcome: a first, playable prototype of the game available for Android platform. At this phase, the game is a stand-alone application partially integrated in the PEGASO ecosystem.

E. Phase 4: Game Mechanics and Usability II Test

At the time of the writing of this paper, this phase is currently on-going. Four countries are involved in the participation to this phase: Italy, Spain, England and Scotland. The results presented here below are related to 10 Italian teenagers (5 girls and 5 boys). Main results:

- Teenagers involved in the study suggested that the possibility of interacting with peers for collaborative or competitive purposes would enhance the engagement toward the game.
- Rather than multiple characters, it would be interesting to have the possibility to customise the main character.
- Having a more varied environment will facilitate the use of the application over a long period of time.

F. Phase 5: Design Iteration and Integration in the PEGASO Companion

The PEGASO Companion [34] is the Android app the will encompass all the PEGASO services (see Figure 2). The main outcome of this phase will be the integration of the serious game in the PEGASO ecosystem and the PEGASO Companion as well as the integration of the most relevant features demanded by the users in the previous phase. Via this integration it will be possible for the PEGASO user to test the "energy" game mechanic that requires the whole PEGASO ecosystem in order to work.

Expected main outcome: The integration of the serious game within the PEGASO Companion and the related apps.

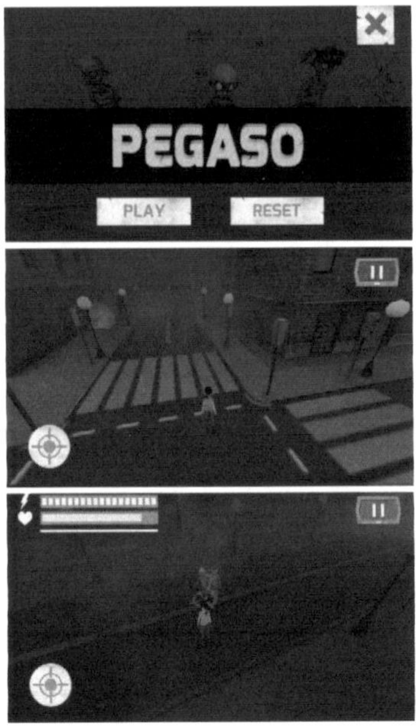

Figure 1. Screenshots from the PEGASO serious game.

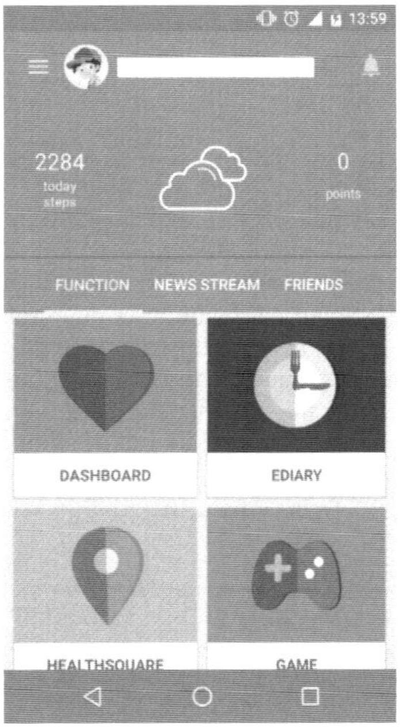

Figure 2. PEGASO Companion UI. From this interface the user can access to the serious game (via the bottom right icon).

In this way, it will be possible to present to the testers the whole PEGASO ecosystem as a unique and integrated environment and evaluate the user experience in the whole PEGASO.

G. Phase 6: Integrated Game Test

In this phase, the user will test the serious game integrated in the whole PEGASO system. The integration in the PEGASO echo-system will allow the user to test different mechanics that would be impossible to evaluate otherwise. For instance, the users will have the possibility to see how their activities in the real world can affect the game and provide a first feedback about the global motivational mechanisms. Additionally, the expected results will address integrations issues and the usability of the PEGASO platform as a unique entity.

H. Phase 7: Preparation for the Pilot

This last design and development phase will be mainly centred on the release of the serious game stable version. It means that this phase will focus on creating mechanisms, contents and procedural solutions that should guarantee the game a life span covering the 6 months duration of the final pilot. Such solutions have to motivate the users to come back playing the game and, at the same time, to stay engaged with the whole PEGASO system.

I. Phase 8: Final Pilot

The final pilot will include about 300 teenagers coming from the same four regions (Italy, Spain, England, and Scotland) for a duration of more than 6 months. Via this pilot, we will estimate the impact of the serious game (and the whole PEGASO ecosystem) on the user behaviours and habits in the medium-long terms.

Figure 3. Users testing the serious game during phase 4.

IV. Preliminary Results

At the time of the writing of this paper, our work is currently between phases 4 and 5 (Figure 3 shows some of the users testing the serious game during the phase 4). Currently, the overall design of the game situates the player's character in a modern city (as depicted in Figure 1), where the player must scavenge for food resources during the day and combine them to obtain abilities which she can use to combat waves of creatures that appear during the night.

The backbone of this idea has been already introduced to teenagers during the first pre-pilot in the Italy (n=10), Spain (n=18) and England (n=16) as mock-ups. The activities consisted in focus groups during which the main planned game mechanics and aesthetics of the design were discussed. As we detail in [33], the first results positively evaluated the acceptability of a mock-up prototype and aesthetical options. At the same time, we collected valuable feedback about usability and suitable improvements.

On the other hand, during the second pre-pilot, we provided a functioning alpha version of the serious game as standalone mobile application (not integrated with the PEGASO Companion and the other apps in the PEGASO ecosystem). Currently, only the data from Italian pilot site are available (n=10, 5 females and 5 males, all aged 15-17 years old). In particular, the aim of the second pre-pilot iteration about the PEGASO serious game is:

- Testing the core game play mechanics (exploring the virtual world and finding food, fighting, unlocking benefits);
- Evaluating the usability of the first prototype.

Together with a couple of questionnaires (SUS [35] and a custom questionnaire), it was decided to conduct a focus group to find out participants' views and overall impressions about the PEGASO serious game.

About the game play mechanics, teenagers provided valuable feedback highlighting the strengths and making suggestions for possible improvements (e.g., add a tutorial the first time that the mini-game about nutrition is played, or make the fight against the enemies more challenging). Similarly, about the game usability they provided suggestions about improvement in the control system and the feedback to the player. Other suggestions were related to the importance of avatar customisation and the possibility to have a competitive or collaborative game modality with the possibility of playing together with friends. The game needs to be variated if have to keep players engaged.

Although the limited findings from this focus group needs to be validated with the results coming from the other focus groups in the other 3 regions (i.e., Catalonia, England and Scotland), globally, these first results show encouraging feedback concerning the generic game mechanics and the game. In particular, the "energy" game mechanic was highly appreciated. At the same time weak points have been highlighted in order to guide future developments. Comparison with testers in the other countries will also allow highlighting possible cultural differences.

V. CONCLUSION AND FUTURE WORK

This paper presents the iterative approach we are using for designing a serious game in the frame of the on-going PEGASO European project. The goal of this serious game is to convey educational contents, foster healthy habits and keep teenagers engaged in the whole PEGASO ecosystem.

The design process that we adopted is user-centred and consists in 8 phases alternating design/development with tests and evaluations with users. In particular, the design steps involving users consist in three short pre-pilots (from 1 day to 2 weeks) and a six-month pilot study.

Currently, we have realised a first functioning prototype of the serious game and tests are ongoing in Italy, Spain and UK (England and Scotland sites). Whilst now working as standalone app, the future work consists in integrate the serious game inside the PEGASO ecosystem and evaluate its impact as tool for behaviour change. In particular, our research will focus on the effect on motivation and engagement derived from linking user's behaviours in the real world with rewards and features in the game world.

ACKNOWLEDGMENT

This work has been supported by the European Commission under the collaborative project PEGASO ("Personalised Guidance Services for Optimising Lifestyle in Teenagers") funded by the European Commission under the Seventh Framework Programme, FP7-ICT-2013-10.

REFERENCES

[1] V. Baert, E. Gorus, T. Mets, C. Geerts, and I. Bautmans, "Motivators and barriers for physical activity in the oldest old: a systematic review." Ageing research reviews 10, no. 4 (2011): 464-474.

[2] D. J. Toobert, L. A. Strycker, R. E. Glasgow, M. Barrera, and J. D. Bagdade, "Enhancing support for health behavior change among women at risk for heart disease: the Mediterranean Lifestyle Trial." Health Education Research 17, no. 5 (2002): 574-585.

[3] M. Grasser and F. Mayer, "Interdisciplinary education in vocational education and training: Supporting joint up thinking—Some insites in lecturing." In Interactive Collaborative Learning (ICL), 2015 International Conference on. IEEE, 2015 pp. 615-619.

[4] M. Prensky, "How to teach with technology: Keeping both teachers and students comfortable in an era of exponential change." Emerging technologies for learning 2, no. 4 (2007): 40-46.

[5] K. Huss et al., "Computer game for inner-city children does not improve asthma outcomes." Journal of Pediatric Health Care 17, no. 2 (2003): 72-78.

[6] K. A. Sward, S. Richardson, J. Kendrick, and C. Maloney, "Use of a web-based game to teach pediatric content to medical students." Ambulatory Pediatrics 8, no. 6 (2008): 354-359. Harvard

[7] P. C. H. Wong et al., "Effects of a 12-week exercise training programme on aerobic fitness, body composition, blood lipids and C-reactive protein in adolescents with obesity." (2008).

[8] M. Prensky, "Computer games and learning: Digital game-based learning." Handbook of computer game studies 18 (2005): 97-122.

[9] J. P. Gee, "What video games have to teach us about learning and literacy." Computers in Entertainment (CIE) 1, no. 1 (2003): 20-20.

[10] G. Sim, S. MacFarlane, and J. Read, "All work and no play: Measuring fun, usability, and learning in software for children." Computers & Education 46, no. 3 (2006): 235-248.

[11] P. Krebs, J. O. Prochaska, and J. S. Rossi, "A meta-analysis of computer-tailored interventions for health behavior change." Preventive medicine 51, no. 3 (2010): 214-221.

[12] D. B. Portnoy, L. A. J. Scott-Sheldon, B. T. Johnson, and M. P. Carey, "Computer-delivered interventions for health promotion and behavioral risk reduction: a meta-analysis of 75 randomized controlled trials, 1988–2007." Preventive medicine 47, no. 1 (2008): 3-16.

[13] M. Maertens, P. Desmet, F. Cornillie, and M. Vandewaetere, "It's all in the game: Assessing the difficulty of math exercises in an educational game." The Biennial Conference of the European Association for Research in Learning and Instruction (EARLI) (2013).

[14] J. Bourgonjon, M. Valcke, R. Soetaert, and T. Schellens, "Students' perceptions about the use of video games in the classroom." Computers & Education 54, no. 4 (2010): 1145-1156.

[15] K. Squire, L. Giovanetto, B. Devane, and S. Durga, "From users to designers: Building a self-organizing game-based learning environment." TechTrends 49, no. 5 (2005): 34-42.

[16] T. Baranowski, R. Buday, D. I. Thompson, and J. Baranowski, "Playing for real: video games and stories for health-related behavior change." American journal of preventive medicine 34, no. 1 (2008): 74-82.

[17] T. M. Connolly, E. A. Boyle, E. MacArthur, T. Hainey, and J. M. Boyle, "A systematic literature review of empirical evidence on computer games and serious games." Computers & Education 59, no. 2 (2012): 661-686.

[18] J. DeShazo, L. Harris, and W. Pratt, "Effective intervention or child's play? A review of video games for diabetes education." Diabetes technology & therapeutics 12, no. 10 (2010): 815-822.

[19] L. Gamberini, G. Barresi, A. Maier, and F. Scarpetta, "A game a day keeps the doctor away: A short review of computer games in mental healthcare." Journal of CyberTherapy and Rehabilitation 1, no. 2 (2008): 127-145.

[20] K. Guse et al., "Interventions using new digital media to improve adolescent sexual health: a systematic review." Journal of Adolescent Health 51, no. 6 (2012): 535-543.

[21] I. Guy, A. Perer, T. Daniel, O. Greenshpan, and I. Turbahn, "Guess who?: enriching the social graph through a crowdsourcing game." In Proceedings of the SIGCHI Conference on Human Factors in Computing Systems. ACM, 2011, pp. 1373-1382

[22] P. M. Kato, "Video games in health care: Closing the gap." Review of General Psychology 14, no. 2 (2010): 113.

[23] H. Kharrazi, A. S. Lu, F. Gharghabi, and W. Coleman, "A scoping review of health game research: past, present, and future." Games for health: Research, Development, and Clinical Applications 1, no. 2 (2012): 153-164.

[24] A. S. Lu, T. Baranowski, D. Thompson, and R. Buday, "Story immersion of videogames for youth health promotion: A review of literature." GAMES FOR HEALTH: Research, Development, and Clinical Applications 1, no. 3 (2012): 199-204.

[25] M. Papastergiou, "Exploring the potential of computer and video games for health and physical education: A literature review." Computers & Education 53, no. 3 (2009): 603-622.

[26] B. A. Primack et al., "Role of video games in improving health-related outcomes: a systematic review." American journal of preventive medicine 42, no. 6 (2012): 630-638.

[27] E. Rahmani and S. A. Boren, "Videogames and health improvement: A literature review of randomized controlled trials." GAMES FOR HEALTH: Research, Development, and Clinical Applications 1, no. 5 (2012): 331-341.

[28] N. Wilkinson, R. P. Ang, and D. H. Goh, "Online video game therapy for mental health concerns: a review." International journal of social psychiatry 54, no. 4 (2008): 370-382.

[29] T. Baranowski et al., "Video game play, child diet, and physical activity behavior change: A randomized clinical trial." American journal of preventive medicine 40, no. 1 (2011): 33-38.

[30] J. Rubin and D. Chisnell, Handbook of usability testing: how to plan, design and conduct effective tests. John Wiley & Sons, 2008. Harvard

[31] I. Dunwell, K. Torrens, S. Clarke, and S. Doukianou, "Game-based lifestyle interventions for adolescents: An evidence-based approach." In Interactive Mobile Communication Technologies and Learning (IMCL), 2014 International Conference on. IEEE, 2014, pp. 269-273.

[32] S. Michie, L. Atkins, and R. West, "The behaviour change wheel: A new method for characterising and designing behaviour change interventions". Implementation Science. 2011.

[33] I. Dunwell, R. Dixon, and D. Morosini, "A mobile serious game for lifestyle change: Conveying nutritional knowledge and motivation through play." In Interactive Mobile Communication Technologies and Learning (IMCL), 2015 International Conference on. IEEE, 2015, pp. 259-263.

[34] S. Carrino, M. Caon, O. A. Khaled, G. Andreoni, and E. Mugellini, "PEGASO: Towards a Life Companion", 16th International Conference on Human-Computer Interaction (HCII2014 Springer International Publishing, 2014, pp. 325-331.

[35] J. Brooke, "SUS-A quick and dirty usability scale." Usability evaluation in industry 189, no. 194 (1996): 4-7.

An Environmental and Wearable System Supporting Monitoring Services at Home for Elderly People

SMARTA project experience in technology reliability and acceptance

Giuseppe Andreoni, Marco Bocciolone, Paolo Perego, Marco Tarabini
Design dept. and Mechanics dept.
Politecnico di Milano
Milan, Italy
e-mail: (giuseppe.andreoni, marco.bocciolone, paolo.perego, marco.tarabini)@polimi.it

Lucia Pigini, Gabriele Bovi, Claudia Panzarino, Valerio Gower, Maurizio Ferrarin
Biomedical Technology Dept. and Cardiology Rehab Dept.
IRCCS Fondazione Don Carlo Gnocchi Onlus
Milan, Italy
e-mail: (lpigini, gbovi, vgower, mferrarin)@dongnocchi.it; claudia.panzarino@gmail.com

Abstract—**This paper presents the system validation of an integrated Personal Health System for supporting elderly remote monitoring and services. Objective of the trial as a whole is the assessment of reliability, effectiveness and usability of the SMARTA system. Outcomes from the validation tests were more than good for the system's reliability and all subjects assessed that they consider the system easy to use and that they will positively consider its introduction in their houses.**

***Keywords-Personal Health systems; Reliability; Usability*.**

I. INTRODUCTION

The SMARTA project (acronym for the Italian title "Sistema di Monitoraggio Ambientale con Rete di sensori e Telemonitoraggio indossabile a supporto di servizi di salute, prevenzione e sicurezza per l'Active Aging", i.e., Environmental Monitoring System through a sensors network and wearable tele-monitoring to support health, prevention and security services for Active Aging)explored the validation and exploitation of innovative technologies about future healthcare services dedicated to elderly for Ambient Assisted Living and for a more accurate, personalized and widespread management of geriatric pathologies. In fact, two billion people will be aged 60 and older by 2050 [1]. This represents both challenges and opportunities. Older people make important contributions to society as family members, volunteers and as active participants in the workforce. The wisdom they have gained through life experience makes them a vital social resource. However, along with these benefits come special health challenges for the 21st century. It is important to prepare health providers and societies to meet the specific needs of older populations [2]. This includes: (1) training for health professionals on old-age care, (2) preventing and managing age-associated chronic diseases, (3) designing sustainable policies on long-term and palliative care, and (4) developing age-friendly services and settings.

In this perspective, the SMARTA project aims at developing and testing a system integrating personal and environmental sensors for the realization of healthcare services for monitoring vital signs, sudden adverse events like falls, promoting exercise and active lifestyle, communicating with caregivers, offering safety services at home (e.g., intrusion detection).

The SMARTA system can be classified into the category of Personal Health Systems (PHS): introduced in late 1990s, they are a recent concept as devices and integrated solutions offering distributed health services by exploiting the innovation in science and technology such as the biomedical, micro- and nano-technologies, and the innovation in the Information and Communication Technologies (ICT). In this specific application, PHS empower elderly in their health management together with supporting their safe home living thanks to continuous monitoring of health related signals and parameters, integrated home automation controls and a web-based assistance. In fact, PHS are empowering systems because they have been designed to place the individual citizen/patient in the center of the healthcare delivery process. They allow citizens/patients to have more awareness and responsibility in managing their own health, and they support the interaction with care providers whenever is necessary.

Among PHS, Wearable Biomedical Systems (WBS) are a specific category and they can be defined as integrated systems on a wearable platform (in the sense of clothing or devices attachable to the human body) and can offer solutions for continuous monitoring by measuring non-invasive biomedical, biochemical and physical parameters. Continuous monitoring with early detection of anomalies has likely the potential to provide patients with an increased level of confidence, which in turn may improve the quality of life. If WBS are an ideal platform for multi-parametric non-intrusive monitoring of health status, their user acceptance and reliability have been investigated in research but rarely in a real scenario. In the frame of the SMARTA project, a deployment of the system into real life has been tested: this activity was approved by Fondazione Don Gnocchi ethical committee and subjects provided and signed the corresponding informed consent.

This paper presents the results of the reliability and user compliance. In section II, the SMARTA system is generally described and the setup of the validation test is presented. Section III introduces the results obtained during the experimental phase, according to overall reliability, wearable ECG reliability and usability, and fall detection system reliability and acceptance. In Section IV, the conclusions of this experimentation are drawn.

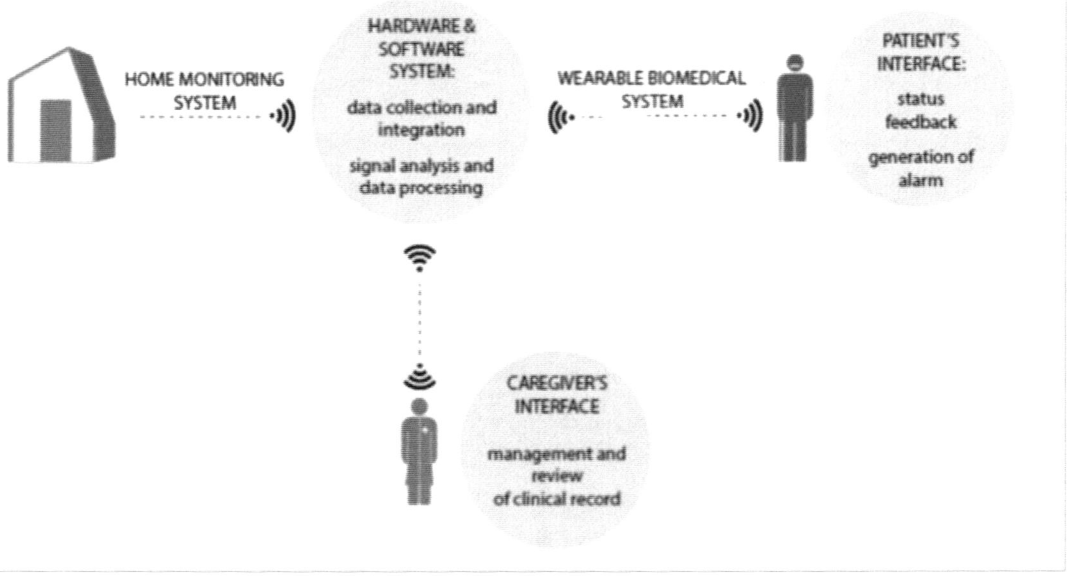

Figure 1. The diagram of the SMARTA system architecture.

II. Materials and Methods

The SMARTA system is the result of the integration of different components developed by the partners of the project. These parts correspond to specific functions of the overall system and are placed on different levels:

- monitoring of vital signs and lifestyle (e.g., no. of steps inside the house/environment);
- supporting the adoption of active lifestyles (e.g., performance monitoring of motor and physiological fitness exercises/prevention) and/or rehabilitation (exercises driven through video and the system is able to record the performance and movements);
- safety system environment (detection of falls, intrusion detection through detection of footsteps on the floor);
- communication with caregivers for continuous assessment and follow up.

These systems send data to a body/home gateway that redirects them to a center which concentrates the data and implements the above services (figure 1). The biomedical monitoring system is composed by wearable and non-wearable sensors. The considered sensors are: pulse-oximeter, glucometer, sphygmomanometer, thermometer, weighting scale, garment provided with sensors for ECG, fall detectors. The system was settled-up at the Smart Home of the "IRCCS S.Maria Nascente" Center, Fondazione Don Gnocchi.

The objectives of the test of the SMARTA system were the following ones. The main goal was the evaluation of the technical reliability of the prototype system in terms of:

- correct recognition and processing of the signals collected through the various system components;
- proper display/data management in the medical record;
- suitable provision or communication to the subject of information related to measures/alarms through the dedicated interface.

This first testing phase was carried out on 10 healthy subjects.

Whenever the technical reliability was good, the second set of evaluations was related to the procedure compliance, system usability and acceptance by the different endusers; in particular, the following elements were considered:

- simulation of the real applicability of the SMARTA system, through usability tests carried out in a controlled home environment;
- assessment of the perception of the SMARTA system in terms of patient's utility, usability, acceptability and attractiveness;
- evaluation of the perception of the SMARTA system in terms of caregiver's usefulness, usability, acceptability and attractiveness.

Thirteen elderly subjects (mean age 66.5 years) affected by cardiologic pathologies were recruited in this second testing phase.

For each trial, a dedicated testing protocol and the related methodology for measuring the proper outcomes and related indexes were defined. This paper is focused on the outcome for (1) the overall system, (2) the wearable monitoring system, and (3) the ground monitoring system.

III. RESULTS AND DISCUSSION

A. *Overall SMARTA system reliability*

The first phase of testing revealed that the SMARTA system has some improvement fields in term of technological stability. In fact, some failed measurements are still present in the system (Table I).

TABLE I. SMARTA RELIABILITY ANALYSIS – ALL SUBJECTS

Reliability	GUI	Measured parameters						
		T	W	AP	SpO2	GI	ECG	TS
% passed	100%	65%	65%	70%	70%	70%	70%	65%

Parameter legend: T = temperature, W = Weight, AP = Arterial Pressure, SpO2 = , Saturation of Peripheral Oxygen GI = glycaemia, ECG = ElectroCardioGram, TS = Tinetti Score.

In particular, the user interface (GUI) is the most stable component (100% of positive results) while during the measurement of the physiological parameters a 30-35% of failed test are shown. These problems were due to some intercommunication bugs distributed in the different hierarchical levels of the system. This test highlighted the need for a further step of tuning to increase the stability and the reliability of the system, even if the performance is generally encouraging.

The system underwent the judgement of thirteen patients in terms of utility (Figure 2), usability (Figure 3), and acceptance (Figure 4). A set of subjective assessment was administered through a dedicated questionnaire to the recruited subjects (Table II). The Likert scale with 7 values (from -3 = fully disagree, to 0 = neutral, up to +3 = fully agree) was chosen for ranking the subjective evaluations.

TABLE II. SMARTA RELIABILITY ANALYSIS - ELDERLY

Parameter/ question no.	Question
Utility	
Q1	Better health status monitoring
Q2	Increased safety at home
Q3	Reduced stress from continuous medical visits
Acceptance	
Q1	Patient privacy compliance
Q2	Comfort of wearable sensors
Q3	Easy to use system
Usability	
Q1	Clarity and immediate understanding of the SMARTA GUI
Q2	Pleasantness of the SMARTA GUI
Q3	Interest in SMARTA system adoption

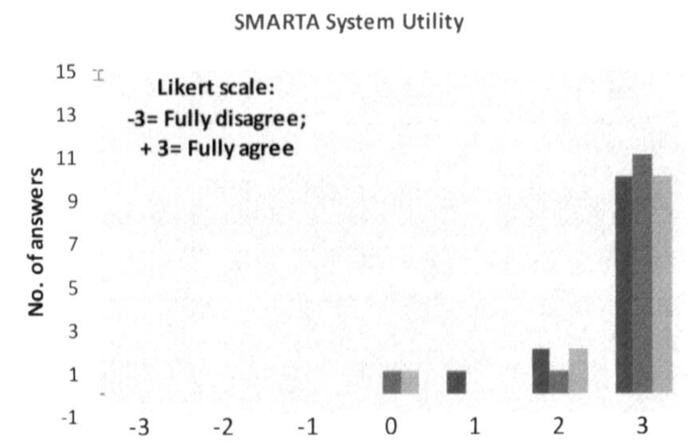

Figure 2. Results of the subjective assessment for the SMARTA system utility.

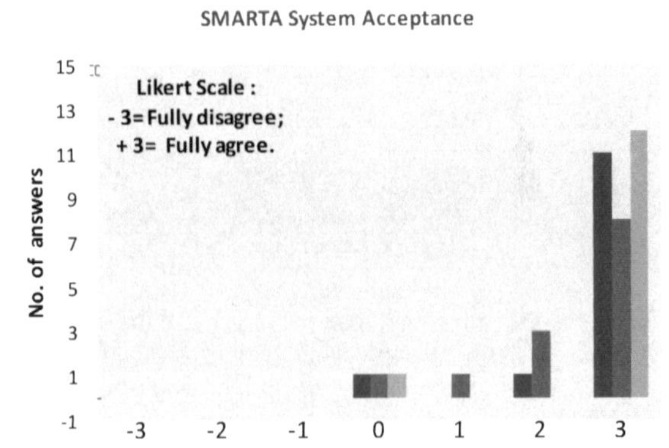

Figure 3. Results of the subjective assessment for the SMARTA system acceptance.

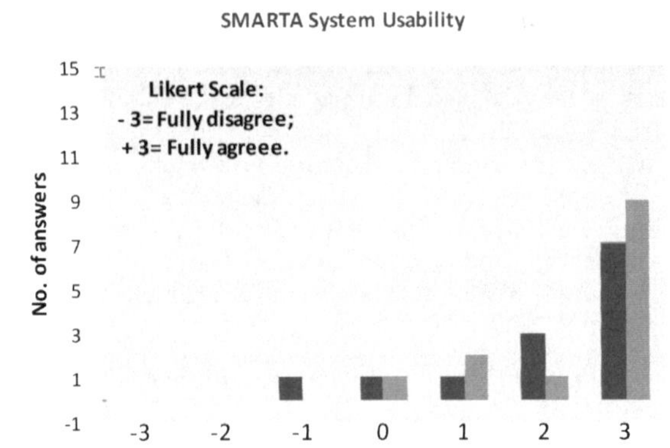

Figure 4. Results of the subjective assessment for the SMARTAsystem usability.

A similar assessment was carried out on the clinical operators that used the system. A set of subjective assessment was administered through another dedicated questionnaire to the four recruited subjects. The three investigated aspects were *utility*, *acceptance*, and *usability* according to the following questions/factors (Table III); the results (Likert scale) are summarized in Table IV.

TABLE III. SMARTA RELIABILITY ANALYSIS - CAREGIVERS

Parameter/ question no.	Question
Utility	
Q1	Reduction of visit number
Q2	Continuous monitoring
Q3	Efficacy of recorded clinical parameters in the personal healthcare folder
Acceptance	
Q1	Patient privacy compliance
Q2	Easy training to use the system
Q3	Technical appropriateness and reliability of the SMARTA system
Usability	
Q1	Clarity of personal healthcare folder consultation
Q2	Clarity of personal healthcare folder setup and management
Q3	Interest in SMARTA system adoption

TABLE IV. SMARTA: RELIABILITY ANALYSIS - CAREGIVERS

Caregivers	Utility			Acceptance			Usability		
	Q1	Q2	Q3	Q1	Q2	Q3	Q1	Q2	Q3
Mean value	2.8	2.8	2.5	3.0	1.8	2.8	1,0	1,8	3.0

The results show that the prototype of the SMARTA system has been highly appreciated by its potential users (both patients and clinicians) due to its potential in terms of improving home care and simultaneous decrease of the workload of caregivers.

Patients consider the interface clear and effective, but further developments should include the implementation of the feedback alarm relating to home automation. Also the management of measurements can be optimized in terms of questions and confirmations to the end-users.

Instead, the dedicated interface to clinical services (medical records), should be partially revised to match more effectively the needs of the operators. Further developments should include a general revision of the GUI and, in particular, an improved version of the management of the patient profile (profile creation and management of custom settings) and of the visualization of the results of the ECG (the track is neither readable nor understandable by patient and their relatives as end-users).

B. Wearable ECG system reliabiliy and usability

The wearable system was a system for non-intrusive monitoring of one ECG lead and trunk actigraphy. The ECG signal is acquired at a sample frequency f_s = 256 Hz with 24 bit resolution. The raw signal and the processed data are stored in an internal flash memory then downloaded through Bluetooth connection. The same device has also a three axes accelerometer used for wearable fall detection. Wearable solutions include two sensing components to be used with the same hardware device for ECG recording and transmission: an adhesive patch embedding the standard silver/silver chloride electrodes, and a garment provided with sensors (a t-shirt for the male patients and a belt for the female subjects as shown in Figure 5). These second elements embed conductive yarns to implement textile electrodes. Being not adhesive but simply adopting the t-shirt/belt elasticity to keep the sensors adhering to the skin and to minimize the skin motion artefacts. Specific attention has been paid to their design and the choice of materials.

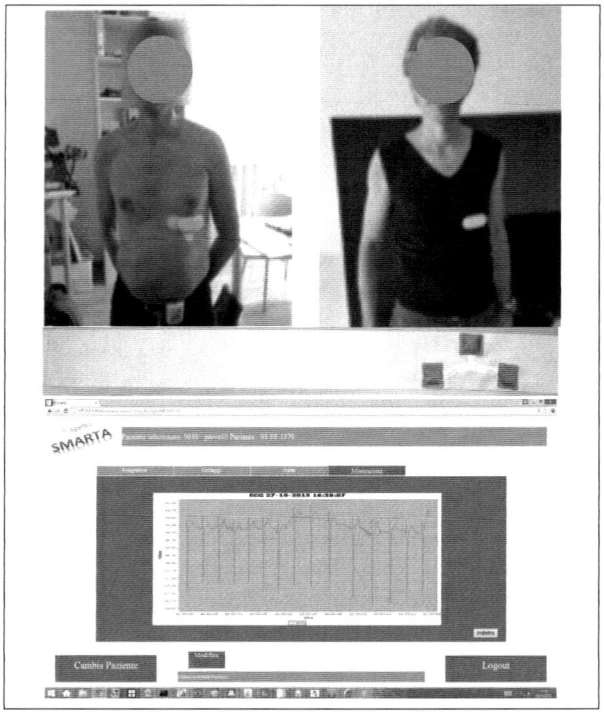

Figure 5. Wearable components of the SMARTA system (the patch, the t-shirt, the belt) and an example of acquired signal.

Preliminary design validation was related to recorded ECG tracks quality compared to the actual one. This analysis demonstrated the clinical suitability of the proposed system. Both solutions provided clinical quality ECG tracks. The validation tests showed also that the sensing patch was preferred for its simplicity in putting it on, while the belt and the t-shirt required more demanding operations. Despite this aspect, prolonged t-shirt or belt wearing was comfortable and did not provoked any skin irritation effect, while adhesive patches usually do. No such tests on the SMARTA patches were carried out during this validation phase.

C. Ground fall detection system reliabiliy and usability

The ground fall detection was set-up with tri-axial MEMS accelerometers [3] located on the ground. The signals were acquired by a Raspberry PI based system and it was analyzed in real time by extracting the RMS and the maximum values of the waveform. An example of the vibration time history is shown in Figure 6.

The vibration measured by the accelerometers was used to estimate if the event can be related to a fall or if it is due to the daily-life activities. The dependence between the measured acceleration and the force generated by the impact was quantified with the ground apparent mass, that was identified with preliminary experiments not described in this work.

The procedure for the fall identification is based on the data fusion at feature level as in [4]. To date, there were no falls and consequently it was not possible to validate the fall detection system. However, results showed that the system are useful to identify the human activities (between 11 and 18.30 in Figure 6) from the period where there was no activity in the room (after 18.30). The vibration spectra evidenced that there were no dominant frequency components up to 50 Hz, coherently with what was evidenced in other kinds of building at the ground floor.

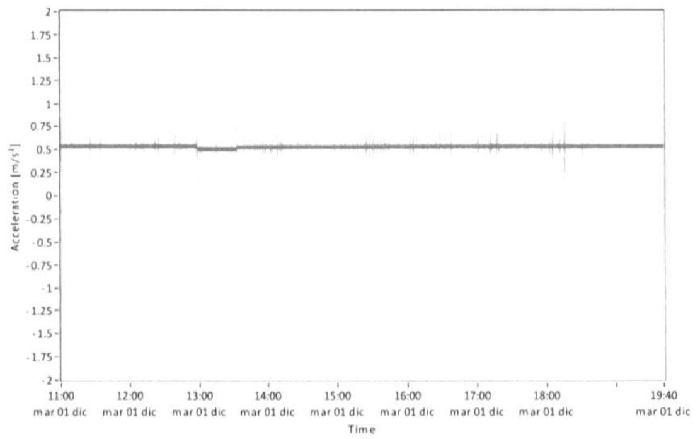

Figure 6. Vibration measured on the floor during the system usability tests.

Experiments performed in laboratory conditions evidenced that the simultaneous use of at least three accelerometers located on the ground allows identifying the position of the event that generated the vibration. The detection is based on a particular triangulation method that uses the wavelet transform to localize the source position; in our case, the triangulation was not effective given the limited bandwidth of the system. The validation of the source localization procedure is deserved to forthcoming studies.

IV. CONCLUSIONS

The SMARTA system demonstrated not only a good technical reliability but also a more than positive usability and acceptance judgement by the end-users. This is a key point for its deployment into real applications. In parallel, the residual problems of communication between the various hierarchical levels of the system, which in particular cases may introduce errors into the proper recording and data storage, were solved and now the robustness of the system ensures its applicability to the patient's home. About the sensors, further developments should be designed with respect to the design and ergonomics of the sensor-shirt and belt to facilitate their use in full autonomy. In doing so, the exploitation of the SMARTA system will aim to bring benefits to citizens and health authorities alike: first, by improving the quality of care for the individuals themselves; second, by containing the rising healthcare costs through the proper and efficient use of technological capabilities.

ACKNOWLEDGMENT

The authors wish to thank all partners of the project (www.smarta-project.it) and specifically: Datamed S.r.l., Flextronics Design S.r.l., Argonet S.r.l., Software Team S.r.l., Electron, Dipartimento di Informatica UNIMI and CoDeBri.

This work was financially supported by a grant from Regione Lombardia ("Avviso pubblico per la realizzazione di progetti di ricerca industriale e sviluppo sperimentale nel settore delle Smart cities and Communities - POR-FESR 2007-2013 asse 1 - Linea di intervento 1.1.1.1. azione E").

REFERENCES

[1] United Nations 2001- World population prospects: the 2000 revision. New York: United Nations.

[2] S. Monfardini, "Can we better manage Unfit older patients?" Cancer Treat Rev, vol. 35, no. 6, 2009, pp. 485-6.

[3] M. Tarabini, B. Saggin, D. Scaccabarozzi, and G. Moschioni, "The potential of micro-electro-mechanical accelerometers in human vibration measurements". Journal of Sound and Vibration, vol. 331, no. 2, 2012, pp. 487-499.

[4] M. Tarabini, G. Moschioni, C. Asensio, D. Bianchi, and B. Saggin, "Unattended acoustic events classification at the vicinity of airports." Applied Acoustics, vol 84, 2014, pp. 91-98.

AUTHOR INDEX

AUTHOR INDEX

International Academy, Research, and Industry Association (IARIA)
412 Derby Way
Wilmington, DE 19810

ISBN 978-1-5108-2311-2
9 781510 823112

CURRAN ASSOCIATES INC.
proceedings
.com